Current Advances and Challenges in Ocean Science—Feature Papers for the Founding of Oceans

Current Advances and Challenges in Ocean Science—Feature Papers for the Founding of Oceans

Editors

Antonio Bode
Michael W. Lomas
Pere Masqué
Diego Macías

MDPI • Basel • Beijing • Wuhan • Barcelona • Belgrade • Manchester • Tokyo • Cluj • Tianjin

Editors

Antonio Bode
Instituto Espanol de
Oceanografia, Centro
Oceanografico de A Coruna
Spain

Michael W. Lomas
Bigelow Laboratory for Ocean
Sciences
USA

Pere Masqué
International Atomic Energy
Agency
Monaco

Diego Macías
Joint Research Centre, European
Commission
Italy

Editorial Office
MDPI
St. Alban-Anlage 66
4052 Basel, Switzerland

This is a reprint of articles from the Special Issue published online in the open access journal *Oceans* (ISSN 2673-1924) (available at: http://www.mdpi.com).

For citation purposes, cite each article independently as indicated on the article page online and as indicated below:

LastName, A.A.; LastName, B.B.; LastName, C.C. Article Title. *Journal Name* **Year**, *Volume Number*, Page Range.

ISBN 978-3-0365-3060-4 (Hbk)
ISBN 978-3-0365-3061-1 (PDF)

Contents

Review

Review of the Scientific and Institutional Capacity of Small Island Developing States in Support of a Bottom-up Approach to Achieve Sustainable Development Goal 14 Targets

Rebecca Zitoun [1,2,3,*], Sylvia G. Sander [4], Pere Masque [4,5,6], Saul Perez Pijuan [7] and Peter W. Swarzenski [4,*]

[1] Department of Botany, University of Otago, Dunedin 9054, New Zealand
[2] National Institute for Water and Atmospheric Research (NIWA)/University of Otago Research Centre for Oceanography, University of Otago, Dunedin 9054, New Zealand
[3] Department of Ocean Systems (OCS), Royal Netherlands Institute for Sea Research (NIOZ), and Utrecht University, 1790 AB Den Burg, The Netherlands
[4] International Atomic Energy Agency, 4a Quai Antoine 1er, 98000 Monaco, Monaco; s.sander@iaea.org (S.G.S.); p.masque-barri@iaea.org (P.M.)
[5] School of Science & Centre for Marine Ecosystems Research, Edith Cowan University, 270 Joondalup Drive, Joondalup, WA 6027, Australia
[6] Departament de Física & Institut de Ciència i Tecnologia Ambientals, Universitat Autònoma de Barcelona, 08193 Bellaterra, Spain
[7] International Atomic Energy Agency, Vienna International Centre, 1400 Vienna, Austria; S.Perez-Pijuan@iaea.org
[*] Correspondence: rebecca.zitoun@outlook.com or rebecca.zitoun@nioz.nl (R.Z.); p.swarzenski@iaea.org (P.W.S.)

Received: 1 June 2020; Accepted: 1 July 2020; Published: 6 July 2020

Abstract: Capacity building efforts in Small Island Developing States (SIDS) are indispensable for the achievement of both individual and collective ocean-related 2030 agenda priorities for sustainable development. Knowledge of the individual capacity building and research infrastructure requirements in SIDS is necessary for national and international efforts to be effective in supporting SIDS to address nationally-identified sustainable development priorities. Here, we present an assessment of human resources and institutional capacities in SIDS United Nations (UN) Member States to help formulate and implement durable, relevant, and effective capacity development responses to the most urgent marine issues of concern for SIDS. The assessment highlights that there is only limited, if any, up-to-date information publicly available on human resources and research capacities in SIDS. A reasonable course of action in the future should, therefore, be the collection and compilation of data on educational, institutional, and human resources, as well as research capacities and infrastructures in SIDS into a publicly available database. This database, supported by continued, long-term international, national, and regional collaborations, will lay the foundation to provide accurate and up-to-date information on research capacities and requirements in SIDS, thereby informing strategic science and policy targets towards achieving the UN sustainable development goals (SDGs) within the next decade.

Keywords: SIDS; UN Member States; sustainable development goal 14; SAMOA pathway; capacity building; scientific infrastructure; IAEA; ocean challenges; nuclear and isotopic techniques

---

1. Introduction

In September 2015, all 193 United Nations (UN) Member States adopted an agenda for sustainable development, including 17 Sustainable Development Goals (SDGs) as an over-arching policy framework through to the year 2030 [1]. The agenda calls for worldwide action among governments, policy makers, private sectors, research communities, and the civil society to actively cooperate to mobilize global efforts around economic, social, and environmental issues in a balanced and integrated manner [1]. SDG 14 (Life Below Water), which focuses on human interactions with the ocean, seas, and marine resources, is critical to an integrated approach to achieve the SDGs agenda [2–4]. SDG 14 is underpinned by seven targets and three means addressing conservation, capacity building, ocean governance, and sustainable use and management of the ocean, seas, and marine resources (Table S1) [1,4]. SDG 14 is thus closely linked with other SDGs, such as Ending Poverty (SDG 1), Ending Hunger (SDG 2), Good Health and Well-Being (SDG 3), Climate Action (SDG 13), and Creating Decent Work and Economic Growth (SDG 8), and has therefore a cross-cutting role in the agenda [2–5].

With the growing awareness of the rapid degradation and over-use of oceans and the importance of healthy oceans and seas to the future of humanity, SDG 14 is gaining increasing attention by governments, industry, scientists, and the public [6]. This awareness led the UN General Assembly to proclaim a Decade of Ocean Science for Sustainable Development, also referred to as the 'Decade', which starts in 2021 and ends in 2030 [6,7]. The primary focus of the Decade is to provide a common framework to support countries in achieving their ocean-related 2030 agenda priorities for sustainable development [7]. The Decade strives to enable UN Member States to (1) work together to protect the health of our shared oceans by coordinating existing programs in ocean science, observations, monitoring, and data exchange [7] and (2) build the scientific and institutional capacity needed to generate scientific knowledge to improve science-based management of our ocean space and resources and thus increase the ability to better inform policy [7].

In the past, interdisciplinary ocean science has made great progress in exploring, describing, understanding, modelling, and enhancing current knowledge of the ocean system due to a myriad of national and international funding efforts and research projects, such as those under the umbrella of the Scientific Committee on Oceanic Research (SCOR; [8]). This knowledge helped to improve and create new technologies, tools, methods, collaborations, partnerships, projects, and scientific infrastructure to achieve a better understanding of the current state of our oceans, adopt appropriate mitigation and management strategies, improve the ability to predict ocean changes in the future, and inform policy and ocean governance [7,8]. However, there are major disparities in the capacity around the world to undertake marine scientific research [7,9] required to achieve national and regional SDG 14 targets. Global ocean science capacities are unevenly distributed, and Small Island Developing States (SIDS; Table S2; [10]) and the Least Developed Countries (LDC; [11]) suffer the most since they are limited in capacity and capability [7,9,12] to tackle global and national ocean-related issues such as climate change, marine pollution, ocean acidification, overfishing, seafood safety and trade, food security, sustainable aquaculture, and degradation and loss of biodiversity and marine habitats [13,14].

The SIDS Accelerated Modalities of Action (SAMOA) Pathway, which is an integral part of the 2030 agenda for the SDGs, reaffirms that SIDS remain a special case for sustainable development, recognizing their ownership and leadership in overcoming their unique challenges but also underscoring the need for partnerships with, and support of, the international community to help the SIDS achieve their SDGs over the next 10 years [15]. The pathway represents the commitments made by 115 SIDS Heads of State and Government as well as high-level representatives and is an official document formally adopted by UN Member States in September 2014. The SAMOA action plan recognizes that SDG 14 in particular is one of the most critical goals for SIDS whose societies, cultures, livelihoods, and economies are inherently linked with healthy, productive, and resilient oceans [7,15].

The success of SDG 14 and the Decade is critically dependent on global capacity building efforts and resource sharing between developed and developing countries [6]. Consequently, supporting SIDS, which frequently have weaker institutional capacity and limited financial resources [13], is indispensable to both individual and collective achievement of SDGs. It is commonly recognized that SIDS have a need for more tertiary education in environmentally relevant fields, technical expertise, training, access to tools, new technology, ocean information, ocean literacy, monitoring networks, data products, and infrastructure and logistics [13]. However, knowledge about their current states of national research capacities and infrastructures and their associated individual capacity building requirements is still scarce. This scarcity of information complicates the efforts of national and international agencies, programs, projects, and collaborations to effectively support SIDS in achieving long-term, sustainable and indeed necessary research capacities and capabilities. The implementation of effective capacity development in SIDS is also hindered by structural hurdles within governmental, national, and regional institutions, i.e., the lack of centralization, as well as the shortage of ownership, leadership, and coordination; the scarcity of active stakeholder participation; the lack of coordination and fragmentation between capacity building efforts at national and international levels; and short-term, project-based approaches of capacity development aids [12,16]. These constraints, among others, prevent fast decision-making and the implementation of ocean science-related issues into national and regional policies and thus exacerbate the challenges of global capacity building efforts and resource sharing [16]. A successful implementation of capacity building in SIDS will thus rely on a holistic approach to address all the gaps in the capacity needs and challenges of SIDS [12,16]. The first key objective for a holistic capacity building process is the development of appropriate human and institutional capacity to engage in (1) the identification of problems and needs in a local context, (2) the collection of scientific information, and (3) the implementation of science-based national plans and international agreements to ensure effective mitigation and adaptation solutions [12,16].

Several regional, national, and international programs and initiatives are already in place to support SIDS to acquire the prerequisite capacity and infrastructure (SDG 14.a) required to understand, monitor, and protect their marine environment in line with nationally identified targets under SDG 14 and the SAMOA Pathway and to fulfil their obligations in the framework of Global Conventions. One such program is the International Atomic Energy Agency's (IAEA's) Technical Cooperation (TC) Programme established in 1957, which among other things, focuses on the safe and peaceful application of nuclear science, technologies, and knowledge to find practical solutions, in this case, for marine environmental challenges. Nuclear and isotopic techniques are essential tools to help identify, monitor, mitigate, and adapt to the effects of sustained climate- and ocean-change such as ocean warming, ocean acidification, biotoxins (e.g., harmful algal blooms), and contaminants (e.g., radionuclides, trace metals, and persistent organic pollutants) [17], and thus directly contribute to SDG 14.1 and 14.3. The TC Programme is the main mechanism through which the IAEA assists and delivers services to its Member States including the development of research capacity to transfer marine technology, in order to improve ocean health and to enhance the contribution of marine biodiversity to the development of developing countries, in particular SIDS and LDC (SDG 14.a). The program helps Member States to build, strengthen, and maintain human expertise and institutional capacities in order to (1) define national needs; (2) make informed decisions on action plans and measures to protect the oceans; (3) assure the sustainable delivery of ecosystem services; and (4) ensure sustainable socioeconomic development [17].

Currently, the IAEA provides technical support to 23 of the 38 SIDS Member States (Table S3) in areas such as the marine environment, health and nutrition, food and agriculture, energy, nuclear knowledge development and management, water and the environment, and industrial applications and radiation technology. Nevertheless, most SIDS still have a low capacity (i.e., limited infrastructure; low technical and institutional capacity; little financial, human, and material resources; and lack of expertise and networks) to conduct research and generate knowledge regarding climate change and marine challenges and thus lack science-based data on which to base policy and create and test necessary adaptation and mitigation strategies [9,18]. To effectively support SIDS in enhancing their capacities to address sustainable marine development challenges that affect them, as highlighted by SDG 14.a, knowledge about their individual research infrastructures and capacities is urgently needed. In particular, in accordance with the statute of the IAEA, it is necessary to assess their research capacities and infrastructure for nuclear and isotopic science, as well as organic and inorganic geochemistry.

The goal of this work was to compile a capacity assessment, which summarizes the existing key research capacities in SIDS UN Member States to help formulate appropriate capacity development responses in marine areas of concern for SIDS. This study focuses primarily on the identification of marine environmental capacities with particular attention to nuclear and isotopic science, as well as inorganic and organic geochemistry facilities. This groundwork is important to ensure that the ocean science community, as well as national and international funding and development agencies, implement relevant and effective capacity development responses in SIDS UN Member States in the coming years to improve SIDS's conditions for sustainable development of the ocean.

2. Overview of Small Island Developing States (SIDS)

SIDS are a distinct group of 58 small island nations, low-lying countries, and territories situated in the tropics and low-latitude sub-tropics [14] across three geographical regions, i.e., the Caribbean; the Pacific; and the Atlantic, Indian Ocean, and South China Seas (AIS) (Figure 1, Figure S1 and Table S2). The SIDS group comprises 38 UN Members and 20 Non-UN Members or Associate Members of Regional Commissions (Figure 1 and Table S2).

2.1. Physiographic and Environmental Characteristics of SIDS

2.1.1. Caribbean SIDS

The Caribbean region contains 29 SIDS, 25 of which lie in the tropics and four are located in the subtropical zone (Figure 1). Their geographical location makes the Caribbean SIDS vulnerable areas to the impact of natural hazards, such as hurricanes, storms, droughts, floods, landslides, volcanic activity, and earthquakes [19]. In total, 26 of the 29 SIDS are classified as islands and three are considered countries (Belize, Guyana, and Suriname). The Caribbean SIDS comprise a total of 1611 islands, islets, and atolls, cover a total land area of 599,105 km^2 with a total coastline of 16,516 km (Table S4; [20]), and comprise a total maritime area of 3,750,019 km^2 [21].

Inland water bodies (lakes, reservoirs, and rivers) cover a total area of 33,429 km^2 (data missing for Guadeloupe and Martinique) with 16 SIDS having no or negligible freshwater sources (Table S4; [20]). The population in the Caribbean SIDS in 2018 amounted to a total of 44.3 million (average population density of 296 people/km^2) with 4.2% of the population living in areas where elevation is below 5 m above sea level (Table S4; [20]). In particular, Suriname and Turks and Caicos Islands have >48% of their population living in areas below 5 m above sea level and thus both SIDS are assumed to be severely affected by future climate change. Further, in the Caribbean SIDS, 4.6% (data missing for Anguilla, Curacao, Guadeloupe, Martinique, Montserrat, and St. Martin) of the total land area is below 5 m of elevation above sea level, with the Bahamas (52.0%), Turks and Caicos Islands (55.9%), and the British Virgin Islands (28.0%) expected to be the most severely affected SIDS by climate change and associated sea level rise in the Caribbean (Table S4; [20]).

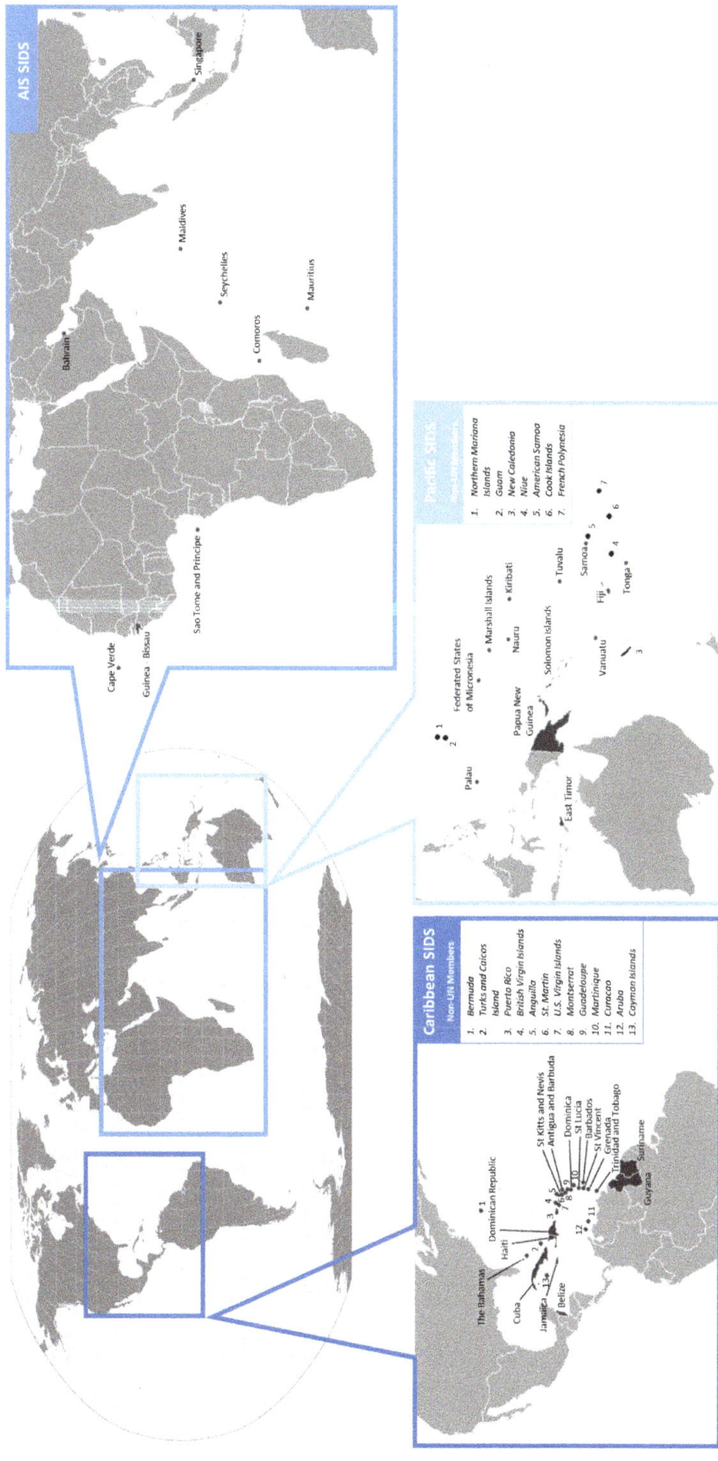

Figure 1. Geographical regions of Small Island Developing States (SIDS). AIS: Atlantic, Indian Ocean, and South China Seas.

An average of 21.6% of the land area in Caribbean SIDS is used for agricultural purposes (i.e., land that is arable, under permanent/temporary crops, or under permanent/temporary pastures) and an average of 40.0% is covered in forest [20]. The coastal ecosystems in the Caribbean region are a mixture of mangrove, sea-grasses, and coral reefs [22], with the Caribbean SIDS containing 5.3% of the world's coral reefs [21]. The Caribbean SIDS have a unique biodiversity and high levels of endemism with a total of 56,236 animal and plant species [23]. However, 3.5% of these species are currently classified as threatened, including a total of 172 bird species, 795 fish species, 100 mammal species, and 878 plant species as of 2018 [20,23]. Natural terrestrial and marine resources of commercial importance are limited, i.e., arable land (e.g., peanuts, tropical fruit), salt, seafood, petroleum, natural gas, timber, hydropower, and elements/minerals (e.g., limestone, cobalt, nickel, copper, gold, silver, phosphate; [19]). With the purpose to conserve the marine and terrestrial biodiversity and natural resources, the Caribbean SIDS protect, on average, 5.7% of their territorial waters and 6.9% of their terrestrial and marine areas as of 2018 [20], with St. Martin conserving as much as 96.4% of its territorial coastal and marine ecosystems.

2.1.2. Pacific SIDS

The Pacific region has 20 tropical SIDS (Figure 1), most of which are exposed to a large number of natural hazards, such as typhoons, storms, droughts, floods, landslides, volcanism, cyclones, earthquakes, tornadoes, and tsunamis [19]. Altogether, the Pacific SIDS have a land area of 555,572 km^2, of which Papua New Guinea comprises 81.5%. The Pacific SIDS contain a total of 5019 islands, islets, and atolls and have a combined coastline of 31,541 km (Table S5; [20]). The territorial waters of the Pacific SIDS cover a total maritime area of 26,911.685 km^2 [21].

Freshwater bodies cover a total area of 11,570 km^2 with 14 of the 20 SIDS having no or negligible freshwater supplies (Table S5; [20]). The resident population of the Pacific SIDS in 2018 amounted to a total of 13.2 million (average population density of 163 people/km^2) with 1.8% of the population living in areas where elevation is below 5 m above sea level [20]. In Pacific SIDS, 0.95% (data missing for Northern Mariana Islands, Cook Islands, Micronesia, and Nuie) of the total land area is below 5 m of elevation above sea level with Kiribati (54.6%), Marshall Islands (43.8%), and Tuvalu (32.6%) being the most severely affected SIDS by climate change and associated sea level rise in the Pacific region (Table S5; [20]).

On average, 23.3% of the land area in the Pacific SIDS is used for agricultural purposes and 53.3% is covered by forest [20]. The Pacific SIDS contain 21.4% of the world's coral reefs and also have the world's highest proportion of endemic species per unit of land area [22]. The biodiversity (a total of 40937 animal and plant species; [23]) is among the most critically threatened with 3.3% of species classified as threatened in 2018 including 195 bird species, 336 fish species, 108 mammal species, and 701 plant species [20,23]. To conserve the ecological biodiversity and the limited natural resources with commercial value (e.g., coconuts, fish, timber, pumice, minerals, hydropower, petroleum, and natural gas; [19]), Pacific SIDS put marine and terrestrial protected areas in place (on average 13.9% of their territorial area). As of 2018, marine protected areas (MPAs) covered on average 13.6% of the territorial waters of the Pacific SIDS with New Caledonia protecting as much as 96.6% of its territorial coastal and marine ecosystems [20].

2.1.3. Atlantic, Indian Ocean, and South China Seas SIDS

The Atlantic, Indian Ocean, and South China Seas (AIS) region consists of nine SIDS covering the tropical, arid, and temperate zones of the world (Figure 1; Table S6). Singapore is a member of the AIS SIDS category, though it is neither a small nor a developing state [18]. The geographical location makes the AIS SIDS high risk areas to the impact of natural hazards including typhoons, storms, droughts, harmattan winds, cyclones, fires, volcanic activity, earthquakes, and tsunamis [19]. The AIS SIDS include eight islands and one country (Guinea-Bissau), which together consist of a total of 1533 islands, islets, and atolls and cover a total land area of 39,248 km^2 with a 3,530 km long coastline (Table S6, [20]).

Guinea-Bissau is the largest SIDS in the AIS region, comprising 72% of the total land area. The maritime claim of each AIS SIDS gives the SIDS a total maritime area of 4,762,621 km^2 [21].

Inland water bodies cover a total area of 8,025 km^2 with six SIDS having no or negligible freshwater supplies (Table S6, [20]). In 2018, the AIS SIDS had a total population of 12.5 million (average population density of 1,487 people/km^2), 13.8% of whom lived in areas where elevation is below 5 m above sea level [20]. The Maldives and Bahrain with 48.2% and 34.3% of their population living in areas < 5 m of elevation above sea level, respectively, are considered the most affected AIS SIDS by climate change. In general, 6.5% (data missing for Guinea-Bissau) of the total land area in AIS SIDS lies below 5 m of elevation, and again, the Maldives (45.5%) and Bahrain (34.0%) are considered the most impacted AIS SIDS regarding future sea level rise (Table S6, [20]).

Agricultural land covers on average 31.5% of the land area in the AIS SIDS, while on average another 33.6% of the land area is covered in forest [20]. The commercially valuable natural resources in the AIS region are limited to oil, natural gas, fish, timber, minerals, hydropower, and coconuts [20]. The AIS SIDS feature numerous marine ecosystems including mangroves and 10% of the world's coral reefs [23]. The AIS region is extremely ecologically diverse and is home to a total of 16486 animal and plant species [23], of which 4.3% are classified as threatened as published by World Bank in 2019 (i.e., 99 bird species, 224 fish species, 59 mammal species, and 329 plant species). As of 2018, the AIS SIDS managed to protect various marine and terrestrial areas amounting to an average of 1.8% of their total territorial area with MPAs covering on average 1.3% of their territorial waters [20].

2.2. Economic Characteristics of SIDS

The Gross Domestic Product (GDP) varies significantly among the SIDS. Singapore had the highest GDP in 2018 with US$ 307.9 billion, while Niue, with only US$ 10 million, had the lowest GDP [19,20]. Singapore, which is relatively rich by developing country standards, is often seen as the exception to the otherwise financially limited SIDS cluster [18]. The average GDP of the SIDS equals US$ 13.9 billion (2018), with the AIS SIDS having the highest average GDP with US$ 40.6 billion (US$ 14940 GDP per capita; AIS excluding Singapore: US$ 7.2 billion GDP and US$ 8735 GDP per capita), followed by the Caribbean SIDS with US$ 13.2 billion (US$ 21111 GDP per capita), and the Pacific SIDS with US$ 2.7 billion (US$ 8079 GDP per capita; Table 1; [19,20]). In total, 83% of SIDS have a GDP lower than the average of US$ 13.9 billion (no data for Guadeloupe and Martinique), and 37% have a GDP lower than US$ 1 billion (Table 1). The growth of most of the Caribbean (average GDP growth of 1.7%) and Pacific SIDS (average GDP growth of 1.6%) economies lagged largely behind the world average (3.0%) in 2018, while the AIS SIDS exceeded the world average with an annual GDP growth of 3.7% (AIS excluding Singapore: 3.8%) in the same year [20].

The tourism industry (direct contribution via the travel and tourism sector and indirect contribution via the services sector, which includes hotels and restaurants) is a vital economic sector for the development of SIDS in terms of GDP earnings and employment (Figure S2; [20,24]). For instance, the Maldives are highly dependent upon revenue gained from tourism, with 38.9% of the island's GDP coming from the travel and tourism sector (Table 1). Other important sectors for the SIDS economies are industry and agriculture (including forestry and fishing) with the latter contributing on average 4.9%, 15.6%, and 11.6% to the GDP in the Caribbean, Pacific, and AIS SIDS, respectively (Table 1 and Figure S2).

Additionally, SIDS rely heavily on trade to drive economic growth [24] with contributions of the export and import of goods and services to the GDP being far higher for SIDS than the global average (Figure S3 and Table 1). The fact that SIDS are dependent on tourism, have a narrow resource base, are usually single commodity exporters which greatly rely on export earnings, and are remote from international markets increases their vulnerability to external economic threats and shocks as well as the impacts of natural disasters (e.g., earthquakes, tsunamis, volcanic eruptions, landslides, tropical storms, floods, and droughts) and climate change [14,24].

Table 1. Economic data of Small Island Developing States (SIDS) [20].

Island State	GDP (US$ Billion, 2018)	GDP Per Capita (US$, 2018)	GDP Growth (annual%, 2017/18)	Travel and Tourism (% of GDP, 2018)	Agriculture, Forestry, and Fishing (% of GDP, 2017/18)	Industry (Including Construction) (% of GDP, 2017/18)	Services (% of GDP, 2017/18)	Exports (% of GDP, 2017/18)	Imports (% of GDP, 2017/18)
Caribbean SIDS									
Anguilla	0.34	-	-	-	-	-	-	-	-
Antigua and Barbuda	1.62	16727.0	4.9	13.1	1.7	19.3	68.0	41.6	46.9
Aruba	2.83	25630.3	1.3	27.6	0.4	16.2	69.6	70.8	75.2
Bahamas	12.16	32217.9	1.4	19.2	1.0	13.2	74.9	33.7	41.8
Barbados	4.85	17949.3	1.0	13.1	1.4	12.7	74.9	38.0	44.9
Belize	1.74	4884.7	3.0	15.0	10.8	13.4	62.9	54.2	57.6
Bermuda	6.43	85748.1	−2.5	5.2	0.7	5.8	-	47.7	29.6
British Virgin Islands	0.94	-	-	33.0	-	-	-	-	-
Cayman Islands	4.44	81124.5	3.0	8.3	0.4	0.0	87.0	-	-
Cuba	102.25	8821.8	1.8	2.6	3.8	24.4	70.8	14.5	11.7
Curacao	5.60	19567.9	−1.7		0.4	18.2	71.7	72.2	85.3
Dominica	0.56	7691.3	0.5	12.3	12.7	12.1	55.1	47.0	71.2
Dominican Republic	77.50	8050.6	7.0	5.4	5.5	27.9	58.9	25.1	30.0
Grenada	1.073	10640.5	4.9	6.9	4.9	13.3	66.7	55.8	52.1
Guadeloupe		-	-	-	-	-	-	-	-
Guyana	3.79	4979.0	3.4	2.7	15.4	31.9	42.0	45.2	60.9
Haiti	8.52	868.3	1.5	3.4	17.7	55.2	24.8	17.1	58.5
Jamaica	13.75	5354.2	1.9	10.5	6.7	20.5	59.2	34.1	48.5
Martinique		-	-	-	-	-	-	-	-
Montserrat	0.07	-	-	-	-	-	-	-	-
Puerto Rico	69.07	31651.3	−4.9	2.5	0.8	50.1	49.6	59.9	45.9
St Martin	0.56	-	-	-	-	-	-	-	-
St. Kitts and Nevis	0.96	19275.4	3.0	6.6	1.2	23.9	63.9	54.1	53.6
St. Lucia	1.69	10566.0	0.6	15.6	2.1	9.7	75.1	36.3	43.7
St. Vincent & Grenadines	0.78	7361.4	2.6	6.0	6.7	15.0	62.3	35.2	55.2
Suriname	4.63	6234.0	2.0	1.3	12.6	32.3	48.7	52.5	38.4
Trinidad and Tobago	23.88	17129.9	0.7	2.8	0.5	39.0	57.1	-	-
Turks and Caicos Islands	0.96	27142.2	5.3	-	0.5	9.7	74.4	-	-
U.S. Virgin Islands	5.18	35938.0	−1.7	18.2	-	-	-	72.5	78.6

Table 1. *Cont.*

Island State	GDP (US$ Billion, 2018)	GDP Per Capita (US$, 2018)	GDP Growth (annual%, 2017/18)	Travel and Tourism (% of GDP, 2018)	Agriculture, Forestry, and Fishing (% of GDP, 2017/18)	Industry (Including Construction) (% of GDP, 2017/18)	Services (% of GDP, 2017/18)	Exports (% of GDP, 2017/18)	Imports (% of GDP, 2017/18)
Caribbean SIDS									
Pacific SIDS									
American Samoa	0.63	11466.7	−5.4	-	-	-	-	57.4	92.7
C. of Northern Marianas	1.24	-	-	-	-	-	-	-	-
Cook Islands	0.25	-	-	-	-	-	-	-	-
F. S. of Micronesia	0.33	3568.3	1.4		27.1	6.1	61.9	26.6	72.4
Fiji	5.11	6267.0	5.0	14.1	10.4	16.1	54.0	-	-
French Polynesia	4.79	14323.8	4.0	-	4.7	-	-	4.9	24.2
Guam	5.79	35712.6	0.2	-	-	-	-	19.3	51.5
Kiribati	0.14	1625.3	2.0	8.7	30.8	12.4	63.4	13.2	92.0
Marshall Islands	0.18	3788.2	2.5	-	16.8	13.9	70.2	31.5	88.8
Nauru	0.12	9888.9	−3.5	-	3.6	5.5	81.6	48.8	101.7
New Caledonia	9.21	12579.6	2.1	-	3.7	19.6		13.1	33.0
Niue	0.01	-	-	-	-	-	-	-	-
Palau	0.20	15859.4	5.0		3.2	8.0	78.4	48.8	75.4
Papua New Guinea	20.69	2730.3	0.4	0.7	17.9	34.7	-	72.2	58.9
Samoa	0.81	4183.4	0.7	-	10.7	21.2	-	31.0	45.4
Solomon Islands	1.22	2137.7	3.4	4.2	35.0	-	-	45.1	53.3
Timor-Leste	2.77	2035.5	2.8	-	10.4	45.8	44.1	61.1	59.9
Tonga	0.43	4364.0	0.3	5.2	17.2	17.3	-	21.5	73.4
Tuvalu	0.03	3700.7	2.5	-	16.5	7.3	-	-	-
Vanuatu	0.86	3123.9	3.2	-	26.6	11.4	-	48.6	49.4
AIS SIDS									
Bahrain	33.81	24050.8	1.8	4.22	0.3	43.4	55.3	88.1	71.1
Cape Verde	1.633	3635.4	5.5	17.66	4.7	19.3	61.1	48.9	67.7
Comoros	0.586	1415.3	2.8	4.25	29.9	12.2	53	11.6	28.5
Guinea-Bissau	1.218	778.0	3.8	-	47.5	12.6	34.8	25.3	32.5
Maldives	4.505	10330.6	6.1	38.92	5.6	12.8	67.4	67.6	72.2
Mauritius	13.413	11238.7	3.8	7.21	2.8	17.6	67.4	40.7	54.2
Sao Tome and Principe	0.41	2001.1	2.7	10.63	11.4	14.8	71.8	-	-
Seychelles	1.825	16433.9	3.6	25.74	2.0	10.9	70.4	87.0	102.1
Singapore	307.9	64581.9	3.1	4.13	0.0	25.2	69.4	176.4	149.8

-: No data available.

Indeed, SIDS are located among the most vulnerable regions in the world in relation to the intensity, frequency, and increasing impact of natural disasters and thus face disproportionately high economic, social, cultural, and environmental consequences [24]. The overall costs of disasters and the economic losses resulting from the negative effects of natural disasters are significant for SIDS and can set back economic development gains by several years [24,25]. For instance, an earthquake in Haiti in 2010 destroyed the equivalent of 120% of Haiti's GDP [20,26], and Hurricane Matthew, which struck the country in 2016, caused damage estimated to equal 32% of Haiti's GDP [26–28]. However, economic losses vary among SIDS, with Caribbean SIDS predicted to face on average the highest annual loss by 2030 as a share of cumulative GDP with 6.3%, followed by the Pacific SIDS with 3.9%, and the AIS SIDS with 0.1% [20,29,30].

The risk of economic breakdown, extensive environmental damage, and disruptions of social and cultural processes in SIDS is only expected to grow owing to climate change [13]. Although the SIDS are among the least responsible for climate change, contributing less than 0.6% (as of 2012) to the global greenhouse gas emissions [20], they are likely to suffer the most from its adverse effects such as sea level rise, extreme storm events and droughts, increasing sea surface temperatures, ocean acidification, coral bleaching, inundation, biodiversity loss, and ecosystem destruction [25]. The socio-economic implications of sea level rise in particular will have drastic negative impacts on all important economic sectors of SIDS including tourism, financial services, agriculture, fisheries, water supply and sanitation, infrastructure, and ecosystem health [25]. For instance, protecting Jamaica's coastline from impacts of a sea-level rise of 1 m is estimated to cost US$ 462 million annually—19% of the country's GDP [13]. Globally, estimated economic stresses due to climate change project losses of US$ 70–100 billion per year through to 2050 [20].

In addition to natural disasters, external economic shocks can also have drastic impacts on financial markets and institutions in SIDS and can cause significant economic, social, and political disruption. For instance, the emergence of the novel coronavirus, COVID-19, highlighted the global economic vulnerability to epidemics and pandemics, placing a disproportionally higher burden on LDC and SIDS due to their dependency on international trade and foreign tourism whose demand has collapsed [31]. On average, trade and tourism (indirectly and directly) account for more than 71% and 30%, respectively, of SIDS GDP, and a decline in tourists by 25% will result in a $7.4 billion or 7.3% fall in SIDS GDP [31]. According to the UN/DESA brief Nr. 64 [31], the projected GDP growth rate of SIDS will likely shrink by 4.7% in 2020, compared to a global contraction of around 3.0%, which will exceed the economic impact of some natural disasters, such as tropical hurricanes. These estimates are, however, rather conservative, and the drop in GDP could be significantly greater in some of the SIDS [31]. The economic consequences of the COVID-19 pandemic will be devastating for many of the SIDS with long lasting effects, which will have major implications for 1) withstanding future natural disasters, and 2) achieving global and national targets within the 2030 agenda and the SDGs [31]. This pandemic underscores the extreme vulnerability of SIDS to global economic shocks and thus further highlights the need of SIDS for international assistance and cooperation.

3. Commonalities of SIDS—Marine Challenges

The data above show that SIDS are geographically dispersed and have diverse ecological and cultural characteristics as well as social and economic structures (i.e., island—non-island, developing—non-developing, states—non-states). Nevertheless, they face similar social, economic, and environmental challenges [14]. Their inherent challenges stem from factors such as their small size, geographical remoteness, narrow resource base, low-lying coastal territories, concentration of the population along coastal zones, growing populations, high volatility of economic growth, dependency on external markets and international assistance, susceptibility to natural disasters (i.e., cyclones and earthquakes), and high vulnerability to the impacts of a changing climate [14,32]. Of particular concern to SIDS are the marine challenges associated with climate change and anthropogenic activity.

Climate change and anthropogenic activities threaten marine ecosystems, economies, and communities on a global scale, but with local consequences. Some impacts are already evident, but other impacts and effects will only become an issue in the coming decades when biological tipping points and thresholds are reached and the resilience of marine ecosystems can no longer be maintained. Climate change and anthropogenic activities are putting the oceans and thus the provision of their ecosystem services at risk. These pressures have a disproportionately greater impact on SIDS, where oceans and marine resources tend to play a fundamental role in financial, social, cultural, and environmental wellbeing [12,14,33]. The most pressing marine issues for SIDS are listed in Table 2, separated into three key themes, with no particular order of priority.

Table 2. List of the most pressing marine issues for Small Island Developing States (SIDS).

Key Area	Marine Challenge
Marine Pollution	Brownification Diverse Contaminants Eutrophication Marine Litter/ Wastewater Microplastics Seafood Safety
Ocean and Climate	Biodiversity Damage and Loss Deoxygenation Harmful Algae Invasive Species Ocean Acidification Ocean Warming Sea Level Rise Severe Weather and Natural Disasters
Marine Resources	Loss of Habitat Exploitation of Marine Resources/Overfishing

All of the current and emerging marine environmental challenges have strong social and economic components, particularly in SIDS which are acutely dependent on oceans. Accordingly, the marine challenges of SIDS should be viewed in a holistic and integrative manner since any changes to any one of these sectors could ultimately lead to the collapse of another [12,25].

However, despite their similar challenges, SIDS do not easily fit into standard response and solution models since each SIDS prioritizes different marine challenges in the context of its unique geophysical characteristics (vulnerabilities), social, economic, and environmental needs, adaptive capacities, and sustainable commitments [12,34,35]. For instance, natural hazards produce widely different outcomes in different countries—while some SIDS seem to cope and adapt fairly well, others suffer tremendously [35]. These differences have to be recognized, understood, and addressed for sustainable development to be suitably realized in SIDS on a national level. Therefore, capacity building, solutions, and adaptive strategies for marine challenges in SIDS need to be custom tailored for each individual SIDS.

4. The Way Forward—Needs to Tackle the Challenges

Despite the specificities of each SIDS, several commonalities regarding possible appropriate and practical responses to address the marine challenges of the 21st century can be identified and are listed below [12,36,37]. These recommendations should be addressed by a broad participation of all relevant stakeholders across different spatial scales and frameworks [9], i.e., international and national governments and agencies, regional and sub-regional organizations, educational and academic entities, and local communities, including multi-stakeholder partnerships, to provide the best chance of long-lasting solutions.

- *Define Indicators*—Identify a manageable set of measurements (proxies and indices), which are globally applicable to evaluate ecosystem health in various regions in order to reduce cost and time; engage government entities, donors, funding agencies, and the public; as well as allow the interoperability of research data. Frameworks such as the Ocean Health Index [38] can be used in SIDS to assess ocean health and thus inform policy and evaluate progress.

- *Establish Protocols and Standards*—Standards and protocols for sample collection, handling, analysis, and data products are needed to obtain high quality data, meet global standards, and allow the interoperability and reproducibility of research data. For this, SIDS can take advantage of repositories such as the Ocean Best Practices System (OBPS; [39]), which includes standard operating procedures, manuals, and method descriptions for most ocean-related sciences and applications.
- *Establish Baseline Measurements*—Establish monitoring programs or engage with existing monitoring programs such as the Global Ocean Acidification Observing Network (GOA-ON; [40]) to quantify baselines and detect changes in the biogeochemistry of water bodies to evaluate ecosystem health. These data are needed to inform policy-makers and thus support better decision making on management strategies.
- *Create Data Reporting Networks*—Create data products, reporting networks, and mechanisms to share scientific findings and lessons. Herein, SIDS can take advantage of the Ocean Data and Information System Catalogue (ODIS, [41]) to find ocean-related web-based data products or participate in existing networks such as the Ocean Acidification International Coordination Centre (OA-ICC; [42]) and the Global Ocean Observing System (GOOS, [43])
- *Improve Local Modelling Projections*—Advance the resolution of existing scenario models to help identify and address local priorities and issues.
- *Advance Scientific Understanding*—Conduct national/regional vulnerability assessments to identify the potential impact of climate change and anthropogenic activity on key ecological, cultural, and economic marine resources and species, as well as the communities that depend on them.
- *Optimize Research*—Enhance the research capacity and infrastructure by developing low-cost instrumentation and secondary standards to reduce costs and help SIDS to gather, access, and use data to take ownership in capacity building activities and decision-making. There is also a necessity to make pre-existing technology (e.g., autonomous underwater vehicles (AUVs); Argo floats; conductivity, temperature, and depth instruments (CTDs); Gliders; Monitoring Buoys etc.) globally available to obtain relevant data for effective adaptation, mitigation, and resilience strategies.
- *Take Meaningful Actions and Implement Solution-based Strategies*—Reduce the causes of climate change and anthropogenic activity by implementing effective mitigation, resilience, and adaptation strategies, tailored to local and regional needs and priorities. Some marine problems can be circumvented with human interventions, e.g., reduction of atmospheric greenhouse gas concentrations, solar radiation management, protection of biota and ecosystems, and manipulation of biological and ecological adaptation [44]. One example of an effective climate change action is championed by the Blue Carbon Initiative [45] which aims to mitigate climate change through the restoration and sustainable use of coastal and marine ecosystems such as mangroves, tidal marshes, and seagrasses. The success of existing and emerging management strategies has to be validated and quantified with scientific means. Networks such as EVALSDGs [46], the International Organization for Cooperation in Evaluation (IOCE; [47]), and Eval4Action [48] can support SIDS to effectively evaluate efforts on the timely delivery of SDGs to guide evidence-based decision-making.
- *Raise Public Awareness*—Inform and educate the wider public about current and emerging local marine challenges and their impacts on social, environmental, and economic security to develop local awareness, expertise, knowledge, and thereby drive action. Initiatives such as UNESCO's Education for Sustainable Development (ESD) program [49] and the Sandwatch program [50] can support regional efforts in SIDS via climate change education.

While some SIDS lack the capacity to act on the above-mentioned recommendations, others such as the Pacific SIDS already have ocean observation processes in place, thanks to national and international support and close collaborations between governments. However, overall, SIDS remain dependent on the support of developed countries to help them establish, gather, access, and use data to build their capacity regarding monitoring activities, sustainable development, and decision-making processes [9,12]. Thus, the increase in national scientific knowledge including long-term education, training, and human resource development, as well as the improvement of national research capacities, and the transfer of marine technology has to be the fundamental purpose of capacity building efforts to help SIDS to help themselves with long-lasting effects. The development of knowledge and skills relevant in the design, development, implementation, management, and maintenance of institutional and scientific infrastructures is key to ensure local ownership and thus the achievement of nationally relevant SDGs. Multilateral organization such as SCOR [51], GEOTRACES [52], GOA-ON [53], and the IAEA [54] are involved in capacity building activities for ocean science and can help SIDS to build capacity via resources, technical guidance, scientific mentorship, training, workshops, summer schools, scholarship, and financial support. Capacity building is also provided by regional actors in SIDS such as the Western Indian Ocean Marine Science Association (WIOMSA; [55]), the Secretariat of the Pacific Regional Environment Programme (SPREP; [56]), and the Caribbean Community Climate Change Centre (CCCCC; [57]), just to name a few. In particular, universities in SIDS will be central players in national capacity building activities, since they are the generators of knowledge, sources of trained personnel, and hubs of innovation [16]. After successful capacity development efforts, effective approaches have to be found and implemented to retain human resources (i.e., trained people) in SIDS. This retention is so far a problem in most developing states, where often people that receive capacity building training leave their institutions and organizations afterwards because of better employment opportunities due to their newly acquired skills. Thus, capacity building is quickly lost again.

For institutional, research, and human capacity building activities to be effective in suitably realizing sustainable development in SIDS on a national level, we believe that it is necessary to take the following factors into account:

- *Coordination and facilitation of capacity building at a national and sub-national level*—To empower national ownership and local leadership, the involvement of stakeholders at the national level is necessary to understand national capacity needs, define and shape the national capacity building agenda, and subsequently guide and coordinate appropriate national capacity efforts. Coordinated governmental bodies, agencies, and institutions, with clearly defined responsibilities and duties, are a requirement to effectively and efficiently guide, navigate, and facilitate the implementation and management of capacity building efforts and SDGs without duplication and fragmentation [16]. Herein, the Ocean Action Hub platform [58] and organizations such as GOA-ON [40] can be used to support SIDS in creating web-based and regional science hubs to facilitate multi-stakeholder engagement in order to address regional challenges, needs, and priorities as a collective.
- *Enhancement of Collaboration and Coordination*—Bridging the gap between scientists, communities, and policy makers is an essential component to increase awareness and understanding of marine risks and to develop mitigation and adaptation responses and resource management strategies targeted to local priorities. Further, there is a need for more sustainable and effective international partnerships and collaborations between and among developed and developing nations to build capacity, transfer technology, link initiatives, share networks, and mobilise resources. International organizations such as the Intergovernmental Oceanographic Commission of UNESCO (IOC-UNESCO; [59]) and The Ocean Foundation [60] can help SIDS to facilitate dialogue among stakeholders and further catalyse partnerships among scientists, policymakers, academia, businesses, industry, and the public.
- *Increase Sustained Financial Support*—Develop coordinated funding strategies and identify existing or potential funding sources that will help complement regional long-term actions such as research, monitoring, and outreach activities. There is a strong need for long-term funding in SIDS to support ongoing costs of analysis centers (e.g., maintenance, staffing, quality control, standard

solutions, and consumables) and to guarantee efficient monitoring efforts and outputs relevant to improve and inform modelling predictions, national strategies, and policy. There are many regional and multi-lateral funding opportunities available for SIDS including those endorsed by the World Bank's Global Environment Facility (GEF), the EU's Advancing Capacity to support Climate Change Adaptation project, the Asian Development Bank, the Adaptation Fund (AF), the Climate Investment Funds (CIF), and most recently the Green Climate Fund (GCF).

Applied and basic marine research plays a strategic and important role in SIDS to move forward and identify, solve, and manage marine problems, particularly those related to SDG 14.1 to 14.5. Capacity building in science is, among others, an essential component in the process of achieving sustainable ocean and coastal development both at national and international levels [36]. However, most SIDS have limited infrastructure; technical and institutional capacity; financial, human, and material resources; as well as a lack of expertise and networks needed to produce scientific information to adequately develop and implement national plans in the context of international agreements [12,36]. To effectively support SIDS to address and overcome nationally-identified marine challenges and promote, develop, and strengthen marine science, sustainable and long-term regional capacity building and research infrastructure programs are key [12,36]. However, for stakeholder efforts to be effective in supporting individual SIDS in enhancing their capacities and capabilities to address their sustainable marine development priorities (developed in line with SDG 14.a), knowledge about their individual capacity building and research infrastructure requirements is necessary. Accordingly, an updated capacity assessment of human resources as well as institutional and technical capacities in each SIDS are needed to create suitable information for future program conception, implementation, execution, and evaluation.

5. Research Infrastructure and Capacities in UN SIDS

The following capacity assessment, which summarizes the existing key capacities in SIDS UN Member States, can be used as a basis to formulate appropriate capacity development responses by the international ocean science community, national governments, as well as funding and development agencies. Data are pooled from online databases including the World Bank [20] for data on national research and education expenditures as well as data on scientific publications, the World Higher Education Database [61] and the Commonwealth Network [62] for data on tertiary education institutions, and the UNESCO Data for Sustainable Development Explorer [63] for data on human resources in research and development. The authors acknowledge the fact that the paper is not a comprehensive analysis of the scientific capacity and capability of SIDS UN Member States in ocean sciences, but is rather a suitable starting point from which it is possible to gain a preliminary and useful insight into current research infrastructures and capacities in SIDS UN Member States. It should also be considered as the initial input of a publicly available and accessible database, future implementation of which we submit as a recommendation. It is also important to bear in mind that the capacity assessment and associated conclusions are limited by the information given on the above-mentioned websites (including the websites of each respective institution, Table S7). Accordingly, the capacity assessment omitted a discussion on research infrastructures and capabilities in SIDS Non-UN Members or Associate Members of Regional Commissions since there are only limited data available.

On average, the government in SIDS UN Member States invests approximately 14.5% of the total government expenditure on general education (European Union, 12.0%; North America 13.4%), and 14.7% of this total government expenditure for general education is invested in tertiary education (European Union, 22.0%; North America 28.0%). In regional terms, most SIDS make similar educational investments, but there are significant differences for some SIDS (Table 3; [20]). Data for research and development expenditure (% of GDP) are only available for some SIDS UN Member States, but usually the amount does not exceed 0.4% of the GDP, except for Singapore (Table 3; [20]), and is thus well below the research and development expenditures (% of GDP) of the global scientific powerhouses, particularly the United States (2.8%), China (2.2%), United Kingdom (1.7%), and Germany (3.0%) [20].

Table 3. Education and research capacity data of UN Small Island Developing States (SIDS) [20].

Island State	Expenditure-on Tertiary Education 2018 (% of Government Expenditure on Education)	Total Government Expenditure on Education 2018 (% of Government Expenditure)	Research and Development Expenditure 2018 (% of GDP)	Scientific and Technical Journal Articles (2016)
Caribbean SIDS				
Antigua and Barbuda	7	6.9	-	8
Bahamas	-	18.9	-	14
Barbados	29	12.9	-	48
Belize	9	21.3	-	6
Cuba	25	-	0.35	1045
Dominica	9	10.5	-	8
Dominican Republic	15	12.6	-	30
Grenada	14	14.9	-	34
Guyana	5	18.3	-	14
Haiti	10	13.1	-	30
Jamaica	20	18.6	0.06	135
St. Kitts and Nevis	15	8.6		21
St. Lucia	0	14.8	0.33	5
St. Vincent & Grenadines	7	18.8	0.12	2
Suriname	-	-	-	21
Pacific SIDS				
Trinidad and Tobago	10	-	-	3
F.S. of Micronesia	23	14.3	-	155
Fiji	9	11.5	-	2
Kiribati	15	22.5	-	1
Marshall Islands	-	-	-	1
Nauru	21	15.5	-	3
Palau	40	-	0.03	54
Papua New Guinea	9	10.5	-	5

Table 3. *Cont.*

Island State	Expenditure-on Tertiary Education 2018 (% of Government Expenditure on Education)	Total Government Expenditure on Education 2018 (% of Government Expenditure)	Research and Development Expenditure 2018 (% of GDP)	Scientific and Technical Journal Articles (2016)
Samoa	14	17.5	-	14
Solomon Islands	4	7.3	-	8
Timor-Leste	22	18.1	-	2
Tonga	0	-	-	14
Tuvalu	6	11.8	-	-
Vanuatu	10	-	-	3
AIS SIDS				
Cape Verde	17	16.4	0.07	8
Guinea-Bissau	4	16.2		12
Sao Tome and Principe	10	18.4	-	1
Comoros	10	15.3	-	5
Maldives	21	11.3	-	5
Mauritius	6	19.6	0.36	156
Seychelles	22	11.7	0.22	12
Singapore	35	20.0	2.22	11254
Bahrain	24	7.2	0.1	211

-: No data available.

The lack of expenditure on research and development and thus the low capacity to do research in Natural Sciences in SIDS UN Member States is also highlighted by the low output of scientific and technical articles (fields: physics, biology, chemistry, mathematics, clinical medicine, biomedical research, engineering and technology, and earth and space sciences; [20]) published in 2016 by SIDS UN Member States—they contributed only 0.6% (Caribbean SIDS: 0.07%, Pacific SIDS: 0.01%, and AIS SIDS: 0.51%; AIS excluding Singapore: 0.02%) of all the scientific and technical articles published worldwide in 2016 (Table 3; [20]).

Data on human resources in SIDS UN Member States are limited and barely up-to-date. Most SIDS suffer from the limitation of a small population and thus their national technical and research capacity is usually very low [38]. For instance, in Bahrain and Papua New Guinea, the number of researchers and technicians engaged in research and development, expressed as per million people, was 367 (2014) and 47 (2016), respectively, while it was 7187 (2014) in Singapore [20]. Consequently, SIDS experience great difficulties in developing local expertise to meet the wide-ranging and growing demands of national and regional marine sustainable development [64].

As of 2019, according to the World Higher Education Database and the Commonwealth Network, there are altogether 277 tertiary education institutions (i.e., universities, polytechnics, and colleges, including associated scientific institutes and centers) in the SIDS UN Member States and only 14 out of these 277 institutions disclose information (on their respective websites) regarding their scientific facilities and capacities. Twelve out of these 14 institutions (five in the Caribbean SIDS and seven in the AIS SIDS, six of which are located in Singapore) report on their facilities, infrastructures, and instrumentation to perform nuclear and isotopic techniques and organic and inorganic geochemistry (Figure 2 and Table S7). However, these values may not reflect the actual overall stage of development of marine science capabilities in SIDS given that the data were obtained from the websites of each respective institution and some of those websites do not disclose information regarding their scientific facilities, capabilities, and capacities or are particularly difficult to navigate. Thus, these figures have to be seen as 'minimum values' since they rely heavily on available statistics and the reporting mechanisms in place, which are usually insufficient in SIDS due to constraints in the human, technical, and financial resources required for generating this kind of information [9]. These values highlight that reporting mechanisms on marine science capacities in SIDS have to be put in place to aid in the development and implementation of relevant, effective, and successful capacity development responses in SIDS in order to improve their conditions for sustainable development of the ocean [9].

Up to 69, 70, and 58 SIDS UN Member State institutions currently teach courses or offer diplomas and certificates in Chemistry, Biology, and Environmental Science, respectively (Figure 2 and Table S7), and thus have the potential to put infrastructure in place (if not already existing) to conduct research necessary to deliver on regional and national environmentally focused SDGs. Furthermore, up to 40%, 21% (10% excluding Singapore) and 11% of Pacific, AIS, and Caribbean SIDS UN Member State institutions already offer courses and diplomas in Marine Science, Oceanography, Aquaculture, Fisheries, and Climate Change (Figure 2 and Table S7), and, therefore, possibly already address, monitor, and tackle national issues identified under SDG 14 (Life Below Water).

The Caribbean SIDS possess the highest tertiary institutional capacities and capabilities in SIDS regions, followed by the AIS and Pacific SIDS. This order might be surprising at first considering that the Caribbean SIDS have the lowest percentage of total government expenditure on tertiary education, but they do have the largest GDP per capita and the highest population (up to 3.5-fold) of the three SIDS regions and thus a higher demand for institutional capacities. The AIS SIDS, even though they have the lowest total population, find themselves ranked in the middle, between Caribbean and Pacific SIDS, regarding their amount of current tertiary institutional capacities and capabilities. This result is largely explained by Singapore, since education has been a key driver in Singapore's economic development over the last decades [65,66]. Singapore's aspirations to be a global education hub lead to large investments in the establishment and expansion of state-of-the-art campuses, centers, research laboratories, and joined degrees with international top-ranking universities [65,66]. The Pacific SIDS have the lowest amount of tertiary institutions, but considering their low GDP (15 times lower in relation to AIS SIDS), their institutional capacities are promising. In addition, up to 40% of the institutions in the Pacific SIDS engage in Marine Science-related topics compared to only 11% in the Caribbean SIDS and 21% (10% excluding Singapore) in the AIS SIDS. This outcome could be attributed to the fact that Pacific SIDS have the largest coastline and maritime area and contain the highest percentage of coral reefs and marine protected areas out of the three SIDS regions and thus have a higher need for stewardship of coastal and marine resources. In general, therefore, it seems that the Pacific SIDS are currently the leaders in ocean science-related research capacities and infrastructures in SIDS regions.

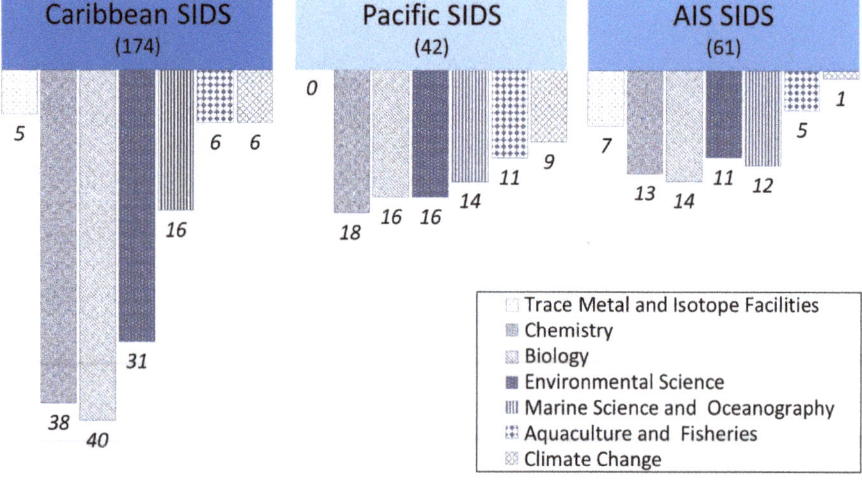

Figure 2. List of existing tertiary institutional capacities in UN Small Island Developing States (SIDS). The numbers in the parentheses at the top represent the number of existing tertiary education institutions in each respective SIDS region. Note that these data are limited by the information given on the website of each respective institution.

6. Conclusion and the Way Forward

The research capacities and capabilities in SIDS are heterogeneous, and some countries will need more help and funding than others to reach international standards. Despite some progress made in the last few years, if SDGs are to be met by SIDS in the next 10 years and if SIDS are to engage productively with international agendas in order to formulate and solve their own priorities it is necessary for SIDS to 1) have an educational shift toward relevant nationally-identified environmental issues, and 2) strengthen research capacities in light of each SIDS's distinctive social, cultural, economic, and environmental challenges. These measures should be supported by international, national, and regional collaborations and partnerships via funds and systematic capacity building which focuses on strengthening national self-determination and ownership [67]. Durable partnerships, based on mutual collaboration, ownership, trust, respect, accountability, and transparency, will be the means for achieving ocean sustainable development via concerted capacity building efforts. However, despite the urgency for SIDS to strengthen their research capacities and infrastructures, ways to achieve this have yet to be fully and critically explored. Nevertheless, there appears to be a consensus that capacity building must include individuals, institutions, and systems that collectively enable effective and sustainable development [16]. Yet, the way forward will still vary for each individual SIDS, dependent on the current state of research and education infrastructure, as well as social, cultural, economic, and environmental priorities.

The importance of relevant and durable research capacity and infrastructure in SIDS has been recognized globally, and regional, national, and international organizations, both past and present, including the UN, have accorded priority to this area, as reflected by SDG 4 (Quality Education). However, the current assessment highlights the fact that there is only limited, if any, up-to-date information publicly available on the number of researchers and technicians engaged in research and development in individual SIDS, and institutional websites commonly lack appropriate information on respective programs and institutional facilities. These limitations make it challenging to conduct a comprehensive analysis of the scientific capacity and capability of SIDS UN Member States to help national and regional policymakers as well as international agencies, programs, and projects to formulate appropriate national and international capacity development responses in, and for, SIDS. The current assessment is a suitable starting point from which it is possible to gain preliminary and useful insight into current research infrastructure and capacities in SIDS UN Member States, but considerably higher amounts of data are necessary to establish a greater degree of understanding of the current state in SIDS—How should we know where to start if we do not know where to start from?

To formulate appropriate national and international capacity development responses (e.g., funding, collaborations, partnerships, and resource management) to achieve SDGs in SIDS, an up-to-date database on educational, institutional, human, and research capacities is necessary. In other words, a bottom up approach is indispensable—a complete database will lay the foundation to provide adequate information on national and regional research capacities, gaps, and requirements. This information in turn will allow regional, national, and international parties to develop and implement durable, relevant, effective, and successful development responses, tailored to local and regional needs and priorities of each SIDS, to deliver on the SDG 14 as individuals and as a collective. To support efforts to create an ocean-related database, the current assessment highlights four key measures:

(1) *Collect data*—Gather more data on human resources, research infrastructures, and institutional capacities and facilities to create regional information baselines. In this connection, it is important that the public websites of institutions have easily-navigated site maps and include up-to-date information on institutional research facilities and capabilities. Regional initiatives (i.e., WIOMSA, SPREP, and CCCCC) will also be key in gathering and updating capacity information outside the tertiary environment and in carrying out activities to improve it.

(2) *Put data networks in place*—Data on research capacities, capabilities, and infrastructures have to be made publicly available and easily accessible via data products and networks to share data and thus inform governments, policymakers, and the international community to support better decision making. The current lack of information sharing and accessibility is supported by the fact that the IAEA's TC Programme supports 23 SIDS, 14, 5 and 4 of which are situated in the Caribbean, Pacific, and AIS region, respectively, but as of yet only 12 SIDS report on available research capacities and infrastructures for nuclear and isotopic science, including organic and inorganic geochemistry.

(3) *Identify metrics*—Reliable assessment procedures and metrics have to be put in place to evaluate the progress of capacity building efforts and serve as an analytical base to understand patterns and determinants of successful capacity building programs [16]. Frameworks such the Capacity Building Initiative for Transparency (CBIT; [68]) under the UNFCCC can help SIDS to track and report progress of existing and future country commitments.

(4) *Increase sustained funding and expertise*—The previously listed measures can only be realized if appropriate access to funding and expertise is put in place. The required financial resources are unlikely to be met from national government budgets since the lack of funding and expertise is the single major impediment in SIDS to sustainably carry out sustainable marine development in the first place [12]. Therefore, the long-term provision of financial resources and capacity building in SIDS must be sought through international and intergovernmental cooperation (lending/funding/donor/aid agencies) to enable SIDS to become equal partners in dealing with global economic and environmental issues [10].

Creating databases on research capacities and infrastructures in SIDS UN Member States will allow these states to develop and successfully implement appropriate development responses to the most urgent present and future national needs on ocean and coastal environmental issues. Long-term, durable, and appropriate capacity responses are decisive for the future of each region [9] and thus should form fundamental strategies to achieve social, cultural, political, economic, and environmentally sustainable development.

It is undeniable that national, regional, and international efforts have already led to an important and considerable improvement in SIDS capacities, both at regional and national levels. However, continued and long-term action is required to cope with the expected increase in the severity of climate-related impacts on ocean systems and services [44,69]. Multilateral programs, such as the Decade and the ongoing IAEA TC Programme, can play a major role in promoting ocean science, catalyzing action, facilitating cooperation, supporting networking forums, and providing capacity development, expertise, and resources in SIDS. The TC Programme for example, can act as a catalyzer, with an advisory, coordinative, and facilitating role to use nuclear science and technologies to address marine challenges, such as coastal zone management, impacts of climate change (e.g., ocean warming and ocean acidification), marine biodiversity loss, seafood safety and trade, food security, water resource management, coastal and marine pollution, harmful algal blooms, marine biotoxins, and sustainable aquaculture [70]. Presently, these marine threats are all areas of concern for SIDS. But more importantly, the IAEA can respond to the need for a database on research capacities and infrastructures in SIDS UN Member States by acting as a hub to develop, facilitate, coordinate, and lead such a data portal, as previously done via the IAEA's Ocean Acidification International Coordination Centre (OA-ICC) after concerns from UN Member States about ocean acidification in 2012. Only a concerted capacity building effort and action by international, national, and regional parties will allow the furthest behind to catch up and become an equal partner in addressing global ocean sustainable development and management.

Supplementary Materials: The following are available online at http://www.mdpi.com/2673-1924/1/3/9/s1, Figure S1: Geographical regions of Small Island Developing States (SIDS), Figure S2: Average composition of the Gross Domestic Product (GDP) of Small Island Developing State (SIDS) regions by sectors as of 2017/2018, Figure S3: Average imports and exports as a percentage of the Gross Domestic Product (GDP) by Small Island Developing State (SIDS) regions, Table S1: Targets of Sustainable Development Goal (SDG) 14—Life Below Water, Table S2: List of Small Island Developing States (SIDS) separated into UN Members and Non-UN Members/Associate Members of Regional Commissions, Table S3: List of Small Island Developing States (SIDS) that participate in the Technical Cooperation (TC) Programme of the International Atomic Energy Agency (IAEA), Table S4: List of physiographic data of Small Island Developing States (SIDS) in the Caribbean region, Table S5: List of physiographic data of Small Island Developing States (SIDS) in the Pacific region, Table S6: List of physiographic data of Small Island Developing States (SIDS) in the Atlantic, Indian Ocean, and South China Seas (AIS) region, Table S7: List of existing tertiary education institutional capacities in UN Small Island Developing States (SIDS).

Author Contributions: Conceptualization, S.P.P., P.W.S. and R.Z.; methodology, R.Z.; formal analysis, R.Z.; investigation, R.Z.; data curation, R.Z.; writing—original draft preparation, R.Z.; writing—review and editing, R.Z., S.P.P, P.M., S.G.S. and P.W.S. All authors have read and agreed to the published version of the manuscript.

Funding: This research was partially funded under the IAEA Interregional Project INT0093 *"Applying Nuclear Science and Technology in Small Island Developing States in Support of the Sustainable Development Goals and the SAMOA Pathway"*.

Acknowledgments: R.Z. thanks W.D.N. Dillon from the University of Otago for his help with the edits of the final draft and for valuable discussions regarding the concepts presented in this manuscript. P.W.S. appreciatively recognizes the continued support provided by the IAEA Environment Laboratories and the IAEA Technical Cooperation Programme. The IAEA is grateful for the support provided to its Environment Laboratories by the Government of the Principality of Monaco. This work contributes to the ICTA "Unit of Excellence" (MinECo, MDM2015-0552).

Conflicts of Interest: The authors declare no conflict of interest. The funders had no role in the design of the study; in the collection, analyses, or interpretation of data; in the writing of the manuscript, or in the decision to publish the results.

References

1. UN. Draft Outcome Document of the United Nations Summit for the Adoption of the Post-2015 Development Agenda. In *Draft Resolution Submitted by the President of the General Assembly, Sixty-Ninth Session, Agenda Items 13 (a) and 115, A/69/L.85, August 12, 2015*; UN: New York, NY, USA, 2015.
2. Schmidt, S.; Neumann, B.; Waweru, Y.; Durussel, C.; Unger, S.; Visbeck, M. *SDG14 Conserve and Sustainably Use the Oceans, Seas and Marine Resources for Sustainable Development*; International Council for Science: Paris, France, 2017.
3. Ntona, M.; Morgera, E. Connecting SDG 14 with the other Sustainable Development Goals through marine spatial planning. *Mar. Policy* **2018**, *93*, 214–222. [CrossRef]
4. Sturesson, A.; Weitz, N.; Persson, Å. SDG 14: Life below Water. A Review of Research Needs. In *Technical Annex to the Formas Report Forskning för Agenda 2030: Översikt av Forskningsbehov och Vägar Framåt*; Stockholm Environment Institute: Stockholm, Sweden, 2018.
5. Singh, G.G.; Cisneros-Montemayor, A.M.; Swartz, W.; Cheung, W.; Guy, J.A.; Kenny, T.A.; McOwen, C.J.; Asch, R.; Geffert, J.L.; Wabnitz, C.C.; et al. A rapid assessment of co-benefits and trade-offs among Sustainable Development Goals. *Mar. Policy* **2018**, *93*, 223–231. [CrossRef]
6. Visbeck, M. Ocean science research is key for a sustainable future. *Nat. Commun.* **2018**, *9*, 1–4. [CrossRef] [PubMed]
7. IOC. *The Science We Need for the Ocean We Want: The United Nations Decade of Ocean Science for Sustainable Development (2021–2030)*; IOC Brochure 2018-7 (IOC/BRO/2018/7 Rev); IOC: Paris, France, 2019; 24p.
8. Urban, E.R.; Bowie, A.R.; Boyd, P.W.; Buck, K.N.; Lohan, M.C.; Sander, S.G.; Schlitzer, R.; Tagliabue, A.; Turner, D. The Importance of Bottom-Up Approaches to International Cooperation in Ocean Science. *Oceanography* **2020**, *33*, 11–15. [CrossRef]
9. Valdés, L. *Global Ocean Science Report: The Current Status of Ocean Science around the World*; UNESCO Publishing: Paris, France, 2017.
10. UN. Small Island Developing States. Available online: https://sustainabledevelopment.un.org/topics/sids/list (accessed on 3 July 2020).
11. UN. LDCs at a Glance. Available online: https://www.un.org/development/desa/dpad/least-developed-country-category/ldcs-at-a-glance.html (accessed on 3 July 2020).
12. Kullenberg, G. Capacity building in marine research and ocean observations: A perspective on why and how. *Mar. Policy* **1998**, *22*, 185–195. [CrossRef]
13. UNFCCC. *Climate Change, Small Island Developing States*; Climate Change Secretariat (UNFCCC): Bonn, Germany, 2005; p. 10.
14. UN-OHRLLS. *Small Island Developing States in Numbers*; Climate Change Edition; UN-OHRLLS: New York, NY, USA, 2015.
15. SAMOA Pathway. SIDS Accelerated Modalities of Action Outcome Statement. 2015. Available online: https://www.un.org/ga/search/view_doc.asp?symbol=A/RES/69/15&Lang=E (accessed on 3 July 2020).
16. Khan, M.; Sagar, A.; Huq, S.; Thiam, P.K. *Capacity building under the Paris Agreement*; European Capacity Building Initiative: Oxford, UK, 2016.
17. IAEA. *How Nuclear Techniques Help Address Environmental Challenges of Island States*; IAEA Brief Environment 2017/4; IAEA Brief Environment: Vienna, Austria, 2017.
18. Baldacchino, G. Seizing history: Development and non-climate change in Small Island Developing States. *Int. J. Clim. Chang. Strateg. Manag.* **2018**. [CrossRef]
19. *The World Factbook*; Central Intelligence Agency: Washington, DC, USA, 2019. Available online: https://www.cia.gov/library/publications/resources/the-world-factbook/index.html (accessed on 1 October 2019).
20. World Bank. *World Development Indicators*; The World Bank Group: Washington, DC, USA, 2019; Available online: data.worldbank.org/indicator/EN.ATM.CO2E.PC (accessed on 1 October 2019).
21. Pauly, D.; Zeller, D. (Eds.) Sea around Us Concepts, Design and Data. 2015. Available online: http://www.seaaroundus.org/ (accessed on 1 October 2019).
22. Al-Jenaid, S.; Chiu, A.; Dahl, A.; Fleischmann, K.; Garcia, K.; Graham, M.; King, P.; Kubiszewsk, I.; McManus, J.; Ragoonaden, S.; et al. *GEO Small Islands Developing States Outlook*; IUCN: Gland, Switzerland, 2014.

23. IUCN. *International Union for Conservation of Nature's Red List of Threatened Species, Version 2019-2*; International Union for Conservation of Nature: Washington, DC, USA, 2019; Available online: https://www.iucnredlist.org (accessed on 1 October 2019).

24. Boto, I.; Biasco, R. *Small Island Economies: Vulnerabilities and Opportunities*; Brussels Rural Development Briefings. A series of meetings on ACP-EU development issues; CTA Brussels Office: Brussels, Belgium, 2012.

25. UNEP. *GEO Small Island Developing States Outlook*; United Nations Environment Programme: Nairobi, Kenya, 2014.

26. Marcelin, L.H.; Cela, T.; Shultz, J.M. Haiti and the politics of governance and community responses to Hurricane Matthew. *Disaster Health* **2016**, *3*, 151–161. [CrossRef] [PubMed]

27. OCHA. Office for the Coordination of Humanitarian Affairs. Haiti: Hurricane Matthew Situation Report no. 19. 2016. Available online: http://reliefweb.int/sites/reliefweb.int/files/resources/ocha_haiti_sitrep_19_02_nov_2016_eng.pdf (accessed on 3 July 2020).

28. Taft-Morales, M. Haiti's Political and Economic Conditions: In Brief. 2017. Available online: https://pdfs.semanticscholar.org/caa7/3af7dddfbd334ee753bf7d3f6ba97781c129.pdf (accessed on 3 July 2020).

29. UNSD. UN Statistics Division. 2019. Available online: https://unstats.un.org/home/ (accessed on 1 October 2019).

30. EM-DAT. The International Disaster Database. 2019. Available online: https://emdat.be/ (accessed on 1 October 2019).

31. UN/DESA. UN/DESA Policy Brief #64: The COVID-19 Pandemic Puts Small Island Developing Economies in Dire Straits. 2020. Available online: https://www.un.org/development/desa/dpad/publication/un-desa-policy-brief-64-the-covid-19-pandemic-puts-small-island-developing-economies-in-dire-straits/ (accessed on 3 July 2020).

32. Wong, P.P. Small island developing states. *Wiley Interdiscip. Rev. Clim. Chang.* **2011**, *2*, 1–6. [CrossRef]

33. CMEP. *Pacific Marine Climate Change Report Card 2018*; Bryony, T., Paul, B., Jeremy, H., Tommy, M., Sylvie, G., Awnesh, S., Gilianne, B., Patrick, P., Sunny, S., Tiffany, S., Eds.; Commonwealth Marine Economies Programme: London, UK, 2018; 12p.

34. Nurse, L.A.; McLean, R.F.; Agard, J.; Briguglio, L.P.; Duvat-Magnan, V.; Pelesikoti, N.; Tompkins, E.; Webb, A. Small islands. In *Climate Change 2014: Impacts, Adaptation, and Vulnerability. Part B: Regional Aspects. Contribution of Working Group II to the Fifth Assessment Report of the Intergovernmental Panel on Climate Change*; Barros, V.R., Field, C.B., Dokken, D.J., Mastrandrea, M.D., Mach, K.J., Bilir, T.E., Chatterjee, M., Ebi, K.L., Estrada, Y.O., Genova, R.C., Eds.; Cambridge University Press: Cambridge, UK; New York, NY, USA, 2014; pp. 1613–1654.

35. Sjöstedt, M.; Povitkina, M. Vulnerability of Small Island Developing States to Natural Disasters: How Much Difference Can Effective Governments Make? *J. Environ. Dev.* **2017**, *26*, 82–105. [CrossRef]

36. Montero, G.G. The Caribbean: Main experiences and regularities in capacity building for the management of coastal areas. *Ocean Coast. Manag.* **2002**, *45*, 677–693. [CrossRef]

37. UN. Report of the Conference of the Parties on Its Tenth Session, Held at Buenos Aires from 6 to 18 December 2004. Available online: https://unfccc.int/resource/docs/cop10/10a01.pdf#page=7 (accessed on 3 July 2020).

38. Ocean Health Index. About Ocean Health Index. Available online: http://www.oceanhealthindex.org (accessed on 3 July 2020).

39. Ocean Best Practices Ocean Best Practices System. Available online: www.oceanbestpractices.org (accessed on 3 July 2020).

40. GOA-ON. About GOA-ON. Available online: http://www.goa-on.org/home.php (accessed on 3 July 2020).

41. ODIS. Ocean Data and Information System "Catalogue of Sources". Available online: https://catalogue.odis.org/ (accessed on 3 July 2020).

42. OA-ICC. IAEA Ocean Acidification International Coordination Centre (OA-ICC) portal for ocean acidification biological response data. Available online: http://oa-icc.ipsl.fr/ (accessed on 3 July 2020).

43. GOOS. Framework for Ocean Observing. Available online: https://www.goosocean.org/ (accessed on 3 July 2020).

44. Gattuso, J.P.; Magnan, A.K.; Bopp, L.; Cheung, W.W.; Duarte, C.M.; Hinkel, J.; Mcleod, E.; Micheli, F.; Oschlies, A.; Williamson, P.; et al. Ocean solutions to address climate change and its effects on marine ecosystems. *Front. Mar. Sci.* **2018**, *5*, 337. [CrossRef]

45. The Blue Carbon Initative. About Blue Carbon. Available online: https://www.thebluecarboninitiative.org/about-blue-carbon (accessed on 3 July 2020).
46. EvalPartners. EvalAgenda2020. Available online: https://www.evalpartners.org/about/about-us (accessed on 3 July 2020).
47. IOCE. About the International Organization for Cooperation in Evaluation (IOCE). Available online: https://www.ioce.net/ (accessed on 3 July 2020).
48. Eval4Action. The Decade of Evaluation for Action. Available online: https://www.eval4action.org/ (accessed on 3 July 2020).
49. UNESCO ESD. Education for Sustainable Development. Available online: https://en.unesco.org/themes/education-sustainable-development (accessed on 3 July 2020).
50. UNESCO. SANDWATCH: A Science Education Scheme through Sustainable Coastal Monitoring. Available online: http://www.unesco.org/new/en/natural-sciences/priority-areas/sids/sandwatch/ (accessed on 3 July 2020).
51. SCOR. Capacity Development. Available online: http://scor-int.org/work/capacity/ (accessed on 3 July 2020).
52. GEOTRACES. GEOTRACES Capacity Building Activities. Available online: https://www.geotraces.org/geotraces-capacity-building-activities/ (accessed on 3 July 2020).
53. GOA-ON. Pier2Peer. Available online: http://www.goa-on.org/pier2peer/pier2peer.php (accessed on 3 July 2020).
54. IAEA. Training Courses. Available online: https://www.iaea.org/services/education-and-training/training-courses (accessed on 3 July 2020).
55. WIOMSA, About Western Indian Ocean Marine Science Association (WIOMSA). Available online: https://www.wiomsa.org/ (accessed on 3 July 2020).
56. SPREP. About The Secretariat of the Pacific Regional Environment Programme (SPREP). Available online: https://www.sprep.org/ (accessed on 1 July 2020).
57. CCCC. The Caribbean Community Climate Change Centre—Empowering People to Act On Climate Change. Available online: https://www.caribbeanclimate.bz/ (accessed on 3 July 2020).
58. UNDP About the Ocean Action Hub. Available online: https://www.oceanactionhub.org/about-ocean-action-hub (accessed on 3 July 2020).
59. UNESCO-IOC. Intergovernmental Oceanographic Commission (IOC). Available online: http://www.unesco.org/new/en/natural-sciences/ioc-oceans/ (accessed on 3 July 2020).
60. The Ocean Foundation. About The Ocean Foundation. Available online: https://oceanfdn.org/ (accessed on 3 July 2020).
61. WHED. IAU the World Higher Education Database (WHED). Available online: https://www.whed.net (accessed on 3 July 2020).
62. Commonwealth Network. Commonwealth in Action. Available online: http://www.commonwealthofnations.org/commonwealth-in-action/ (accessed on 3 July 2020).
63. UNESCO. Data for the Sustainable Development Goals. Available online: http://uis.unesco.org/ (accessed on 3 July 2020).
64. UNEP. Progress in the implementation of the programme of action for the sustainable development of small island developing States. Report of the Secretary-General–Addendum. In Proceedings of the Commission on Sustainable Development, Sixth Session, New York, NY, USA, 20 April–1 May 1998; UNEP: Nairobi, Kenya, 1989.
65. Luke, A.; Freebody, P.; Shun, L.; Gopinathan, S. Towards research-based innovation and reform: Singapore schooling in transition. *Asia Pac. J. Educ.* **2005**, *25*, 5–28. [CrossRef]
66. Olds, K. Global assemblage: Singapore, foreign universities, and the construction of a "global education hub". *World Dev.* **2007**, *35*, 959–975. [CrossRef]
67. Crossley, M.; Holmes, K. Challenges for educational research: International development, partnerships and capacity building in small states. *Oxf. Rev. Educ.* **2001**, *27*, 395–409. [CrossRef]
68. CBIT. The Capacity Building Initiative for Transparency (CBIT) Global Coordination Platform. Available online: https://www.cbitplatform.org/about (accessed on 3 July 2020).

69. Gattuso, J.-P.; Magnan, A.; Billé, R.; Cheung, W.W.L.; Howes, E.L.; Joos, F.; Allemand, D.; Bopp, L.; Cooley, S.R.; Eakin, C.M.; et al. Contrasting futures for ocean and society from different anthropogenic CO_2 emissions scenarios. *Science* **2015**, *349*, aac4722. [CrossRef] [PubMed]
70. Betti, M. ASLO 2009: The Marine Environment Laboratories of the International Atomic Agency, Monaco (IAEA-MEL). *Limnol. Oceanogr. Bull.* **2008**, *17*, 88–89. [CrossRef]

Technical Note

Oceanographic Determinants of the Abundance of Common Dolphins (*Delphinus delphis*) in the South of Portugal

Joana Castro [1,2,*], Ana Couto [2,3], Francisco O. Borges [2], André Cid [1], Marina I. Laborde [1,2], Heidi C. Pearson [4] and Rui Rosa [2]

[1] AIMM—Associação para a Investigação do Meio Marinho, 1500-399 Lisboa, Portugal; andre.cid@aimmportugal.org (A.C.); marina.laborde@aimmportugal.org (M.I.L.)

[2] MARE—Marine and Environmental Sciences Centre, Faculdade de Ciências da Universidade de Lisboa, 1749-016 Lisboa, Portugal; anasscouto@gmail.com (A.C.); franciscoomcborges@gmail.com (F.O.B.); rarosa@fc.ul.pt (R.R.)

[3] CIBIO/InBIO—Universidade do Porto, 4485-661 Porto, Portugal

[4] University of Alaska Southeast, Juneau, AK 99801, USA; hcpearson@alaska.edu

* Correspondence: jmadeiracastro@gmail.com

Received: 5 July 2020; Accepted: 25 August 2020; Published: 28 August 2020

Abstract: Off mainland Portugal, the common dolphin (*Delphinus delphis*) is the most sighted cetacean, although information on this species is limited. The Atlantic coast of Southern Portugal is characterized by an intense wind-driven upwelling, creating ideal conditions for common dolphins. Using data collected aboard whale-watching boats (1929 sightings and 4548 h effort during 2010–2014), this study aims to understand the relationships between abundance rates (AR) of dolphins of different age classes (adults, juveniles, calves and newborns) and oceanographic [chlorophyll a (*Chl-a*) and sea surface temperature (SST)] variables. Over 70% of the groups contained immature animals. The AR of adults was negatively related with *Chl-a*, but not related to SST values. The AR of juveniles was positively related with SST. For calves and newborns, although the relationship between SST and AR is similar to that observed for juveniles, the effect could not be distinguished from zero. There was no relationship between *Chl-a* levels and the AR of juveniles, calves and newborns. These results corroborate previous findings that common dolphins tend to occur in highly productive areas demonstrating linkages between their abundance and oceanographic variables, and that this region may be a potential nursery ground.

Keywords: ecology; oceanography; Portugal; abundance rate; nursery; common dolphin; *Delphinus delphis*

---

1. Introduction

The abundance and distribution of cetaceans is influenced by a series of oceanic and environmental variables [1,2]. Several studies worldwide have demonstrated these relationships, and have shown that cetacean movements vary within and between species [2] and are influenced by numerous variables, including sea-surface temperature [3,4], salinity [3], depth [5], seabed gradient [6], thermocline [7], oxygen minimum layer [8] and prey availability [9]. As top predators, cetacean distribution is closely related to the distribution of their prey [10–12], which in turn can be affected by upwelling systems [13–15] and SST [1]. Thus, in order to understand the factors driving cetacean distribution, insight into these environmental variables is needed.

The Iberian Peninsula constitutes an excellent scenario for conducting ecological niche studies of small cetaceans, because it constitutes a transition area between two distinct environments (i.e., the Mediterranean Sea and the North Atlantic Ocean), and thus exhibits a high level of habitat complexity [16,17]. The coastal region off southwest Iberia is part of the North Atlantic eastern boundary, which encompasses the northern branch of the Canary/Iberian Eastern Boundary Upwelling System (EBUS)—one of the world's foremost productive marine ecosystems [18] and the western part of the northern margin of the Gulf of Cadiz [19,20] (Cape St. Vicente point—Sagres, see Figure 1). This region is characterized by an intense wind-driven upwelling season spanning from March to October, and by an upwelling center in the area of Cape St. Vicente [21]. These features restrict sardines (*Sardina pilchardus*), one of the main prey species of small cetaceans (e.g., common dolphins) off Portugal [22,23], to coastal waters.

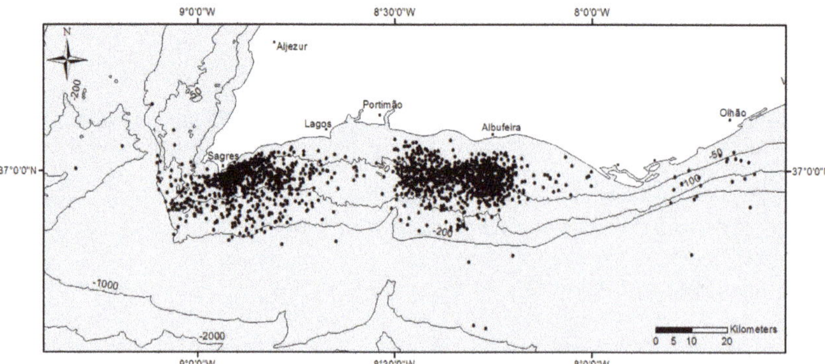

Figure 1. Map of the study area in southern Portugal with bathymetric lines in meters. Each black dot represents one common dolphin (*Delphinus delphis*) sighting.

The south coast of the Iberian Peninsula has been the target of many studies that have revealed a rich cetacean biodiversity [24–26]. However, knowledge concerning the distribution and abundance of cetaceans off the Atlantic coast of the Iberian Peninsula remains scarce and often limited to specific areas [17,24,25,27]. In particular, few studies have examined cetacean occurrence in the waters off southern Portugal [17]. The common dolphin (*Delphinus delphis*) is the most abundant cetacean species in this area [17], although very little is known regarding their biology and ecology, including their abundance and distribution. Further, there is limited information regarding the effect of the aforementioned environmental variables on common dolphin distribution [17]. Common dolphins occur in most coasts of the world, and mainly in the continental shelf but can be found in all depth ranges [28–31]. The common dolphin is described as a highly gregarious species, forming groups of up to several hundreds or even thousands of individuals, although presenting a basic social unit of between 20 to 30 individuals per group [32]. Generally regarded as an opportunistic feeder [10,33], this species is considered a preferential ichtyophageous in the South of Portugal [34], in the Bay of Biscay [35,36] and in the Mediterranean [37].

The apparent lack of studies on cetaceans off southern Portugal could be related to the logistical constraints of performing dedicated cetacean surveys. Such surveys are demanding in terms of the financial resources, time, and personnel required, and thus researchers often utilize alternative study approaches, such as the use of platforms of opportunity (e.g., whale-watching vessels [31,38]). Although there are inherent limitations to this approach (e.g., the potential interaction between the vessel and cetaceans, and the influence of sea state and wind direction on the sighting ability, distribution and behaviour of the animals [38]), platforms of opportunity can yield valuable information regarding cetacean ecology that would otherwise be unobtainable.

To better understand the role of environmental variables on the ecology of common dolphins off southern Portugal, systematic data on their distribution together with oceanographic data must be collected [2]. Therefore, the main goal of this study was to model the abundance of common dolphins in the southern waters of Portugal in relation to oceanographic variables, specifically chlorophyll a (*Chl-a*) and SST, using data collected aboard platforms of opportunity. Ultimately, this study contributes to a better understanding of patterns in the occurrence and habitat use of common dolphins off the Portuguese southern coast.

2. Material and Methods

For the purpose of this study, data collection took place over the span of five years (between April 2010 and November 2014) aboard 11 whale-watching boats from a total of seven companies, which have the common dolphin as one of its target species. Observations occurred along the south coast of Portugal, between the areas of Cape St. Vicente and Olhão (Figure 1) where random transects were conducted, at a maximum distance of 25 nautical miles from shore (the boats used were not allowed to go further offshore due to permits and boat class), and occurred between April and November of each year.

All whale-watching trips occurred between 9 a.m. and 5 p.m., lasting for 90 min each. The number of trips per day varied according to the availability of tourists, the sea state, meteorological conditions and when such conditions were deemed by the whale watching companies as being safe to conduct the activity. During whale-watching trips, two trained observers continuously searched for common dolphins in the region 180° ahead of the vessel. For each common dolphin sighting, GPS position, date and time, bathymetry, group size and composition were recorded. The length of the whale-watching boats was between 7 and 10 m, and bathymetry kept between 30 and 250 m. A group was defined as any pod of dolphins observed in apparent association and moving in the same direction, usually engaged in similar activities [39]. Groups were documented if present at a distance of 300 m to 30 m from the boat. Group composition was described according to the number of adults: individuals between 180 and 220 cm in length, apparently fully grown and physically mature; juveniles: immature animals not yet fully grown, but larger than calves; calves: animals $\leq \frac{1}{2}$ the length of an adult, travelling alongside an adult; and newborns: animals with the same size or less than calves, exhibiting foetal folds and travelling alongside an adult [40]. Immature animals were defined as all individuals that did not appear fully grown (ca < 180 cm).

Sampling effort was measured as the number of hours spent searching for dolphins. Abundance rate (AR) was calculated as the monthly rate of the number of individual common dolphins to the hours of sampling (effort). Only sightings observed from 0 to 4 according to the Beaufort Scale were considered. Sightings observed above Beaufort 4 were discarded.

Concerning the oceanographic variables, L3 satellite-derived monthly night time sea surface temperature (SST) and *Chl-a* data were acquired from Moderate-resolution Imaging Spectroradiometer (Aqua-MODIS; http://oceancolor.gsfc.nasa.gov/), with a spatial resolution of approximately 4 km. The monthly average of each variable was calculated for the area covered by the surveys (Lat: 36.5–37.5, Lon: −10, −7 decimal degrees), before being used in the models.

Sighting data were analyzed according to specific age classes (adult, juvenile, calf and newborn). For each age class, generalized additive models (GAMs) with a lognormal distribution were used to estimate relationships between the AR and environmental variables (SST and *Chl-a*). GAMs are semi-parametric extensions of generalized linear models (GLMs) with the ability to deal with non-linear relationships between response and explanatory variables [41]. As non-linear relationships were expected a priori, GAMs were preferred over GLMs. Year was included as a random effect to account for temporal variance that cannot be explained by the environmental variables.

For all variables, a thin plate regression spline was used. Model selection was done automatically by including an additional penalty for each smooth, so that if smoothing parameters for a term tend to infinity, the term will be selected out of the model [42]. Statistical analysis was performed using the "mgcv" [41] package in "R" (version 3.6.1, R Development Core Team).

3. Results

Sampling effort totalled 4548 h (Table 1), and a total of 1929 common dolphin groups were sighted. Groups composed only of adults corresponded to 27.52%, while the rest of the sighted groups contained adults and immature animals (72.48%). Calves and/or newborns were present in 42.5% of the sightings.

On average, group size was 31 ± 51.6 individuals (ranging from 1 to 1000 individuals), with groups consisting, on average, of 4.9 ± 8.55% juveniles, 3.0 ± 6.15% calves, and 0.8 ± 3.03% newborns. All age classes were observed in every year sampled, and overall the abundance rates did not differ between years (F = 0.51, p = 0.728; Figure 2).

There was variation in oceanographic variables according to month and year (see Supplementary Figure S1). SST tends to increase during the spring and summer months, reaching higher values towards the end of the summer (months of August and September) depending on the year, while *Chl-a* tends to decrease. For adults, the effect of *Chl-a* was significantly different from zero (which indicates no effect) (p value < 0.05), with a negative relationship between abundance rate and *Chl-a* observed. There was no effect of SST on the abundance rate of adults (Figure 3).

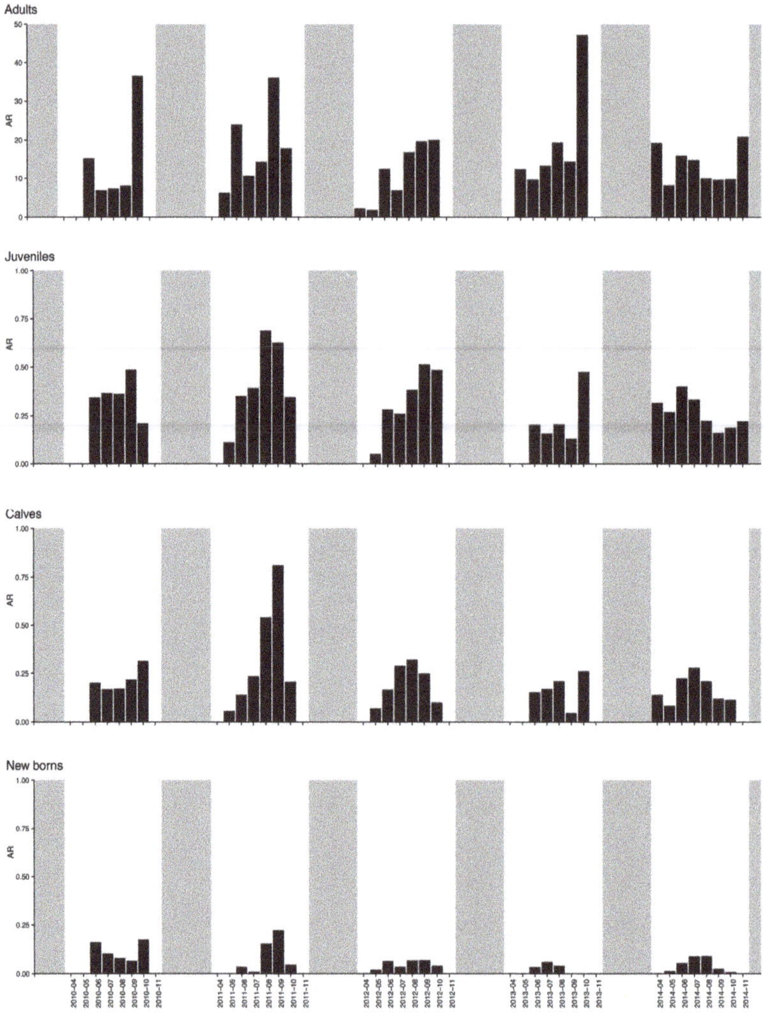

Figure 2. Monthly abundance rates (AR) of common dolphin (*Delphinus delphis*) adults, juveniles, calves and newborns during the study sampling period (April 2010 to November 2014). Grey X-axis breaks represent months with no sampling due to unfavourable winter conditions.

Table 1. Summary of the number of sightings and sampling hours (effort) per month (April to November), during the sampling period of 2010–2014. Grey areas represent months with no sampling due to unfavorable winter conditions.

	April	May	June	July	August	September	October	November	Total
No. sightings 2010	0	0	29	145	164	72	27	0	437
Effort (hours) 2010	0	0	49.5	508.5	658.5	216	28.5	0	1461
No. sightings 2011	13	7	17	64	95	62	31	0	289
Effort (hours) 2011	21	18	28.5	102	174	76.5	43.5	0	463.5
No. sightings 2012	1	8	24	71	119	43	27	0	293
Effort (hours) 2012	13.5	99	78	195	250.5	72	49.5	0	757.5
No. sightings 2013	0	15	43	144	168	67	58	0	495
Effort (hours) 2013	9	19.5	117	280.5	271.5	130.5	42	0	870
No. sightings 2014	17	27	91	120	93	31	34	2	415
Effort (hours) 2014	28.5	70.5	147	192	276	156	121.5	4.5	996

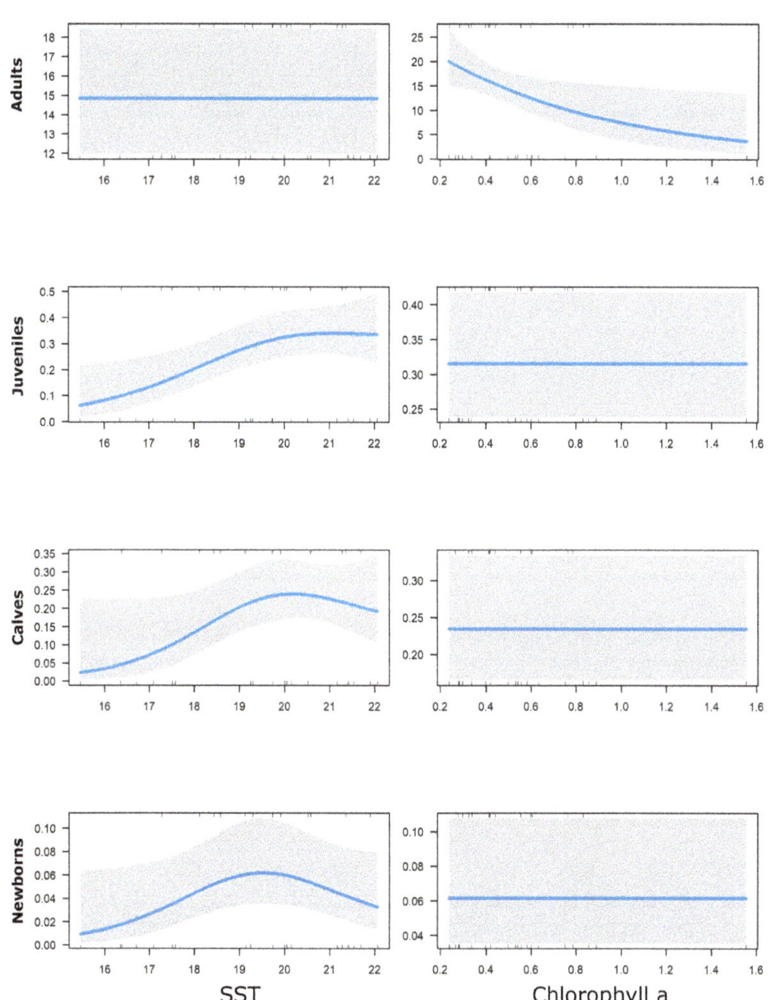

Figure 3. Model predictions for each age class.

For juveniles, the effect of SST was significantly different from no effect (p value < 0.05), with a positive relationship between abundance rate and SST observed. There was no effect of *Chl-a* on the abundance rate of juveniles (Figure 3). Although similar trends between abundance rate of calves and newborns and SST were observed (Figure 3), there was no evidence that the observed trends were different from zero (no effect) with a *p*-value higher than 0.05 (Figure 3). There was no effect of *Chl-a* on the abundance rate of calves and newborns (Figure 3).

4. Discussion

This study utilized whale watching boats as platforms of opportunity, providing support for its effectiveness as a cetacean research technique. However, it is important to point out some of the limitations that these platforms present when compared with dedicated scientific surveys, namely, in this study, the unsystematic sampling effort that can lead to biased results [17].

Common dolphins are considered to be opportunistic feeders [10,33], thus it is reasonable to expect that their abundance is related to *Chl-a* concentration, which is often used as a proxy for prey abundance and distribution [24]. Previous studies conducted off the Portuguese coast considered that chlorophyll concentration was one of the most important variables associated with this species' distribution [17,43].

In our study, increasing values of *Chl-a* were related to lower AR of adults. It is likely that the reduced ARs could be due to the succession from primary to secondary consumers (i.e., the change in the local trophic web to larger predators, preying upon primary consumers) in response to the increase in microalgae abundance [43]. As such, the presence of potential prey was likely still relatively reduced when *Chl-a* values were higher.

Our results showed a positive relationship between the AR of juveniles and SST values and suggest a similar relationship for calves and newborns. This is consistent with other common dolphin studies, in which groups containing immature individuals (juvenile, calf or newborn) tend to be more frequently associated with warmer waters [44].

The majority of the groups sighted in this study were composed of both adults and immature individuals (this represented 72.48% of the total number of sightings), and the presence of calves and/or newborns (accounting for 42.5%) was of particular relevance. This strongly suggests the importance of this area as a potential nursery ground for common dolphins and is consistent with other studies [31]. Furthermore, several studies indicate that water temperature influences cetaceans births [45–47] and, for example, in bottlenose dolphins (*Tursiops truncatus*) a peak of births usually coincides with a peak in water temperature [45,47]. SST values for the Portuguese coast are highest during the period of intense upwelling activity in southern Portugal (i.e., between March and October [48]). According to Cañadas and Vázquez [49], SST plays a significant role in common dolphin distribution and density and a strong change in SST may affect how this species distributes and their abundance, at least at a local level.

As with previous studies [1,44,50], our results suggest that common dolphin abundance is linked to not only "extrinsic" factors (in this case chlorophyll concentration and SST) but also to "intrinsic" factors such as the presence of juveniles, calves and newborns. In order to improve our knowledge on common dolphin ecology, it is essential to develop more studies to comprehend how these influences work.

5. Conclusions

Our study has shown that the South of Portugal is an important area for the common dolphin, particularly as a potential nursery ground, and that the AR of this species are associated, to some extent, with oceanographic variables such as SST and chlorophyll. Conservation strategies should consider these relationships, especially with respect to changing conditions due to climate change. For example, changes in upwelling and ocean current oscillation patterns could lead to decreased prey availability, justifying the need for increased conservation and management action. While the common dolphin is currently not considered to be endangered or threatened in Portugal (according to the International

Union for Conservation of Nature [51]), it is important to use a proactive approach and manage this species for "robustness" [52] in order to increase resiliency to future environmental or anthropogenic changes. For effective conservation action, further research is needed to better understand the current human uses of these coastal waters in which common dolphins occur.

Supplementary Materials: The following are available online at http://www.mdpi.com/2673-1924/1/3/12/s1. Figure S1. Monthly variation of the environmental variables examined: Sea Surface Temperature (SST) and Chlorophyll a (Chla). Grey X-axis breaks represent months with no sampling due to unfavourable winter conditions and thus were excluded from analysis. Figure S2. Model validation plots of the four categories: Adults, Juveniles, Calves and Newborns.

Author Contributions: Conceptualization, J.C. and A.C. (André Cid); Methodology, J.C. and A.C. (André Cid); Software, A.C. (Ana Couto), M.I.L. and F.O.B.; Validation, J.C., A.C. (Ana Couto) and F.O.B.; Formal Analysis, A.C. (Ana Couto), M.I.L. and F.O.B.; Investigation, J.C. and A.C. (André Cid); Resources, J.C.; Data Curation, J.C. and A.C. (Ana Couto); Writing—Original Draft Preparation, J.C., F.O.B. and A.C. (Ana Couto); Writing—Review and Editing, H.C.P. and R.R.; Visualization, J.C., A.C. (Ana Couto) and F.O.B.; Supervision, H.C.P. and R.R.; Project Administration, J.C.; Funding Acquisition, J.C., A.C. (André Cid), M.I.L., F.O.B. and R.R. All authors have read and agreed to the published version of the manuscript.

Funding: This research was funded by Associação para a Investigação do Meio Marinho (AIMM) and Fundação para a Ciência e a Tecnologia (FCT), through Programa Investigador FCT 2013 granted to RR. FCT also supported this study through (i) the strategic project UID/MAR/04292/2019 granted to MARE and (ii) PhD grant to J.C. (SFRH/BD/134156/2017), M.I.L. (SFRH/BD/84297/2012) and F.O.B. (SFRH/BD/147294/2019).

Acknowledgments: The authors would like to thank the following whale-watching companies for their support during this study: Cape Cruiser, AlgarExperience, Dolphins Driven, Dream Wave, X Ride, EcoDolphin and Formosa Mar. A special thanks to all AIMM volunteers who participated in the data collection.

Conflicts of Interest: The authors have no conflict of interest to declare.

References

1. Neumann, D.R. The activity budget of free-ranging common dolphins (*Delphinus delphis*) in the Northwestern Bay of Plenty, New Zealand. *Aquat. Mamm.* **2001**, *27*, 121–136.
2. Hastie, G.; Swift, R.; Slesser, G.; Thompson, P.; Turrel, W. Environmental models for predicting oceanic dolphin habitat in the Northeast Atlantic. *ICES J. Mar. Sci.* **2005**, *62*, 760–770. [CrossRef]
3. Selzer, L.A.; Payne, P.M. The distribution of white-sided (*Lagenorhynchus acutus*) and common dolphins (*Delphinus delphis*) vs. environmental features of the continental shelf of the Northeastern United States. *Mar. Mammal Sci.* **1988**, *4*, 141–153. [CrossRef]
4. Baumgartner, M.F.; Mullin, K.D.; May, L.N.; Leming, T.D. Cetacean habitats in the Northern Gulf of Mexico—Statistical data included. *Fish. Bull.* **2001**, *99*, 219–239. [CrossRef]
5. Gowans, S.; Whitehead, H. Distribution and habitat partitioning by small odontocetes in the gully, a submarine canyon on the scotian shelf. *Can. J. Zool.* **1995**, *73*, 1599–1608. [CrossRef]
6. Baumgartner, M.F. The distribution of Risso's dolphin (*Grampus griseus*) with respect to the Physiography of the Northern Gulf of Mexico. *Mar. Mammal Sci.* **1997**, *13*, 614–638. [CrossRef]
7. Reilly, S.B. Seasonal changes in distribution and habitat differences among dolphins in the Eastern Tropical Pacific. *Mar. Ecol. Prog. Ser.* **1990**, *66*, 1–11. [CrossRef]
8. Polacheck, T. Relative abundance, distribution and inter-specific relationship of cetacean schools in the Eastern Tropical Pacific. *Mar. Mammal Sci.* **1987**, *3*, 54–77. [CrossRef]
9. Cockcroft, V.G.; Peddemors, V.M. seasonal distribution and density of common dolphins *Delphinus delphis* off the south-east coast of Southern Africa. *S. Afr. J. Mar. Sci.* **1990**, *9*, 371–377. [CrossRef]
10. Young, D.D.; Cockcroft, V.G. Diet of common dolphins *Delphinus delphis* off the south-east coast of Southern Africa: Opportunism or specialization? *J. Zool.* **1994**, *234*, 41–53. [CrossRef]
11. Amaral, A.R.; Beheregaray, L.B.; Bilgmann, K.; Freitas, L.; Robertson, K.M.; Sequeira, M.; Stockin, K.A.; Coelho, M.M.; Möller, L.M. Influences of past climatic changes on historical population structure and demography of a cosmopolitan marine predator, the common dolphin (genus *Delphinus*). *Mol. Ecol.* **2012**, *21*, 4854–4871. [CrossRef] [PubMed]

12. Henderson, E.E.; Forney, K.A.; Barlow, J.P.; Hildebrand, J.A.; Douglas, A.B.; Calambokidis, J.; Sydeman, W.J. Effects of fluctuations in sea-surface temperature on the occurrence of small cetaceans off Southern California. *Fish. Bull.* **2014**, *112*, 159–177. [CrossRef]

13. Ballance, L.T.; Pitman, R.L.; Fiedler, P.C. Oceanographic influences on seabirds and cetaceans of the eastern tropical pacific: A review. *Prog. Oceanogr.* **2006**, *69*, 360–390. [CrossRef]

14. Bilgmann, K.; Möller, L.M.; Harcourt, R.G.; Gales, R.; Beheregaray, L.B. Common dolphins subject to fisheries impacts in southern australia are genetically differentiated: Implications for conservation. *Anim. Conserv.* **2008**, *11*, 518–528. [CrossRef]

15. Valdez, F.P.; Corrigan, S.; Möller, L.; Bilgmann, K.; Beheregaray, L.B.; Allen, S. Fine-Scale genetic structure in short-beaked common dolphins (*Delphinus delphis*) along the East Australian current. *Mar. Biol.* **2010**, *158*, 113–126. [CrossRef]

16. Alcaraz, D.; Paruelo, J.; Cabello, J. Identification of current ecosystem functional types in the Iberian Peninsula. *Glob. Ecol. Biogeogr.* **2006**, *15*, 200–212. [CrossRef]

17. Moura, A.E.; Sillero, N.; Rodrigues, A. Common dolphin (*Delphinus delphis*) habitat preferences using data from two platforms of opportunity. *Acta Oecologica* **2012**, *38*, 24–32. [CrossRef]

18. Bograd, S.J.; Rykaczewski, R.R.; García-Reyes, M.; Sydeman, W.J.; Bakun, A.; Miller, A.J.; Black, B.A. Anticipated effects of climate change on coastal upwelling ecosystems. *Curr. Clim. Chang. Rep.* **2015**, *1*, 85–93. [CrossRef]

19. Cravo, A.; Relvas, P.; Cardeira, S.; Rita, F.; Madureira, M.; Sánchez, R. An upwelling filament off southwest iberia: Effect on the chlorophyll a and nutrient export. *Cont. Shelf Res.* **2010**, *30*, 1601–1613. [CrossRef]

20. Cravo, A.; Relvas, P.; Cardeira, S.; Rita, F. Nutrient and chlorophyll a transports during an upwelling event in the NW margin of the gulf of cadiz. *J. Mar. Syst.* **2013**, *128*, 208–221. [CrossRef]

21. García Lafuente, J.; Ruiz, J. The gulf of cádiz pelagic ecosystem: A review. *Prog. Oceanogr.* **2007**, *74*, 228–251. [CrossRef]

22. Silva, M.A.; Sequeira, M. Patterns in the mortality of common dolphins (*Delphinus delphis*) on the portuguese coast, using stranding records, 1975–1998. *Aquat. Mamm.* **2005**, *29*, 88–98. [CrossRef]

23. Beare, D.; Pierce, G.J.; Patterson, I.A.P.; Reid, R.J.; Learmonth, J.A.; Reid, D.G.; Santos, M.B.; Ross, H.M. Variability in the diet of harbor porpoises (*Phocoena phocoena*) in scottish waters 1992–2003. *Mar. Mammal Sci.* **2006**, *20*, 1–27. [CrossRef]

24. Cañadas, A. Towards Conservation of Dolphins in the Alborán Sea. Hacia La Conservación de Los Delfines En El Mar de Alborán. Ph.D. Thesis, Universidad Autónoma de Madrid, Madrid, Spain, 2006.

25. De Stephanis, R.; Cornulier, T.; Verborgh, P.; Sierra, J.S.; Gimeno, N.P.; Guinet, C. Summer spatial distribution of cetaceans in the strait of gibraltar in relation to the oceanographic context. *Mar. Ecol. Prog. Ser.* **2008**, *353*, 275–288. [CrossRef]

26. Verborgh, P.; De Stephanis, R.; Pérez, S.; Jaget, Y.; Barbraud, C.; Guinet, C. Survival rate, abundance, and residency of long-finned pilot whales in the strait of gibraltar. *Mar. Mammal Sci.* **2009**, *25*, 523–536. [CrossRef]

27. Brito, C.; Vieira, N.; Sá, E.; Carvalho, I. Cetaceans' occurrence off the west central portugal coast: A compilation of data from whaling, observations of opportunity and boat-based surveys. *Environ. Res.* **2009**, *2*, 2–5.

28. Forcada, J.; Hammond, P. Geographical variation in abundance of striped and common dolphins of the western mediterranean. *J. Sea Res.* **1998**, *39*, 313–325. [CrossRef]

29. Peddemors, V.M. Delphinids of Southern Africa: A review of their distribution, status and life history. *J. Cetacean Res. Manag.* **1999**, *1*, 157–165.

30. Cañadas, A.; Sagarminaga, R.; García-Tiscar, S. Cetacean distribution related with depth and slope in the mediterranean waters off Southern Spain. *Deep. Res. Part I Oceanogr. Res. Pap.* **2002**, *49*, 2053–2073. [CrossRef]

31. Evans, P.G.H.; Hammond, P.S. Monitoring cetaceans in european waters. *Mamm. Rev.* **2004**, *34*, 131–156. [CrossRef]

32. Evans, P. *Dolphins*; Whittet Books Limited: London, UK, 1994.

33. Gannier, A. Les Cétacés de Méditerranée Nord-Occidentale: Estimation de Leur Abondance et Mise En Relation de La Variation Saisonnière de Leur Distribution Avec l'écologie Du Milieu. Ph.D. Thesis, Ecole Pratique des Hautes Etudes, Montpellier, France, 1995.

34. Santos, M.; German, I.; Correia, D.; Read, F.; Martinez Cedeira, J.; Caldas, M.; López, A.; Velasco, F.; Pierce, J. Long-Term variation in common dolphin diet in relation to prey abundance. *Mar. Ecol. Prog. Ser.* **2013**, *481*, 249–268. [CrossRef]

35. Pusineri, C.; Magnin, V.; Meynier, L.; Spitz, J.; Hassani, S.; Ridoux, V. Food and feeding ecology of the common dolphin (*Delphinus delphis*) in the oceanic northeast atlantic and comparison with its diet in neritic areas. *Mar. Mammal Sci.* **2007**, *23*, 30–47. [CrossRef]
36. Spitz, J.; Mourocq, E.; Léauté, J.-P.; Quéro, J.-C.; Ridoux, V. Prey selection by the common dolphin: Fulfilling high energy requirements with high quality food. *J. Exp. Mar. Biol. Ecol.* **2010**, *390*, 73–77. [CrossRef]
37. Gannier, A. Present distribution of common dolphin *Delphinus delphis* in French mediterranean and adjacent waters as obtained from small boat surveys. *Aquat. Conserv. Mar. Freshw. Ecosyst.* **2018**, 1–8. [CrossRef]
38. Vinding, K.; Bester, M.; Kirkman, S.P.; Chivell, W.; Elwen, S.H. The Use of Data from a platform of opportunity (Whale Watching) to study coastal cetaceans on the southwest coast of South Africa. *Tour. Mar. Environ.* **2015**, *11*, 33–54. [CrossRef]
39. Shane, S.H. Behavior and ecology of the bottlenose dolphin at Sanibel Island, Florida. *Bottlenose Dolphin* **1990**, 245–265.
40. Neumann, D.R.; Orams, M.B. Behaviour and ecology of common dolphins (*Delphinus delphis*) and the impact of tourism in mercury bay, North Island, New Zealand. *Sci. Conserv.* **2005**, 5–40.
41. Wood, S.N. *Generalized Additive Models: An Introduction with R*, 2nd ed.; CRC Press: New York, NY, USA, 2017. [CrossRef]
42. Marra, G.; Wood, S.N. Practical variable selection for generalized additive models. *Comput. Stat. Data Anal.* **2011**, *55*, 2372–2387. [CrossRef]
43. Kämpf, J.; Chapman, P. *Upwelling Systems of the World*; Springer: Cham, Switzerland, 2016. [CrossRef]
44. Stockin, K.A.; Pierce, G.J.; Binedell, V.; Wiseman, N.; Orams, M.B. Factors affecting the occurrence and demographics of common dolphins (*Delphinus* sp.) in the Hauraki Gulf, New Zealand. *Aquat. Mamm.* **2008**, *34*, 200–211. [CrossRef]
45. Würsig, B. Occurrence and group organization of Atlantic bottlenose porpoises (*Tursiops truncatus*) in an argentine bay. *Biol. Bull.* **1978**, *154*, 348–359. [CrossRef]
46. Evans, P.G.H. *The Natural History of Whales and Dolphins*; Christopher Helm Mammal Serie: London, UK, 1987.
47. Bearzi, G.; Notarbartolo-Di-Sciara, G.; Politi, E. Social ecology of bottlenose dolphins in the Kvarneric (Northern Adriatic Sea). *Mar. Mammal Sci.* **1997**, *13*, 650–668. [CrossRef]
48. Baptista, V.; Silva, P.L.; Relvas, P.; Teodósio, M.A.; Leitão, F. Sea surface temperature variability along the portuguese coast since 1950. *Int. J. Climatol.* **2018**, *38*, 1145–1160. [CrossRef]
49. Cañadas, A.; Vázquez, J.A. Common dolphins in the alboran sea: Facing a reduction in their suitable habitat due to an increase in sea surface temperature. *Deep Sea Res. Part II Top. Stud. Oceanogr.* **2017**, *141*, 306–318. [CrossRef]
50. Cañadas, A.; Hammond, P.S. Abundance and habitat preferences of the short-beaked common dolphin *Delphinus delphis* in the southwestern mediterranean: Implications for conservation. *Endanger. Species Res.* **2008**, *4*, 309–331. [CrossRef]
51. Hammond, P.S.; Bearzi, G.; Bjorge, A.; Forney, K.; Karczmarski, L.; Kasuya, T.; Perrin, W.F.; Scott, M.D.; Wang, J.Y.; Wells, R.S.; et al. *Delphinus delphis*. In *The IUCN Red List of Threatened Species*; International Union for Conservation of Nature and Natural Resources: Gland, Switzerland, 2008; p. e.T6336A12649851. [CrossRef]
52. Würsig, B.; Reeves, R.R.; Ortega-Ortiz, J.G. Global climate change and marine mammals. In *Marine Mammals*; Evans, P.G.H., Raga, J.A., Eds.; Springer US: Boston, MA, USA, 2002; pp. 589–608. [CrossRef]

Article

Climate and Local Hydrography Underlie Recent Regime Shifts in Plankton Communities off Galicia (NW Spain)

Antonio Bode *, Marta Álvarez, Luz María García García, Maria Ángeles Louro, Mar Nieto-Cid, Manuel Ruíz-Villarreal and Marta M. Varela

Instituto Español de Oceanografía, Centro Oceanográfico de A Coruña, 15001 A Coruña, Spain;
marta.alvarez@ieo.es (M.Á.); luz.garcia@ieo.es (L.M.G.G.); gelines.louro@ieo.es (M.Á.L.);
mar.nietocid@ieo.es (M.N.-C.); manuel.ruiz@ieo.es (M.R.-V.); marta.varela@ieo.es (M.M.V.)
* Correspondence: antonio.bode@ieo.es; Tel.: +34-981205362

Received: 15 June 2020; Accepted: 23 September 2020; Published: 25 September 2020

Abstract: A 29-year-long time series (1990–2018) of phyto- and zooplankton abundance and composition is analyzed to uncover regime shifts related to climate and local oceanography variability. At least two major shifts were identified: one between 1997 and 1998, affecting zooplankton group abundance, phytoplankton species assemblages and climatic series, and a second one between 2001 and 2002, affecting microzooplankton group abundance, mesozooplankton species assemblages and local hydrographic series. Upwelling variability was relatively less important than other climatic or local oceanographic variables for the definition of the regimes. Climate-related regimes were influenced by the dominance of cold and dry (1990–1997) vs. warm and wet (1998–2018) periods, and characterized by shifts from low to high life trait diversity in phytoplankton assemblages, and from low to high meroplankton dominance for mesozooplankton. Regimes related to local oceanography were defined by the shift from relatively low (1990–2001) to high (2002–2018) concentrations of nutrients provided by remineralization (or continental inputs) and biological production, and shifts from a low to high abundance of microzooplankton, and from a low to high trait diversity of mesozooplankton species assemblages. These results align with similar shifts described around the same time for most regions of the NE Atlantic. This study points out the different effects of large-scale vs. local environmental variations in shaping plankton assemblages at multiannual time scales.

Keywords: phytoplankton; zooplankton; time series; regime shift; climate; nutrients

---

1. Introduction

Ecosystem regime shifts, or "sudden, dramatic, long-lasting changes in ecosystem structure and function" [1], represent changes in state that may be irreversible. Earlier theories implied that the ecosystem alternated between two stable states depending on the intensity of the attractor variables and feedback mechanisms, while recent studies reveal rather more complex stepwise changes [2]. Most of the existing studies point to climatic oscillations as major drivers of these shifts [3–5]. However, shifts may imply non-linear responses to many drivers [6], including local conditions and anthropogenic factors, such as fisheries [7]. Recognizing shifts is important when interpreting past changes and predicting future states (e.g., those related to climate change), as they provide the basis for management and the development of conservation policies [8,9].

Regime shifts in marine ecosystems are increasingly reported as the time series of observations reach several decades. For instance, there are examples of shifts in ecosystem components such as plankton [10–12], fish populations [7,13], and whole food webs [7,14–16]. Recent evidence points to a quasi-synchronicity in these changes, particularly those observed near the turn of the 21st century [2,3,7,8,17], with local variations imposed on large-scale, climatic drivers.

Recent plankton regime shifts were described for several regions of the NE Atlantic. Major shifts were observed in the late 1980s, as illustrated by the analysis of observational time series from the North Sea and nearby regions [3,4,17], but soon more examples were provided for southern areas, such as the English Channel [6,18,19], the Bay of Biscay [14,15], and the Iberian shelf [20,21]. All these studies reported a major reorganization of plankton assemblages but there was a large variability in the potential drivers identified, from climate-related factors such as the North Atlantic Oscillation [3,6,7,10] to local factors operating at decadal time scales [12]. While both factors are currently known to operate synergistically (e.g., [2]), one of the limitations for separating climatic from local effects is the availability of observations collected over decadal time periods.

Previous studies of the year-to-year changes in the pelagic ecosystems of the northwestern Iberian shelf and southern Bay of Biscay were limited by the availability of long time series. Consequently, they focused on stationary responses to climate [22–26] and only recently were regime shifts recognized in plankton [20,21,27] and fish [9]. Because of the major importance of the seasonal upwelling as inputs of nutrients for primary production in this region [28,29], these studies dedicated specific efforts to finding predictable deterministic functions of the ecosystem response to variations in upwelling intensity or frequency. For instance, dramatic changes in primary production and phytoplankton composition were predicted from decreases in upwelling [25]. However, both the results of continued analysis of observational time series [27] and current estimates of future changes in upwelling [30] do not support these predictions. This discrepancy suggests that both climate-driven and local environmental factors would probably contribute to the observed long-term patterns. In this context, the identification of regime shifts may shed new light on the distinction among climate and local factors affecting the structure of marine ecosystems in this region.

The objective of the present analysis is to illustrate major regime shifts in plankton assemblages in the Galician coast (NW Spain) since 1990 in relation to shifts in climate and local oceanography. The resulting regimes are compared to those described in other regions of the NE Atlantic with common patterns highlighted.

2. Materials and Methods

2.1. Data Series

This study was made on annually resolved series of climatic and meteorological variables, and local oceanographic and plankton data collected between 1990 and 2018 (Supplementary Tables S1 and S2). Oceanographic and plankton data were provided by the time series project "Series temporales de oceanografía en el norte de España" (RADIALES) [31]. In this study, we used data from Station E2CO (43°25.32′ N, 8°26.22′ W, 80 m deep) located in shelf waters off A Coruña (NW Spain).

Several climatic indices were compiled to represent global and large-scale regional variability and climatic teleconnections: the North Atlantic Oscillation (NAO), the Atlantic Multidecadal Oscillation (AMO), the index of Global Average Temperature and Sea Surface Temperature Anomalies (GLOTI), the Pacific Decadal Oscillation (PDO), the Arctic Oscillation Index (AO), the El Niño Index (ENN), and the Number of Sun Spots (SUNSP). Monthly values of climatic series were obtained from the National Oceanographic and Atmospheric Administration (NOAA Earth System Research Laboratory of the United States [32], except for NAOs, which were obtained from the Climate Research Unit of the University of East Anglia [33]. Values of each series were averaged for each year between 1990 and 2018 (or the latest date available). In the case of NAO, summer (July, August) and winter (December to March) averages were also computed. All of these indices were reported to co-vary to some degree, with long-term variability in plankton in previous studies on the North Atlantic (e.g., [15,16,19,34]).

Meteorological variables included precipitation in A Coruña (PCO) provided as six-hourly values by the Agencia Española de Meteorología (AEMET) [35]. For this study, annual means and sd values of PCO were computed along with those for the spring season (March–May), as previous studies in the region indicated covariations in precipitation with plankton [21,27]. Wind forcing was expressed by the Upwelling Index (UI), or Ekman transport derived from surface winds in a cell of $1° \times 1°$ centered at $44°$ N, $9°$ W ([36]). Annual series of UI values were obtained from the six-hourly values reported by the Instituto Español de Oceanografía [37]. Four different series of UI were used: annual mean and sd series, and annual mean of positive (upwelling) or negative (downwelling) monthly values.

Local observations at Station E2CO were represented by monthly values of temperature, salinity, inorganic nutrients, dissolved oxygen, chlorophyll-a, primary production and particulate organic matter measured in water samples collected with Niskin bottles at 7 depth levels (5 levels in case of primary production). Details of the sampling and analysis of these variables are provided elsewhere [27]. To represent the vertical variability, three different integrations were made: means for the surface layer (0–5 m) and means and sd for the whole water column. These series of environmental variables will be subsequently termed as hydrography series (Supplementary Table S1). The original raw data for these series are available from the PANGAEA repository [38,39]. Plankton data at Station E2CO included phytoplankton, microzooplankton (40–200 μm) and mesozooplankton (>200 μm). In all series, only the periods with continuous and comparable data were considered for the subsequent analysis, thus avoiding possible methodological bias introduced by changes in the taxonomy experts' identifications. Phytoplankton data were counts of cells in several taxonomic categories obtained at 3–7 depth levels (see details in [40]). For this study, annual mean values were computed for the 0–40 m depth layer from monthly values. Taxa included major groups such as diatoms, dinoflagellates, Cryptophyceae, small flagellates and other groups with lower abundance (Supplementary Table S1) recorded between 1990 and 2017, but also detailed counts for several species or genera consistently recognized between 1990 and 2010 (Supplementary Table S2). Microzooplankton series contained abundance data from monthly samples collected by vertical hauls of a Bongo-type net (50 cm in diameter, 40-μm mesh), subsequently filtered through a 200-μm screen and preserved in buffered formaldehyde (4% final concentration). Individuals were identified with a stereomicroscope and the abundance of major taxonomic groups recorded between 1990 and 2015 (Supplementary Table S1). Mesozooplankton was collected by double oblique tows of a Juday–Bogorov net (50 cm in diameter, 200 μm mesh size) between the surface and the bottom ([23]). The abundance of major taxonomic groups was recorded between 1990 and 2018, and also detailed counts by species or genera consistently recognized between 1994 (1990 for Cladocera) and 2018 (Supplementary Table S2). The original raw data for plankton series are available from the PANGAEA repository (phytoplankton [41]microzooplankton [42]; mesozooplankton [43]).

2.2. Data Preparation

All series were converted to annual means to filter seasonal variability and focus on multiannual variability. Previous studies stressed that a large source of variability was seasonal, i.e., six and or 12-month periodicity, particularly for meteorological and plankton series (e.g., [44]). Large outliers (i.e., exceeding 3 sd) were removed and replaced by the equivalent value up to 3 times the series sd. All series were tested for normal distribution (Shapiro–Wilk test), autocorrelation and trends (Mann–Kendall test). No significant autocorrelation at the annual time scale was found for any of the series. Mean and sd values and significant trends were provided in Supplementary Tables S1 and S2. Transformations were applied where required for normality (logarithmic and fourth-root transformations). Finally, all annual series were expressed as 3-year running means of standardized anomalies from the overall series mean.

2.3. Statistical Analysis

The variability of each type of series (climate, hydrography, phytoplankton, micro- and mesozooplankton) was summarized by Principal Component Analysis (PCA). A previous selection of series was made for each type to avoid large correlations of related variables (Supplementary Figure S1). The first components of each PCA were subsequently analyzed for regime shift using the Rodionov test [45]. Regime shifts were identified at $p < 0.05$ using a cut-off length of 10 years and red noise correction for autocorrelation.

The identification of regimes in plankton species assemblages was made by applying Non-Metric Multidimensional Analysis (MDS) on species series. The main regimes suggested by the ordination were confirmed with cluster analysis (Euclidean distance, paired group clustering). The most discriminating species for the regimes were identified with the procedure Similarity Percentages (SIMPER) [46]. Plankton regimes were related to regime shifts identified from PCA variables by projecting the principal components for each series with the two first coordinates of the MDS.

The statistical packages PRIMER 6.0 [46] and PAST 4.0 [47] were used for data transformation and analysis.

3. Results

3.1. Multiannual Variability of Series

Large year-to-year variability was observed in all series (e.g., selected series in Figure 1), with significant trends in some of them (see Supplementary Tables S1 and S2), which suggests multiannual phases. For instance, there was a significant increase in climatic indices, such as AMO from 1990 to mid-2000, and also rises in GLOTI and SUNSP (Supplementary Table S1). In other cases, there were quasi-decadal oscillations, as observed for UI. In the case of hydrography series, there was a significant increase in phosphate and silicate but a decrease in the mean oxygen concentrations after 2000 (Supplementary Table S1). Moreover, there was a significant rise in primary production throughout the series. Phytoplankton series showed a significant decrease in the total abundance of diatoms, while overall abundance did not show a clear trend (Figure 1). Microzooplankton abundance displayed wide oscillations in total abundance, mainly due to copepods, while other groups (e.g., medusae and Siphonophora) became less abundant. In turn, mesozooplankton showed a clear increase in the abundance of almost all groups, but particularly of copepods, Bivalvia, Euphausiacea, Cladocera, Gastropoda and Chaetognatha, after 2000 (Supplementary Table S1). All series showed relative maxima and minima clustered in periods of several years (Figure 2). For instance, climate series showed positive anomalies at the beginning and end of the 1990s, but some series also showed these anomalies after 2012 (e.g., NAO, AMO, PDO, AO). Most hydrography series changed from negative to positive anomalies around 2002. However, phytoplankton series did not show a clear trend, while microzooplankton series showed a general dominance of positive anomalies after 2002, and almost all mesozooplankton series concentrated positive anomalies after 1998.

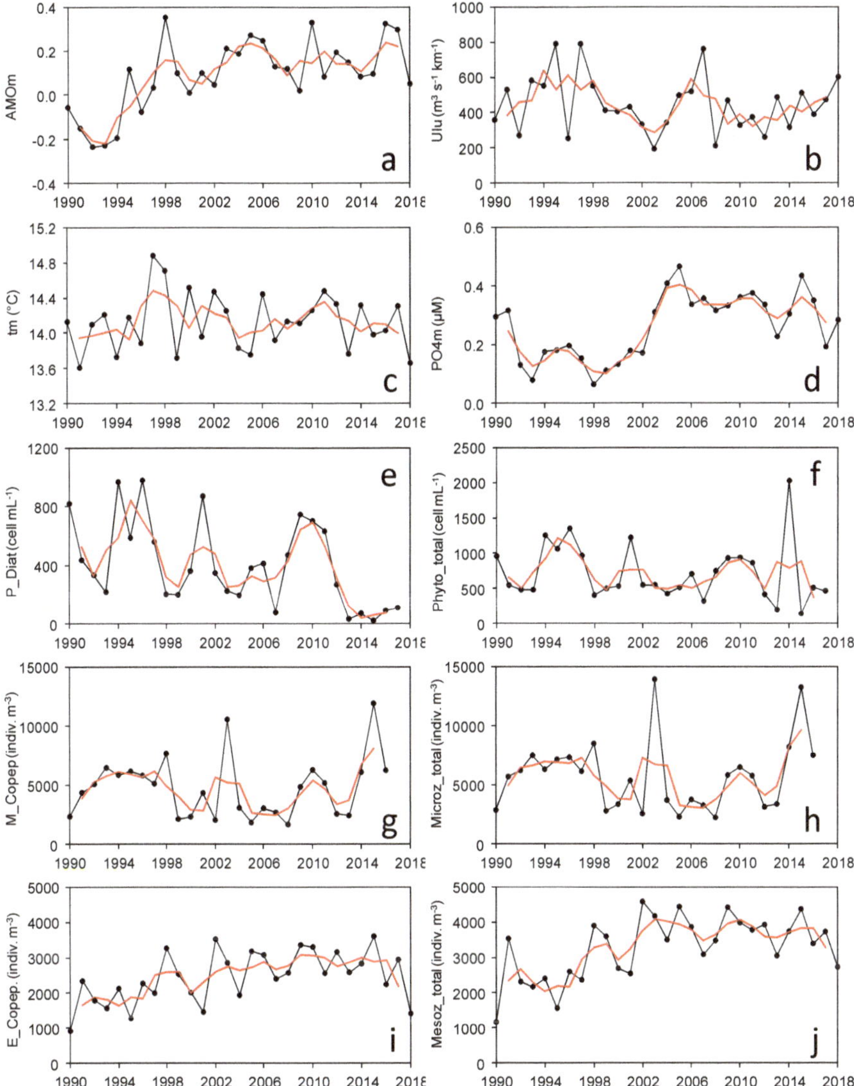

Figure 1. Distribution of annual means of example series for each variable category (climate, hydrography, phytoplankton, microzooplankton, and mesozooplankton). (**a**): North Atlantic Oscillation (NAOm, relative units), (**b**): positive monthly values of Upwelling Index (UIu, m^3, s^{-1}, km^{-1}), (**c**): water column temperature (tm, °C), (**d**): water column phosphate concentration (PO4m, µM), (**e**): photic zone diatom abundance (P_Diat, cell mL^{-1}), (**f**): photic zone total phytoplankton abundance (P_Phyto_total, cell mL^{-1}), (**g**): water column integrated microzooplanktonic copepod abundance (M_Cope, indiv. m^{-3}), (**h**): water column integrated total microzooplankton abundance (Microz_total, indiv. m^{-3}), (**i**): water column integrated mesozooplanktonic copepod abundance (E_Cope, indiv. m^{-3}), (**j**): water column integrated total mesozooplankton abundance (Mesoz_total, indiv. m^{-3}). Red lines indicate 3-year running means.

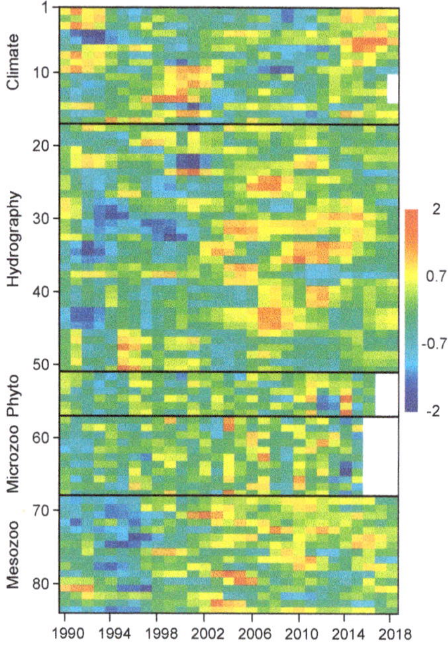

Figure 2. Distribution of anomalies from the overall mean for climate, hydrography and plankton series employed in this study. Each series was prefiltered (±3 sd), transformed for normality, normalized, standardized and smoothed using a 3-year running mean (see Methods). Series identification is shown in Supplementary Table S1.

3.2. Regime Shifts

The two first components of the PCA accounted for 56.08–66.37% of the total variance for the different series types (Supplementary Table S3). For climatic–meteorological series, the highest correlations with the first component (PC1) included the summer NAO, El Niño index (positive) and AMO, GLOTI and the mean precipitation at A Coruña (negative). For hydrography series, phosphate and silicate concentrations and primary production were the main variables with positive values of correlation with PC1, while oxygen and particulate organic carbon and nitrogen had the highest values of negative correlation. The abundance of flagellates, Cryptophycea and other taxa were the main contributors to the positive values of the PC1 for phytoplankton series, while diatoms and dinoflagellates contributed to negative values. Copepods generally dominated the positive correlations with PC1 for both micro- and mesozooplankton series, the latter showing negative correlations with other groups, such as Foraminifera and Annelida.

Most of the PC1 for each series type showed significant regime shifts (Figure 3). In the case of climate series, the shift was between 1997 and 1998 ($p < 0.01$), while it was between 2001 and 2002 ($p < 0.001$) for hydrography series. No shift was found for phytoplankton series. In contrast, microzooplankton showed shifts between 2003 and 2004 ($p < 0.05$) and between 2012 and 2013 ($p < 0.05$). One shift between 1997 and 1998 was determined for mesozooplankton series ($p < 0.001$). Few additional shifts were found for the second principal component (PC2) of all series (Supplementary Table S3, Supplementary Figure S2). For instance, there was a shift in climate PC2 between 2013 and 2014, related to the positive correlations of this component with GLOTI and the negative correlations with SUNSP. Mesozooplankton PC2 (positively correlated with Cnidaria and negatively with Euphausiacea) also showed shifts between 2006 and 2007, and between 2016 and 2017.

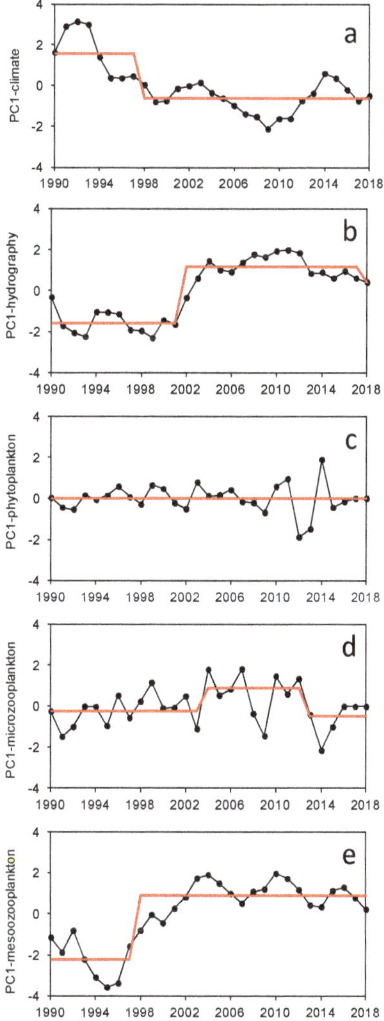

Figure 3. Time series of the first principal component of the Principal Component Analysis (PCA) on each type of series. (**a**): Climate series, (**b**): hydrography series, (**c**): phytoplankton series, (**d**): microzooplankton series; (**e**): mesozooplankton series. The red line indicates the weighed means of the regimes determined with the Rodionov test (see Methods).

3.3. Regimes in Plankton Species Assemblages

Detailed taxa series showed different patterns among groups of taxa: (Figure 4 and Supplementary Table S2). In the case phytoplankton, positive anomalies prevailed either at the beginning or at the end of the series. Those prevailing mainly at the beginning were mostly diatoms (e.g., *Chaetoceros socialis, Ch. gracilis, Ch. decipiens, Lauderia annulata, Skeletonema costatum*) but also taxa from Dictyochophyceae (*Octactis speculum*), dinoflagellates (*Tripos lineatus*) or Euglenophyceae (*Eutreptiella* spp.). Those prevailing at the end were dinoflagellates (e.g., *Gyrodinium spirale, Heterocapsa niei*) but also diatoms (e.g., *Pseudo-nitzschia delicatissima, Navicula transitans*, other *Chaetoceros* and *Nitzschia* spp., *Thalassiossira levanderi*) and Dictyochophyeace (*Dictyocha fibula*). For mesozooplankton, positive anomalies clustered before or after approximately 2002.

Before 2002, the dominant taxa were Cladocera (*Penilia avirostris*) and copepods (e.g., *Centropages chierchiae, Ctenocalanus vanus, Temora spinifera, T. longicornis*). After 2002, there was a change in the dominant taxa of Cladocera (*Evadne spinifera*) and copepods (e.g., *Centropages typicus, Oithona nana, Corycaeus* spp.).

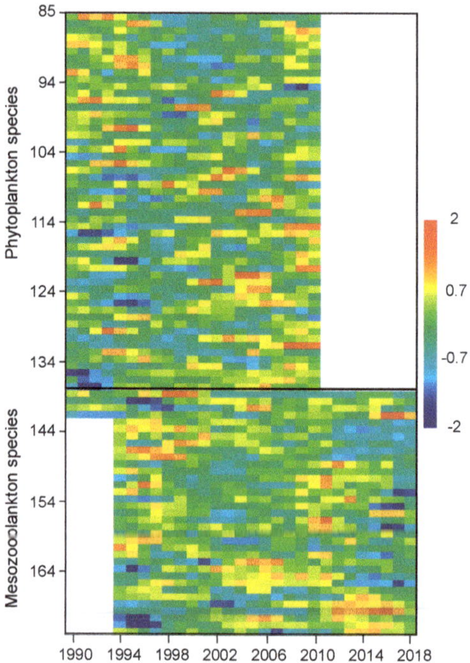

Figure 4. Distribution of anomalies from the overall mean for series of phytoplankton and mesozooplankton species considered in this study. Each series was prefiltered (±3 sd), transformed for normality, normalized, standardized and smoothed using a 3-year running mean (see Methods). Series identification is shown in Supplementary Table S2.

For both phytoplankton and mesozooplankton, two main temporal regimes were identified by MDS and cluster analysis (Figure 5 and Supplementary Figure S2). Phytoplankton showed a first regime for the period 1990–1996, correlated positively with climate PC1 and negatively with the first component of the other series. The second phytoplankton regime (1997–2010) was negatively correlated with climate PC1 and positively with the first component of the other series. Diatoms (*N. transitans, P. delicatissima, Chaetoceros* spp.), dinoflagellates (*Torodinium robustum, H. niei*), and Dictyochophyceae (*D. fibula*) were the main taxa that increased from the first to the second regime (Table 1). In turn, taxa with decreasing abundances were mainly diatoms (*L. annulata, S. costatum*, several spp. of *Chaetoceros*).

Mesozooplankton regimes spanned the periods 1994–2001 (with a positive correlation with PC1 of climate, and a negative correlation with the other series, except phytoplankton) and 2002–2018 (with negative correlations with climate PC1 and positive correlations with other series, except phytoplankton). The main discriminating taxa for these periods were those increasing their abundance between regimes, such as some copepods (*Oithona similis, O. nana*) and Cladocera (*Podon intermedius, Evadne nordmani*), and those decreasing, such as the copepods *C. vanus, T. stylifera, T. longicornis*, or the Cladocera *P. avirostris* (Table 2).

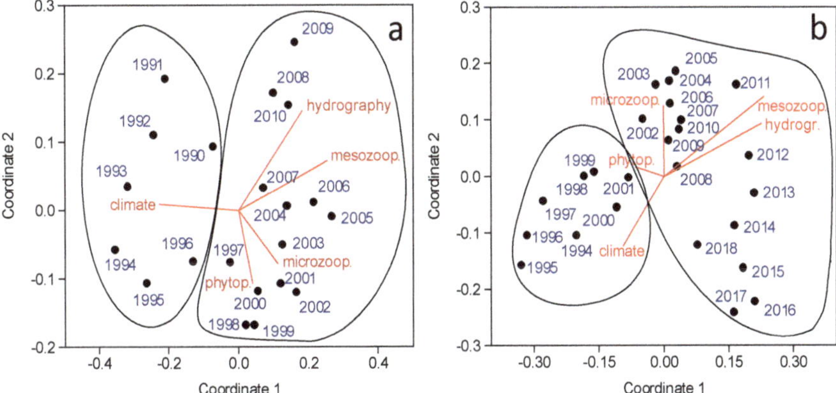

Figure 5. Plot of the values of the 2 first coordinates of the Non-Metric Multidimensional Analysis (MDS) on the annual series of phytoplankton (**a**) or mesozooplankton taxa (**b**). The correlation of the first principal component of the PCA on each type of series is indicated by the length of the vectors. The solid lines enclose the main periods detected by cluster analysis (see Methods).

Table 1. Main phytoplankton taxa contributing to the separation between periods identified by MDS and cluster analysis. The mean values of abundance (cells mL^{-1}), squared Euclidean distance (D^2), the index of discrimination consistency (D^2/sd), and the percent contribution (Contr. %) of each taxa to the separation determined by the Similarity Percentages (SIMPER) procedure between periods are given. Relatively low (blue) or high (red) mean abundances are indicated by shading. Series numbers as in Supplementary Table S2.

Series No.	Series	Taxonomic Group	1990–1997 Mean Abund.	2000–2010 Mean Abund.	D^2	D^2/sd	Contr. %
116	*Navicula transitans*	Diatoms	0.33	0.82	2.64	1.61	5.30
126	*Pseudo-nitzschia delicatissima*	Diatoms	3.00	12.93	2.14	1.16	4.31
108	*Chaetoceros* spp.	Diatoms	6.15	30.46	2.09	1.16	4.21
136	*Torodinium robustum*	Dinoflagellates	0.20	0.50	1.88	0.93	3.78
138	*Heterocapsa niei*	Dinoflagellates	4.05	8.26	1.71	0.90	3.43
97	*Lauderia annulata*	Diatoms	4.11	1.08	1.60	1.20	3.21
101	*Skeletonema costatum*	Diatoms	42.56	3.21	1.57	1.04	3.16
120	*Nitzschia* spp.	Diatoms	0.77	1.08	1.52	0.93	3.06
110	*Dictyocha fibula*	Dictyochophyceae	0.05	0.16	1.29	1.02	2.58
85	*Bacteriastrum delicatulum*	Diatoms	1.11	0.04	1.22	1.05	2.45
106	*Chaetoceros decipiens*	Diatoms	1.15	0.29	1.18	1.04	2.38
123	*Prorocentrum balticum*	Dinoflagellates	1.03	4.47	1.16	0.85	2.32
132	*Thalassiosira levanderi*	Diatoms	0.49	20.64	1.15	0.68	2.31
137	*Gyrodinium spirale*	Dinoflagellates	0.23	0.39	1.12	0.94	2.25
122	*Proboscia alata*	Diatoms	0.87	2.05	1.11	0.73	2.22
92	*Chaetoceros socialis*	Diatoms	252.08	131.83	1.07	0.80	2.16
91	*Chaetoceros gracilis*	Diatoms	3.56	2.30	1.05	0.84	2.11

Table 2. Main mesozooplankton taxa contributing to the separation between periods identified by MDS and cluster analysis. The mean values of abundance (cells mL^{-1}), squared Euclidean distance (D^2), the index of discrimination consistency (D^2/sd), and the percent contribution (Contr. %) of each taxa to the separation determined by the SIMPER procedure between periods are given. Relatively low (blue) or high (red) mean abundances are indicated by shading. Series numbers as in Supplementary Table S2.

			1994–2001	2002–2018			
Series No.	Series	Taxonomic Group	Mean Abund.	Mean Abund.	D^2	D^2/sd	Contr. %
173	*Oithona similis*	Copepoda	34.12	133.02	2.80	1.04	7.63
172	*Oithona nana*	Copepoda	8.48	27.82	2.27	0.92	6.18
139	*Penilia avirostris*	Cladocera	0.44	0.02	1.87	1.28	5.10
140	*Podon intermedius*	Cladocera	66.51	126.59	1.72	0.86	4.68
141	*Evadne nordmani*	Cladocera	25.89	83.63	1.43	1.01	3.89
144	*Ctenocalanus vanus*	Copepoda	28.37	14.54	1.39	1.21	3.78
148	*Temora stylifera*	Copepoda	18.45	4.75	1.37	1.08	3.74
164	*Oncaea media*	Copepoda	238.55	438.45	1.33	0.75	3.62
168	*Pseudocalanus elongatus*	Copepoda	134.46	152.58	1.29	1.09	3.50
170	*Centropages typicus*	Copepoda	2.87	9.23	1.26	1.20	3.42
171	*Corycaeus* spp.	Copepoda	0.12	1.17	1.24	0.84	3.39
147	*Temora longicornis*	Copepoda	74.14	47.80	1.21	0.95	3.29

4. Discussion

4.1. Regime Shifts in NW Iberia

This study reports, for the first time, multiple regime shifts in plankton assemblages off NW Iberia for the last 29 years. The lack of a long enough series may explain why previous studies in this region focused on continuous trends [22,23,25], or only on some components such as zooplankton [20–22], phytoplankton [40], or primary production [24,27]. The new approach revealed a differential influence of climatic and local conditions on plankton assemblages, thus providing a general framework for the changes previously identified not only in this region, but also in the whole NE Atlantic. For instance, our results agree with the observed reorganization of marine ecosystems related to major climate and oceanographic changes after 1990 [7,8,34].

Considering only the first principal component of each series (there were less significant shifts when using the second component), two major types of regime shifts were found in this study. First, the shift between 1997 and 1998 was related to climate and meteorological variables (Figure 6a). Climatic indices for the 1990–1997 period were characterized by relatively low values of AMO, but high values of NAO, ENN and SUNSP, associated to a predominating cold and dry conditions over the NE Atlantic [48–50], as indicated by the relatively low precipitation recorded in A Coruña during the spring in this period. Conversely, warm and wet conditions prevailed in the period 1998–2018, when a positive phase of AMO started [51]. In this analysis, the various metrics of the upwelling influence did not show particular correlations with the shift. This result may appear unexpected because of the dependence of the wind regime on major climatic conditions, such as NAO [52]. However, most of the changes in upwelling intensity off Galicia were observed at the seasonal scale (e.g., [44]). For instance, modeling studies found differential changes in upwelling-favorable winds in spring–summer vs. autumn–winter seasons [30], changes that could be largely buffered at the annual scale used in this study. This fact does not imply that upwelling is unimportant in regional oceanography, but simply indicates that its significance is mainly manifested at seasonal and lower time scales, as shown by previous studies [25,29,44].

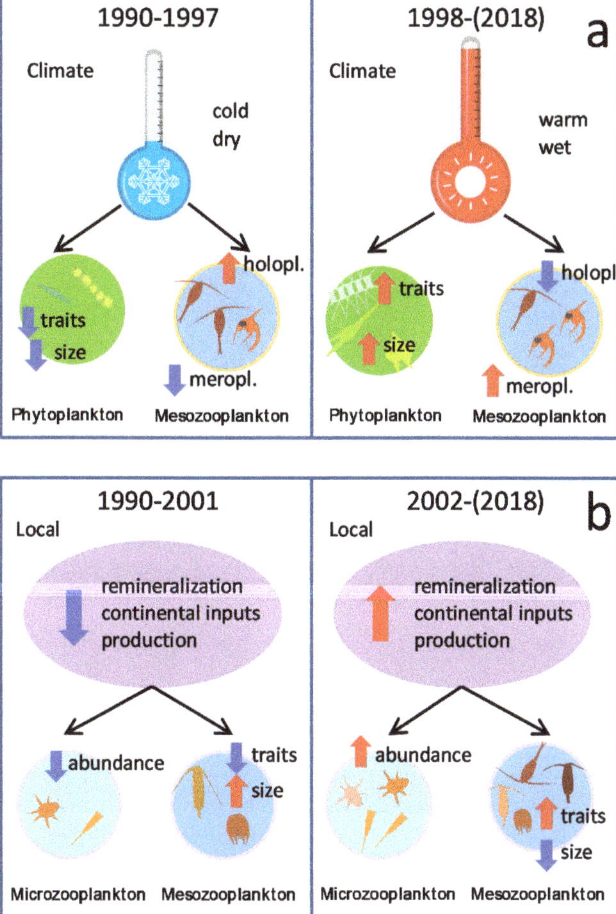

Figure 6. Schema of the main properties of the different plankton regimes identified and related to climatic (**a**) or local oceanography conditions (**b**). Relative increases (decreases) are indicated by red (blue) arrows.

Second, another type of regime shift was found between 2001 and 2002. In this case, the shift was observed in local oceanographic conditions (Figure 6b). The first period (1990–2001) was characterized by relatively low concentration values of some nutrients, such as phosphate and silicate, and high dissolved oxygen and particulate organic matter, compared with converse conditions in the second period (2002–2018). This suggests either a major oceanographic change (e.g., in the water masses) or enhanced remineralization after 2001. For the Eastern North Atlantic Central Waters (ENACW), the main water mass affected by coastal upwelling in this region, increased inputs of saline and warm subtropical waters in the NE Atlantic have been attributed to negative values of NAO, while cooler and less saline waters prevailed in years of high positive NAO values [53]. These extremes correspond to the southern and northern components of ENACW [54], with the former having a higher proportion of nutrients derived from in situ remineralization than the latter, being more dependent on the mixing depth during winter [55]. This would explain the lower oxygen and organic matter content of waters observed at St. E2CO after 2002, along with changes in the proportion of inorganic nutrients (Figure 2).

However, the characteristics of water over the shelf are also affected by processes other than upwelling. For instance, the input of continental waters may alter both thermohaline properties and the nutrient content of shelf waters, particularly in this region because of the high concentrations of some nutrients, such as phosphate and silicate ([28]), both showing a relative excess after 2002 (Figure 2). Such inputs are favored by increased precipitation [56] and a previous study on the primary production series at St. E2CO pointed out the association of rainy periods during the spring, high phosphate concentrations and increased production [27]. Similarly, the analysis of a multidecadal series of zooplankton and precipitation from Vigo (ca. 1° S of Sta. E2CO) found a significant shift in zooplankton assemblages after 2001 related to changes in local precipitation [21]. It is probable that both mechanisms (i.e., change in water masses and variations in runoff) contributed to the observed changes in local oceanography.

Detecting regime shifts, such as those outlined above, is very dependent on the length and quality of the series. One major recommendation is the use of multivariate series, instead of single series, as the former can evidence the interaction of multiple variables [57]. However, even when we have used a collection of multivariate series consistently recorded for nearly three decades, there are several issues that could affect their interpretation. Methodological variations along the recording period can be generally solved through corrections and calibration. For instance, changes in the method for analyzing chlorophyll after 2000 required the application of a correction factor to earlier measurements [27]. In other cases, such as the changes in the experts' identifications of plankton, there were different solutions. These changes did not affect the abundance counts of major taxa; however, some differences were found in the recognition of phytoplankton (but not zooplankton) species between the different teams of participating experts. For this reason, the analysis of regime shifts in plankton used major taxa (e.g., diatoms, copepods), while the analyses of assemblages were restricted to the species unequivocally recognized for all experts for the longest period available in each case. In addition, there were statistical constraints, mainly related to the significance of the shifts and the length of the regimes than can be detected using a particular time series. The use of principal components, rather than individual variables, allowed for the extraction of the main signal of abrupt change in different types of series and for analyzing their relationships [14,15]. The significance of the shifts in the components was analyzed using a sequential test that enhances its detection at the ends of time series [45]. In addition, the error in the identification due to serial correlation was reduced by estimating and removing the red noise from the series [58]. However, it must be noted that we only analyzed shifts based on abrupt changes in the mean value of the component series, and that different regimes may also result from shifts in the variance and frequency structure [59]. Finally, the series must express a significant proportion of regional variability that can be distinguished from local causes. This is the case of the series observed at the station used in this study, which is well connected to the dynamics of the upwelling transition zone on the NW Iberian coast [20,22–24].

4.2. Shifts Related to Climate

The start of a warm period of AMO in 1995 was identified as one of the main drivers of regime shift in many components of the marine ecosystems of the NE Atlantic [31]. Climate-related regime shifts between 1997 and 1999 were reported for plankton in the North Sea [16] and in the Bay of Biscay [15], but were also found in other ecosystem components as pelagic and demersal fish in the North Sea [34] and in Galicia [9], and in birds and cetaceans in the Bay of Biscay [14]. Other studies of plankton series in the NE Atlantic also reported major shifts in this period, but attributed them mainly to changes in local oceanography [60–62]. Local conditions also affected plankton regimes detected in this study, as evidenced by their correlations with PC1 of the hydrography series (Figure 5). In any case, even when the exact identification of timing in these multiannual shifts varied according to the series selected and the strength of the coupling between climatic and regional or local oceanographic variables (e.g., [15]), there were similar characteristics shared for the different regimes in all zones.

In this study, the phytoplankton assemblages for the regime prior to 1997 were characterized mainly by chain-forming diatom species, of small individual sizes, and of cylindrical or oval cellular shape. In contrast, after 1997, the characterizing species were from more diverse groups, in addition to diatoms (as dinoflagellates and Dyctiochophyeae), had, in general, larger individual sizes, and more diverse cellular shapes. These characteristics can be assimilated with life traits [60,61] illustrating the different compositions of the assemblages for each regime (Figure 6). Similar results were reported for phytoplankton regimes in the North Sea, where an enlargement in niche size and shifts in niche position were described after 1997, and the dominance of typical cold water species (e.g., *S. costatum*) in the previous regime [62]. However, despite sharing some characteristics (e.g., dominance of diatoms in cold, low NAO periods), local changes in the coastal regions may be different, as illustrated by the increase in the prymnesiophyte *Phaeocystis globosa* after 1999 on the Dutch coast ([16]) and the relative decrease in dinoflagellates in other areas [63]. The species composition for the regimes described here was advanced in a previous analysis which was part of the same series, highlighting the diversity of trends in the annual series of individual species [40], and the rapid response of small-sized species (particularly diatoms) to cold water, high-nutrient conditions [61]. A sharp increase in primary production and chlorophyll in the studied station after 1998 was also reported [27].

The climate-related regimes for zooplankton were expressed mainly as a change in the abundance ratio of major groups (Figure 3), rather than individual species. However, climatic effects may have favored the composition of the first regime (1994–2001) for zooplankton species, as suggested by the position of PC1 for climate in Figure 5. Prior to 1998, there was a dominance of holoplanktonic groups (e.g., copepods) over those of meroplankton (e.g., larvae of bivalve and gasteropod mollusks), and a converse situation thereafter. Pronounced changes were also reported for zooplankton in other regions in the NE Atlantic, mainly for copepods. For instance, after 1997, there was a reduction in the total abundance of copepods, due mainly to a decrease in neritic species and a relative increase in small-sized, subtropical species in the North Sea [63,64] and in the Bay of Biscay [15]. All these changes were linked to an enhanced input of warm ocean waters in the period dominated by positive AMO and decreasing NAO conditions, thus pointing to a major role of climate-induced oceanographic processes in the control of zooplankton assemblages, at least at the level of major taxonomic groups. As similar and concurrent shifts were found in upper trophic levels, such as planktivorous fish [7,9], birds and cetaceans [14], these changes are indicative of a major role of bottom-up control of the corresponding regimes described for plankton. A top-down control by their predators (or trophic cascade) cannot be excluded, but this mechanism is more likely to operate in semi-enclosed seas [16].

4.3. Shifts Related to Local-Oceanography

Major shifts in plankton were found for most of the NE Atlantic between 1999 and 2002. Most of them were related to local and sub-regional changes in hydrography and nutrients, but, in contrast to climate-related shifts, all cases reported preferentially affected coastal zones. Changes in phytoplankton included an increase in the most recent regime of at least some species of diatoms (e.g., *Pseudo-nitzschia* spp.), dinoflagellates, including in other groups (e.g., *Phaeocystis globosa*) in areas of the North Sea [15,16,64], and also in the English Channel [6,18,19], with noticeable reductions (extensions) in spring (autumn) blooms. There were also reported shifts in zooplankton in this period, mostly related with the increase in coastal locations of small copepods (e.g., *Oithona*, *Oncaea*), a decrease in Calanoida, particularly in those of relatively large size (e.g., *Paracalanus*, *Pseudocalanus*, *Temora*), and a general increase in species diversity in areas of the North Sea and the English Channel [6,18,19,64]. In southern regions, such as the Bay of Biscay [26] and Galicia [21] there was an increase in total abundance and biomass. In all reported cases, these shifts were mainly related to local and regional hydrographic conditions.

In our study, the observed shifts in plankton after 2001 were attributed mainly to changes in nutrients and primary production (Figure 6). These shifts appeared in the zooplankton taxa but not in the phytoplankton ones. In the case of microzooplankton, there was a general increase in abundance for all taxa after 2001; however, such an increase was not detected for mesozooplankton general taxa, but for individual species instead. The species increasing their abundance displayed a larger variety of life traits and smaller size than those of the previous regime. The former included ambush-feeding omnivores

(*O. similis*, *O. nana*, *O. media*), filter-feeding herbivores (*Pseudocalanus elongatus*, *Podon intermedius*, *Evadne nordmanni*), omnivores (*C. typicus*), and ambush-feeding carnivores (*Corycaeus* spp.), according to general trophic classifications (e.g., [65]). Those species with a higher abundance prior to 2002 were, in general, of larger size and with more specialized feeding behaviors, such as filter-feeding herbivores (*P. avirostris*, *T. stylifera*, *T. longicornis*) or omnivores (*C. vanus*). Similar variations in traits were described for zooplankton series in the North Pacific after the 1997–1998 El Niño [65], thus suggesting that functional diversity may represent a better tracer of community shifts than taxonomic diversity, at least in some ecosystems.

5. Conclusions

The shift in plankton regimes around the turn of the century described in this study implied an increase in complexity of the ecosystem, with a higher number of traits, small-sized species, and a likely increase in diversity. The fact that some of the shifts were mainly related to climate, while others were concurrent with changes in local oceanographic variables, may be due to time lags in the response of the ocean, but also to the particular conditions of the studied sites. For instance, most of the shifts related to changes in currents, nutrients or salinity occurred in coastal zones. This local effect appears to be a characteristic of the response of marine ecosystems to climate change (e.g., [66]), in contrast with terrestrial ecosystems where the effect of climate change may be more direct ([5]). Trophic cascades, however, cannot be discarded as they have been shown to trigger regime shifts in semi-enclosed seas ([16]). In addition, the diversity of variables affecting these shifts supports the hypothesis of a predominant role of non-stationary effects of the climate on oceanographic variables, including plankton ([6]). In this way, the response of the ecosystem would be triggered mainly by the variance in the climate rather than by specific changes in the mean value of a particular index. The recognition of non-stationarity in climate responses would have major implications for future models of climate change, which are currently driven by stationary dynamics (e.g., [67]).

The described regimes appeared synchronized across the NE Atlantic ([7]), and even at larger spatial scales ([17]), supporting the major role of warming and of variations in atmospheric pressure fields on ocean currents and concomitant effects on ecosystems, even when modulated by local factors. The consideration of these biological teleconnections would affect our understanding of the relative importance of climatic vs. local factors in marine ecosystems. Future projections of ecosystem change and management policies (such as the European Marine Strategy Framework Directive) may benefit from considering the diversity of life traits, as a measure of ecosystem complexity, non-stationary effects of climate, and teleconnections across the ocean. All these factors, however, cannot be assessed without the continuation of long-term, multidisciplinary time series of observations, such as the ones used in this analysis.

Supplementary Materials: The following are available online at http://www.mdpi.com/2673-1924/1/4/14/s1, Figure S1: Pearsons' r correlations between series within each type, Figure S2: Time series of the second principal component of the PCA on each type of series. Figure S3: Cluster of annual series of phytoplankton or mesozooplankton taxa. Table S1: List of original series along with mean, sd and trends. Table S2: List of original series of plankton taxa along with mean, sd and trends. Table S3: Loadings of the first (PC1) and second (PC2) principal components of the PCA for each type of series.

Author Contributions: Conceptualization, A.B.; methodology, A.B., M.Á., M.Á.L., M.R.-V., M.M.V.; formal analysis, A.B.; investigation, L.M.G.G.; data curation, A.B., M.Á., M.Á.L., M.R.-V., M.M.V.; writing—original draft preparation, A.B.; writing—review and editing, A.B., M.Á., L.M.G.G., M.Á.L., M.N.-C., M.R.-V., M.M.V.; visualization, A.B., M.Á.L., M.N.-C.; project administration, A.B.; funding acquisition, A.B., M.R.-V. All authors have read and agreed to the published version of the manuscript.

Funding: This research was funded by the RADIALES project of the Instituto Español de Oceanografía (IEO), with additional funds from the MarRisk project (Interreg POCTEP Spain–Portugal; grant number: 0262 MARRISK 1 E), as well as from the Contrato-Programa Axencia Galega de Innovación (GAIN)-IEO grant (grant number: IN607A 2018/2) of the Axencia Galega de Innovación (GAIN, Xunta de Galicia, Spain).

Acknowledgments: We are grateful for the dedicated work of the many ship crews, technicians, and scientists that made the collection of the time series observations employed in this study possible. Particularly, we thank David Marcote and Ángel F. Lamas (CTD), Nicolás González, Rosario Carballo, and Mónica Castaño (nutrient analysis), Fernando Rozada (POM), Manuel Varela and Jorge Lorenzo (phytoplankton), and Maite Álvarez-Ossorio and Elena Rey (zooplankton) for the initiation and maintenance of the series.

Conflicts of Interest: The authors declare no conflict of interest. The funders had no role in the design of the study; in the collection, analyses, or interpretation of data; in the writing of the manuscript, or in the decision to publish the results.

References

1. Möllmann, C.; Folke, C.; Edwards, M.; Conversi, A. Marine regime shifts around the globe: Theory, drivers and impacts. *Phil. Trans. R. Soc. B* **2015**, *370*. [CrossRef]
2. Conversi, A.; Dakos, V.; Gardmark, A.; Ling, S.; Folke, C.; Mumby, P.J.; Greene, C.; Edwards, M.; Blenckner, T.; Casini, M.; et al. A holistic view of marine regime shifts. *Phil. Trans. R. Soc. B* **2015**, *370*. [CrossRef]
3. Reid, P.C.; Edwards, M.; Beaugrand, G.; Skogen, M.; Stevens, D. Periodic changes in the zooplankton of the North Sea during the 20th century linked to oceanic inflow. *Fish. Oceanogr.* **2001**, *12*, 260–269. [CrossRef]
4. Beaugrand, G. Long-term changes in copepod abundance and diversity in the North-East Atlantic in relation to fluctuations in the hydroclimatic environment. *Fish. Oceanogr.* **2003**, *12*, 270–283. [CrossRef]
5. Hallett, T.B.; Coulson, T.; Pilkington, J.G.; Clutton-Brock, T.H.; Pemberton, J.M.; Grenfell, B.T. Why large-scale climate indices seem to predict ecological processes better than local weather. *Nature* **2004**, *430*, 71–75. [CrossRef]
6. Molinero, J.C.; Reygondeau, G.; Bonnet, D. Climate variance influence on the non-stationary plankton dynamics. *Mar. Environ. Res.* **2013**, *89*, 91–96. [CrossRef]
7. Alheit, J.; Gröger, J.; Licandro, P.; McQuinn, I.H.; Pohlmann, T.; Tsikliras, A.C. What happened in the mid-1990s? The coupled ocean-atmosphere processes behind climate-induced ecosystem changes in the northeast atlantic and the mediterranean. *Deep-Sea Res. II* **2019**, *159*, 130–142. [CrossRef]
8. Möllmann, C.; Conversi, A.; Edwards, M. Comparative analysis of european wide marine ecosystem shifts: A large-scale approach for developing the basis for ecosystem-based management. *Biol. Lett.* **2011**, *7*, 484–486. [CrossRef]
9. Kraberg, A.C.; Wiltshire, K.H. Regime shifts in the marine environment: How do they affect ecosystem services. In *The Mediterranean Sea: Its History and Present Challenges*; Goffredo, S., Dubinsky, Z., Eds.; Springer Science + Business Media: Dordrecht, The Netherlands, 2014; pp. 499–504.
10. Beaugrand, G. The North Sea regime shift: Evidence, causes, mechanisms and consequences. *Prog. Oceanogr.* **2004**, *60*, 245–262. [CrossRef]
11. McQuatters-Gollop, A.; Vermaat, J.E. Covariance among North Sea ecosystem state indicators during the past 50 years—Contrasts between coastal and open waters. *J. Sea Res.* **2011**, *65*, 284–292. [CrossRef]
12. Peperzak, L.; Witte, H. Abiotic drivers of interannual phytoplankton variability and a 1999–2000 regime shift in the North Sea examined by multivariate statistics. *J. Phycol.* **2019**, *55*, 1274–1289. [CrossRef] [PubMed]
13. Cabrero, Á.; González-Nuevo, G.; Gago, J.; Cabanas, J.M. Study of sardine (*Sardina pilchardus*) regime shifts in the Iberian Atlantic shelf waters. *Fish. Oceanogr.* **2019**, *28*, 305–316. [CrossRef]
14. Hemery, G.; D'Amico, F.; Castege, I.; Dupont, B.; D'Elbee, J.; Lalanne, Y.; Mouches, C. Detecting the impact of oceano-climatic changes on marine ecosystems using a multivariate index: The case of the Bay of Biscay (North Atlantic-European ocean). *Glob. Chang. Biol.* **2008**, *14*, 27–38. [CrossRef]
15. Luczak, C.; Beaugrand, G.; Jaffré, M.; Lenoir, S. Climate change impact on Balearic shearwater through a trophic cascade. *Biol. Lett.* **2011**, *7*, 702–705. [CrossRef] [PubMed]
16. Pershing, A.J.; Mills, K.E.; Record, N.R.; Stamieszkin, K.; Wurtzell, K.V.; Byron, C.J.; Fitzpatrick, D.; Golet, W.J.; Koob, E. Evaluating trophic cascades as drivers of regime shifts in different ocean ecosystems. *Phil. Trans. R. Soc. B* **2015**, *370*. [CrossRef]
17. Kröncke, I.; Neumann, H.; Dippner, J.W.; Holbrook, S.; Lamy, T.; Miller, R.; Padedda, B.M.; Pulina, S.; Reed, D.C.; Reinikainen, M.; et al. Comparison of biological and ecological long-term trends related to northern hemisphere climate in different marine ecosystems. *Nat. Conserv.* **2019**, *34*, 311–341. [CrossRef]

18. Hernández-Fariñas, T.; Soudant, D.; Barillé, L.; Belin, C.; Lefebvre, A.; Bacher, C. Temporal changes in the phytoplankton community along the French coast of the Eastern English Channel and the Southern Bight of the North Sea. *ICES J. Mar. Sci.* **2014**, *71*, 821–833. [CrossRef]

19. Reygondeau, G.; Molinero, J.C.; Coombs, S.; MacKenzie, B.R.; Bonnet, D. Progressive changes in the western English Channel foster a reorganization in the plankton food web. *Prog. Oceanogr.* **2015**, *137*, 524–532. [CrossRef]

20. Bode, A.; Alvarez-Ossorio, M.T.; Miranda, A.; Ruiz Villarreal, M. Shifts between gelatinous and crustacean plankton in a coastal upwelling region. *ICES J. Mar. Sci.* **2013**, *70*, 934–942. [CrossRef]

21. Buttay, L.; Miranda, A.; Casas, G.; González-Quirós, R.; Nogueira, E. Long-term and seasonal zooplankton dynamics in the northwest Iberian shelf and its relationship with meteo-climatic and hydrographic variability. *J. Plankton Res.* **2016**, *38*, 106–121. [CrossRef]

22. Valdés, L.; López-Urrutia, A.; Cabal, J.; Alvarez-Ossorio, M.; Bode, A.; Miranda, A.; Cabanas, M.; Huskin, I.; Anadón, R.; Alvarez-Marqués, F.; et al. A decade of sampling in the Bay of Biscay: What are the zooplankton time series telling us? *Prog. Oceanogr.* **2007**, *74*, 98–114. [CrossRef]

23. Bode, A.; Alvarez-Ossorio, M.T.; Cabanas, J.M.; Miranda, A.; Varela, M. Recent trends in plankton and upwelling intensity off Galicia (NW Spain). *Prog. Oceanogr.* **2009**, *83*, 342–350. [CrossRef]

24. Bode, A.; Anadón, R.; Morán, X.A.G.; Nogueira, E.; Teira, E.; Varela, M. Decadal variability in chlorophyll and primary production off NW Spain. *Clim. Res.* **2011**, *48*, 293–305. [CrossRef]

25. Pérez, F.F.; Padin, X.A.; Pazos, Y.; Gilcoto, M.; Cabanas, M.; Pardo, P.C.; Doval, M.D.; Farina-Bustos, L. Plankton response to weakening of the Iberian coastal upwelling. *Glob. Chang. Biol.* **2010**, *16*, 1258–1267. [CrossRef]

26. González-Gil, R.; González Taboada, F.; Höffer, J.; Anadón, R. Winter mixing and coastal upwelling drive long-term changes in zooplankton in the Bay of Biscay (1993-2010). *J. Plankton Res.* **2015**, *37*, 337–351. [CrossRef]

27. Bode, A.; Álvarez, M.; Ruíz-Villarreal, M.; Varela, M.M. Changes in phytoplankton production and upwelling intensity off A Coruña (NW Spain) for the last 28 years. *Ocean Dyn.* **2019**. [CrossRef]

28. Alvarez-Salgado, X.A.; Beloso, S.; Joint, I.; Nogueira, E.; Chou, L.; Pérez, F.F.; Groom, S.; Cabanas, J.M.; Rees, A.P.; Elskens, M. New production of the NW Iberian shelf during the upwelling season over the period 1982–1999. *Deep Sea Res.* **2002**, *49*, 1725–1739. [CrossRef]

29. Beca-Carretero, P.P.; Otero, J.; Land, P.E.; Groom, S.; Alvarez-Salgado, X.A. Seasonal and inter-annual variability of net primary production in the NW Iberian margin (1998–2016) in relation to wind stress and sea surface temperature. *Prog. Oceanogr.* **2019**, *178*. [CrossRef]

30. Casabella, N.; Lorenzo, M.N.; Taboada, J.J. Trends of the Galician upwelling in the context of climate change. *J. Sea Res.* **2014**, *93*, 23–27. [CrossRef]

31. Available online: http://www.seriestemporales-ieo.net (accessed on 24 September 2020).

32. Available online: https://psl.noaa.gov/gcos_wgsp/Timeseries/ (accessed on 30 March 2020).

33. Available online: https://crudata.uea.ac.uk/cru/data/nao/ (accessed on 25 March 2020).

34. Edwards, M.; Beaugrand, G.; Helaouet, P.; Alheit, J.; Coombs, S. Marine ecosystem response to the Atlantic multidecadal oscillation. *PLoS ONE* **2013**, *8*, e57212. [CrossRef]

35. Available online: https://opendata.aemet.es/ (accessed on 15 March 2020).

36. González-Nuevo, G.; Gago, J.; Cabanas, J.M. Upwelling index: A powerful tool for marine research in the nw iberian upwelling system. *J. Oper. Oceanogr.* **2014**, *7*, 45–55. [CrossRef]

37. Available online: http://www.indicedeafloramiento.ieo.es/ (accessed on 15 March 2020).

38. Available online: https://doi.org/10.1594/PANGAEA.885413 (accessed on 24 September 2020).

39. Available online: https://doi.pangaea.de/10.1594/PANGAEA.919082 (accessed on 24 September 2020).

40. Bode, A.; Estévez, M.G.; Varela, M.; Vilar, J.A. Annual trend patterns of phytoplankton species abundance belie homogeneous taxonomical group responses to climate in the NE Atlantic upwelling. *Mar. Environ. Res.* **2015**, *110*, 81–91. [CrossRef] [PubMed]

41. Available online: https://doi.org/10.1594/PANGAEA.908815 (accessed on 24 September 2020).

42. Available online: https://doi.pangaea.de/10.1594/PANGAEA.919081 (accessed on 24 September 2020).

43. Available online: https://doi.pangaea.de/10.1594/PANGAEA.919080 (accessed on 24 September 2020).

44. Nogueira, E.; Perez, F.F.; Rios, A.F. Seasonal patterns and long-term trends in an estuarine upwelling ecosystem (Ria de Vigo, NW Spain). *Estuar. Coast. Shelf Sci.* **1997**, *44*, 285–300. [CrossRef]

45. Rodionov, S.N. A sequential algorithm for testing climate regime shifts. *Geophys. Res. Lett.* **2004**, *31*. [CrossRef]
46. Clarke, K.R.; Gorley, R.N. *Primer V6: User Manual/Tutorial*; PRIMER-E Ltd.: Plymouth, UK, 2006; 192p.
47. Hammer, Ø.; Harper, D.A.T.; Ryan, P.D. Past: Paleontological statistics software package for education and data analysis. *Palaeontol. Electron.* **2001**, *4*, 9.
48. Folland, C.K.; Knight, J.; Linderholm, H.; Fereday, D.; Ineson, S.; Hurrell, J.W. The summer North Atlantic Oscillation: Past, present and future. *J. Clim.* **2009**, *22*, 1082–1103. [CrossRef]
49. López-Parages, J.; Rodríguez-Fonseca, B.; Dommenget, D.; Frauen, C. ENSO influence on the North Atlantic European climate: A non-linear and non-stationary approach. *Clim. Dyn.* **2016**, *47*, 2071–2084. [CrossRef]
50. O'Reilly, C.H.; Woollings, T.; Zanna, L. The dynamical influence of the atlantic multidecadal oscillation on continental climate. *J. Clim.* **2017**, *30*, 7213–7230. [CrossRef]
51. Knudsen, M.F.; Seidenkrantz, M.-S.; Jacobsen, B.H.; Kuijpers, A. Tracking the atlantic multidecadal oscillation through the last 8000 years. *Nat. Commun.* **2011**, *2*. [CrossRef]
52. Marshall, J.; Kushnir, Y.; Battisti, D.; Chang, P.; Czaja, A.; Dickson, R.; Hurrell, J.; McCartney, M.; Saravanan, R.; Visbeck, M. North Atlantic climate variability: Phenomena, impacts and mechanisms. *Int. J. Climatol.* **2001**, *21*, 1863–1898. [CrossRef]
53. Prieto, E.; Gonzalez-Pola, C.; Lavin, A.; Holliday, N.P. Interannual variability of the Northwestern Iberia deep ocean: Response to large-scale North Atlantic forcing. *J. Geophys. Res. Oceans* **2015**, *120*, 832–847. [CrossRef]
54. Rios, A.F.; Pérez, F.F.; Fraga, F. Water masses in the upper and middle North Atlantic Ocean East of Azores. *Deep Sea Res.* **1992**, *39*, 645–658. [CrossRef]
55. Pérez, F.F.; Mouriño, C.; Fraga, F.; Rios, A.F. Displacement of water masses and remineralization rates off the Iberian peninsula by nutrient anomalies. *J. Mar. Res.* **1993**, *51*, 869–892. [CrossRef]
56. Otero, P.; Ruiz-Villarreal, M.; Peliz, A. River plume fronts off NW Iberia from satellite observations and model data. *ICES J. Mar. Sci.* **2009**, *66*, 1853–1864. [CrossRef]
57. deYoung, B.; Jarre, A. Regime shifts: Methods of analysis. In *Encyclopedia of Ocean Sciences*; Cochran, J.K., Bokuniewicz, H., Yager, P., Eds.; Academic Press: London, UK, 2019; pp. 489–492.
58. Rodionov, S.N. Use of prewhitening in climate regime shift detection. *Geophys. Res. Lett.* **2006**, *33*. [CrossRef]
59. Liu, Q.; Wan, S.; Gu, B. A review of the detection methods for climate regime shifts. *Discrete Dyn. Nat. Soc.* **2016**, *2016*. [CrossRef]
60. Litchman, E.; Klausmeier, C.A. Trait-based community ecology of phytoplankton. *Annu. Rev. Ecol. Evol. Syst.* **2008**, *39*, 615–639. [CrossRef]
61. Otero, J.; Bode, A.; Álvarez-Salgado, X.A.; Varela, M. Role of functional traits variability in the response of individual phytoplankton species to changing environmental conditions in a coastal upwelling zone. *Mar. Ecol. Prog. Ser.* **2018**, *596*, 33–47. [CrossRef]
62. Freund, J.A.; Grüner, N.; Brüse, S.; Wiltshire, K.H. Changes in the phytoplankton community at Helgoland, North Sea: Lessons from single spot time series analyses. *Mar. Biol.* **2012**, *159*, 2561–2571. [CrossRef]
63. Alvarez-Fernandez, S.; Lindeboom, H.; Meesters, E. Temporal changes in plankton of the North Sea: Community shifts and environmental drivers. *Mar. Ecol. Prog. Ser.* **2012**, *462*, 21–38. [CrossRef]
64. Boersma, M.; Wiltshire, K.H.; Kong, S.M.; Greve, W.; Renz, J. Long-term change in the copepod community in the Southern German Bight. *J. Sea Res.* **2015**, *101*, 41–50. [CrossRef]
65. Pomerleau, C.; Sastri, A.R.; Beisner, B.E. Evaluation of functional trait diversity for marine zooplankton communities in the Northeast subarctic Pacific ocean. *J. Plankton Res.* **2015**, *37*, 712–726. [CrossRef]
66. Dippner, J.W.; Kornilovs, G.; Junker, K. A multivariate Baltic Sea environmental index. *Ambio* **2012**, *41*, 699–708. [CrossRef] [PubMed]
67. Chust, G.; Allen, J.I.; Bopp, L.; Schrum, C.; Holt, J.; Tsiaras, K.; Zavatarelli, M.; Chifflet, M.; Cannaby, H.; Dadou, I.; et al. Biomass changes and trophic amplification of plankton in a warmer ocean. *Glob. Chang. Biol.* **2014**, *20*, 2124–2139. [CrossRef] [PubMed]

 *oceans*

Article

Influence of Seawater Ageing on Fracture of Carbon Fiber Reinforced Epoxy Composites for Ocean Engineering

Antoine Le Guen-Geffroy [1], Peter Davies [1,*], Pierre-Yves Le Gac [1] and Bertrand Habert [2]

[1] IFREMER Centre Bretagne, Marine Structures Laboratory, 29280 Plouzané, France;
leguengeffroy.antoine@gmail.com (A.L.G.-G.); Pierre.Yves.Le.Gac@ifremer.fr (P.-Y.L.G.)

[2] Direction Générale de l'Armement, IP/MCM/PMA, PC62, 60 boulevard du Général Valin,
75015 Paris, France; bertrand.habert@intradef.gouv.fr

* Correspondence: peter.davies@ifremer.fr

Received: 29 August 2020; Accepted: 22 September 2020; Published: 27 September 2020

Abstract: Carbon fiber reinforced composite materials are finding new applications in highly loaded marine structures such as tidal turbine blades and marine propellers. Such applications require long-term damage resistance while being subjected to continuous seawater immersion. However, few data exist on which to base material selection and design. This paper provides a set of results from interlaminar fracture tests on specimens before and after seawater ageing. The focus is on delamination as this is the main failure mechanism for laminated composites under out-of-plane loading. Results show that there are two contributions to changes in fracture toughness during an accelerated wet ageing program: effects due to water and effects due to physical ageing. These are identified and it is shown that this composite retains over 70% of its initial fracture properties even for the worst case examined.

Keywords: composite; delamination; seawater; immersion; ageing

---

1. Introduction

Fiber reinforced composites are widely used in the marine industry [1,2]. Glass fiber reinforced polyesters have been used in pleasure boat and military vessel construction for over 50 years, but higher performance carbon fiber reinforced composites are finding increasing applications. These include racing yachts [3], offshore applications [4,5] and marine renewable energy structures. The latter include tidal turbine blades [6], up to 10 m long, and the latest generation of offshore wind turbine blades, over 100 m long [7]. These must perform under severe loading conditions while keeping costs low. As a result, there is increasing interest in the long-term behavior of carbon fiber composites produced by non-aerospace manufacturing routes; infusion and low pressure injection methods are of particular interest.

Despite exceptional in-plane properties, an inherent weakness of laminated composites is their sensitivity to interlaminar crack propagation or delamination due to out-of-plane loading. This has been the subject of a large number of papers since the 1970s [8], and specific experimental techniques have been developed to characterize the resistance to interlaminar fracture [9]. Mode I tests (crack opening under tensile loads) were developed first, and an ISO standard procedure was agreed in the late 1990s [10]. Mode II loading (in-plane shear) is more difficult to apply and various test configurations have been proposed [11,12], with the end loaded split (ELS) test method now an ISO standard [13]. It is also important to be able to characterize the interactions between mode I and mode II so various mixed mode loading tests have been developed. The most popular uses the mixed mode bending (MMB) set-up, a specially designed fixture allowing different combinations of mode I and mode II to be applied by changing the fixture loading arm geometry [14]. This is an ASTM standard [15]. The output values from these tests are critical strain energy release rates, G_c, which quantify the energy necessary to either initiate or propagate delaminations. Originally used to compare materials on a consistent basis, they have become more useful in recent years as modeling techniques have been developed which use these values; in particular, cohesive zone methods (CZM) require G_c input values [16,17].

The main particularity of marine applications is temporary or permanent exposure to seawater. Many polymers are sensitive to moisture so the influence of seawater on all matrix dominated composite properties needs to be studied. For laminated composites with fibers aligned in the 0° direction, this includes off-axis tension, axial and off-axis compression, in-plane shear and all the through-thickness properties. The latter are particularly difficult to measure as laminated composites are usually quite thin and, in many cases, only the apparent interlaminar shear (ILSS) can be obtained, using short beam shear specimens (ASTM D2344). An alternative approach is to test specimens with predefined defects, thin films which are implanted during manufacture at the specimen mid-thickness. These can then be subjected to different loading types in order to measure fracture energies or critical strain energy release rates, G_c.

The resistance to interlaminar crack propagation is one of the matrix and fiber/matrix interface dominated properties which would be expected to be affected by water ingress. There have been a number of previous studies of the influence of water on delamination toughness, but the majority were focused on glass reinforced composites as this was the preferred reinforcement for marine applications. Marom summarized early work, stating that the short-term effect of water is to increase fracture toughness (plasticization), while in the long term, a deterioration effect will prevail (hydrolysis, fiber/matrix debonding) [18]. One of the aims of durability studies is to establish when this transition occurs.

Due to the more limited use of carbon composites in marine applications in the past compared to aerospace structures, there are considerably fewer papers on the delamination in water of these materials. Table 1 shows a brief overview of some results published over the last 40 years. These results concern carbon fiber reinforced epoxies and PolyEtherEtherKetone (PEEK) composites. The latter is a highly durable semi-crystalline thermoplastic.

Table 1. Previous work on the influence of wet ageing on interlaminar fracture of carbon reinforced composites.

Loading Mode	Material	Ageing Condition. Starter Crack Type	Influence of Water on Fracture Toughness	Reference
I (Double Torsion)	Carbon/Epoxy	Water 70 °C. Saw cut + razor	No effect	Thomson and Broutman, 1982 [19]
I/II	Carbon/Epoxy	50% RH, Wet. PTFE film	$G_c\uparrow$ for mode I dominated. No effect on G_c for shear dominated tests	Russell and Street, 1985 [20]
I, II	Carbon/Epoxy	50% RH, Wet. Mode I precrack	$G_{Ic}\uparrow$, $G_{IIc}\downarrow$ for wet	Garg and Ishai, 1985 [21]
I, II	Carbon/Epoxy, Carbon/PEEK	Water 22, 77, 100 °C Unaged, half-saturated, fully saturated. Foil film starter	Epoxy: $G_{Ic}\uparrow$, $G_{IIc}\downarrow$ PEEK: G_{Ic} no effect, $G_{IIc}\downarrow$	Selzer and Friedrich, 1995 [22]
II	Carbon/Epoxy	Various RH conditions. Film	$G_{IIc}\downarrow$	Zhao, Gaedke, 1996 [23]
	Carbon/Epoxy	Dry, 50% RH, Wet Precrack	$G_{Ic\ initiation}$ no effect, $G_{Ic\ propagation}\downarrow$ for wet $G_{IIc}\downarrow$ for wet	Chou I, 1998 [24]
I, II, I/II	Carbon/Epoxy	Moisture saturated. Polyimide film	G_{Ic} no effect, G_c mixed and mode II $\downarrow$ for wet	Asp, 1998 [25]
II	Carbon/Epoxy	Saturated distilled water. Film and precrack	$G_{IIc}\downarrow$ for wet	Landry et al., 2012 [26]
I	Carbon/Toughened epoxy adhesively bonded	Fresh and sea water, 70, 82 °C. Precrack	$G_{Ic}\downarrow$	Couture, 2013 [27]
I/II	Carbon/Epoxy	Saturated distilled water 70 °C. Precrack	$G_{Ic}\uparrow$ $G_{I/IIc}\downarrow$ for wet	LeBlanc et al., 2015 [28]

For mode I fracture, both increases and decreases in G_{Ic} after water saturation have been reported, while the effect of water absorption on the mode II and mixed mode I and II tends to reduce fracture toughness. The majority of the previous work in this area has been limited to either pure mode I or mode II loading; to the authors' knowledge, there have been very few studies in which the effects of natural seawater ageing on carbon/epoxy marine composites have been studied over the full range of mode I, mode II and mixed mode I/II loading. This is the aim of the present work.

2. Materials

The composite material studied here is a carbon fiber reinforced epoxy. Both fibers and matrix are commercially available products. The epoxy resin pre-polymers are diglycidyl ether of bisphenol F (DGEBF, ≥50%), diglycidyl ether of bisphenol A (DGEBA, ≥10%) and 1,6 hexa diglycidyl ether (≥10%). This epoxy resin is commercialized by the French manufacturer Sicomin® under the SR8100 product name. The hardener was also supplied by Sicomin®, reference SD4772. This amine-based hardener is designed for manufacturing of thick composite parts under anaerobic processes such as resin transfer molding (RTM) and infusion processes. The carbon fibers used were T700 Torayca® 12 K, 1600 tex, standard modulus unidirectional reinforcement. For both ease of handling and ease of resin impregnation, these were woven in the perpendicular direction with 68 tex glass fibers, representing a 13 to 590 glass to carbon weight ratio, resulting in a mass per unit area of 603 g/m^2.

Six-ply thick 500×500 mm^2, unidirectional composite plates were manufactured using the vacuum infusion process, of thickness 3.4 ± 0.1 mm. In order to perform fracture tests, a 12 μm thick by 80 mm wide sheet of polytetrafluoroethylene (PFTE) was placed at mid-thickness at each side of the plate perpendicular to the carbon fiber direction before infusion.

The composites plates were cured at ambient temperature for 24 h and post-cured for 8 h at 80 °C, followed by a post-cure of 2 h at 120°C. This final cure was necessary to completely cure and stabilize the resin, to prevent any residual curing during seawater ageing. Using TGA (thermogravimetric analysis under a nitrogen atmosphere) and a helium pycnometer to measure density, fiber and void volume ratios were calculated. The fiber volume ratio obtained from these measurements was estimated to be around $62 \pm 3\%$ with $0.8 \pm 0.3\%$ void volume ratio. The material's glass transition temperature (Tg) was measured using differential scanning calorimetry to be around 75 ± 2 °C. The overall quality of panels was checked using ultrasonic C-scan and very low attenuation was found, confirming good internal quality. Interlaminar shear strength was measured using ASTM D2344 to be 64 ± 3 MPa.

3. Experimental Methods and Data Analysis

This study focuses on the effect of long-term seawater ageing on the fracture properties of the composite. It is first necessary to establish how long the specimens need to saturate and the water temperature which can be used to accelerate water ingress. A preliminary weight gain study on the water ageing behavior of $50 \times 50 \times 3$ mm^3 coupons showed Fickian behavior of the composite. This means that there is a stable and reversible state that is reached after a certain time, representing the fully water-saturated state. Therefore, in this study, only the fully dried, fully saturated and re-dried after full saturation states were examined. These three conditions will be referred to as unaged (fully dried), saturated (fully saturated) and saturated then dried (re-dried after full saturation).

Seawater ageing tanks were used to age composite samples. These 180 L capacity insulated tanks were filled with natural seawater pumped directly from the Brest estuary. This seawater was continuously renewed (volume replaced every 24 h). The temperature was controlled to within ±2 °C (Figure 1) and monitored continuously.

Based on results from the preliminary study, shown in Figure 2, the water temperature for specimen ageing was chosen to be 60 °C. This temperature corresponds to the highest possible accelerating temperature with respect to the material's dry Tg of 75 °C. The time to saturation of specimens at this temperature was found to be 15 weeks.

Figure 1. Seawater ageing tanks.

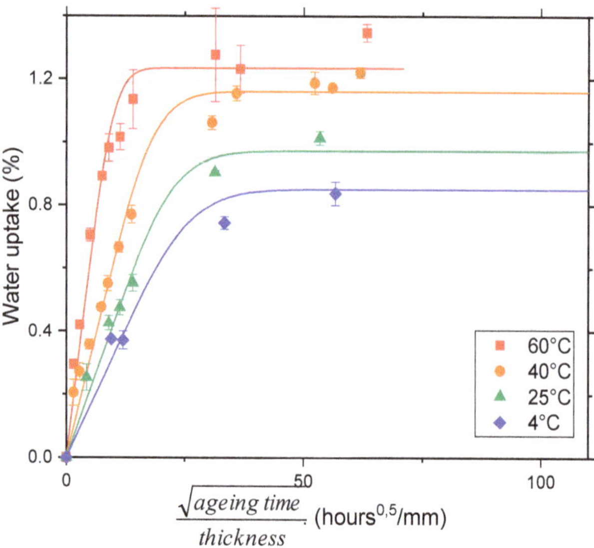

Figure 2. Preliminary water uptake results for the composite in seawater at different ageing temperatures. Lines show respective Fickian model fit. Error bars show standard deviations.

All specimens were dried prior to either testing (unaged condition) or immersion. Some mode I and mode II specimens were also dried after reaching saturation, in order to examine the reversibility of water effects, by placing them in an oven at 60 °C in desiccators, also for 15 weeks, their weight being controlled frequently in order to ensure that they were fully dried.

The interlaminar fracture properties of the composite were studied using different test methods: pure mode I delamination, pure mode II delamination and combinations of mode I and mode II loading. For all the tests, an Instron™ 5966 electromechanical test machine was used, with an Instron™ 2580-500N load cell. Controlled laboratory environment conditions of 21 ± 1 °C and 50 ± 5% relative humidity were maintained throughout the tests.

3.1. Mode I Delamination: Double Cantilever Beam (DCB)

The double cantilever beam (DCB) test method was chosen in order to test the mode I fracture toughness. These tests were performed based on the ISO 15024 standard test method [9]. Specimens were cut to 20 by 150 mm^2 dimensions. The compliance calibration (CC) data reduction method was chosen to calculate the mode I energy release rate, G_{Ic} in J/m^2, as

$$G_{Ic} = \frac{nP\delta}{2ba} \tag{1}$$

In this equation, n is the slope of the compliance logarithm log(C) versus crack length logarithm log(a). P is the load in N at the opening displacement δ in mm, b is the specimen's width in mm and a is the crack length in mm. In order to measure crack length, one of the specimen edges was painted using a white Posca$^®$ pen. This particular marker pen has many advantages such as being water-based, preventing unwanted degradation as opposed to alcohol-based paints, providing a matt finish with good opacity, thus increasing contrast with the crack, and, finally, it has no mechanical resistance so it does not affect the load measurements. A camera was placed facing the white-painted edge of the specimens and linked to a PC to record the images together with corresponding load and displacement values using a data acquisition card. The crack length value was interpolated between the initial crack length and the final crack length using the compliance. The validity of this interpolation was checked by comparing the calculated crack lengths and the optically measured crack lengths, which showed good agreement. The crosshead displacement rate was 1 mm/min, in order to follow the crack propagation accurately.

Results from all these delamination tests include both initiation and propagation values. The former can be defined by the values on the force versus displacement corresponding to either beginning of non-linearity (NL) or a 5% change in slope (5%) and using the initial starter crack length. Both were calculated here. As recommended by the standard, two cases can be identified. The first corresponds to initiation from the insert film: a short natural crack is created (5 to 10 mm from the film insert) and then the specimen is unloaded. The second initiation value is determined when the small precrack created by this first initiation cycle is reloaded. Loading is then continuous and propagation values are also calculated, using the load and displacement corresponding to each calculated crack length once the crack has started to advance along the specimen.

3.2. Mode II Delamination: Calibrated Edge Loaded Split (C-ELS)

The mode II delamination properties were evaluated using the ISO calibrated edge loaded split (C-ELS) standard test method [13]. This particular test method was chosen over other mode II configurations [11,12] due to better crack stability and longer crack propagation length. The specimen's width and length were also 20 by 150 mm^2, similar to the DCB dimensions, which facilitates specimen preparation. As with mode I tests, the loading rate was 1 mm/min. The pure mode II energy release rate, G_{IIc} in J/m^2, was calculated using the Corrected Beam Theory with the effective crack length (CBTE) data reduction method:

$$G_{IIc} = \frac{9P^2 a_e^2}{4b^2 h^3 E_{1f}}, \text{ with} \tag{2}$$

$$a_e = \left[\frac{1}{3} \left\{ 2bCh^3 E_{1f} - \left(L_1 + \Delta_{clamp} \right)^3 \right\} \right]^{\frac{1}{3}} \tag{3}$$

In this equation, in addition to the equivalent parameters from Equation (1), a_e is the effective crack length in mm, h is the half-thickness of the specimen in mm, E_{1f} is the flexural modulus that can be determined from three point bending test or from clamp calibrating the ELS test in MPa, C is the compliance (displacement/force), L_1 is the free length, meaning the distance from the load point to the crack tip in mm, and finally, Δ_{clamp} is the clamp correction determined through the clamp calibration procedure detailed in [13].

3.3. Mode I/II Delamination: Mixed Mode Bending (MMB)

The third test procedure was also based on a standardized test, the mixed mode bending (MMB) [15]. This test method uses a special lever and base fixture that combines a three-point bending end notched flexure with a DCB test. By adjusting the position of the load point on the lever, different mode I/II combinations are obtained, as shown in Figure 3.

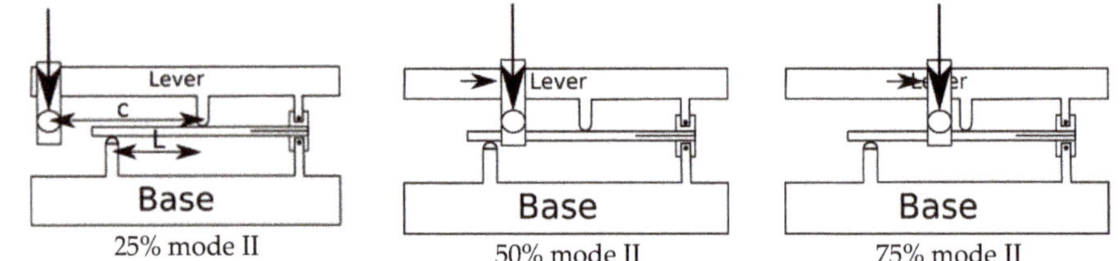

Figure 3. Different MMB test configurations.

The mode I and mode II energy release rates are calculated separately as follows:

$$G_I = \frac{12P^2(3c-L)^2}{16b^2h^3L^2E_{1f}}(a+\chi h)^2 \tag{4}$$

$$G_{II} = \frac{19P^2(c+L)^2}{16b^2h^3L^2E_{1f}}(a+0.42\chi h)^2, \text{ with} \tag{5}$$

$$\chi \equiv \sqrt{\frac{E_{11}}{11G_{13}}\left(3-2\left(\frac{\Gamma}{1+\Gamma}\right)\right)} \text{ and } \Gamma \equiv 1.18\frac{\sqrt{E_{11}E_{22}}}{G_{13}} \tag{6}$$

In all these equations, in addition to the terms from previous equations and starting from the calculation of the mode I energy release rate G_I, c is the lever length, as shown in Figure 3 in mm, L is the half-span length, also shown in Figure 3 in mm, χ is the crack length correction factor, E_{11} is the longitudinal modulus measured experimentally at 97 GPa, E_{22} is the transverse modulus, measured experimentally to be 7.0 GPa and, finally, G_{13} is the shear modulus estimated to be 8.5 GPa. As for the previous test methods, the loading rate was 1 mm/min. Crack length was also measured using a camera and by painting one of the specimen's edges, together with numerical interpolation. The results from the MMB test are plotted as the total energy release rate $G = G_I + G_{II}$ versus the mode II contribution $\frac{G_{II}}{G}$, as suggested by Benzeggagh and Kenane [29]. In this study, three ratios of mode mixity were tested, corresponding to mode II percentages of 25, 50 and 75%.

Data were then fitted using the Benzeggagh–Kenane (B-K) criterion using the following equation:

$$G_T = G_{Ic} + \left(G_{IIc} - G_{Ic}\right)\left(\frac{G_{II}}{G}\right)^n \tag{7}$$

The parameter n of the B-K criterion was found using a least squares method. This parameter reflects the influence of the mode II loading on the global energy release rate. A higher value of n indicates that the mode II plays a greater role in the total energy release rate, but the value is only used as a material characterization parameter here.

In order to obtain reliable propagation values, a particular data reduction method was developed to process the R-curves (G versus crack length). The curve was first limited to only include data for crack increases of 0.1 mm. This allowed a global average to be calculated without artificially weighting the value. Then, peaks and troughs were automatically retrieved from the curve using an Excel® routine. An example of an R-curve with peaks and troughs is shown in Figure 4.

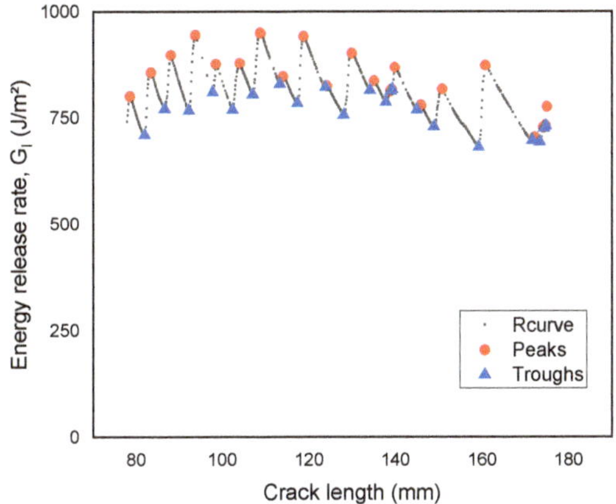

Figure 4. Example of peaks and troughs retrieved from R-curve.

4. Results

4.1. Water Uptake

Figure 5 shows weight gain measurements for three mode I fracture test specimens, first for immersion to saturation in seawater at 60 °C and then for drying.

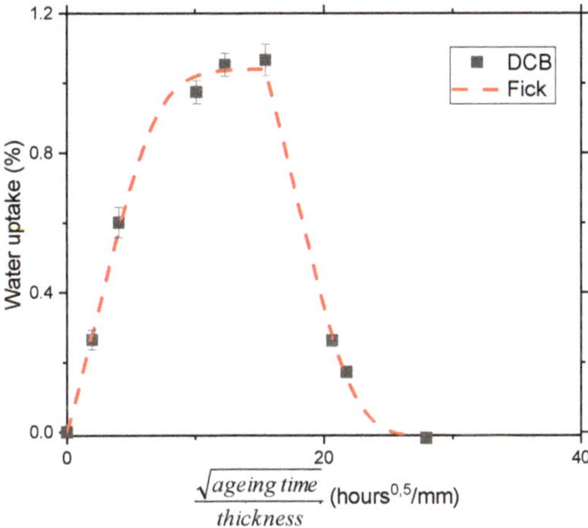

Figure 5. Mode I specimen water uptake after aging in natural seawater at 60°C and drying at 60 °C.

This plot shows similar weight gain kinetics and saturation values to those of the square coupon samples (Figure 2) and indicates that the weight gain is reversible after drying.

4.2. Mode I Fracture

Figure 6 shows examples of load-displacement plots for unaged mode I and mode II specimens. The propagation is unstable. One can note the difference in load level required to propagate cracks for these two tests. For both, precracked initial crack length was around 70 mm and final crack length was around 110 mm for DCB and 130 mm for ELS.

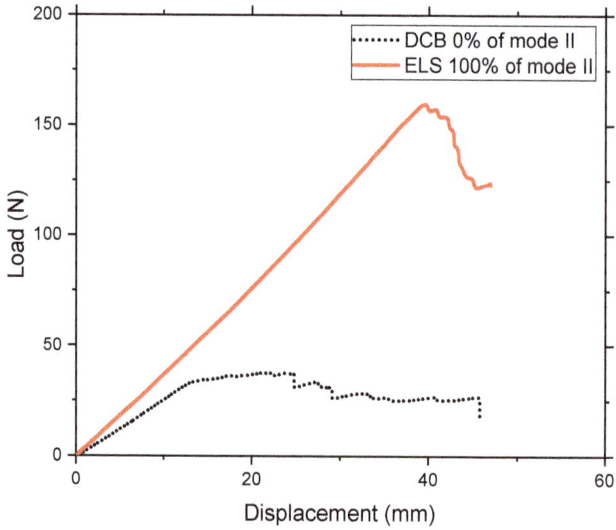

Figure 6. Load-displacement plot, unaged mode I and mode II specimens.

An example of an R-curve (G_I versus crack length) for the composite under mode I loading in the three ageing conditions is shown in Figure 7. The trend of the R-curves is globally stable with crack length but one can clearly observe local peaks and troughs for the unaged material, often attributed to a stick-slip behavior. This has been explained previously in terms of a local crack blunting mechanism [30]. Five specimens were tested for each condition, but only four results were retained for the saturated and dried conditions due to experimental incidents (debonding of load blocks during the tests). The dried condition showed less variability in propagation compared to the two other ageing conditions. These values are summarized with initiation values in Table 2.

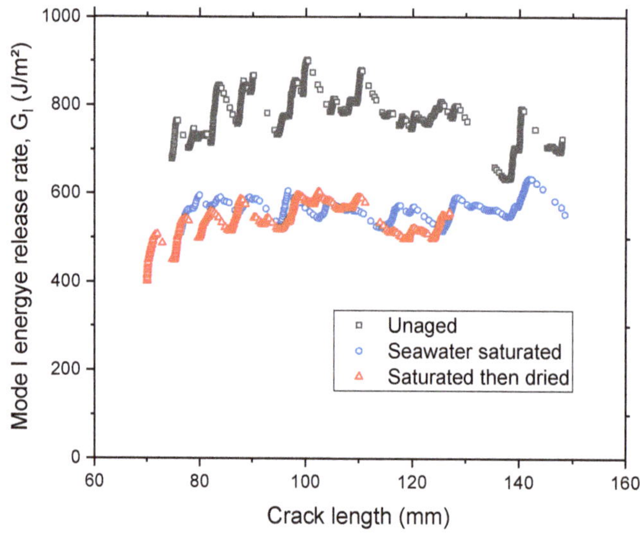

Figure 7. Examples of R-curves for unaged, seawater saturated and saturated then dried mode I specimens.

Table 2. Initiation and propagation values for all mode I tests; mean (standard deviation) values in J/m^2 (arrows show trends compared to the previous ageing condition).

	Initiation from Film		Initiation from Precrack		Propagation		
Criterion	**NL**	**5%/Max.**	**NL**	**5%/Max**	**Mean Entire R-Curve**	**Peak**	**Valley**
Unaged, Dry	281 (55)	353 (19)	388 (80)	527 (79)	799 (89)	835 (93)	722 (77)
Saturated	↘246 (67)	↘293 (25)	↘277 (13)	↘432 (11)	↘600 (105)	↘616 (96)	↘597 (102)
Re-dried after saturation	↘219 (13)	↘272 (48)	↗355 (60)	↗468 (42)	↘564 (57)	↘570 (55)	↘559 (55)

There is some scatter, but it is possible to obtain an average propagation energy release rate of around 800 J/m^2 for unaged, 600 J/m^2 for seawater saturated and 560 J/m^2 for re-dried specimens. Overall, for all the fracture parameters calculated from mode I loading tests, saturation in seawater at 60 °C reduces mode I fracture resistance by around 30% and this change is not reversible after drying.

4.3. Mode II Fracture

Figure 8 shows examples of the R-curves (G_{II} versus crack length) for the composite under mode II loading. One can see the difference in shape of the R-curves compared to those of the DCB test. For mode II, the R-curves are smooth compared to the mode I propagation; therefore, no "Peak" nor "Trough" values can be calculated. It is also worth noting the higher values of energies involved in mode II delamination compared to mode I. Table 3 shows initiation and propagation values for all mode II tests.

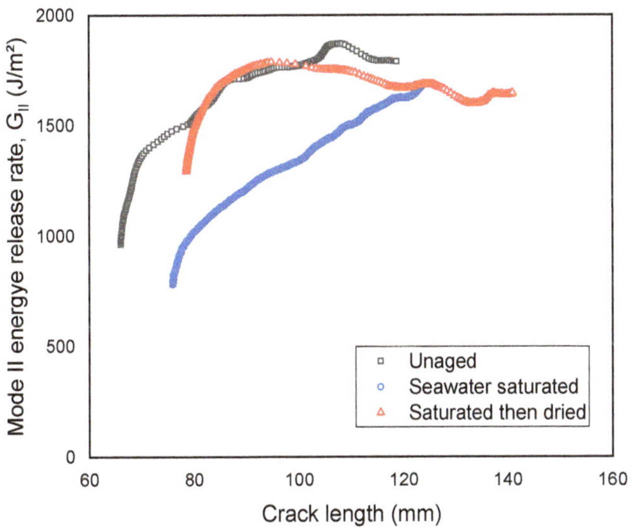

Figure 8. Examples of R-curves for unaged, saturated and saturated then dried mode II specimens.

Table 3. Initiation and propagation values for all mode II tests; mean (standard deviation) values in J/m^2 (arrows show trends compared to the previous ageing condition).

	Initiation from Film		Initiation from Precrack		Propagation
Criterion	NL	5%/Max.	NL	5%/Max	Mean Entire R-Curve
Unaged, Dry	738 (234)	1046 (216)	1078 (316)	1455 (253)	1790 (410)
Saturated	↘562 (97)	↘713 (93)	↘1032 (176)	↘1255 (261)	↘1378 (388)
Re-dried after saturation	↗728 (124)	↗958 (80)	↘993 (163)	↗1374 (236)	↗1475 (255)

There is a trend showing a decrease in G$_{II}$ after seawater ageing, but there is significant variability in the results and differences are within one standard deviation. This will be discussed further below. Values after drying tend to increase but do not return to unaged values. As with mode I, the standard deviation seems to be lower in the final dried ageing condition.

4.4. Mixed Mode Fracture

Figure 9 shows an example of the load–displacement plots from mixed mode MMB tests for the three mode ratios. Here, again, the increasing mode II contribution corresponds to higher loads during the test.

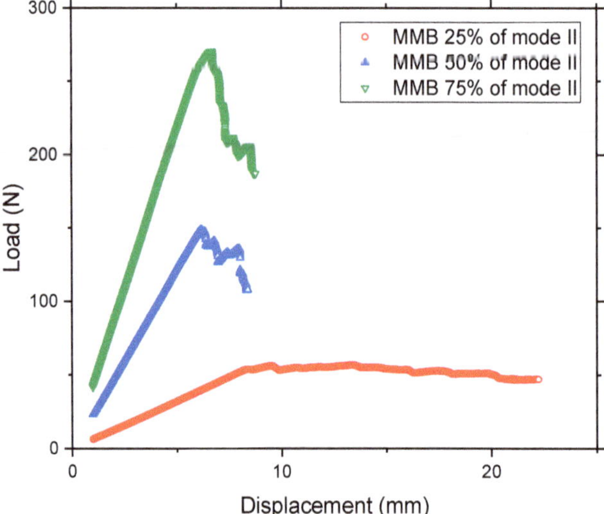

Figure 9. Load-displacement plots, 3 mixed mode ratios.

The R-curves of the MMB tests at the two ageing conditions and for the three mode mixities is shown in Figure 10.

These plots reveal one of the difficulties with MMB tests, namely that the propagation length is quite short (20–30 mm at most) compared to the mode I tests. As a result, the R-curves tend to increase and do not reach a plateau value. In this case, it is interesting to compare initiation values. For the 75% mode II case, the R-curves are more stable. Table 4 shows these initiation and propagation values for the three mixed mode tests.

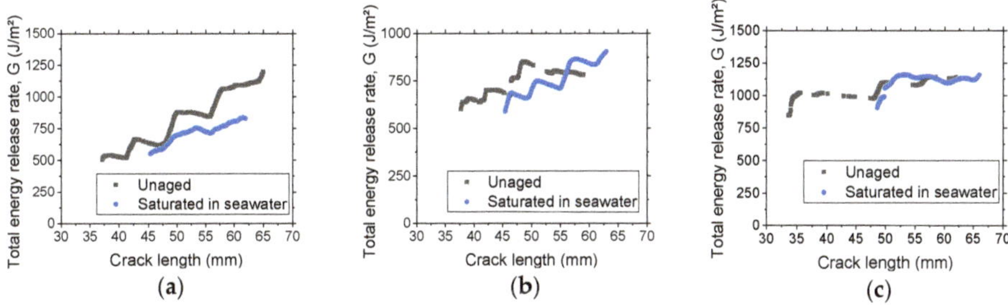

Figure 10. Examples of R-curves for unaged, and saturated mixed mode specimens, (25 (**a**), 50 (**b**) and 75% (**c**) of mode II).

Table 4. Initiation and propagation values for mixed mode (I/II) tests.

25% Mode II	Initiation from Film		Initiation from Precrack		Propagation		
Criterion	NL	5%/Max.	NL	5%/Max	Global mean	Peak	Valley
Unaged, Dry	348 (134)	449 (148)	373 (163)	473 (192)	818 (315)	913 (339)	859 (333)
Saturated	↘174 (48)	↘271 (46)	↘280 (29)	↘416 (33)	↘712 (183)	↘764 (175)	↘728 (168)
50% Mode II	**Initiation from Film**		**Initiation from Precrack**		**Propagation**		
Criterion	NL	5%/Max.	NL	5%/Max	Global mean	Peak	Valley
Unaged, Dry	544 (36)	752 (52)	517 (48)	670 (132)	795 (134)	813 (148)	780 (148)
Saturated	↘303 (52)	↘403 (40)	↘421 (58)	↘576 (66)	↘708 (134)	↘748 (139)	↘706 (143)
75% Mode II	**Initiation from Film**		**Initiation from Precrack**		**Propagation**		
Criterion	NL	5%/Max.	NL	5%/Max	Global mean	Peak	Valley
Unaged, Dry	591 (162)	783 (135)	562 (109)	759 (111)	942 (113)	1020 (112)	935 (97)
Saturated	↘491 (13)	↘716 (75)	↗676 (141)	↗867 (110)	↗1063 (144)	↗1134 (111)	↗1093 (88)

Once again, the overall trend indicates a decrease due to seawater ageing for the mode I dominated fractures, but for the 75% mode II fracture, saturation of specimens tends to result in a slightly higher energy release rate.

5. Discussion

- **Influence of an accelerated wet ageing protocol on mixed mode failure criteria**

The influence of seawater ageing on initiation and propagation values of strain energy release rate is shown in Figure 11.

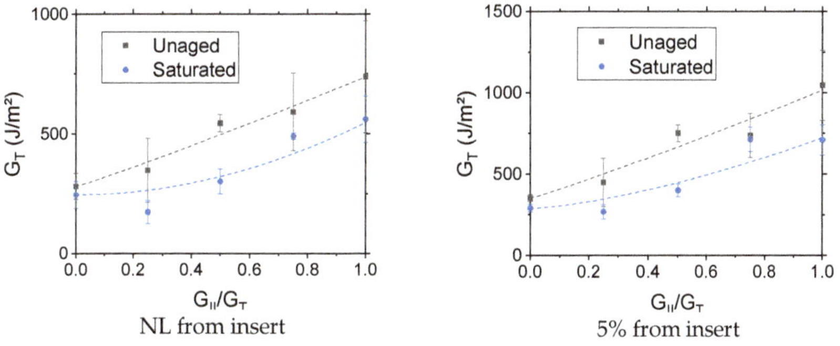

Figure 11. *Cont.*

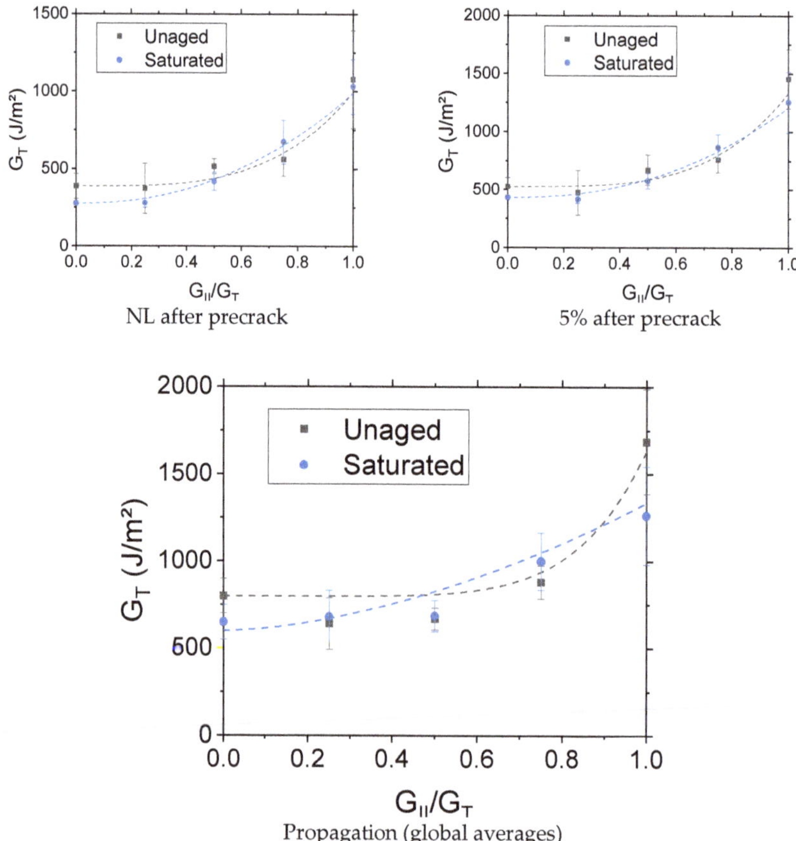

Figure 11. Summary of results for initiation and propagation; lines are B-K fits.

First, it should be noted that the interlaminar fracture toughness values measured here are in a similar range to published values for marine carbon/epoxy specimens. For example, Baral et al. studied a range of high modulus composites used for racing yachts and found mode I initiation values in the range of 200–300 J/m² with propagation values up to 700 J/m² [31]. Mode II fracture energies for carbon/epoxies are typically three times higher than those found under mode I loading [32], with this ratio decreasing as mode I toughness increases.

Second, it may be noted that, globally, the influence of seawater saturation on the interlaminar fracture behavior of this carbon/epoxy composite is small given the error bars. The results have been analyzed using the Benzeggagh and Kenane representation (B-K) [29] and this provides a reasonably good fit to all data. The n-parameters obtained for the unaged material are compared with published values for a glass/epoxy and carbon/peek composite in Table 5. It is apparent that, for values measured from the insert film, there is a stronger influence of the mode II loading component after saturation, while, for values from a precrack and propagation values, this mode II influence is lower after saturation. No published values were found for this parameter after seawater saturation.

Table 5. Mixed mode B-K failure criteria parameter n.

n	Unaged	Seawater Saturated
NL film	1.085	↗1.972
5% film	1.079	↗1.458
NL precrack	3.588	↘2.215
5% precrack	3.793	↘2.300
Propagation	6.189	↘1.693
Glass/epoxy [29]	2.6	-
Carbon/PEEK [31]	2.284	-

- **Influence of physical ageing**

An important feature of the values of G_{Ic} after drying, shown in Table 2, and, to a lesser extent, mode II values (Table 3) is that they did not return to their initial unaged values, even though the water ingress was shown to be reversible (Figure 5). This suggests that the material has undergone some permanent changes. This may indicate permanent damage, in the form of fiber/matrix debonding. Scanning electron microscopy was performed on unaged and aged fracture surfaces, but no significant differences were noted (no evidence of cleaner fibers after ageing, for example), so images are not shown here. Another possible explanation, which has received little attention in the published literature to date, is that the material undergoes physical ageing during immersion and drying at 60 °C. Physical ageing (PA) is a gradual change towards a more stable state of a polymer which has undergone rapid cooling below its glass transition temperature. Heating to a temperature close to the Tg will accelerate this change, related to molecular rearrangements, and it is usually accompanied by a reduction in tensile strain at failure [33,34]. At lower temperatures, these changes will take longer. Previous work on this same epoxy resin without fiber reinforcement [35] showed such an embrittlement with physical ageing, with faster property changes at closer to Tg temperatures. This is a reversible process; complete recovery of initial properties was noted after rejuvenation, a thermal treatment above Tg.

In order to examine the hypothesis that PA occurred during immersion and drying, a second series of mode I and mode II tests was performed on samples which had either been rejuvenated before testing or placed in conditions encouraging PA. The rejuvenation procedure involved heating above the Tg for a short time, then quenching in 15 °C water (samples were placed in sealed bags to prevent any contact with the water). Physically aged specimens were heated in an oven at 60 °C for 3 weeks. This temperature and ageing time were found in the previous study on the resin [35] and were estimated to provide conditions for a relatively complete PA. Tables 6 and 7 show the results from the mode I and mode II fracture tests on these conditioned specimens.

Table 6. Influence of physical ageing on energy release rate in mode I.

Criterion	Initiation from Film		Initiation from Precrack		Propagation		
	NL	5%/Max.	NL	5%/Max	Mean Entire R-Curve	Peak	Valley
Rejuvenated	190 (57)	359 (120)	497 (21)	816 (14)	800 (103)	832 (98)	744 (76)
Physically aged in air	←189 (57)	↘295 (80)	↘386 (77)	↘632 (95)	↘621 (57)	↘645 (57)	↘603 (53)

These results indicate that PA does have an impact on the energy release rate for both mode I and mode II for the majority of the initiation and propagation values. The energy release rates decrease with PA. The reduction is significant for mode I and less important for mode II loading. The magnitude is similar to the reductions noted after seawater ageing, suggesting that PA, due to the increase in temperature imposed by the accelerated ageing protocol, is the main mechanism acting to reduce delamination resistance here.

Table 7. Influence of physical ageing on energy release rate in mode II.

Criterion	Initiation from Film		Initiation from Precrack		Propagation
	NL	5%/Max.	NL	5%/Max	Mean Entire R-Curve
Rejuvenated	546 (90)	720 (95)	683 (131)	971 (116)	1275 (121)
Physically aged in air	↗584 (51)	↗768 (24)	↘596 (20)	↘912 (23)	↘1119 (98)

From this complementary study, it is clear that accelerated ageing of carbon/epoxy composites with low Tg resin can lead to complex test results. Nevertheless, it should be emphasized that the overall effect is limited; once it has been quantified, it is easy to design structures so that early failure can be avoided, at least under quasi-static loadings.

- **Applicability of test results**

The data shown in this paper were analyzed using an approach based on a number of assumptions. First, the equations employed assume linear elastic fracture mechanics. The specimens were dimensioned following the standards to achieve this, and unloading indicated no permanent residual displacement which would have indicated widespread damage in the specimen arms.

Second, physical ageing effects due to the extended period at 60 °C also contribute to the final value and result in a lower delamination resistance. This contribution will also occur if ageing and drying are performed at lower temperatures and in real marine structures but will require longer times to appear. The ageing protocol adopted here thus provides a conservative value of delamination resistance. Further tests would be required at different times and temperatures to establish the physical ageing kinetics in these composites, as was done for the resin alone previously [35]. Given the fact that water ingress can also reduce the Tg of epoxy matrices due to plasticization [36–38], physical ageing could also finally appear at ageing temperatures considered as too far from Tg to happen in relevant timescales.

Third, there has been considerable discussion in the published literature on the values to be used in modeling. Tests for each loading condition yield several results here, with initiation from an implanted film, from a precrack and average values for a propagating crack. In addition, different initiation criteria and data analyses are provided by the standard test procedures. In the present case, up to seven values are defined. This multiplicity of values arose due to both the integration of different national protocols into one ISO document and the need to highlight the specimen geometry dependence of propagation values caused by fiber bridging and multiple cracking. Given that the same specimen geometry was used throughout this study, it is interesting to compare the ratio of results from five values before and after water ageing. Figure 12 shows these ratios.

There is a change in the influence of the water ageing protocol as the mode II contribution increases, with the effect becoming positive at high mode II ratio, but the results from the ELS test are in contradiction with this trend.

Pure mode II tests show considerably higher scatter than those from 75% mode II, which makes it difficult to formulate conclusions regarding ageing effects. Development of a standard test to measure mode II delamination resistance has proved difficult. Initially, the end notch flexure specimen appeared suitable [39,40], but the unstable nature of propagation during this test was not satisfactory. Tests promoting stable propagation, using 4-point ENF (End Notched Flexure) [41] and end loaded split specimens such as those tested here [13] were therefore preferred as they allow multiple values to be measured on each specimen. However, these and all mode II configurations suffer from the influence of friction between the propagating crack surfaces. As a result, some authors have questioned whether a pure mode II test is possible [42]. For ageing studies, this also introduces an additional unknown factor: the influence of water in the saturated material on friction effects.

An alternative pragmatic approach to examining the influence of in-plane shear loading is to apply a mode II dominated mixed mode test, as this involves a small opening component which limits crack surface contact. In the present case, such an approach appears more consistent with mixed mode results for lower mode II loads. If propagation results are needed and the short MMB propagation distance is a constraint, some additional tests with an alternative fixture such as the asymmetric DCB proposed by Vanderkley [43] may be useful.

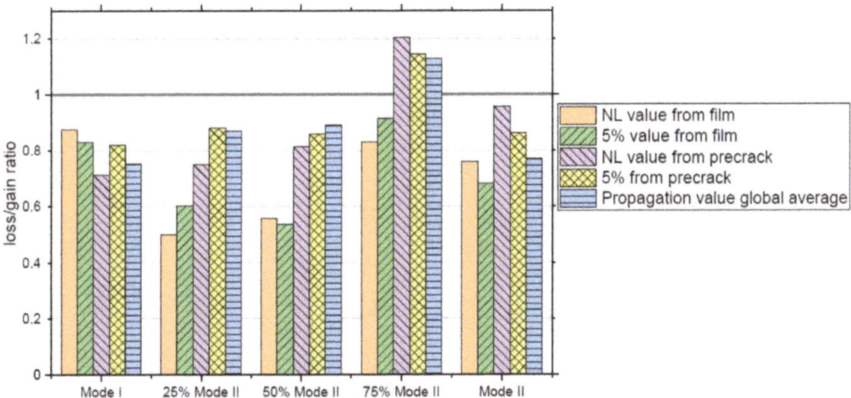

Figure 12. Ratios of water saturated/unaged fracture resistance values.

6. Conclusions

This paper presents a new set of results from a detailed study of the influence of seawater immersion on the interlaminar fracture resistance of a carbon/epoxy composite. It is shown that care is needed to dissociate changes due to water ingress from those associated with physical ageing due to polymer chain reorganization. The latter tend to reduce fracture toughness; the mechanism is slow at temperatures well below the glass transition temperature, but when raised temperatures are used to accelerate water ingress, these will also accelerate physical ageing. Specific tests were performed to quantify the latter and it appears to be the main mechanism contributing to changes in fracture resistance with this accelerated ageing protocol. It is therefore strongly recommended that for marine composites subjected to ageing at temperatures approaching their Tg separate tests be performed to dissociate these two effects.

However, even accounting for this second ageing mechanism, the interlaminar fracture behavior of this composite is quite stable, indicating that it is a good contender for marine applications where out-of-plane loading occurs.

Author Contributions: Funding acquisition, P.D. and B.H.; investigation, A.L.G.-G.; methodology, A.L.G.-G., P.D., P.-Y.L.G. and B.H.; supervision, P.D. and P.-Y.L.G.; writing—original draft, A.L.G.-G.; writing—review and editing, P.D., P.-Y.L.G. and B.H. All authors have read and agreed to the published version of the manuscript.

Funding: This research was funded by DGA and IFREMER.

Acknowledgments: The authors are grateful to the DGA and IFREMER for co-funding this study. The technical assistance of colleagues at the IFREMER Centre (Luc Riou, Mickael Premel-Cabic, Nicolas Lacotte and Mael Arhant) is gratefully acknowledged.

Conflicts of Interest: The authors declare no conflict of interest.

References

1. Smith, C.S. *Design of Marine Structures in Composite Materials*; Elsevier Applied Science: London, UK, 1990.
2. Graham-Jones, J.; Summerscales, J. (Eds.) *Marine Applications of Advanced Fibre-reinforced Composites*; Woodhead Publishing: Cambridge, UK, 2016.
3. Marsh, G. Fifty years of reinforced plastic boats. *Reinf. Plast.* **2006**, *50*, 16–19. [CrossRef]

4. Ochoa, O.; Salama, M.M. Offshore composites: Transition barriers to an enabling technology. *Compos. Sci. Technol.* **2005**, *65*, 2588–2596. [CrossRef]

5. Melot, D. Present and Future Composites Requirements for the Offshore Oil and Gas Industry. In *Durability of Composites in a Marine Environment 2*; Davies, P., Rajapakse, Y.D.S., Eds.; Springer: Dordrecht, The Netherlands, 2017; pp. 151–172.

6. Bir, G.S.; Lawson, M.; Li, Y. Structural Design of a Horizontal-Axis Tidal Current Turbine Composite Blade. In Proceedings of the ASME 30th International Conference on Ocean, Offshore, and Arctic Engineering, Rotterdam, The Netherlands, 19–24 June 2011.

7. Ennis, B.L.; Kelley, C.L.; Naughton, B.T.; Norris, R.E.; Das, S.; Lee, D.; Miller, D.A. *Optimized Carbon Fiber Composites in Wind Turbine Blade Design*; SANDIA Report SAND2019-14173; Sandia Labs: Albuquerque, NM, USA, November 2019.

8. Friedrich, K. (Ed.) *Application of Fracture Mechanics to Composite Materials*; Elsevier: Amsterdam, The Netherlands, 1989.

9. Moore, D.R.; Williams, J.G.; Pavan, A. *Fracture Mechanics Testing Methods for Polymers, Adhesives and Composites*; Elsevier: Amsterdam, The Netherlands, 2001.

10. ISO 15024:2001. *Fibre-Reinforced Plastic Composites—Determination of Mode I Interlaminar Fracture Toughness, GIC, for Unidirectionally Reinforced Materials*; ISO: Geneva, Switzerland, 2001.

11. Davies, P.; Sims, G.D.; Blackman, B.R.K.; Brunner, A.J.; Kageyama, K.; Hojo, M.; Tanaka, K.; Murri, G.; Rousseau, C.; Gieseke, B.; et al. Comparison of test configurations for the determination of GIIc: Results from an international round robin. *Plast. Rubber Compos.* **1999**, *28*, 432–437. [CrossRef]

12. Davidson, B.D.; Sun, X. Effects of Friction, Geometry, and Fixture Compliance on the Perceived Toughness from Three-and Four-Point Bend End-Notched Flexure Tests. *J. Reinf. Plast. Compos.* **2005**, *24*, 1611–1628. [CrossRef]

13. ISO 15114:2014. *Fibre-Reinforced Plastic Composites—Determination of the Mode II Fracture Resistance for Unidirectionally Reinforced Materials Using the Calibrated End-Loaded Split (C-ELS) Test and an Effective Crack Length Approach*; ISO: Geneva, Switzerland, 2014.

14. Reeder, J.; Crews, J.J. The mixed-mode bending method for delamination testing. *AIAA J.* **1989**, *28*, 1270–1276. [CrossRef]

15. ASTM D6671. *Standard Test Method for Mixed Mode I-Mode II Interlaminar Fracture Toughness of Unidirectional Fiber Reinforced Polymer Matrix Composites*; ASTM: Philadelphia, PA, USA, 2001.

16. Alfano, G.; Crisfield, M.A. Finite element interface models for the delamination analysis of laminated composites: Mechanical and computational issues. *Int. J. Numer. Methods Eng.* **2001**, *50*, 1701–1736. [CrossRef]

17. Blackman, B.R.K.; Hadavinia, H.; Kinloch, A.J.; Williams, J.G. The use of a cohesive zone model to study the fracture of fibre composites and adhesively-bonded joints. *Int. J. Fract.* **2003**, *119*, 25–46. [CrossRef]

18. Marom, G. Environmental Effects on Fracture Mechanical Properties of Polymer Composites. In *Application of Fracture Mechanics to Composite Materials*; Friedrich, K., Ed.; Elsevier: Amsterdam, The Netherlands, 1989; pp. 397–424.

19. Thomson, K.W.; Broutman, L.J. The effect of water on the fracture surface energy of fiber-reinforced composite materials. *Polym. Compos.* **1982**, *3*, 113–117. [CrossRef]

20. Russell, A.; Street, K. Moisture and Temperature Effects on the Mixed-Mode Delamination Fracture of Unidirectional Graphite/Epoxy. In *Delamination and Debonding of Materials*; ASTM STP 876; Johnson, W., Ed.; ASTM International: Philadelphia, PA, USA, 1985; pp. 349–370.

21. Garg, A.; Ishai, O. Hygrothermal influence on delamination behavior of graphite/epoxy laminates. *Eng. Fract. Mech.* **1985**, *22*, 413–427. [CrossRef]

22. Selzer, R.; Friedrich, K. Influence of water up-take on interlaminar fracture properties of carbon fibre-reinforced polymer composites. *J. Mater. Sci.* **1995**, *30*, 334–338. [CrossRef]

23. Zhao, S.; Gaedke, M. Moisture effects on Mode II delamination behavior of carbon/epoxy composites. *Adv. Compos. Mater.* **1996**, *5*, 291–307. [CrossRef]

24. Chou, I. Effect of fiber orientation and moisture absorption on the interlaminar fracture toughness of CFRP laminates. *Adv. Compos. Mater.* **1998**, *7*, 377–394. [CrossRef]

25. Asp, L.E. The effects of moisture and temperature on the interlaminar delamination toughness of a carbon/epoxy composite. *Compos. Sci. Technol.* **1998**, *58*, 967–977. [CrossRef]

26. Landry, B.; La Plante, G.; Le blanc, L.R. Environmental effects on mode II fatigue delamination growth in an aerospace grade carbon/epoxy composite. *Compos. Part A Appl. Sci. Manuf.* **2012**, *43*, 475–485. [CrossRef]

27. Couture, A.; Laliberte, J.; Li, C. Mode I Fracture Toughness of Aerospace Polymer Composites Exposed to Fresh and Salt Water. *Chem. Mater. Eng.* **2013**, *1*, 8–17.

28. Le Blanc, L.; La Plante, G.; Li, C. Moisture effects on mixed-mode delamination of carbon/epoxy composites, in Design, Manufacturing and Applications of Composites. In *Proceedings of the Tenth Joint Canada-Japan Workshop on Composites, Vancouver, BC, Canada, 19–21 August 2014*; DEStech Publications, Inc.: Lancaster, PA, USA, 2015; p. 115.

29. Benzeggagh, M.; Kenane, M. Measurement of mixed-mode delamination fracture toughness of unidirectional glass/epoxy composites with mixed-mode bending apparatus. *Compos. Sci. Technol.* **1996**, *56*, 439–449. [CrossRef]

30. Kinloch, A.J.; Williams, J.G. Crack blunting mechanisms in polymers. *J. Mater. Sci.* **1980**, *15*, 987–996. [CrossRef]

31. Baral, N.; Davies, P.; Baley, C.; Bigourdan, B. Delamination behaviour of very high modulus carbon/epoxy marine composites. *Compos. Sci. Technol.* **2008**, *68*, 995–1007. [CrossRef]

32. Reeder, J.R. *An Examination of Mixed Mode Delamination Failure Criteria*; NASA Technical Memo 104210: Hampton, VA, USA, 1992.

33. Struik, L.C.E. *Physical Aging in Amorphous Polymers and Other Materials*; Elsevier: Amsterdam, The Netherlands, 1978.

34. Odegard, G.M.; Bandyopadhyay, A. Physical aging of epoxy polymers and their composites. *J. Polym. Sci. Part B Polym. Phys.* **2011**, *49*, 1695–1716. [CrossRef]

35. Le Guen-Geffroy, A.; Le Gac, P.-Y.; Habert, B.; Davies, P. Physical ageing of epoxy in a wet environment: Coupling between plasticization and physical ageing. *Polym. Degrad. Stab.* **2019**, *168*, 108947. [CrossRef]

36. El Yagoubi, J.; Lubineau, G.; Saghir, S.; Verdu, J.; Askari, A. Thermomechanical and hygroelastic properties of an epoxy system under humid and cold-warm cycling conditions. *Polym. Degrad. Stab.* **2014**, *99*, 146–155. [CrossRef]

37. Morgan, R.J.; O'Neal, J.E. The Durability of Epoxies. *Polym. Technol. Eng.* **1978**, *10*, 49–116. [CrossRef]

38. Nogueira, P.; Torres, A.; Abad, M.-J.; Cano, J.; Barral-Losada, L.F.; Ramírez, C.; López, J.; López-Bueno, I. Effect of water sorption on the structure and mechanical properties of an epoxy resin system. *J. Appl. Polym. Sci.* **2001**, *80*, 71–80. [CrossRef]

39. Barrett, J.; Foschi, R. Mode II stress-intensity factors for cracked wood beams. *Eng. Fract. Mech.* **1977**, *9*, 371–378. [CrossRef]

40. Carlsson, L.; Gillespie, J.; Pipes, R. On the Analysis and Design of the End Notched Flexure (ENF) Specimen for Mode II Testing. *J. Compos. Mater.* **1986**, *20*, 594–604. [CrossRef]

41. Martin, R.H.; Davidson, B. Mode II fracture toughness evaluation using a four point bend end notched flexure test. In *Proceedings of the 4th International Deformation and Fracture of Composites Conference (DFC4), London, UK, 24–26 March 1997*; Institute of Materials: London, UK, 1997; pp. 243–252.

42. O'Brien, T.K.; O'Brien, T. Composite Interlaminar Shear Fracture Toughness, GIIc: Shear Measurement or Sheer Myth? In *Composite Materials: Fatigue and Fracture*; Volume 7th, Bucinell, R., Ed.; ASTM STP 1330; ASTM International: West Conshohocken, PA, USA, 1998; pp. 3–18.

43. Vanderkley, P.S. Mode I-Mode II Delamination Fracture Toughness of a Unidirectional Graphite/Epoxy Composite. Master's Thesis, Texas A & M University, College Station, TX, USA, 1981.

Article

Seasonal and Inter-Annual Variability of the Phytoplankton Dynamics in the Black Sea Inner Basin

Svetla Miladinova [1],*, Adolf Stips [1], Diego Macias Moy [2] and Elisa Garcia-Gorriz [1]

[1] Joint Research Centre (JRC), European Commission, 21027 Ispra, Italy; Adolf.STIPS@ec.europa.eu (A.S.); Elisa.GARCIA-GORRIZ@ec.europa.eu (E.G.-G.)

[2] Instituto de Ciencias Marinas de Andalucía (ICMAN), Consejo Superior de Investigaciones Científicas (CSIC), 11510 Puerto Real, Spain; diego.macias@icman.csic.es

* Correspondence: svetla.miladinova@ext.ec.europa.eu

Received: 18 August 2020; Accepted: 14 October 2020; Published: 20 October 2020

Abstract: We explore the patterns of Black Sea phytoplankton growth as driven by the thermohaline structure and circulation system and the freshwater nutrient loads. Seasonal and inter-annual variability of the phytoplankton blooms is examined using hydrodynamic simulations that resolve mesoscale eddies and online coupled bio-geochemical model. This study suggests that the bloom seasonality is homogeneous across geographic locations of the Black Sea inner basin, with the strongest bloom occurring in winter (February–March), followed by weaker bloom in spring (April–May), summer deep biomass maximum (DBM) (June–September) and a final bloom in autumn (October–November). The winter phytoplankton bloom relies on vertical mixing of nitrate from the intermediate layers, where nitrate is abundant. The winter bloom is highly dependent on the strength of the cold intermediate layers (CIL), while spring/summer blooms take advantage of the CIL weakness. The maximum phytoplankton transport across the North Western Shelf (NWS) break occurs in September, prior to the basin interior autumn bloom. Bloom initiation in early autumn is associated with the spreading of NWS waters, which in turn is caused by an increase in mesoscale eddy activity in late summer months. In summary, the intrusion of low salinity and nitrate-rich water into the basin interior triggers erosion of the thermocline, resulting in vertical nitrate uplifting. The seasonal phytoplankton succession is strongly influenced by the recent CIL disintegration and amplification of the Black Sea circulation, which may alter the natural Black Sea nitrate dynamics, with subsequent effects on phytoplankton and in turn on all marine life.

Keywords: Black Sea; biogeochemical modelling; seasonal phytoplankton and inter-annual variation

---

1. Introduction

Phytoplankton blooms [1] are identifiable signals of the annual growth activity in pelagic systems. Typical Black Sea seasonal phytoplankton dynamics are characterised by a major phytoplankton bloom in winter (February) followed by a smaller spring (March–April) bloom [2–5]. It has been found that windy, cold winters lead to an abundance of phytoplankton biomass (PHY) or chlorophyll-a (CHLa) concentrations in winter-spring, due to enhanced vertical mixing and stronger upwelling. Thus, the inter-annual variability of phytoplankton blooms has been associated with the winter severity. In summer the phytoplankton growth on the surface decays because of nitrate depletion, and autumn bloom follows between September and November. During the late autumn and winter-spring blooms, PHY reaches a maximum in the upper mixed layer. From June to October, the maximum of PHY (so-called DBM) appears at the subsurface layer, associated with the uppermost pycnocline. The February–April blooms are clearly detected by both types of CHLa measurement modalities used: in-situ and satellite [2,5,6]. Satellite data showed that the autumn blooms are more abundant than the spring blooms. The trend towards milder winters has been cited as the reason for the disappearance of the spring bloom in the open Black Sea [7]. It has been assumed that in years with temperatures in line with the inter-annual average, CHLa peaks in winter and the spring peak is absent, whereas in cold years, the relatively low CHLa in winter is followed by a spring bloom. Specifically, a decrease in the frequency of cold winters results in changes in the timing of the winter-early spring bloom, which either weakened or disappeared altogether depending on local meteorological and oceanographic conditions [5]. It is believed that the open Black Sea experiences a bloom cycle with an autumn CHLa peak and summer minimum, but no spring bloom on the surface in an atypical year [8]. Several authors suggested a possible relationship between cold intermediate layers (CIL) thickness and phytoplankton blooms [5,6,8]. There is an established intensification of the autumn blooms over the last 20 years [9]. In general, there is no clear consensus in the literature concerning the exact seasonality of PHY/CHLa dynamics observed in the Black Sea.

The inconsistency between measurements is noticeable when comparing in-situ and remote sensing studies in the Black Sea. For example, it is suggested that CHLa density is overestimated for the open sea by up to a factor of four, when computed using the global SeaWiFS chlorophyll model, when compared to direct in-situ measurements [10]. The difference between satellite-based estimates and in-situ measurements is even larger for the coastal regions [11]. The analysis of MODIS-Aqua data for 2003–2013 and SeaWiFS data for 1998–2007 indicates that Oceancolor Level 2 and 3 chlorophyll data contain systematic inconsistencies in the Black Sea basin, which are related to the effect of cloud shadows [12]. In summary, the seasonal CHLa density estimation from satellite observations appears to be less reliable for the Black Sea, and consequently so are the conclusions for seasonal phytoplankton and inter-annual evolution in the literature, which are solely based on satellite data.

During the last two decades, the Black Sea ecosystem evolution has been studied by using a variety of models with different levels of complexity [13–16]. Even though the existing studies have provided deep knowledge on the Black Sea plankton dynamics and certain important North Western Shelf (NWS) processes, the current understanding of the mechanisms that control the seasonality and magnitude of the phytoplankton blooms is still incomplete.

The impact of mesoscale (or sub-mesoscale) physics on both hydrodynamics and phytoplankton production has been discussed [17,18]. The largest mesoscale eddies (MEs) in the Black Sea emerge from instabilities of horizontally sheared motions, particularly in the boundaries of the Rim Current. The MEs appears to be a possible mechanism for offshore transport of nutrients. The distribution of hydro-chemical properties in the upper layer seems to be sensitive to external pressures originating from regional weather variability [19,20] and anthropogenic inputs [21,22]. It is important to recognise the contribution of the mesoscale circulations in supporting offshore phytoplankton production. In addition, the ability of nutrient-rich NWS waters to enhance phytoplankton blooms should be evaluated.

The present study motivation is to address the gaps in the literature regarding forcing and dynamics of plankton seasonality in the basin of the Black Sea. To achieve this goal, the study exploits a previously developed hydrodynamic model to perform reliable long-term simulations of the Black Sea dynamics [23]. The paper aims to explore the seasonality and inter-annual variation of the phytoplankton bloom in the Black Sea inner basin. The effect of CIL thickness and temperature on seasonal phytoplankton blooms is studied in detail. We examine the dynamics of the Black Sea surface circulation system and cross-shelf transport to propose a mechanism of initiation of the early autumn bloom.

2. Methods and Data

2.1. Geographical Area of Interest

When identifying the sources and levels of nitrate, the Black Sea basin can be divided into two main parts. One part includes the NWS and South Western Self (SWS), which receive nitrate coming from the Danube and other big rivers flowing into the NWS during a large part of the year. It is assumed that phosphate is limiting the phytoplankton growth in these regions, due to excess of inorganic nitrogen exceeding the stoichiometric amount of phosphate [19]. In the rest of the Black Sea, however, nitrate is the limiting nutrient for phytoplankton growth [11]. The nitrate concentration is less than 0.1 mmol N m^{-3} in the surface waters of the inner sea for most of the year except for periods of strong vertical winter mixing and infrequent intrusion from coastal regions [24,25]. The euphotic zone in the Black Sea interior typically extends to a depth of about 30–45 m [26], in which the concentration of oxygen is high (more than 100 mmol O$_2$ m^{-3}), while nutrients and organic matter concentrations vary seasonally [25]. The depth distribution profile of nitrate is characterised by a subsurface maximum located below the euphotic zone (~55 m depth), following a rapid decrease of nitrate with increasing depth, and absence of nitrate in the deep sea. That intermediate layer (located between 30–150 m depth) containing a high concentration of nitrate (0.5–5 mmol N m^{-3}) is called the nitrate storage henceforth.

The accumulation of nitrate in intermediate water layers can be explained by its production, namely, nitrification and remineralisation, as well as the very slow uptake rates. It is further assumed that there are external sources of nitrate contributing to the inner basin concentration, such as river-based and atmospheric input. For instance, 20% of nitrates consumed by phytoplankton in the open waters are of allochthonous origin [27]. Moreover, based on satellite observation, it is thought that anthropogenic forces on the NWS shelf did not significantly influence the ecosystem of the deep part of the Black Sea during the 1970s to mid-1980s [21].

The Black Sea is characterised by large spatial heterogeneity of bio-hydro-chemical characteristics [19,23,28]. The main processes governing nutrient spreading are large-scale horizontal currents and mesoscale eddy fields, in conjunction with vertical turbulent mixing during phases of convection in winter. One general hypothesis is the strengthening of the basin-scale cyclonic circulation in winter and the intensification of anticyclone activity in the warm summer–autumn months [23,29,30]. Typically, in late summer and early autumn, an abundance of anticyclonic eddies is formed between the shelf edge and the Rim Current.

We consider three deep regions: The deep basin, of depth greater than 1500 m, is divided into three regions (WEST, CENTRAL and EAST). The basin separation is introduced for model validation purposes. It is assumed that the layer located between 200 and 1500 m in depth covers the shelf-break (SB) and continental slope. The southern and south-western shelf-break areas (SWSB) comprise the southern part of this strip. The NWS is less than 200 m deep and is located above 43.5° N, while the SWS is the area below 43.5° N. This study is focused on the phytoplankton succession in the inner Black Sea basin (limited by the 1500 m isobath in Figure 1).

2.2. Hydrodynamic Model

A hydrodynamic model, which represents the mesoscale circulations and thermohaline structure in the Black Sea for a continuous multi-decadal period, without any relaxation towards external fields, is used for this study [23]. This 3D hydrodynamic model involves two modules: GETM [31] and GOTM. It is initialised on a high resolution 2 × 2 min latitude–longitude horizontal grid. The meteorological forcing from the European Centre for Medium Range Weather Forecast (ECMWF) [32], based on 6-hourly records, has been applied (ERA-Interim project). Monthly mean freshwater input has been estimated using the values from the Global Runoff Data Centre (GRDC) runoff [33]. The model is initialised using temperature and salinity 3D fields coming from the MEDAR/MEDATLAS II project [34]. The 3D hydrodynamic model has been successfully applied to study the long term (1960–2015) thermohaline structure and circulation in the Black Sea previously [23,35].

2.3. Lower Trophic Level Model

The flow fields from the hydrodynamic model are used to predict the evolution of the low trophic level components of the food chain in the Black Sea ecosystem [36]. A nitrate-based biogeochemical model has been implemented, taking inspiration from the models already existing in the literature [37,38]. The model is purposefully adapted for the Black Sea basin, where a unique ecosystem exists. It represents the classical omnivorous food-web with 12 state variables (Figure 2). These include two phytoplankton size groups–small (P_S) and large (P_L), four zooplankton groups, including micro and meso-zooplankton (Z_S and Z_L), non-edible dinoflagellate species as Noctiluca (Z_N) and gelatinous zooplankton species Mnemiopsis and Beroe Ovata (Z_G). All plankton biomasses are expressed in nitrogen units, since nitrogen is considered to be the most important limiting nutrient for the interior Black Sea ecosystem [39]. Nitrogen is represented by two inorganic nutrients—nitrate (N_n) and ammonium (N_a) and is included in the particulate organic material—detritus (D). The Black Sea Ecosystem Model (BSEM) model is further modified in Reference [40] where the phosphate (PO) state variable is added in order to consider phosphorus as limiting the phytoplankton production rate together with nitrogen and photosynthetically available radiation. Other state variables incorporate dissolved oxygen (DO) and hydrogen sulphide (HS). The values of the coefficients in the BSEM model equations are obtained either from the literature or from fitting the model to measured concentration profiles in the Black Sea basin. Nevertheless, the updated BSEM parameters (Tables S1 and S2) do not differ substantially from their original values [38]. The Supplementary file includes all model equations describing the evolution of the state variables and the values of model parameters.

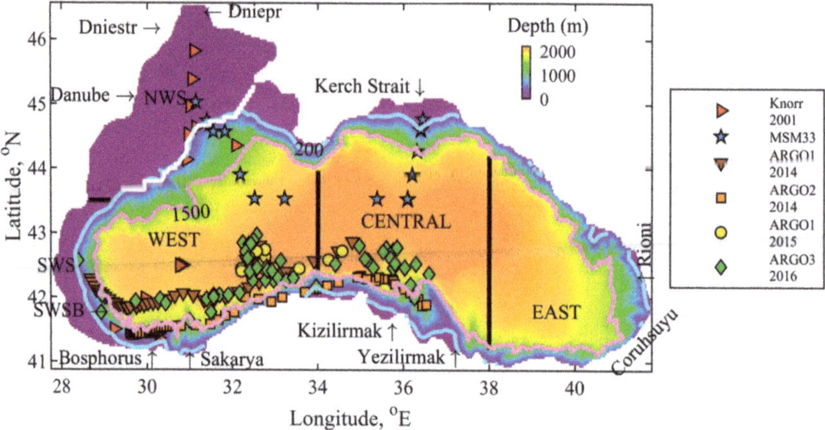

Figure 1. Bathymetry and location map of the Black Sea, North Western Shelf (NWS) and South Western Self (SWS) regions, and the main rivers. The 1500 m isobath is given in magenta and the 200 m isobath in cyan. The area between these two isobaths is considered herein as a shelf-break region. The boundaries of the three deep basin compartments 'WEST', 'CENTRAL' and 'EAST' are shown. The southernmost region, characterised by water depth inferior to 1500 m, is called the southern shelf. Symbols show the R/V Knorr and MSM33 routes and position of ARGO's drift stations (Table 1). The NWS boundary, where the offshore/onshore exchanges are estimated, is denoted by a white tracer.

River nutrient load data is issued from the SESAME and PERSEUS projects [41]. It consists of systematic nutrient measurements at the Danube discharge sections from the 1990s until 2008. Climatological mean Danube nutrient loads (averaged over the 1990–2008 period) are used from the beginning of 2009 for the Danube. For the other model rivers (Figure 1) data is not available since the 1990s, thus, climatological mean data is used in all cases to fill in the gaps. Model sensitivity analysis shows that initial nutrient concentration in the intermediate and deep layers is a very important part of the model setup, since it can support phytoplankton growth, even without loading from the rivers. Thus, the nutrient climatology for the Black Sea, coming from [34], is used as initial conditions for nitrate, ammonium and phosphate. This dataset reflects the main features known from observations, such as the existence of the intermediate layer in the basin interior of high nitrate concentration (nitrate storage) [25,42]. At depths greater than 115 ± 15 m, ammonium, phosphate and hydrogen sulphide concentrations increase. Hydrogen sulphide is zero in the upper 90 m, and then increases linearly to 500 mmol HS m^{-3} at the seabed level. Dissolved oxygen decreases linearly from 340 mmol O$_2$ m^{-3} to 0 in the upper 150 m and is set to zero further below [43,44]. All other BSEM model state variables are set to 0.0225 mmol N m^{-3} and vertically uniformity over the entire water column is assumed. This assumption is based on the fact that equilibrium structures do not depend on the initial conditions and are emergent properties of the model dynamics. A comparison of the atmospheric and riverine inputs for the Black Sea revealed that atmospheric-dissolved, inorganic nitrogen ranged between 4% and 13% of the total inorganic nitrogen, while the atmospheric-dissolved, inorganic phosphorus fluxes had significantly higher contributions with values ranging from 12% to 37% of the total inorganic phosphorus [45]. According to Reference [46], the atmospheric input of oxidised nitrogen can reach 13% of the total inorganic nitrogen input of the Danube. Such a flux can sustain up to 15% of the new phytoplankton production. We considered constant atmospheric surface fluxes of 0.083 mmol N m^{-2}d^{-1} nitrate, 0.0225 mmol N m^{-2}d^{-1} ammonium, and 0.0052 mmol P m^{-2}d^{-1} phosphate. These values are in close agreement with the lower bounds of the experimental data presented in Reference [45].

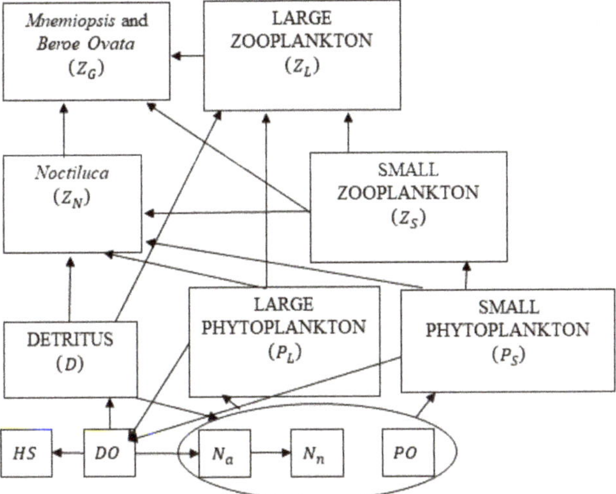

Figure 2. Schematic representation of the Black Sea Ecosystem Model (BSEM) model structure that includes the basic omnivorous food web and its interactions with the gelatinous carnivore predators *Mnemiopsis, Beroe Ovata* and *Noctiluca*.

The issue of what constitutes a bloom is more than a biomass issue [1]. It is assumed in Reference [47] that the total carbon biomass of non-blooming events was normally distributed with a seasonal mean described by a periodic spline function. Blooms were identified as significant deviations above this pattern. In this study, we use a criterion for the bloom definition by considering an occurrence of a bloom when the actual PHY (chlorophyll-a concentration) in the basin is larger than the mean average for the entire analysed period. We are not dealing merely with the exponential growth part of the phytoplankton curve, but are looking at the long-term, seasonal dynamic of that curve.

PHY (mg C m^{-3}) is calculated as the sum of small and large phytoplankton, which is converted from mmol N m^{-3} to mg C m^{-3} using the 6.625 Redfield ratio. Then, in order to compare with observational data, the stoichiometric conversion of mmol N m^{-3} to mg CHLa m^{-3} is done. Despite the existing underlying complexity of this conversion [48,49], the last ratio is usually taken as a constant, which is assumed to be 1 [13,14,50].

2.4. Model Validation Statistics

Model validation statistics used in this study are the average absolute error (*AAR*) and Pearson correlation coefficient (*R*), displayed with their 95% confidence intervals. Model skills (*Mskill*) are evaluated using root mean square error (*RMSE*) and standard deviation (*std*), namely:

$$Mskill = 1 - RMSE/std \text{ (observational data)}.$$

A value close to one indicates a close match between observations and model prediction. A value close to zero indicates that the mechanistic model has the poor predictive capacity and performs, as well as a simple, data-driven statistical model would. Values inferior to zero indicate that the observation means would be a better predictor than the mechanistic model results. Statistical measures, such as the normalised root mean squared difference (*RMSD*) and normalised standard deviation (model statistics are divided by the observational ones) are used in Section 3 for model validation (Taylor diagram).

2.5. Validation Data

Several different CHLa datasets of the Black Sea surface and water column are used in this study (Table 1). The Copernicus dataset is averaged and reprocessed with the BSAlg algorithm [51]. TheRegAlg time series of CHLa from 1999 to 2008 are redrawn from Reference [4]. In-situ vertical profiles of CHLa were collected during the R/V Knorr 2001 research cruise to the Black Sea (23 May to 10 June 2001). The station locations for the R/V Knorr 2001 are shown in Figure 1. They are well situated to study WEST, SWSB, and NWS-SB regions of the Black Sea [52]. Refer to Figure 1 for the location of the regions of interest. MSM33 denotes sampling conducted from the 11 November to the 2 December 2013 during cruise 33 of the German R/V Maria S. Merian [53]. The cruise itinerary covered the north-western and north-eastern shelf-break and the central deep Black Sea (Figure 1). In-situ vertical profiles of CHLa from the profiling float PROVOR II (ARGO) is available several (three to ten) times a month from 2014 to 2016 [54]. Locations of ARGO's data stations can be seen in Figure 1, while the time coverage is given in Table 1.

Table 1. The chlorophyll-a (CHLa) datasets names and coverage areas over space and time.

Short Name	Type of Dataset	Resolution	Coverage	References
Copernicus	Satellite (Copernicus OCTAC)	1 × 1 km, 1-day average	1998–2017, Surface of the entire basin (a lot of missing data in December and January)	[54]
RegAlg	Satellite (SeaWiFS)	4 × 4 km, daily	1999–2008, Mean surface data for several particular regions	[4]
Knorr 2001	In-situ	Stations	23 May to 10 June 2001 Vertical profiles in WEST, SWSB (south-western shelf-break areas) and NWS	[52]
MSM33	In-situ	Stations	11 November to 2 December 2013	[53]
ARGO1	In-situ	Profiling float	2014, Vertical profiles in WEST and CENTRAL 2015, in WEST	[54]
ARGO2	In-situ	Profiling float	1 January to 18 October 2014 Vertical profiles in SS	[54]
ARGO3	In-situ	Profiling float	2016, Vertical profiles in WEST and CENTRAL	[54]

3. Validation of the Phytoplankton Distribution Model

3.1. Validation against In-Situ Vertical Profile Data

As introduced above, the study focuses on the phytoplankton bloom dynamics and the physical factors governing its life cycle in the Black Sea inner basin. Before addressing this topic, we assess the model's ability to capture the amplitudes and phases of known seasonal blooms, as well as CHLa seasonal cycles for the different study areas, located in the Black Sea basin. The in-situ measurements are used to assess the deep CHLa maximum (DCM) over a short period of time, although the data coverage from in-situ observations remains sparse in the Black Sea.

A comparison between the Knorr 2001 CHLa (mg CHLa m^{-3}) vertical profiles and the modelled CHLa (mg CHLa m^{-3}) concentrations are shown in Figure 3. Based on different hydrodynamic conditions, the vertical CHLa profiles are sorted into three groups, namely, for WEST, SWSB and NWS-SB. This categorisation is necessary because a few CHLa samples exist in each of these areas (locations of the sample stations are marked with red symbols in Figure 1). There is no data for the CENTRAL, EAST or SWS regions. In the WEST Black Sea waters (Figure 3a), the CHLa concentration profile exhibits a quasi-regular change below the thermocline, where the CHLa value increases to a maximum below the surface and then sharply decreases to zero. Hardly any abundance of CHLa is predicted by our model below 50 m depth, in accordance with in-situ data [2]. Our model results suggest a local maximum at about 20–25 m depth, while the in-situ data shows a maximum at about 15–20 m. In the NWS-SB area, the empirically measured CHLa profiles (Figure 3b) show a lack of obvious pattern: No noticeable maximum, a uniform concentration in the upper 20 m, and then a systematic decrease to zero. Our CHLa vertical profiles are in good agreement with the observational data in the aforementioned studies. In Figure 3c, are shown vertical CHLa profiles for the Knorr 2001 measurements, gathered by stations located in the SWSB. This is a hydrodynamically dynamic area, which is strongly influenced by the Rim Current speed. The DCM modelled by the BSEM is located at 20 m below the surface, while the measured DCM oscillates between 10 and 20 m below the surface. The individual CHLa model results from the three areas (WEST, NWS-SB and SWSB) are compared with the corresponding Knorr 2001 data in Figure 4a in the shape of a Taylor diagram. The best correlation is found for NSW-SB area, where the correlation coefficient is ~0.85, and the normalised RMSD is the lowest. Because of the limited amount of Knorr 2001 data, the model performance is evaluated over the merged datasets. In particular, the AAE = 0.151 ± 0.034 mg CHLa m^{-3} and Mskill = 0.266. The discrepancy between our model and observed values is small, and the correlation between them is moderate, *Mskill* is greater than zero, which tells us that the model can describe the average empirical CHLa vertical profile. Figure 4b shows the comparison between MSM33 surface CHLa and modelled surface CHLa concentration. The correlation is good, yet the calculated values are lower than both in-situ measurements. A possible explanation includes the assumed N:CHLa ratio in this study.

3.2. Validation against Remote Sensing Data

In Figure 5, we compare different surface CHLa datasets for the period 1999–2008. The time period is chosen to correspond to the RegAlg data availability. BSEM model results are compared with two data sets based on satellite observations. We decided to compare eight-day mean surface values of the CHLa because the most important events in the seasonal phytoplankton succession, such as blooms, often occur over a short time period, on the time scale of a week. The AAE and R coefficients, quantifying similarities between different time series are presented in Table 2. The agreement between the BSEM model and RegAlg survey is the highest, while there is no correspondence with the Copernicus survey data. This is not surprising because both satellite datasets disagree completely on the frequency and amplitude of the blooms. Interestingly, an agreement between BSEM and RegAlg in the CENTRAL Black Sea region is best. Modelled bloom amplitudes in WEST and CENTRAL regions are of the same magnitude, while in the EAST they are slightly smaller. When analysing CHLa seasonality in the three different regions, we can conclude that patterns are similar across different regions of the Black Sea. More details about PHY seasonal and inter-annual variability in the entire inner basin are given in Sections 4.1 and 4.2.

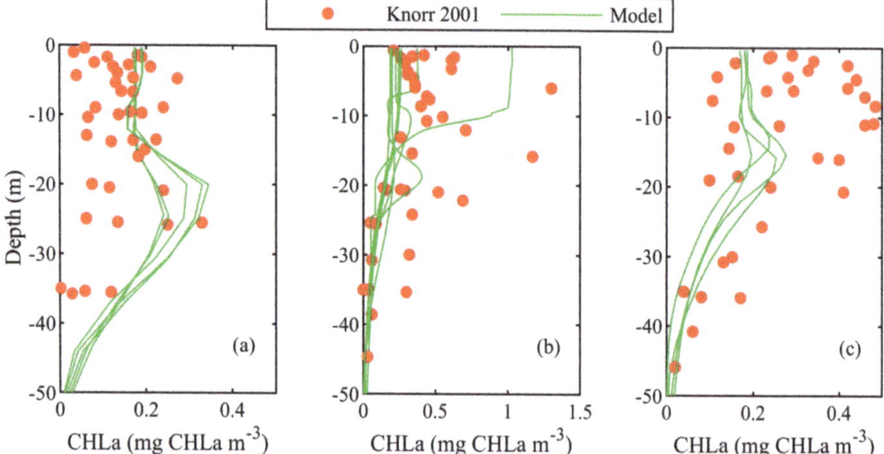

Figure 3. A comparison of CHLa (mg CHLa m^{-3}) profiles, acquired empirically by Knorr 2001 R/V (red symbols) and BSEM predictions (the lines represent modelled vertical profiles in the stations of Knorr 2001 R/V) in (**a**) WEST, (**b**) NWS-SB and (**c**) SWSB.

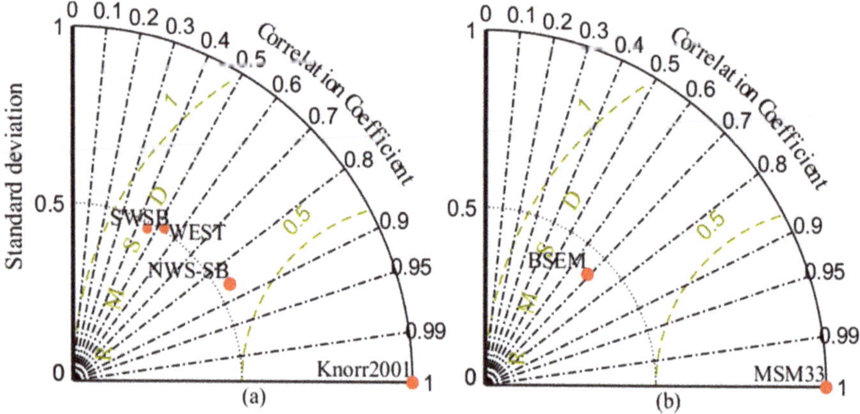

Figure 4. Taylor diagram comparing modelled and in-situ CHLa (mg CHLa m^{-3}). (**a**) vertical profiles acquired by Knorr 2001 R/V and BSEM model predictions for the stations distributed in the WEST, NWS-SB and SWSB and (**b**) surface CHLa acquired by MSM33 and BSEM predictions.

3.3. Model Validation against Argo Floats Empirical Data

The high spatial and temporal coverage of satellite observations makes them extremely suitable for testing model simulations. However, it is clear from the above results that using only satellite CHLa data to validate model estimates has its limitations. Model validation requires comparisons with in-situ observations, too. Now, we focus on the comparison between the CHLa concentration, as measured by the ARGO floats and computed by the BSEM model. For this reason, simulated daily CHLa values are extracted at the ARGO locations (Figure 1) and averaged for the upper 5 m Black Sea depth. Then, results for all datapoints in a month are further averaged temporally. ARGO-floats acquired measurements are also averaged monthly. Comparison of the BSEM forecast with in-situ data of ARGO1, 2 and 3 (Table 1) is presented in Figure 6. The ARGO1 starts its water trajectory in January 2014 from the western shelf-break (Figure 1). Until November, the float crosses the WEST region, and reaches the CENTRAL Black Sea region in December. The ARGO2 float travels along

Anatolian coast, through southern shelf-break. In 2015 the ARGO1 resided almost entirely in the WEST region, while the ARGO3 trajectory for 2016 covers the SWSB, WEST and CENTRAL regions. In all considered regions, the surface CHLa profile has a U-shape, with a maximum in winter and minimum in summer. This finding is clearly evident in both the ARGO floats data and in the BSEM model forecast. We find that the similarities between the BSEM model results and ARGO data remarkable, because during 2014–2016 the model is constrained only with climatological nutrient data for all considered rivers. Moreover, the model parameters are calibrated to capture the long-term ecosystem evolution rather than to represent a particular year, and the model is not related to any external data during the entire model run. The BSEM model was not calibrated to represent the specifics of the taxonomic composition that occur in a given year, however it has capabilities to correctly identify the overall phytoplankton patterns on both a yearly and long-term basis. Finally, the statistical analysis performed for the merged ARGO time series and corresponding BSEM model outputs indicates $AAE = 0.31 \pm 0.08$ mg CHLa m^{-3}, $R = 0.86$ (0.74, 0.93) and *Mskill* = 0.42.

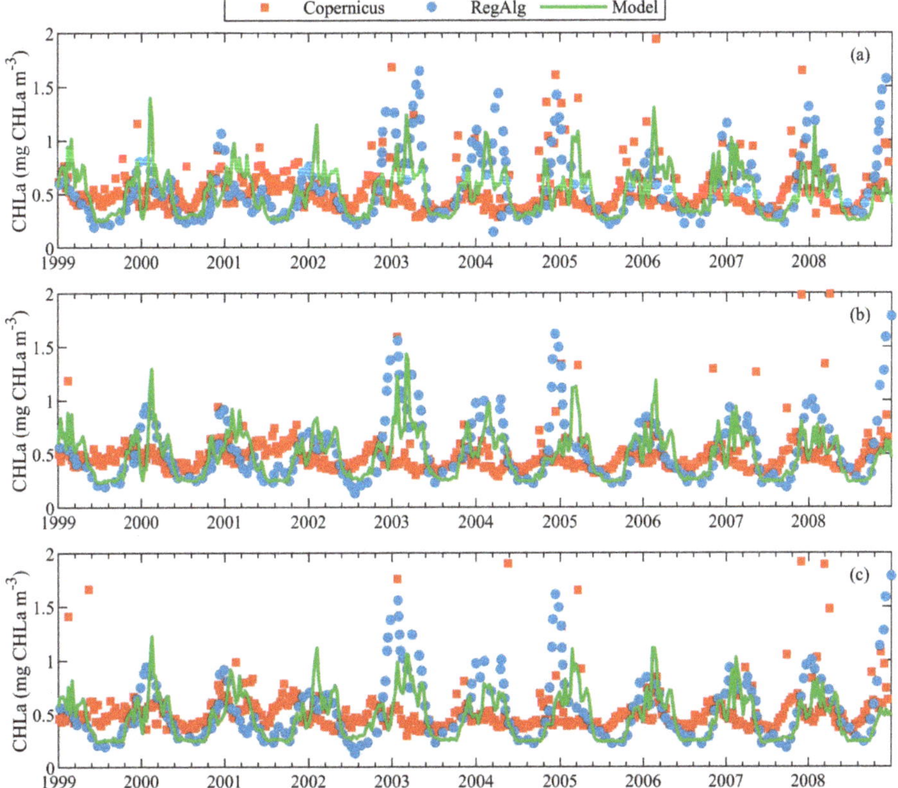

Figure 5. A comparison between eight-day mean surface CHLa (mg CHLa m^{-3}) measurements and model predictions, within the different regions: (**a**) WEST; (**b**) CENTRAL; and (**c**) EAST. Green lines represent surface CHLa from BSEM model, blue points–RegAlg and red squares–Copernicus. Time series are given for the area mean, averaged over a specified sub-region.

Table 2. The average absolute error (*AAE*) (mg CHLa m^{-3}) and R between pairs of different time series of eight-day mean surface CHLa over 1999–2008 in WEST, CENTRAL and EAST Black Sea regions (Figure 5). *AAE* error (mg CHLa m^{-3}) and 95% confidence intervals (in parentheses), in the case when *p*-value < 0.05 are shown.

Region	WEST		CENTRAL		EAST	
Data sources	*AAE* (±error)	*R* (interval)	*AAE* (±error)	*R* (interval)	*AAE* (±error)	*R* (interval)
RegAlg Copernicus	0.275 ± 0.014	0.08 (0.05, 0.11)	0.267 ± 0.019	-	0.259 ±0.032	-
RegAlg BSEM	0.197 ± 0.009	0.44 (0.42, 0.47)	0.158 ± 0.008	0.58 (0.56, 0.60)	0.151 ± 0.008	0.56 (0.54, 0.59)
Copernicus BSEM	0.229 ± 0.013	0.08 (0.05, 0.11)	0.220 ±0.017	-	0.234 ± 0.031	-

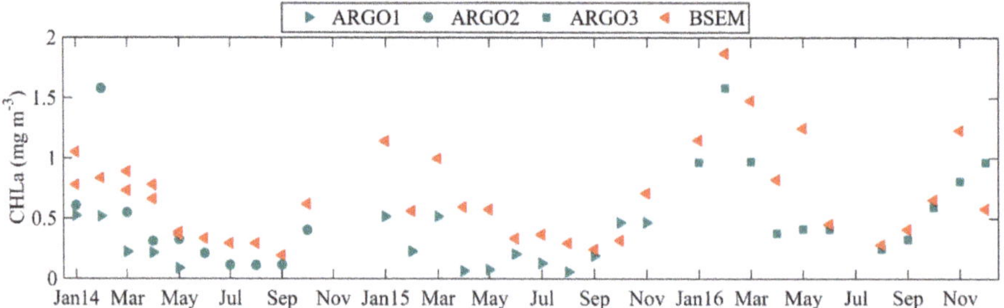

Figure 6. Comparison of surface monthly CHLa concentrations determined by our proposed BSEM model and data gathered by ARGO floats. Model results are extracted at ARGO's geographical location points (Figure 1).

4. Results

4.1. Seasonal Variability of PHY

We consider herein a year as *'cold'* when the Black Sea's winter (December–March) surface temperature (SST) in the inner basin (with depth > 1500 m) is less than the long-term mean winter SST, 8.3 °C [23]. On the contrary, when the winter SST is equal to or greater than 8.3 °C, the year is considered as *'warm'*. So, for the current simulation period, considering 1998 to 2017, the years 1998, 2000, 2002–2006, 2008, 2012 and 2017 are considered *'cold'* (10 years) and the rest *'warm'* (10 years).

Given that this study aims to better describe the seasonal cycle and multi-annual variation of phytoplankton through the Black Sea, in Figure 7 are plotted the daily mean surface PHY (mg C m^{-3}) model predictions, averaged over the whole deep basin (with depth > 1500 m). The annual mean surface PHY is shown with a dashed line. All annual mean values are close to the overall mean value (38.88 mg C m^{-3}) for the whole simulation period. According to our bloom definition, the most abundant surface blooms take place in winter and early spring, while the surface bloom in summer is absent. The autumn bloom is clearly visible in our simulations, yet its amplitude is less than the corresponding winter bloom amplitude and of similar size to the spring bloom. Stronger winter blooms are calculated for the years with winter SST below the average.

In Figure 8 are given climatological vertical distributions of (a–b) sigma density (kg m^{-3}), (c–d) nitrate (mmol N m^{-3}), (e–f) PHY (mg C m^{-3}) and (g–h) detritus (mmol N m^{-3}) averaged over the WEST region (Figure 1) from December to November in *'cold'* and *'warm'* years, as extracted from the BSEM output. First, we compute the seasonal cycle (365 days—each average of the 10-year period for *'cold'* and *'warm'* years, respectively). The seasonal cycle of the water column in the WEST region is then calculated by averaging the time series for all WEST model grid boxes. The WEST region is used as an example when comparing the mechanisms behind the blooms in both warm and cold periods. We find that the mechanisms governing blooms are closely related to the complexity of the physics in this deep-sea region. Note that the PHY profiles are plotted from the surface to 50 m depth, while the other variables are plotted down to 150 m depth. Typically, from mid-November to the end of February nitrate in the upper part of the storage (~60 m depth) rises and is utilised for biological production. Consistent with previous studies, simulations show that in *'warm'* winters the entrainment of denser water into the mixed layer is lower (Figure 8a,b). During warmer winters, vertical surface mixing is reduced. The lower entrainment in warm winters results in a shallow winter mixed layer, which reduces the lift-up of nutrients from the pycnocline to the mixed layer (Figure 8c,d). Conversely, in *'cold'* years, a stronger vertical mixing occurs in December through to February. Of note, in Figure 8c, the mechanism of the uplifting of nitrate from the storage and the increase of nitrate concentration in the mixed layer above 1 (mmol N m^{-3}) can be seen.

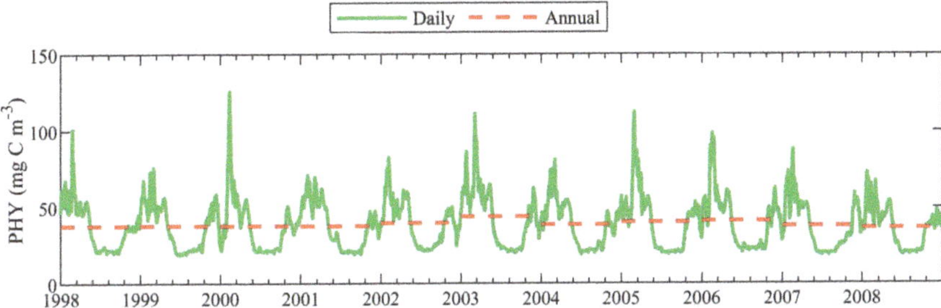

Figure 7. Daily mean surface phytoplankton biomass (PHY) (mg C m^{-3}) model prediction averaged over the whole deep basin (with depth > 1500 m). The annual mean surface PHY (mg C m^{-3}) is shown with a dashed line.

The winter blooms in the WEST Black Sea region utilise the nitrate lifted from the storage and/or coming from the NWS. Generally, winter blooms start in January and attain maxima in February or the first half of March. The phytoplankton grows abundantly in the upper 15–25 m and decreases with depth (Figure 8e,f). A small amount of phytoplankton is mixed into the deeper water and is lost from the production regions of the water column. The BSEM model forecasts a winter bloom peak in January or February in *'warm'* years and in February or March during *'cold'* years. A lower phytoplankton bloom in December-January in the *'cold'* years is compensated by higher bloom in the following two months (Figure 8e). The sunlight availability in December-January is an important factor in shaping the winter bloom. The winter surface cooling and strong winds that carry phytoplankton to depths below the euphotic zone have weaker negative effects on the phytoplankton abundance. Two distinct peaks (in January and February) are noticeable for the *'warm'* years, followed by less pronounced blooms in March. Spring blooms are of smaller magnitude than the winter ones, and they attain a maximum in the second half of April for *'warm'* years and at the end of April/beginning of May for *'cold'* years. In summary, our results show, in agreement with the existing literature [5,55], that the lower winter nutrient supply from the storage in *'warm'* years is a key factor in shaping the winter Black Sea phytoplankton bloom. In addition, our model clearly indicates the presence of both winter and spring blooms regardless of local meteorological and oceanographic conditions, and that the winter bloom in *'cold'* years is more intense.

During summer months, DBM occurs in the thermocline at ~25–35 m depth and lasting June to September, fuelled by the nitrate in the storage. At the surface, PHY reaches a minimum in summer months. In the *'warm'* summers, the DBM occurs in a slightly deeper location, and the PHY increases. An autumn bloom usually develops in October or November. Then, in December and in the first half of January, the bloom can decline, owing to low light intensity. The autumn peak occurs earlier and by the end of October in *'cold'* years.

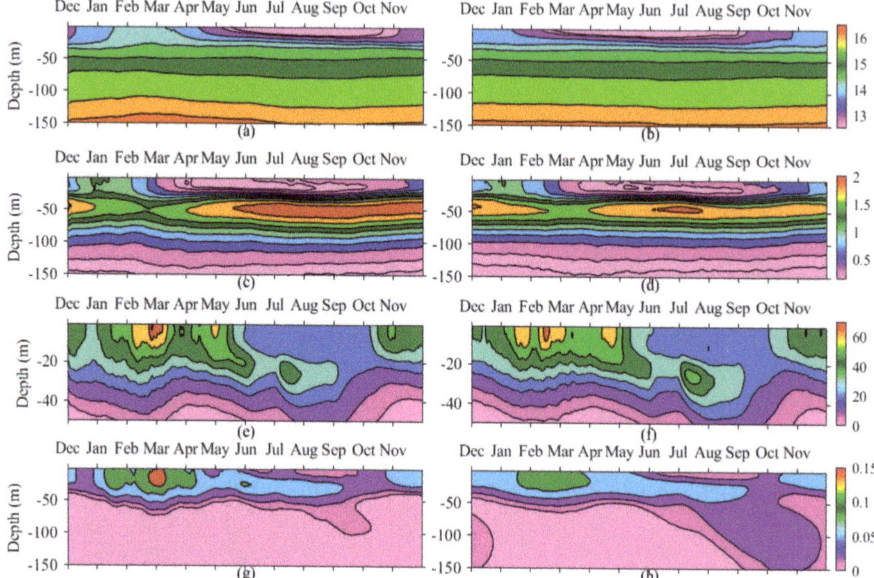

Figure 8. Climatological daily mean vertical distribution in the WEST region for: (**a**) Sigma density (kg m^{-3}) in 'cold' years and (**b**) in 'warm' years; (**c**) nitrate (mmol N m^{-3}) in 'cold' years and (**d**) in 'warm' years; (**e**) PHY (mg C m^{-3}) in 'cold' years and (**f**) in 'warm' years; and (**g**) detritus (mmol N m^{-3}) in 'cold' years and (**h**) in 'warm' years.

The detritus is mainly mineralised in the euphotic zone except for the amount produced by the summer phytoplankton growth in the thermocline (Figure 8g,h). The most obvious reason for detritus sinking in late summer is the lack of oxygen for mineralisation. The deep winter mixing causes deeper penetration of oxygen in the upper halocline, which can be used in late summer for detritus decomposition. For example, BSEM model simulations for the deep basin oxygen predict concentrations of over 50 (mmol O$_2$ m^{-3}) at about 100 m depth in the summers of 2003, 2006 and 2012 and close to zero in 2001 and 2011. The weaker winter, mixing in *'warm'* years, results in a loss of subsurface organic matter in summer. At the end of February and beginning of March, the detritus in the upper 50 m reaches a maximum for *'cold'* years. The nitrate storage is refilled to a greater extent in *'cold'* years, which translates into a higher winter-spring detritus concentration.

4.2. Inter-Annual Variability of Plankton Blooms

Seasonality of the blooms is linked to the nitrate storage content, as well as to nutrient mixing during winter convection, as well as to the suppressed vertical mixing during the summer thermocline months. However, it is not fully understood how the nitrate storage is formed and how it correlates with the CIL dynamics. The CIL is defined in the literature as an intermediate layer with a temperature of less than 8 °C [56], while we assume it to be an intermediate layer, characterised by temperatures of less than 8.35 °C [35].

The Black Sea vertical thermohaline structure consists of a typical two-layer stratified system. The relatively low salinity surface water layer, which is approximately 100 m thick, is separated from the underlying water body by a permanent pycnocline located between sigma density $\sigma_t \sim 14.5$ and 16.5 kg m^{-3} (upper halocline). The strong vertical density gradients hinder the transport of nitrate from the halocline to the euphotic zone. Such transport of nitrate generally occurs in the deep basin by mechanisms of convection and advection in winter. The halocline acts not only as a barrier to vertical transport during these periods of thermocline presence, but it is also an essential nitrate source (Figure 8a–d).

Figure 9 presents the vertical distribution of the inner basin mean temperature (°C), PHY (mmol C m^{-3}) and nitrate (mmol N m^{-3}) in the basin interior from the year 1998 to 2017. The vertical temperature profiles illustrate the CIL evolution in the inner basin (of depth > 1500 m) through the simulation period. From the model results, we can conclude that the storage is thicker and contains more nitrate when the CIL volume increases and its temperature decreases (Figure 9). That usually happens in *'cold'* years, when the strong vertical winter mixing gives rise to strong phytoplankton growth in winter-spring, followed by the refilling of the nitrate storage with the regenerated nitrate. So, the CIL volume and temperature are important in regulating nitrate storage in the Black Sea waters. However, in several WARM years (e.g., 2010, 2011, 2015 and 2016) we found that the volume and/or nitrate content of the storage increase although the CIL is almost eroded. It is interesting to note that the storage has been substantially refilled in the years with high Danube discharge (e.g., 2010, 2015 and 2016). In 2010 the Danube discharge was extremely high, while 2015 and 2016 are the last two years of the four-year high Danube discharge period (discharges are significantly above the long-term mean levels). Thus, it turns out that both winter vertical mixing and Danube discharge are important for nitrate storage refilling.

Seasonal and inter-annual variation of the inner basin PHY is visible in Figure 9. The PHY reaches a maximum in winter with one exception in the year 2003 when it is shifted towards early spring. When comparing the winter-spring blooms in 1998–2009 and 2010–2017, there can be seen a tendency towards a smaller winter peak for the second period.

It is well known that the winter nitrate uplifting from the storage to the euphotic zone supports the winter-spring blooms. However, the storage in the years preceding the strong winter-spring blooms does not always contain high levels of nitrate (e.g., in 2002, 2015 and 2016). This is in accordance with [57], displaying that during severe winters, the content of nutrients in the CIL is not a reliable indicator of the bloom intensity. The convective mixing during the winter season is a dominant mechanism.

4.3. The Role of CIL

To study the impact of CIL evolution on nitrate content in the intermediate layers, the upper and lower CIL boundaries are presented together with the nitrate content in the pycnocline in Figure 10. The isotherm criteria for the CIL recognition (8.35 °C) is applied herein to identify its upper and lower boundary in the warm season. For example, the upper boundary is located at the first grid box below the surface where the temperature is lower than 8.35 °C. The lower boundary is located at the first grid box where the temperature becomes higher than 8.35 °C. Furthermore, CIL properties are averaged for the warm season (May-September) and over the deep basin (depth > 1500 m). During this season the CIL experiences less variation, due to a strong vertical stratification. Typically, the CIL upper boundary is about 25 m, however, it can be found exceeding 30 m depths in *'warm'* years as a result of surface water warming. At the same time, the lower CIL boundary becomes shallower because of less intensive CIL refilling (see the red dashed line in Figure 9). In contrast, the depth of the CIL increases significantly in *'cold'* years, mainly due to an intensified CIL refilling. The cold layer, defined as the water body having a temperature less than 8.35 °C, almost vanishes in 2016. The nitrate concentration in the storage is integrated along a vertical column, from 25 m depth (the upper boundary of the nitrate storage) to the seabed and then averaging is also performed across the entire warm months and over the entire extent of the deep basin. The annual nitrate variation simulated here is in accordance with the CIL variation (Table 3). For example, a significant positive correlation is found between the nitrate content and the upper/lower boundaries of CIL, while the correlation between nitrate content and CIL temperature is negative. Thus, the stronger the CIL, the higher the nitrate content of the pycnocline. Recently, the CIL is weak and eroded, and it is not surprising that the nitrate content of storage tends to decline (Figure 9).

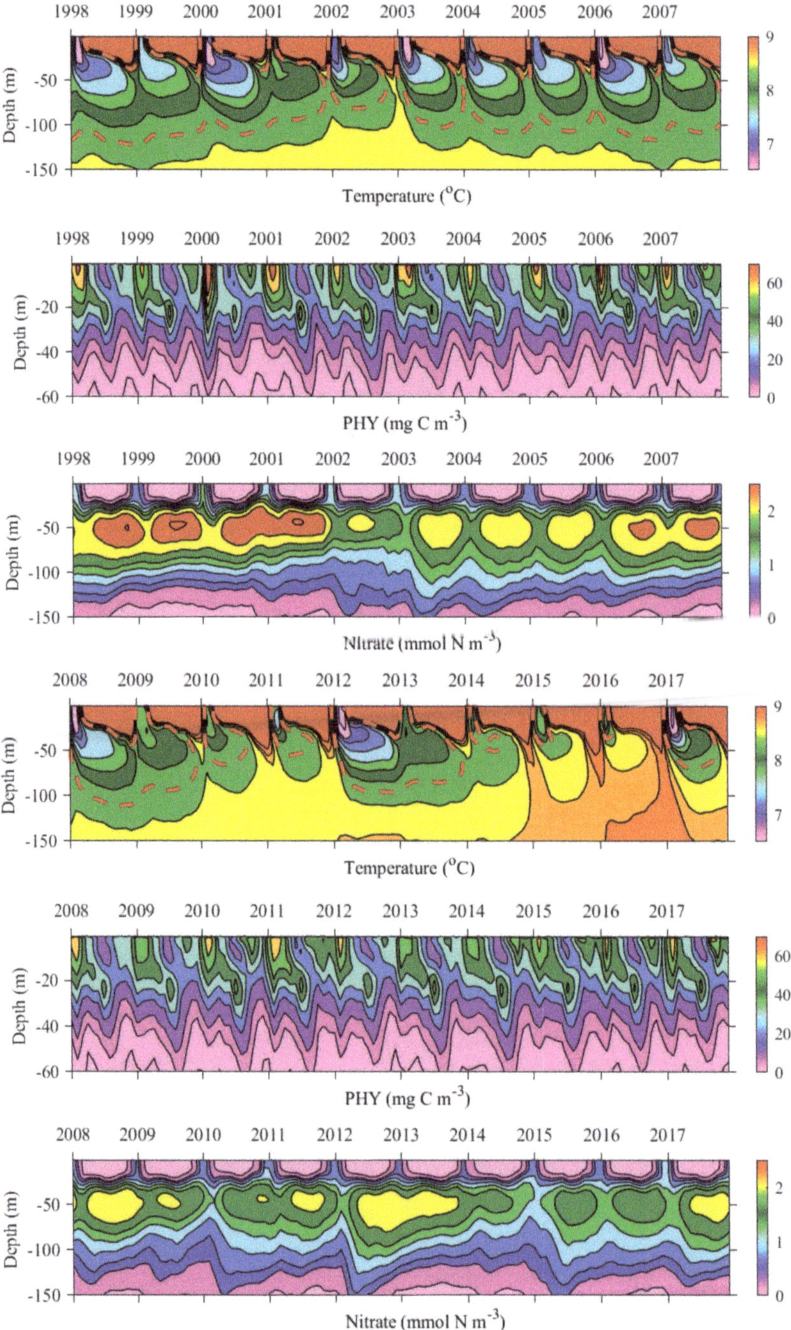

Figure 9. Deep basin monthly modelled vertical contours of temperature (°C), PHY (mg C m^{-3}) and nitrate (mmol N m^{-3}) from the years 1998 to 2017. To facilitate reading, the temperature colour bar is set in the range 6.5–9 °C and the 8.35 °C isotherm is shown with a red dashed line.

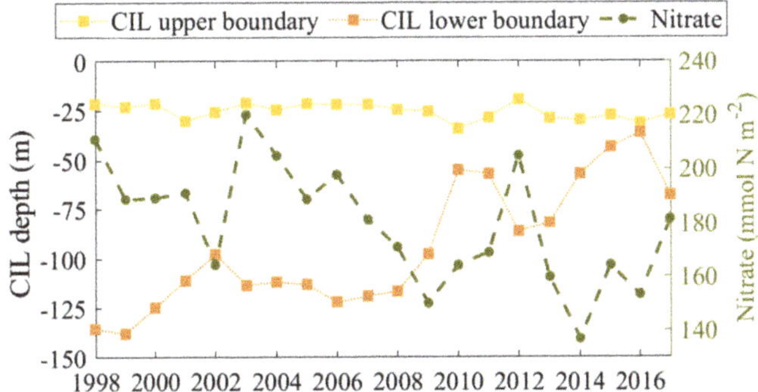

Figure 10. Mean cold intermediate layers (CIL) upper and lower boundary (m) in the deep Black Sea basin, averaged over the warm season (May-September) and vertically integrated nitrate concentration in the pycnocline (right axis, in units of mmol N m^{-2}).

Table 3. Correlation matrix over 1998–2017 between mean CIL properties, vertically integrated PHY (mg C m^{-2}) and Nitrate (mmol N m^{-2}) in the pycnocline. It includes the R-coefficient (p-value < 0.05 for all correlations) and the 95% confidence interval.

	Winter PHY	Spring PHY	Summer PHY	Autumn PHY	Nitrate in the Storage
CIL properties			R (interval)		
Temperature	−0.59 (−0.82, −0.20)	0.75 (0.45, 0.89)	0.67 (0.33, 0.86)	0.65 (0.07, 0.77)	−0.76 (−0.9, −0.48)
Lower boundary	0.64 (0.28, 0.84)	−0.65 (−0.85, −0.29)	−0.69 (−0.87, −0.35)	−0.51 (−0.78, −0.09)	0.65 (0.3, 0.85)
Nitrate in the storage	0.72 (0.41, 0.88)	-	−0.54 (−0.79, −0.12)	-	1

Winter, spring, summer and autumn PHY (mg C m^{-2}) in the inner Black Sea basin are shown in Figure 11. They represent vertically integrated PHY values for the upper 50 m and are consequently averaged over the deep basin for January–March (winter PHY), April–May (spring PHY), June-September (summer PHY) and October–November (autumn PHY), respectively. Comparing the PHY successions in the deep Black Sea basin over the seasons, we highlight that the highest PHY is in the winter and is followed by the spring/summer PHY (both of similar magnitude). The seasonality found here agrees with previous studies [2,58]. The difference between the strength of winter and spring/summer PHY is reduced in recent years, for example, after 2009. The weakest value of PHY occurs in the autumn, however it increases in '*warm*' years. A relationship between CIL properties and winter PHY is established. Like the nitrate content of pycnocline, the winter PHY is positively affected by the CIL strength (Table 3). The convective mixing during the winter season is a dominant mechanism for the CIL refilling, and essentially supports the nitrate lifting from the nitrate storage. On the contrary, the spring, summer and autumn PHY successions are favoured by the CIL weakness. A possible explanation includes an elevation of the CIL position toward the euphotic zone during periods of weak CIL. In this way, the nutrient storage is lifted-up, and the phytoplankton growth is enhanced. Note that autumn PHY does not correlate with nitrate content in the storage (Table 3), however the autumn PHY increases in periods of high Danube discharge. The factors influencing the increase of autumn PHY are presented and discussed in Sections 4.4 and 4.5.

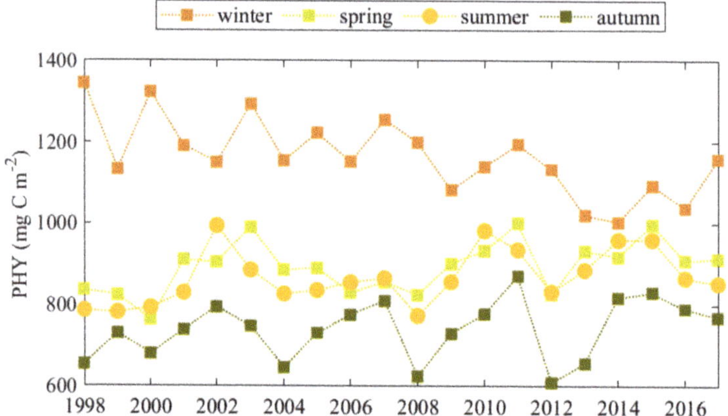

Figure 11. Vertically integrated modelled PHY (mg C m^{-2}) in the inner Black Sea basin—average PHY for January–March (winter), April–May (spring), June–September (summer) and October–November (autumn).

4.4. Mechanism behind Autumn Blooms in the Black Sea

The next topic of study is on the mechanisms behind the previously detected earlier autumn phytoplankton blooms. We found that the autumn surface bloom is always present it the Black Sea deep basin (Figure 7). The bloom usually begins in the second half of October (Figure 8c,d), when the PHY reaches values higher than 39 mg C m^{-3}. However, in several years the phytoplankton blooms earlier in October (e.g., in 2011 among others, Figure 9). We set out to investigate the processes which are responsible for these premature blooms in autumn. Erosion of the thermocline and vertical mixing, due to surface cooling and strong winds typically start at the end of November, hence, they cannot be directly responsible for the early autumn bloom observed in References [9,55,59].

Figure 12 illustrates possible mechanisms for the early autumn bloom. We decided to illustrate the onset of bloom on 9 October 2004, due to the better quality of the satellite image on this day. The corresponding vertical profiles of nitrate in two specific geographic locations, indicated by A and B in Figure 12a, are plotted in Figure 12b. A daily nitrate distribution is included because nitrate is quickly utilised at the surface for biological production and disappears in days (Figure 12a). Figure 12c shows the SeaWiFS CHLa image, acquired on the same day [60], while the surface CHLa field and the surface currents, predicted by our modelling, are presented in Figure 12d. Previous modelling results [23,61] have shown that many MEs grow in late summer at spatial scales of 10–100 km and timescales of one month or more. The monthly velocity field given in Figure 12d could not capture all MEs which arose in a selected day. The western part of the Rim Current disintegrates into several vigorous cyclonic eddies.

On 9 October 2004, the Danube water transport is Northbound, then Eastbound and then flows offshore by the AE1 anticyclonic ME. Another part of the freshwater plume is partially locked by AE2 eddy, and then the plume splits toward west and east. The eastern part is mixed with the inner surface waters by the help of two energetic cyclonic eddies. One of them is located on the shelf-break, while another is in the inner basin. The western part of the plume moves along the west coast in a south-western direction. The spreading of the NWS waters towards the Anatolian coast is suppressed by AE3 eddy, which returns some of the water to the NWS circulation before reaching the Anatolian coast. At the Anatolian coast, the plume is transported to the western gyre by several anticyclonic MEs, two of them are AE4 and AE5. Then, the existence of cyclonic MEs inside the western gyre helps to further transport the bio-chemical matter (nitrate and biological material).

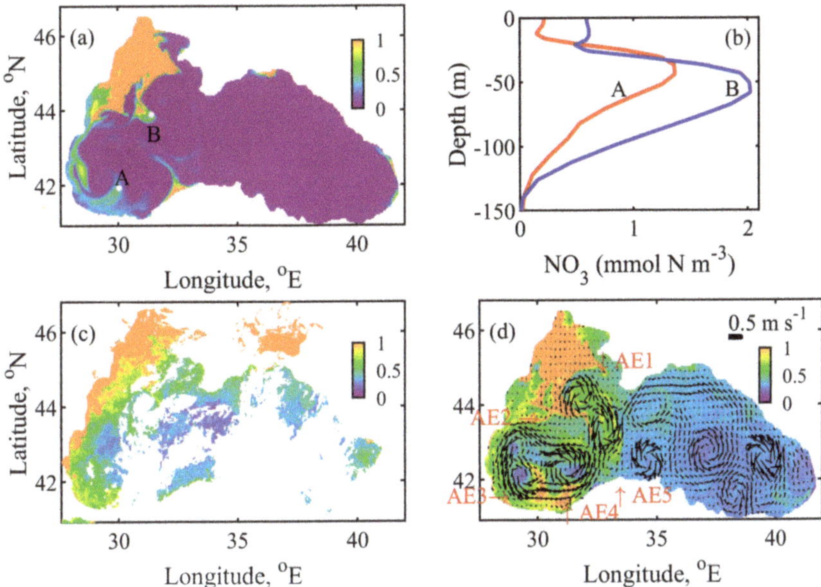

Figure 12. Daily mean simulation results on 9 October 2004 (**a**) surface nitrate (mmol N m^{-3}); (**b**) vertical nitrate profiles at two points (locations are marked by 'o's in (**a**)): A (31.633° E, 42.667° N)—red line and B (31.467° E, 43.667° N)—blue line; (**c**) satellite image of CHLa (mg CHLa m^{-3}) and (**d**) CHLa fields and currents, the colour bar represents the CHLa (mg CHLa m^{-3}), the arrows show the speed (m s^{-1}), and direction of the currents, where the length of the 0.5 m s^{-1} arrow is shown for scale, AE1–AE5 denote the locations of the calculated anticyclonic MEs. To facilitate reading, the nitrate colour bar is set in the range 0–1 mmol N m^{-3}, while the CHLa colour bar is set in the range 0–1 mg CHLa m^{-3}.

BSEM model results successfully reproduce the position and timing of the surface phytoplankton bloom caused by the NWS spreading (Figure 12c,d). SeaWiFS's image contains many clouds, so the spread of CHLa from the south shelf-break into the inner basin is difficult to track. The NWS plume intrusion into the inner basin, induced by the MEs, can be seen to some extent in Figure 12c. In summary, the NWS waters were observed to mix laterally with the offshore waters through the shelf-break, due to the NWS anticyclonic eddies in our simulations. The cyclonic current of the west gyre transports nitrate and biological substances from the NWS to the southern part, where they are laterally mixed. Vertical nitrate profiles in Figure 12b show the existence of lateral surface mixing before the onset of vertical mixing. Cyclonic MEs in the inner basin contributes not only to the lateral mixing process, but also to the vertical uplift of nitrates to the surface.

Another example of the onset of early autumn bloom is given in Figure 13. On 9 October 2011, the hydrodynamic conditions are more complicated, since the Danube waters are mainly carried to the north, partially move along the south coast and trapped offshore by the anticyclonic MEs. In the year 2004, the early October surface bloom decays in a week, and the next bloom starts in late November, after the onset of vertical mixing, due to surface cooling. In 2011, the penetration of fresher NWS water offshore and amplified mesoscale circulation triggered the vertical mixing and lift-up nitrate from the storage. Thus, blooming, beginning in the first half of October, lasts until late December (Figure 9). Penetration of the nitrate-rich peripheral water into the interior part of the sea, provoking one of the most abundant autumn blooms throughout the 20-year simulation period. A substantial infiltration of the peripheral water into the inner basin in autumn is associated with the typical late summer Rim Current disintegration (Figure 13b). In summary, the intensification of the autumn blooms can be attributed to a combination of several mechanisms–strengthening of the Rim Current in winter-spring prevents lateral mixing of NWS waters into the inner basin; intensification of the mesoscale and sub-mesoscale eddies in autumn provides favourable conditions for nutrient and biological substance transport from the periphery to the basin interior. The existence and timing of NWS break transport regarding plankton bloom dynamics of the Black Sea are discussed in the next and final subsection.

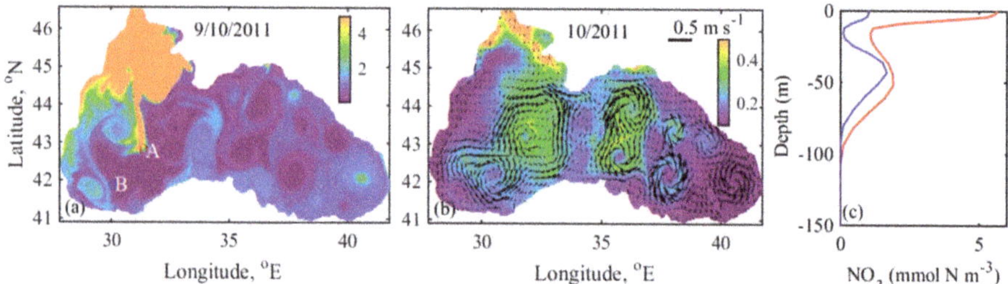

Figure 13. (**a**) Daily mean surface nitrate field on 9 October 2011 (mmol N m^{-3}); (**b**) monthly mean surface CHLa fields and currents averaged over upper 5 m in October 2011, the colour bar represents the CHLa (mg CHLa m^{-3}), the arrows show the speed (m s^{-1}), and direction of the currents, where the length of the 0.5 m s^{-1} arrow is shown for scale; (**c**) vertical nitrate profiles at the two points, shown in (**a**) A (red line) and B (blue line). To facilitate reading, the nitrate colour bar is set in the range 0–5 mmol N m^{-3}, while the CHLa colour bar is set in the range 0–0.5 mg CHLa m^{-3}.

4.5. Mass and Volume Transport across the NWS Break

In Section 4.4, we showed that basin circulation supports lateral mixing in autumn. We will now provide supporting evidence for the strong influx of nitrate and biomass into the deep basin in September, which can fuel blooms of phytoplankton in October. Usually, the water masses coming from the NWS do not penetrate into the interior of the sea, where the salinity is higher, but are transported from the periphery of the Rim Current along the coast. A tendency of decreasing Danube plume transport, southward along the coastline and increasing transport to the north and north-eastern parts of the NWS and then to the southwest is established in Reference [61]. Moreover, it is shown that the river-borne substances achieve maximum concentrations in the inner basin in September. Following [62], we estimate the transport of water between the NWS and the Black Sea interior along with a boundary contouring the 200 m isobath, and including a segment connected to the coast of less than 20 m in depth (Figure 1, white path). For each day, the total transport is calculated by integrating the individual daily-averaged fluxes through the model grid boxes along the basin interior boundary in the horizontal plane and down to the 20 m layer in the vertical plane. The NWS freshwater plume usually forms a current, which is confined to the upper 25 m [63]. Offshore and onshore transports of nitrate and PHY through this boundary are also calculated. The water volume calculations across the NWS break show the presence of weak seasonal variation of the offshore/onshore transport (Figure 14a). Both offshore and onshore transports are plotted as positive values in Figure 14. The highest water volume, leaving and entering the NWS across the considered boundary occurs in September. We estimate that monthly offshore transport exceeds by about 20–30 km^3 the onshore transport. Additionally, offshore and onshore transports display a similar seasonal variation.

Both offshore and onshore transport of nitrate (kton N) reach local maxima in September and then in December (Figure 14b), and minima in spring. A well-defined seasonal cycle is evident for the transport of nitrate and PHY (Figure 14b,c). The offshore nitrate transport is twice as large as the onshore transport. Interestingly, both nitrate transports are almost zero in April, indicating the depletion of nitrate in upper 20 m after the winter bloom.

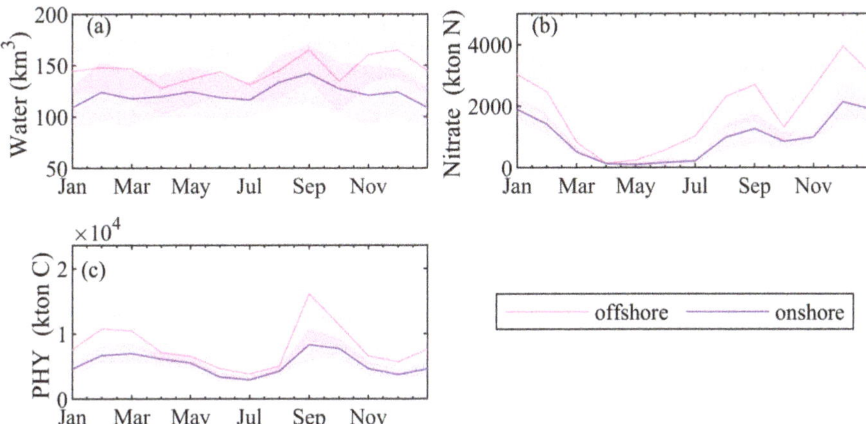

Figure 14. Calculated monthly transport across the NWS of (**a**) water volume (km^3); (**b**) nitrate (kton N) and (**c**) PHY (kton C). Mean offshore transport is presented by a pink line, while mean onshore transport by a purple line. Shaded areas represent 95% confidence intervals of the mean.

The greatest seasonal fluctuations are predicted for the PHY (kton C) transport (Figure 14c). The maximum PHY transport occurs in September, prior to the autumn bloom, coinciding with the phytoplankton bloom on the NWS during this time. The smaller second peak of the PHY transport occurs in February and corresponds to the winter bloom on the NWS. Note that in September, the offshore PHY transport is over twice as the size of the onshore transport. The enhanced nitrate and PHY transport from the NWS to the inner basin in September can support early autumn blooms if the system circulation favours lateral transport.

5. Discussions and Conclusions

A modelling study of the phytoplankton evolution in the Black Sea inner basin during 1998–2017 is presented. The model is based on the coupling of GETM/GOTM and BSEM models. It reproduces the observed seasonal and annual growth of the phytoplankton adequately. An important factor influencing the discrepancy between the results of the BSEM model and the observed data is the conversion factor between N/C and CHLa. The C:CHLa ratio, within the mixed layer in the oligo/meso/eutrophic waters of the tropical Atlantic Ocean is estimated to be 145, 96 and 37, respectively [48]. Based on research conducted in the Black Sea from 2000 to 2011, the seasonal evolution of the C:CHLa ratio and its spatial variability in surface water layer (0–0.5 m) are analysed in Reference [64]. The maximum ratio is observed in summer months (~300 C:CHLa) and minimum value of ~50 in winter. Intermediate C:CHLa values are detected in spring and autumn. The main reasons for the temporal variability of the organic carbon to CHLa ratio are the variability in abundance of light, the seasonal variability in phytoplankton size, as well as in its taxonomic composition. To better compare the model (PHY) and the data (CHLa), an approach describing the dynamics of the C:CHLa ratio as a function of temperature, daily exposure and nutrient-restricted growth rate should be applied [49]. In its present configuration, the BSEM model simulates the phosphate limitation implicitly by linking it to nitrate dynamics using a fixed N:P ratio (Redfield). Future improvements of the model could include an explicit representation of the phosphate path through different state variables to adopt a flexible N:P ratio according to recent approaches [65,66]. More observational data of the river nutrient loads are required in order to better understand the observed CHLa dynamics in the Black Sea.

The simulations illustrate the remarkable complexity of the physical and biological mechanisms in the Black Sea, and at the same time, reveal some intriguing regularities. Despite its decreasing tendency, the winter phytoplankton bloom is still the most abundant seasonal bloom. Winter biomass production is closely related to the CIL dynamics and is essentially supported by the nitrate lifting from the storage in winter. Spring/summer blooms increase in *'warm'* years, when the Danube discharge is high or when the CIL is weak. Both spring and summer blooms show an increasing tendency over the simulation period. Model results suggest that the timing of autumn phytoplankton bloom in the inner basin is related to high fluxes of nitrate and PHY across the NWS break in September. Mechanisms responsible for extensive autumn blooms include amplification of the Danube discharge, weakness of the CIL, and the abundance of energetic cyclonic and anticyclonic mesoscale eddies in late summer–autumn. Generally, they induce horizontal transport of nitrate and biological matter and generate autumn phytoplankton blooms, which consequently spread over the basin. Our future studies will address the relationship between physical structures of oceanic fronts and biological activity in order to quantify the role of eddies for biology.

Supplementary Materials: The following are available online at http://www.mdpi.com/2673-1924/1/4/18/s1, BSEM model equations; Table S1. Food preference coefficients of the predator groups on the prey groups; Table S2. BSEM input parameters.

Author Contributions: S.M.: Conceptualization, Methodology, Software, Writing—original draft. A.S.: Supervision, Project administration, Funding acquisition, Writing—review & editing. D.M.M.: Conceptualization, Methodology, Writing—review & editing. E.G.-G.: Data curation, Visualization, Software. All authors have read and agreed to the published version of the manuscript.

Funding: This research received no external funding.

Acknowledgments: We thank to "The Global Runoff Data Centre, 56068 Koblenz, Germany" for the Danube daily discharge rates. Special thanks go to the GETM/GOTM/FABM developers for providing and maintaining the model. Proofreading and fruitful discussions with Peter D. Marinov are gratefully acknowledged.

Conflicts of Interest: The authors declare no conflict of interest.

References

1. Smayda, T.J. What is a bloom? A commentary. *Limnol. Oceanogr.* **1997**, *42*, 1132–1136. [CrossRef]
2. Vedernikov, V.I.; Demidov, A.B. Vertical Distribution of Primary Production and Chlorophyll during Different Seasons in Deep Regions of the Black Sea. *Oceanology* **1997**, *37*, 376–384.
3. Nezlin, N.P. Seasonal and interannual variability of remotely sensed chlorophyll. In *The Handbook of Environmental Chemistry*; Kostianoy, G., Kosarev, A.N., Eds.; Springer: Berlin/Heidelberg, Germany, 2006; pp. 333–349.
4. Finenko, Z.Z.; Suslin, V.V.; Kovaleva, I.V. Seasonal and long-term dynamics of the chlorophyll concentration in the Black Sea according to satellite observations. *Oceanology* **2014**, *54*, 596–605. [CrossRef]
5. Oguz, T.; Cokacar, T.; Malanotte-Rizzoli, P.; Ducklow, H.W. Climatic warming and accompanying changes in the ecological regime of the Black Sea during 1990s. *Glob. Biogeochem. Cycles* **2003**, *17*. [CrossRef]
6. Oguz, T.; Dippner, J.W.; Kaymaz, Z. Climatic regulation of the Black Sea hydro-meteorological and ecological properties at interannual-to-decadal time scales. *J. Mar. Syst.* **2006**, *60*, 235–254. [CrossRef]
7. Oguz, T. Long-Term Impacts of Anthropogenic Forcing on the Black Sea Ecosystem. *Oceanography* **2005**, *18*, 112–121. [CrossRef]
8. McQuatters-Gollop, A.; Mee, L.D.; Raitsos, D.E.; Shapiro, G.I. Non-linearities, regime shifts and recovery: The recent influence of climate on Black Sea chlorophyll. *J. Mar. Syst.* **2008**, *74*, 649–658. [CrossRef]
9. Mikaelyan, A.S.; Chasovnikov, V.K.; Kubryakov, A.A.; Stanichny, S.V. Phenology and drivers of the winter–spring phytoplankton bloom in the open Black Sea: The application of Sverdrup's hypothesis and its refinements. *Prog. Oceanogr.* **2017**, *151*, 163–176. [CrossRef]
10. Oguz, T.; Ediger, D. Comparision of in situ and satellite-derived chlorophyll pigment concentrations, and impact of phytoplankton bloom on the suboxic layer structure in the western Black Sea during May-June 2001. *Deep. Res. Part II Top. Stud. Oceanogr.* **2006**, *53*, 1923–1933. [CrossRef]
11. BSC2008 State of Environment of the Black Sea (2001–2006/7). Available online: http://www.blacksea-commission.org/_publ-SOE2009.asp (accessed on 14 September 2020).
12. Kubryakov, A.A.; Stanichny, S.V.; Zatsepin, A.G.; Kremenetskiy, V.V. Long-term variations of the Black Sea dynamics and their impact on the marine ecosystem. *J. Mar. Syst.* **2016**, *163*, 80–94. [CrossRef]

13. Oguz, T.; Ducklow, H.; Malanotte-Rizzoli, P.; Tugrul, S.; Nezlin, N.P.; Unluata, U. Simulation of annual plankton productivity cycle in the Black Sea by a one-dimensional physical-biological model. *J. Geophys. Res. Ocean.* **1996**, *101*, 16585–16599. [CrossRef]

14. Grégoire, M. Modeling the nitrogen cycling and plankton productivity in the Black Sea using a three-dimensional interdisciplinary model. *J. Geophys. Res. Ocean.* **2004**, *109*, C05007. [CrossRef]

15. Oguz, T.; Salihoglu, B. Simulation of eddy-driven phytoplankton production in the Black Sea. *Geophys. Res. Lett.* **2000**, *27*, 2125–2128. [CrossRef]

16. Tsiaras, K.P.; Kourafalou, V.H.; Davidov, A.; Staneva, J. A three-dimensional coupled model of the Western Black Sea plankton dynamics: Seasonal variability and comparison to Sea WiFS data. *J. Geophys. Res. Ocean.* **2008**, *113*. [CrossRef]

17. Zatsepin, A.G. Observations of Black Sea mesoscale eddies and associated horizontal mixing. *J. Geophys. Res.* **2003**, *108*, 3246. [CrossRef]

18. Kubryakov, A.A.; Bagaev, A.V.; Stanichny, S.V.; Belokopytov, V.N. Thermohaline structure, transport and evolution of the Black Sea eddies from hydrological and satellite data. *Prog. Oceanogr.* **2018**, *167*, 44–63. [CrossRef]

19. Konovalov, S.K.; Murray, J.W. Variations in the chemistry of the Black Sea on a time scale of decades (1960–1995). *J. Mar. Syst.* **2001**, *31*, 217–243. [CrossRef]

20. Oguz, T.; Velikova, V. Abrupt transition of the northwestern Black Sea shelf ecosystem from a eutrophic to an alternative pristine state. *Mar. Ecol. Prog. Ser.* **2010**. [CrossRef]

21. Kideys, A.E. ECOLOGY: Enhanced: Fall and Rise of the Black Sea Ecosystem. *Science (80-)* **2002**, *297*, 1482–1484. [CrossRef]

22. Oguz, T.; Fach, B.; Salihoglu, B. Invasion dynamics of the alien ctenophore Mnemiopsis leidyi and its impact on anchovy collapse in the Black Sea. *J. Plankton Res.* **2008**, *30*, 1385–1397. [CrossRef]

23. Miladinova, S.; Stips, A.; Garcia-Gorriz, E.; Macias Moy, D. Black Sea thermohaline properties: Long-term trends and variations. *J. Geophys. Res. Ocean.* **2017**, *122*, 5624–5644. [CrossRef] [PubMed]

24. Ærtebjerg, G.; Carstensen, J. Indicator Fact Sheet: (WEU4) Nutrients in Coastal Waters. Available online: https://www.eea.europa.eu/data-and-maps/indicators/nutrients-in-coastal-waters-1/ (accessed on 14 September 2020).

25. Tugrul, S.; Basturk, O.; Saydam, C.; Yilmaz, A. Changes in the hydrochemistry of the Black Sea inferred from water density profiles. *Nature* **1992**, *359*, 137–139. [CrossRef]

26. Churilova, T.; Suslin, V.; Krivenko, O.; Efimova, T.; Moiseeva, N.; Mukhanov, V.; Smirnova, L. Light absorption by phytoplankton in the upper mixed layer of the Black Sea: Seasonality and parametrization. *Front. Mar. Sci.* **2017**, *4*, 1–14. [CrossRef]

27. McCarthy, J.J.; Yilmaz, A.; Coban-Yildiz, Y.; Nevins, J.L. Nitrogen cycling in the offshore waters of the Black Sea. *Estuar. Coast. Shelf Sci.* **2007**, *74*, 493–514. [CrossRef]

28. Yunev, O.A.; Carstensen, J.; Moncheva, S.; Khaliulin, A.; Aertebjerg, G.; Nixon, S. Nutrient and phytoplankton trends on the western Black Sea shelf in response to cultural eutrophication and climate changes. *Estuar. Coast. Shelf Sci.* **2007**. [CrossRef]

29. Oguz, T.; Latun, V.S.; Latif, M.A.; Vladimirov, V.V.; Sur, H.I.; Markov, A.A.; Özsoy, E.; Kotovshchikov, B.B.; Eremeev, V.V.; Ünlüata, Ü. Circulation in the surface and intermediate layers of the Black Sea. *Deep. Res. Part I* **1993**, *40*, 1597–1612. [CrossRef]

30. Korotaev, G.; Oguz, T.; Nikiforov, A.; Koblinsky, C. Seasonal, interannual, and mesoscale variability of the Black Sea upper layer circulation derived from altimeter data. *J. Geophys. Res. Ocean.* **2003**, *108*. [CrossRef]

31. GETM, 3D General Estuarine Transport Model. Available online: http://www.getm.eu/ (accessed on 1 October 2020).

32. ECMWF, European Centre for Medium Range Weather Forecast. Available online: http://www.ecmwf.int (accessed on 10 October 2020).

33. GRDC, Global Runoff Data Centre. Available online: http://www.bafg.de/GRDC (accessed on 10 October 2020).

34. MEDAR/MEDATLAS II Project. Available online: http://www.ifremer.fr/medar (accessed on 10 October 2020).

35. Miladinova, S.; Stips, A.; Garcia-Gorriz, E.; Macias Moy, D. Formation and changes of the Black Sea cold intermediate layer. *Prog. Oceanogr.* **2018**, *167*, 11–23. [CrossRef]

36. Miladinova, S.; Stips, A.; Garcia-Gorriz, E.; Macias Moy, D. *Modelling Toolbox 2: The Black Sea Ecosystem Model*; Publications Office of the European Union: Luxemburg, 2016.

37. Oguz, T.; Ducklow, H.W.; Malanotte-Rizzoli, P.; Murray, J.W.; Shushkina, E.A.; Vedernikov, V.I.; Unluata, U. A physical-biochemical model of plankton productivity and nitrogen cycling in the Black Sea. *Deep Res. Part I Oceanogr. Res. Pap.* **1999**. [CrossRef]

38. Oguz, T. Modeling aggregate dynamics of transparent exopolymer particles (TEP) and their interactions with a pelagic food web. *Mar. Ecol. Prog. Ser.* **2017**, *582*, 15–31. [CrossRef]

39. Oguz, T.; Merico, A. Factors controlling the summer Emiliania huxleyi bloom in the Black Sea: A modeling study. *J. Mar. Syst.* **2006**, *59*, 173–188. [CrossRef]

40. Miladinova, S.; Adolf, S.; Diego, M.M.; Elisa, G.G. *Revised Black Sea Ecosystem Model*; Publications Office of the European Union: Luxemburg, 2017.

41. Ludwig, W.; Dumont, E.; Meybeck, M.; Heussner, S. River discharges of water and nutrients to the Mediterranean and Black Sea: Major drivers for ecosystem changes during past and future decades? *Prog. Oceanogr.* **2009**, *80*, 199–217. [CrossRef]

42. Tuğrul, S.; Murray, J.W.; Friederich, G.E.; Salihoğlu, I. Spatial and temporal variability in the chemical properties of the oxic and suboxic layers of the black sea. *J. Mar. Syst.* **2014**, *135*, 29–43. [CrossRef]

43. Codispoti, L.A.; Friederich, G.E.; Murray, J.W.; Sakamoto, C.M. Chemical variability in the Black Sea: Implications of continuous vertical profiles that penetrated the oxic/anoxic interface. *Deep Res. Part A* **1991**, *38*, S691–S710. [CrossRef]

44. Baştürk, Ö.; Tuğrul, S.; Konovalov, S.; Salihoğlu, İ. Variations in the Vertical Structure of Water Chemistry within the Three Hydrodynamically Different Regions of the Black Sea. In *Sensitivity to Change: Black Sea, Baltic Sea and North Sea*; Springer: Dordrecht, The Netherlands, 1997; pp. 183–196.

45. Koçak, M.; Mihalopoulos, N.; Tutsak, E.; Violaki, K.; Theodosi, C.; Zarmpas, P.; Kalegeri, P. Atmospheric deposition of macronutrients (dissolved inorganic nitrogen and phosphorous) onto the Black Sea and implications on marine productivity. *J. Atmos. Sci.* **2016**, *73*, 1727–1739. [CrossRef]

46. Kubilay, N.; Yemenicioglu, S.; Saydam, A.C. Airborne material collections and their chemical composition over the Black Sea. *Mar. Pollut. Bull.* **1995**, *30*, 475–483. [CrossRef]

47. Carstensen, J.; Klais, R.; Cloern, J.E. Phytoplankton blooms in estuarine and coastal waters: Seasonal patterns and key species. *Estuar. Coast. Shelf Sci.* **2015**, *162*, 98–109. [CrossRef]

48. Finenko, Z.Z.; Hoepffner, N.; Williams, R.; Piontkovski, S.A. Phytoplankton carbon to chlorophyll a ratio: Response to light, temperature and nutrient limitation. *Mar. Ecol. J.* **2003**, *II*, 40–64. [CrossRef]

49. Cloern, J.E.; Grenz, C.; Vidergar-Lucas, L. An empirical model of the phytoplankton chlorophyll: Carbon ratio-the conversion factor between productivity and growth rate. *Limnol. Oceanogr.* **1995**, *40*, 1313–1321. [CrossRef]

50. Karl, D.M.; Knauer, G.A. Microbial production and particle flux in the upper 350 m of the Black Sea. *Deep. Res. Part A* **1991**, *38*, S921–S942. [CrossRef]

51. Kopelevich, O.V.; Sheberstov, S.V.; Yunev, O.; Basturk, O.; Finenko, Z.Z.; Nikonov, S.; Vedernikov, V.I. Surface chlorophyll in the Black Sea over 1978–1986 derived from satellite and in situ data. *J. Mar. Syst.* **2002**, *36*, 145–160. [CrossRef]

52. Knorr 2001, NATO 2001 Black Sea Expedition. Available online: https://www.ocean.washington.edu/cruises/Knorr2001/ (accessed on 10 October 2020).

53. Kaiser, D.; Konovalov, S.; Schulz-Bull, D.E.; Waniek, J.J. Organic matter along longitudinal and vertical gradients in the Black Sea. *Deep Res. Part I Oceanogr. Res. Pap.* **2017**, *129*, 22–31. [CrossRef]

54. CMEMS, Copernicus Marine Environment Monitoring Service. Available online: https://marine.copernicus.eu/ (accessed on 10 October 2020).

55. Mikaelyan, A.S.; Kubryakov, A.A.; Silkin, V.A.; Pautova, L.A.; Chasovnikov, V.K. Regional climate and patterns of phytoplankton annual succession in the open waters of the Black Sea. *Deep Res. Part I Oceanogr. Res. Pap.* **2018**, *142*, 44–57. [CrossRef]

56. Filippov, D.M. *Water Circulation and Structure of the Black Sea*; Nauka: Moscow, Russia, 1968.

57. Mikaelyan, A.S.; Zatsepin, A.G.; Chasovnikov, V.K. Long-term changes in nutrient supply of phytoplankton growth in the Black Sea. *J. Mar. Syst.* **2013**, *117–118*, 53–64. [CrossRef]

58. Yunev, O.A.; Vedernikov, V.I.; Basturk, O.; Yilmaz, A.; Kideys, A.E.; Moncheva, S.; Konovalov, S.K. Long-term variations of surface chlorophyll a and primary production in the open Black Sea. *Mar. Ecol. Prog. Ser.* **2002**, *230*, 11–28. [CrossRef]
59. Silkin, V.A.; Pautova, L.A.; Giordano, M.; Chasovnikov, V.K.; Vostokov, S.V.; Podymov, O.I.; Pakhomova, S.V.; Moskalenko, L.V. Drivers of phytoplankton blooms in the northeastern Black Sea. *Mar. Pollut. Bull.* **2019**, *138*, 274–284. [CrossRef]
60. NASA Ocean Color. Available online: http://oceancolor.gsfc.nasa.gov/ (accessed on 10 October 2020).
61. Miladinova, S.; Stips, A.; Macias Moy, D.; Garcia-Gorriz, E. Pathways and mixing of the north western river waters in the Black Sea. *Estuar. Coast. Shelf Sci.* **2020**, *236*, 106630. [CrossRef]
62. Zhou, F.; Shapiro, G.; Wobus, F. Cross-shelf exchange in the northwestern Black Sea. *J. Geophys. Res. Ocean.* **2014**, *119*, 2143–2164. [CrossRef]
63. Oguz, T.; Malanotte-Rizzoli, P.; Aubrey, D. Wind and thermohaline circulation of the Black Sea driven by yearly mean climatological forcing. *J. Geophys. Res.* **1995**, *100*, 6845. [CrossRef]
64. Stelmakh, L.V.; Gorbunova, T.I. Carbon-to-chlorophyll-a ratio in the phytoplankton of the Black Sea surface layer: Variability and regulatory factors. *Ecol. Montenegrina* **2018**, *17*, 60–73. [CrossRef]
65. Galbraith, E.D.; Martiny, A.C. A simple nutrient-dependence mechanism for predicting the stoichiometry of marine ecosystems. *Proc. Natl. Acad. Sci. USA* **2015**, *112*, 8199–8204. [CrossRef] [PubMed]
66. Macias, D.; Huertas, I.E.; Garcia-Gorriz, E.; Stips, A. Non-Redfieldian dynamics driven by phytoplankton phosphate frugality explain nutrient and chlorophyll patterns in model simulations for the Mediterranean Sea. *Prog. Oceanogr.* **2019**, *173*, 37–50. [CrossRef] [PubMed]

Article

Application of Photo-Identification and Lengthened Deployment Periods to Baited Remote Underwater Video Stations (BRUVS) Abundance Estimates of Coral Reef Sharks

Mauvis Gore [1,2], **Rupert Ormond** [1,2,3,*], **Chris Clarke** [4], **Johanna Kohler** [1,2,5], **Catriona Millar** [1,2] **and Edward Brooks** [6]

[1]   Marine Conservation International, South Queensferry, Edinburgh EH30 9WN, UK; mauvis.gore.mci@gmail.com (M.G.); johanna.k.kohler@gmail.com (J.K.); catrionamillar@hotmail.co.uk (C.M.)

[2]   Centre for Marine Biology and Biodiversity, Heriot-Watt University, Edinburgh EH14 4AS, UK

[3]   Faculty of Marine Science, King Abdulaziz University, Jeddah 21589, Saudi Arabia

[4]   Marine Research Facility, North Obhur, Jeddah 23817-4981, Saudi Arabia; chris@danahdivers.com

[5]   Cayman Islands Department of Environment, P.O. Box 10202, Georgetown, Grand Cayman KY1-1002, Cayman Islands

[6]   Cape Eleuthera Island School, PO Box EL-26029, Rock Sound, Eleuthera, Bahamas; eddbrooks@islandschool.org

*   Correspondence: rupert.ormond.mci@gmail.com; Tel.: +44-1680-300-073

Received: 18 June 2020; Accepted: 23 October 2020; Published: 3 November 2020

Abstract: Baited Remote Underwater Video Stations (BRUVS) are widely used for monitoring relative abundances of fishes, especially sharks, but only the maximum number of individuals seen at any one time (MaxN) is usually recorded. In both the Cayman Islands and the Amirante Islands, Seychelles, we used photo-ID to recognise individual sharks recorded on BRUVS videos. This revealed that for most species the actual numbers of separate individuals (IndN) visiting the BRUVS were significantly higher than MaxN, with, for example, ratios of IndN to MaxN being 1.17 and 1.24 for Caribbean reef, *Carcharhinus perezi*, and nurse, *Ginglymostoma cirratum*, sharks in the Cayman Islands, and 2.46 and 1.37 for blacktip reef, *C. melanopterus*, and grey reef, *C. amblyrhynchos*, sharks, respectively, in the Amirantes. Further, for most species, increasing the BRUVS deployment period beyond the 60 min normally used increased the observed IndN, with more than twice as many individuals in the Cayman Islands and >1.4 times as many individuals in the Amirantes being recorded after 120 min as after 60 min. For most species, MaxN and IndN rose exponentially with time, so data from different deployment periods cannot reliably be compared using catch-per-unit-effort (CPUE) calculated as catch-per-unit-time. In both study areas, the time of first arrival of individuals varied with species from <1 min to >2 h. Individually identifiable sharks were re-sighted after up to 429 days over 10 km away in the Cayman Islands and 814 days over 23 km away in the Amirantes, demonstrating that many individuals range over considerable distances. Analysis of Cayman re-sightings data yielded mean population estimates of 76 ± 23 (SE) and 199 ± 42 (SE) for *C. perezi* and *G. cirratum*, respectively. The results demonstrate that, for sharks, the application of both photo-identification and longer deployment periods to BRUVS can improve the precision of abundance estimates and provide knowledge of population size and ranging behaviour.

Keywords: capture–recapture; Cayman Islands; Seychelles; monitoring; endangered species; maximum number of individuals; photo-identification; mark-recapture; movement ecology

1. Introduction

Globally, many species of shark have suffered a dramatic decline in abundance over the past 30 years [1–3], principally as a result of by-catch and finning in response to the demand for shark fin by the Far-East restaurant trade [4,5]. As a consequence, 11 shark species are now considered Critically Endangered, 15 Endangered and 48 Vulnerable [6]. Most large-bodied sharks are threatened or endangered [2,7], while even once common reef species have become scarce in many areas with, for example, the long-established shark fishery in the Seychelles recording a significant decrease in catch per effort from 16 to 6 kg day^{-1} between 1992 and 2016 [8].

As a result, monitoring the population abundance of sharks is much needed [9]. Baited Remote Underwater Video Stations (BRUVS), underwater video cameras set to record the fishes attracted to bait, is now the method most frequently used for surveying and monitoring sharks in particular [10–12]. Stereo-BRUVS, in which two cameras are focused towards the bait, can provide further data by enabling both the distance to a fish and its length to be determined [13], although the associated equipment costs are necessarily higher if similar numbers of units are to be deployed. Cappo et al. [14] have reviewed BRUVS techniques for camera, bait, and orientation of bait and camera, as well as potential auditory, olfactory and behavioural cues, while Harvey et al. [15] concluded that BRUVS provide a statistically robust and cost-effective method of monitoring diverse assemblages of fishes in a number of habitat types.

BRUVS are not the only visually-based method regularly employed for monitoring fishes. Several forms of Underwater Visual Census (UVC) are widely used by divers [16–18]. However, remote video has the advantage of being practical in conditions where divers cannot easily operate, as in very deep water or rough conditions, and of being able to detect species, such as large sharks, which are often too scarce or too wary to be regularly sighted by divers [13]. Video methods are only fully effective in reasonably clear water, and baited video will only detect species that either are resident close to the BRUVS, or passing through, such as roaming schools of carangids, or like sharks have a well-developed olfactory sense enabling them to home-in on the bait [19].

To date, the most commonly used measure of the numbers of fishes coming to the BRUVS has been the maximum number of fishes recorded in a single frame during the viewing period (referred to as MaxN or Nmax). While simple for an observer to record, this measure suffers a number of limitations, most obviously that the observed MaxN may be much lower than the total number of separate individual fish attracted to the bait. Harvey et al. [15] discussed alternative approaches, including the use of cumulative MaxN (recorded each time MaxN is updated for each species). Campbell et al. [20] compared MaxN with MeanCount (the mean number of fishes observed in a series of frames or stills). They concluded that these two measures were comparable for abundance estimation, although MeanCount resulted in less precision for all species analysed. Similarly, Schobernd et al. [21] compared the advantages of using MaxN and MeanCount in both simulations and laboratory experiments, as well as in empirical data. They concluded that an advantage of MeanCount is that it is linearly related to actual abundance, whereas MaxN is not.

Understanding the accuracy and precision of monitoring and surveying methods is important for interpreting any apparent change or lack of change in species number or relative abundance. A weakness of the above counting methods is that they do not record whether different individuals are visiting a BRUVS at different times, as opposed to a single individual making repeat visits. Nor do they detect when the same individual visits two or more different BRUVS. This limitation is most obvious with less abundant but not uncommon species, such as many reef sharks and rays, in which different individuals may visit the same BRUVS, but at different times.

To address this issue, studies on other taxa have experimented with photo-identification as a tool for distinguishing different individuals. Stobart et al. [22] used BRUVS to record spiny lobster, *Palinurus elephas*, and compared three measures of abundance: MaxN, mean MaxN (using 5 min sampling periods) and the true number of individuals per recording based on individual identification (referred to in the present study as IndN). The authors concluded that while all three measures could distinguish areas of contrasting population density, MaxN and MeanN were not appropriate for documenting changes in population abundance because they were liable to sample saturation. That is, an increase in population abundance may not result in any increase in the MaxN. More recently, Irigoyen et al. [23] used three relative abundance indices in a survey of broadnose sevengill shark, *Notorynchus cepedianus*: Nmax, NmaxIND (cumulative number of different sharks and similar to IndN) and Nocc (total number of occurrences in a BRUVS session). The index NmaxIND appeared to show a greater abundance of sharks than Nmax. Sherman et al. [24] used photo-identification to determine the true numbers of two species of ray visiting BRUVS and found that the actual abundances were for *Neotrygon orientalis* 2.4, and for *Taeniura lymma* 1.1, times greater than the recorded MaxN values. However, to date this approach has not been generally applied to BRUVS-based monitoring of sharks, in part because the earlier cameras used were of lower resolution, so that it was not feasible to distinguish between most individuals of the same species.

Photo-identification has been found to be practical in some other studies of sharks. Population estimates of basking shark, *Cetorhinus maximus* [25], and great white shark, *Carcharhinus carcharius* [26], have been made based on individual recognition through quality images of dorsal fins. Photo-identification based on images captured by divers or fishers has also been used in ecological studies of blacktip reef shark, *C. melanopterus* [27], Greenland shark, *Somniosus microcephalus* [28], nurse shark, *Ginglymostoma cirratum* [29], sicklefin sharks, *Negaprion acutidens* [30], broadnose sevengill shark, *Notorynchus cepedianus* [23], and whale shark [31,32], as well as flapper skate, *Dipturus intermedius* [33]. In addition, BRUVS have been used to monitor the presence and size [34], as well as site fidelity [11], of individually identified sharks. Here, the feasibility of applying photo-identification methodology to the video footage recorded by BRUVS is assessed to monitor more effectively the abundance of coral reef shark species and test whether the data can also be used to generate additional information on population size or ranging behaviour.

The deployment period used to record individuals visiting a BRUVS (i.e., the effective length of deployment) has also varied between studies. The majority of previous studies have used deployment periods of 60 min (for example [24,35,36]), often because this was the maximum recording time that could be guaranteed by the batteries in older cameras. Some studies used shorter times, such as 30 min ([19] fish species), while Brooks et al. ([10] shark species) used 90 min, Harasti et al. ([34] white shark) 5 h and Stobart et al. ([22] lobster) 7 h. After using BRUVS to record coral reef fishes in the Hawaiian Islands, Asher [37] concluded that while a short deployment of 20 min was sufficient to capture fast-reacting species and residents, longer periods of 60 min were required to capture macropiscivores, including sharks. Our experience suggested that additional individual and species of shark could arrive at a BRUVS even after this time. A study by Torres et al. [38] exceeded other deployment times in using a duration of 24 h for sharks in the western Mediterranean. As a result of the wide range of deployment times used in different studies, the effect of deployment period on the numbers of individuals and species of shark recorded at a site was also tested in the present study.

In the present study, the BRUVS video recordings obtained during two shark conservation projects, one in the Cayman Islands (western Caribbean) and the other in the Amirante Islands (south-western Seychelles) were reviewed. Photo-identification methodology was applied to distinguish the separate individuals visiting BRUVS over different periods. The aims of the study were to assess whether (a) the use of photo-identification can provide better estimates of the numbers of sharks being recorded, (b) the approach can be used to generate estimates of population size by applying mark–recapture analyses to the data, and (c) the method can also be used to provide information about the movements of individuals. In addition, (d) it was investigated whether for reef sharks the numbers of species and individuals recorded varied significantly with BRUVS deployment period. We found that photo-identification could enhance the value of data obtained in relation to all three questions. However, the present study is not intended to compare abundances between areas or habitats, or across years (hence locations of individual stations are not included); those analyses are being reported on separately.

2. Materials and Methods

2.1. Study Areas

The study was based on data collected from two widely separated island groups, one in the western Atlantic and the other in the Indian Ocean. The Cayman Islands are located in the North-West Caribbean Sea (Figure 1a) and include three islands, Grand Cayman, Little Cayman and Cayman Brac. These islands are emergent sections of the Cayman Ridge which runs adjacent to the 7 km deep Cayman Trench. Sampling was conducted around the largest island, Grand Cayman (in a study area bounded by Latitudes 19.25° N and 19.42° N and Longitudes 81.05° W and 81.44° W), and also around Little Cayman (in a study area bounded by Latitudes 19.64 and 19.73° N and Longitudes 79.95 and 80.13° W), 130 km to the northeast. Both islands have well-developed fringing coral reefs beyond which a narrow coastal shelf drops steeply to very deep water. A series of marine parks and conservation/replenishment areas, occupying about 25% of the 160 km long coastline, were designated over 30 years ago. Shark abundance, although a little higher than on many other Caribbean Islands, was still considered low in relation to relatively pristine reef areas [39]. In consequence sharks were given full protection throughout Cayman waters in 2015 [40].

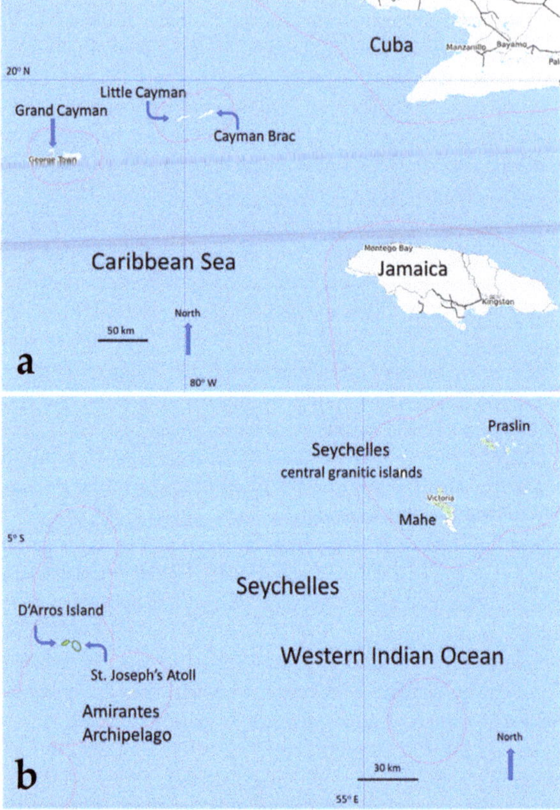

Figure 1. Locations of study areas: (**a**) the three islands constituting the Cayman Islands in the Caribbean Sea, and (**b**) D'Arros Island and St. Joseph's Atoll in the Amirante Islands, Seychelles, in the Indian Ocean. Data analysed for this study resulted from deployments of BRUVS at varied intervals of 0.5–2 km around the whole of both Grand Cayman and Little Cayman in the Cayman Islands, and D'Arros Island and St. Joseph's Atoll in the Seychelles. The faint red lines around each island indicate the extent of territorial waters [41].

The Amirante Islands in the south-western part of the Seychelles, in the western Indian Ocean (Figure 1b), are a series of widely separated small islands and coral reefs sitting on the 180 km by 40 km Amirante Bank, beyond which the plateau gives way to very deep water. The islands include D'Arros Island (53.29° E, 5.41° S) and the adjacent St. Joseph Atoll (53.33° E, 5.43° S), around both of which the data used here were collected (the combined study area being bounded by Latitudes 5.27° S and 5.47° S and Longitudes 53.17° E and 53.38° E). D'Arros and the outer faces of St. Joseph Atoll are surrounded by well-developed fringing reefs that slope down from a reef crest to the seabed at 15 to 20 m, while the atoll itself encloses a 3.5 km wide sandy lagoon, 1–6 m deep. Adjacent bank areas (5–35 m deep) that were also surveyed have a mainly sandy seabed, which in its shallower parts are colonised by dense seagrass beds (mostly *Thalassodendron ciliatum*) and occasional coral patches. D'Arros and St. Joseph Atoll are more remote from significant human populations than the Cayman Islands, and the adjacent reefs have been protected from fishing for almost three decades; in contrast, however, more distant bank areas are regularly fished, including by targeted long-lining for sharks. Nevertheless, sharks are more abundant in the Amirante Islands than in the Cayman Islands, and a comparison between the two areas provided an indication of how different levels of abundance influence the value of applying photo-identification to BRUVS recordings. The precise locations of sampling points are not relevant to the analysis, and for both ethical and legal reasons are not included in this report.

2.2. BRUVS

In both study areas the BRUVS body was built from an open weave plastic crate, upturned and weighted down by a series of dive weights attached close to the rim (now the lower side). The bait arm was a 1.5 m long PVC tube fixed to the top of the crate, with the bait enclosed in a double-layered plastic mesh bag placed at the far end of the bait arm (Figure 2). The amount of bait used was standardised at 300 g of mackerel, *Scomber scombrus*, in the Cayman Islands and bonito, *Euthynnus affinis*, in the Amirante Islands. A camera was fixed on the crate next to the base of the bait arm and pointed towards the bait bag. The cameras used were models 3 and 4 Silver GoPros, which recorded in high definition (HD) and could be fitted with double battery packs, allowing recording periods of 2+ h. The availability of HD cameras was key to the present work, as the high quality of images extracted from video permitted the detection of fine detail on the individual sharks.

Figure 2. Image of a BRUVS as used in the present study, showing the open weave plastic body, a high-definition camera (GoPro Hero 3 or 4) under a roll cage, a bait arm with outer bait bag (lower left) and lines (upper right) leading to a surface marker buoy. (Photo with permission by James Lea).

BRUVS were normally deployed in both study areas on the fore-reef at depths of 10–18 m, but at 2–4 m deep in the lagoon of St. Joseph Atoll and in the Cayman Islands' Sounds (equivalent to lagoons). Deployment by boat took place between 07:35 and 15:34 h in the Amirante Islands and 08:26 and 15:04 in the Cayman Islands. BRUVS were positioned on the seabed in open spaces among corals on reefs, or in gaps within seagrass beds. Appropriate placement of the BRUVS was achieved with the aid of either a viewing scope or of a snorkeler who signalled when the BRUVS was over a suitable area. In both locations BRUVS were deployed in sets of four, one at each of four stations that were located at variable intervals of 500 m to 2 km in a line along each section of coast (i.e., coasts orientated in a particular direction); the arrangement thus approximated a stratified random sampling design. In repeat deployments over successive years BRUVS were placed at the same stations at the same sites. Survey work was undertaken when the Douglas sea scale was ≤ 4 and wave or swell height ≤ 1.5 m. Following retrieval of the BRUVS, memory cards were removed from the cameras and the files copied to computers on which the videos could be viewed. Visual analysis was carried out for the full length of each video, with a minimum of 25% of each video being checked by a second researcher to ensure that comparable standards were being observed. For each shark that was detected, the time(s) of arrival in view and loss from view were recorded on each occasion, along with species, sex, maturity and size, as estimated by comparison with the size of the bait bag and graduated marks on the bait arm [42].

In the Cayman Islands BRUVS were deployed once or twice a year, in or around June and December, and for present purposes data from 2015 to 2017 have been analysed. In the Amirante Islands BRUVS were deployed once per year, in or around February or March, and for this study data from 2015 to 2017 were similarly selected.

2.3. Photo-Identification

To facilitate individual recognition, so far as possible detailed information was recorded on the appearance of each newly observed shark. The characteristics recorded included the shapes of fins and other body parts (if these appeared in anyway atypical), and any variation in natural patterning or colouration, as well as the shape and location of any naturally acquired markings, such as wounds, injuries or scars (Figures 3 and 4). Details were recorded of distinguishing features on both sides of the body, so that with few exceptions individuals could be recognised on viewing from either side. Acquired marks could be temporary, as with minor abrasions or injuries, but most used in individual identification were more permanent, as when part of a fin was missing or torn. Evidence from longer-term studies suggested that while many non-fatal injuries heal, scars and serious fin injuries in sharks can remain evident for many years [25,26].

Figure 3. *Cont.*

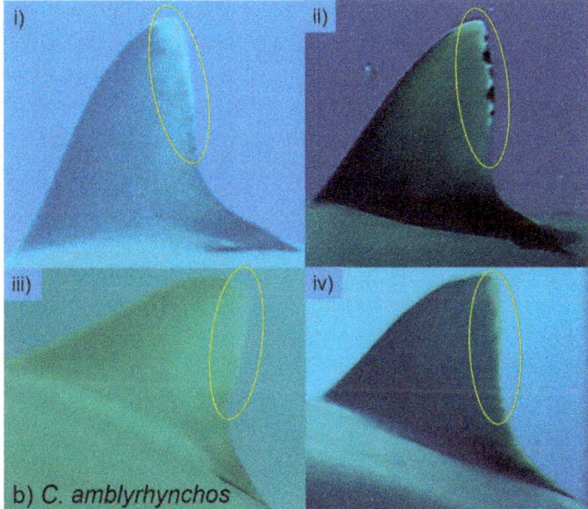

Figure 3. Examples of variations in natural patterning and injuries on dorsal fins of sharks recorded as screen grabs from BRUVS videos. (**a**) Differences in natural patterning on right and left sides of first dorsal fins of two different blacktip reef sharks, *Carcharinus melanopterus*, (first shark i and ii, second shark iii and iv), and (**b**) dorsal fins of four different grey reef sharks, *C. amblyrhynchos*. The yellow circles highlight the location of distinctive features; the yellow lines are added to allow variation in the shape of the black fin tip to be seen more clearly. Note that the different sides of an individual blacktip reef shark's dorsal fin tend to be similar, but are not identical to each other.

Figure 4. Examples of variation in natural markings (circled) on different shark species taken from BRUVS videos of two different Caribbean reef sharks, *Carcharinus perezi* (top), a nurse shark, *Ginglymostoma cirratum*, (centre), and a set of four images (bottom) showing distinctive features on different individual whitetip reef sharks, *Triaenodon obesus*.

Features observed on individual sharks were marked on a standardised cartoon (sketch-like drawing) for each shark (see [25] for further discussion of the method and for examples). Characteristics recorded on each cartoon were then entered into a computer database, along with the details of each deployment and selected images showing the individual's distinctive features. The cartoons and database could be interrogated for individual characteristics to facilitate matching with any other recorded individual. Potential matches were then either confirmed or rejected by comparison of the images or video clips of each shark. Potential matches were independently verified by a second researcher.

2.4. IndN and MaxN

Data on the numbers of sharks were used to compare the results of using two different methods for recording the numbers of sharks of each species attracted to each BRUVS. First, MaxN, the greatest number of sharks of a given species observed in a single frame during a set period, was determined. Second, photo-identification was used to determine so far as possible IndN, the total number of different individuals of each species visible at any time in the field of view. In practice, not all the sharks detected on screen approached the BRUVS close enough for the individual to be identified. Individuals that lacked distinguishing features enabling them to be recognised in the long term, but that could still be separated from other sharks seen on the same recording (for example, if the shark was a different size), were added to the tally. Nevertheless, on some occasions two or more similar individuals may not have been seen clearly enough to distinguish, and on others not every shark attracted may have entered the field of view. Thus, in some instances, the IndN recorded will have been less than the actual number of individuals visiting.

Catch per unit effort (CPUE) has been used to compare the numbers of sharks visiting different BRUVS [10], but this metric can be misleading because, as evident from our results, the numbers of new sharks arriving at a BRUVS tends to decline with time. That is, the plot of the accumulative number of individuals against time is asymptotic. Therefore, longer deployments may generate lower CPUEs, even if more sharks are recorded. For this reason, the values of MaxN and IndN were compared with respect to three different deployment periods commencing from the time when the BRUVS arrived at the seabed; these were 60, 120 and 150 min. This comparison was undertaken to investigate the effects of using longer deployment periods by determining if these result in the detection of more individuals and/or more species. Deployment period was the term used to describe the recording time on the seabed.

2.5. Statistical Analysis and Mark–Recapture Population Estimates

A series of tests were applied to the data to determine statistical significance as detailed in the relevant parts of the Results section; these include Friedman ANOVA, Wilcoxon matched pairs test, ANOVA, paired t and post-hoc LSD tests. The statistical package used in the study for parametric and non-parametric tests (described in Results) was Statistica 64 V. 12.

Although large numbers of BRUVS videos were analysed (>500) and a large number of individually identifiable sharks recorded, the numbers of individuals re-sighted on successive campaigns were too low for open population mark–recapture models to be employed, such as the Schnabel and Jolly–Seber methods which we have applied to basking shark in Scotland [25]. However, the numbers of re-sightings in the Cayman Islands for six surveys over three years were high enough for successive population estimates of two species to be made using the Chapman derivative of the basic Lincoln–Peterson estimator [43,44], which is more robust for small sample sizes [45]. For each "re-capture" occasion t, the population size N_t was calculated where n_t = the number of sharks sighted on sample occasion t, r_t = the number of marked sharks re-sighted in sample n_t, and m_t = the number of marked sharks at the beginning of sample occasion t:

$$N_t = [(m_t + 1) * (n_t + 1)/(r_t + 1)] - 1 \tag{1}$$

Assuming that k_1 number of sharks were initially marked (i.e., identified) at sample occasion 1 and that $k_2 = n_2 - r_2$ was the number of unmarked (previously unknown) sharks sighted by the end of sample occasion 2, then m_t, the number of effectively marked sharks at the next sample occasion t, equals $\sum k_t$. The total population size ($\overline{N}$) was estimated as the mean across sample occasions s [41,42] according to:

$$\overline{N} = \sum Nt/(s - 1) \tag{2}$$

3. Results

A total of 661 BRUVS were deployed, 451 in the Cayman Islands (at 92 locations) and 210 BRUVS in the Amirante Islands (at 60 locations). These BRUVS recorded a total of six species in the Cayman Islands and eight in the Amirante Islands (Table 1), representing two orders (Carcharhiniformes and Orectoloboformes) and three families (Carcharhinidae, Ginglymostomatidae and Sphyrnidae) of sharks.

Table 1. The numbers of individually identifiable sharks of different species (i.e., those carrying distinguishing features enabling them to be recognised again as individuals) and the corresponding percentages these represent of all records of that species observed on video recordings from Baited Remote Underwater Video Stations (BRUVS) (a) in the Cayman Islands (451 BRUVS) and (b) in the Amirante Islands (210 BRUVS).

Area	Scientific Names	English Names	Number of Individually Identifiable Sharks	% Recognisable Sharks of All Records of the Species
(a) Cayman Islands	*Carcharhinus limbatus*	blacktip	8	60
	Carcharhinus perezi	Caribbean reef	135	54.4
	Galeocerdo cuvier	tiger	4	100
	Ginglymostoma cirratum	nurse	201	82.8
	Negaprion brevirostris	lemon	14	66.7
	Sphyrna mokarran	great hammerhead	4	66.7
(b) Amirante Islands	*Carcharhinus albimarginatus*	silvertip	4	16.7
	Carcharhinus amblyrhynchos	grey reef	96	77.9
	Carcharhinus leucas	bull	13	9.1
	Carcharhinus melanopterus	blacktip reef	322	85.3
	Nebrius ferrugineus	tawny nurse	44	78.3
	Negaprion acutidens	sicklefin	65	55.3
	Sphyrna mokarran	great hammerhead	1	33.3
	Triaenodon obesus	whitetip reef	30	63.9

3.1. Recognisability of Individual Sharks

In most species the majority of sharks observed could be recognised as individuals (Table 1), although the proportions of sharks observed for which sufficient features were evident to permit this varied from 100% for tiger shark, *Galeocerdo cuvier*, to 9.1% for bull shark, *Carcharhinus leucas*. In some species, individuals were relatively easy to distinguish. For example, the dorsal fins of *C. melanopterus* (85.3% individually identifiable) are capped by an area of black pigmentation, the exact shape of which varies markedly both between individuals and between opposite sides of the same individual (Figure 3). Similarly, whitetip reef sharks, *Triaenodon obesus*, (63.9% individually identifiable) have spot patterns on the body that are unique to individuals (Figure 4). Other species, such as *C. amblyrhynchos*, (77.9% individually identifiable) and *G. cirratum* (82.8% individually identifiable) (Figure 4), required more careful examination of the video footage to discern individually distinctive markings. *C. leucas* and silvertip, *C. albimarginatus*, sharks were the most difficult to identify as individuals on recordings since they tended not to approach the camera, making detection of individual features often impracticable. The mean percentage of distinguishable sharks across species was 60% and the overall percentage of recognisable individuals (irrespective of species) was 66%.

3.2. Comparison of IndN with MaxN

For all shark species combined in both the Cayman Islands and the Amirante Islands, the mean IndN (1.88, 2.71, 2.67) was significantly greater than mean MaxN (1.20, 1.57, 1.51) for each of the three deployment periods (60, 120 and 150 min, respectively) (Table A1). Considering the species separately, in the Cayman Islands for both *C. perezi* and *G. cirratum*, mean IndN was significantly greater than mean MaxN for each deployment period (60, 120 and 150 min) (Figure 5, Table A1), the overall ratio of IndN to MaxN being 1.17 for *C. perezi*, and 1.24 for *G. cirratum*. Similarly, in the Amirante Islands, mean IndN was significantly greater than mean MaxN for all three deployment periods for *C. amblyrhynchos*, *C. melanopterus* and *N. acutidens*, but only for 120 min for tawny nurse shark, *Nebrius ferrugineus*, (Figure 6, Table A1). The overall ratios of IndN to MaxN were 2.46 for *C. melanopterus*, and 1.37 for *C. amblyrhynchos*. For other species, the sample sizes were too low for significant differences to be detected (Figures 5 and 6). For all species in which there were significant differences between IndN and MaxN, the ratio of IndN to MaxN increased with time, as it did also for *C. limbatus* and *T. obesus*, in which differences were not significant (Table A2).

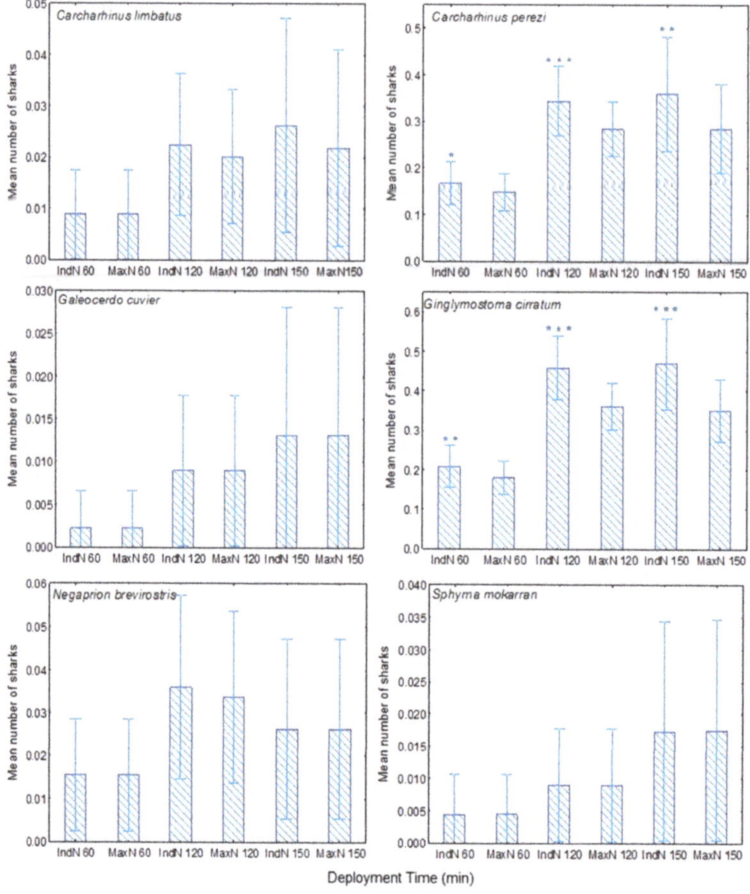

Figure 5. Cayman Islands: The mean number (bar) and 95% confidence interval (error bar) of individually identified sharks observed for deployment periods of 60, 120 and 150 min for IndN compared with the same periods for MaxN. Significant differences found between pairs of IndN and MaxN using paired *t* tests are noted by ★ for $p \leq 0.05$, ★★ for $p \leq 0.001$ and ★★★ for $p \leq 0.0001$ (see Table A1).

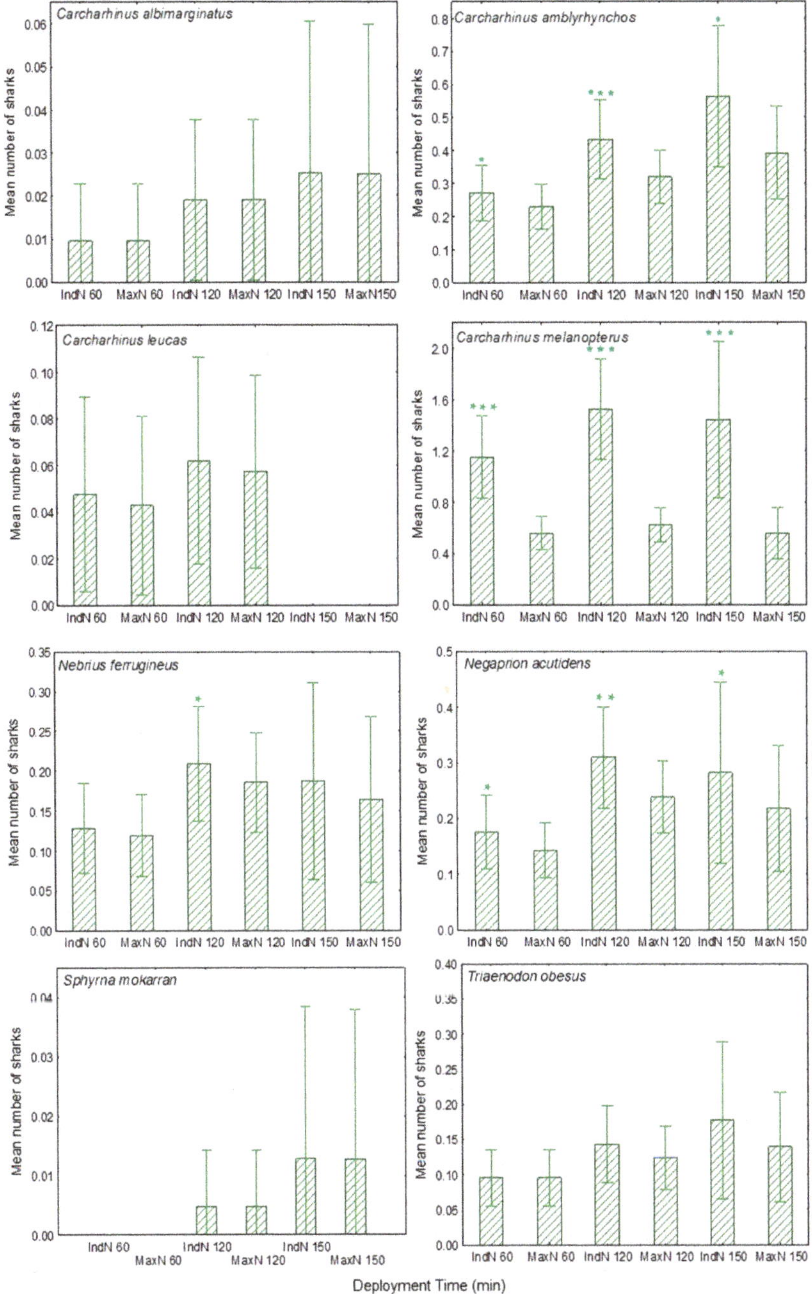

Figure 6. Amirante Islands: The mean number (bar) and 95% confidence interval (error bar) of individually identified sharks observed for deployment periods of 60, 120 and 150 min for IndN compared with the same periods for MaxN. Significant differences found between pairs of IndN and MaxN using paired *t* tests are noted by ★ for $p \leq 0.05$, ★★ for $p \leq 0.001$ and ★★★ for $p \leq 0.0001$ (see Table A1).

The number of BRUVS deployments in which the application of individual identification revealed the presence of more sharks than would otherwise have been evident (i.e., where IndN > MaxN) is shown in Table 2. While the use of photo-identification revealed the presence of additional sharks in the more abundant species (three in the Cayman Island and six in the Amirante Islands), this was not the case for the less common species.

Table 2. Proportion of BRUVS deployments in which IndN was greater than MaxN in (a) the Cayman Islands (n = 451) and (b) the Amirante Islands (n = 210). The proportion as a percentage of BRUVS is the percentage of BRUVS with sharks in which IndN was greater than MaxN.

Area	Species	No. of BRUVS with Sharks	BRUVS IndN > MaxN as%
(a) Cayman Islands	*Carcharhinus limbatus*	8	0
	Carcharhinus perezi	85	20
	Galeocerdo cuvier	4	0
	Ginglymostoma cirratum	131	22.1
	Negaprion brevirostris	11	9.1
	Sphyrna mokarran	4	0
(b) Amirante Islands	*Carcharhinus albimarginatus*	1	0
	Carcharhinus amblyrhynchos	57	33.3
	Carcharhinus leucas	9	11.1
	Carcharhinus melanopterus	80	66.3
	Nebrius ferrugineus	33	15.2
	Negaprion acutidens	46	28.3
	Sphyrna mokarran	1	0
	Triaenodon obesus	27	7.4

3.3. Effect of Deployment Period for All Species Combined

The mean IndN and MaxN of all species combined after different recording times in each of the two study areas are significantly different in each case (Table 3). In both the Cayman Islands and the Amirante Islands there was a significant increase in IndN and in MaxN both from 60 to 120 min, and from 120 to 150 min in the Amirante Islands (Table A3).

Table 3. Comparison of IndN and MaxN for all shark species combined ± standard deviation (S.D.) recorded on BRUVS in (a) Cayman Islands (BRUVS n = 109) and (b) Amirante Islands (BRUVS n = 55) for 60, 120 and 150 min deployment periods. Only cases where BRUVS ran for 150 min were used to compare across the three periods with Friedman ANOVA χ^2, p = probability value, df = 2.

Study Area	Deployment Period (min)	IndN			MaxN		
		Mean Number & S.D. of All Sharks	χ^2	p	Mean Number & S.D. of All Sharks	χ^2	p
(a) Cayman Islands	60	0.78 *0.91*			0.72 *0.82*		
	120	1.81 *1.38*	137.9	0.00001	1.44 *0.95*	119.2	0.00001
	150	1.96 *1.5*			1.56 *1.08*		
(b) Amirante Islands	60	2.45 *2.71*			1.45 *1.24*		
	120	3.73 *3.05*	72.8	0.00001	2.13 *1.28*	54.3	0.00001
	150	3.84 *3.04*			2.16 *1.23*		

Overall, using IndN there were 2.1 times as many sharks recorded after 120 min as after 60 min in the Cayman Islands, and 1.4 times as many recorded in the Amirante Islands after 120 min than after 60 min. Similarly, there were 1.6 and 1.4 as many individual sharks recorded after 150 min as after 120 min in the two areas, respectively.

3.4. Effect of Deployment Period for Separate Shark Species

The mean IndN (i.e., the mean number of separate individuals visiting a BRUVS) was significantly different over the periods 60, 120 and 150 min for *C. limbatus*, *C. perezi*, *G. cirratum* and lemon shark, *Negaprion brevirostris*, in the Cayman Islands, and for *C. amblyrhynchos*, *C. melanopterus*, *N. ferrugineus*, *N. acutidens* and *T. obesus* in the Amirante Islands (Table 4). Among these species, *C. perezi*, *G. cirratum*, *C. amblyrhynchos*, *C. melanopterus* and *N. acutidens* showed a significant increase in mean number from 60 to 120 min, while *C. perezi* and *G. cirratum* also showed an increase from 120 to 150 min (Table A4).

Table 4. Comparison of mean ± standard deviation (S.D.) values of IndN and MaxN for each shark species for BRUVS deployment periods of 60, 120 and 150 min for (a) the Cayman Islands and (b) the Amirante Islands. Shown too are the results of a Friedman ANOVA (χ^2, p = probability value, n = number of BRUVS in sample; df = 2) for the differences in the means across the three deployment periods. Post hoc results are available in Table A4.

Area	Shark Species	Period min	IndN Mean	IndN S.D.	χ^2 / p / n	MaxN Mean	MaxN S.D.	χ^2 / p / n
	Carcharhinus limbatus	60	0.5	0.55	6	0.6	0.55	4
		120	1		0.049	1		0.1
		150	1		6	1		5
	Carcharhinus perezi	60	0.78	0.88	54.5	0.64	0.7	50.6
		120	1.67	1.22	0.00001	1.28	0.88	0.00001
		150	1.82	1.35	45	1.43	1.02	47
(a) Cayman Islands	Galeocerdo cuvier	60	0.33	0.58	4	0.33	0.58	4
		120	1		0.1	1		0.1
		150	1		3	1		3
	Ginglymostoma cirratum	60	0.6	0.75	79.9	0.55	0.63	67.6
		120	1.43	1.03	0.00001	1.1	0.6	0.00001
		150	1.57	0.93	70	1.19	0.46	69
	Negaprion brevirostris	60	0.33	0.52	8	0.33	0.52	8
		120	1		0.018	1		0.018
		150	1		6	1		6
	Sphyrna mokarran	60	0.5	0.58	4	0.5	0.58	4
		120	1		0.1	1		0.1
		150	1		4	1		4
	Carcharhinus albimarginatus	60	0.5	0.71	?	0.5	0.71	2
		120	1		0.4	1		0.4
		150	1		2	1		2
	Carcharhinus amblyrhynchos	60	0.69	0.62	18.7	0.69	0.62	18.7
		120	1.12	0.59	0.00009	1.12	0.59	0.00009
		150	1.19	0.49	26	1.19	0.49	26
(b) Amirante Islands	Carcharhinus melanopterus	60	1.23	1.07	14	1.23	1.07	14
		120	1.47	0.86	0.0009	1.47	0.86	0.0009
		150	1.47	0.86	30	1.47	0.86	30
	Nebrius ferrugineus	60	0.8	0.79	8	0.8	0.79	8
		120	1.3	0.48	0.018	1.3	0.48	0.018
		150	1.3	0.48	10	1.3	0.48	10
	Negaprion acutidens	60	0.64	0.63	14	0.64	0.63	14
		120	1.21	0.43	0.0009	1.21	0.43	0.0009
		150	1.21	0.43	14	1.21	0.43	14
	Triaenodon obesus	60	0.64	0.5	8	0.64	0.5	54.3
		120	1		0.018	1		0.00001
		150	1		11	1		11

The mean MaxN was significantly different across different deployment periods for *C. perezi*, *G. cirratum* and *N. brevirostris* in the Cayman Islands and for *C. amblyrhynchos*, *C. melanopterus*, *N. ferrugineus*, *N. acutidens* and *T. obesus* in the Amirante Islands (Table 4). Post-hoc tests indicated that in the Amirante Islands, *C. amblyrhynchos*, *C. melanopterus* and *N. acutidens* all showed a significant increase in mean MaxN from 60 to 120 min, as well as *G. cirratum* in the Cayman Islands, while *C. perezi* showed significant increases in mean MaxN both from 60 to 120 and from 120 to 150 min (Table A4).

3.5. Arrival Times at BRUVS

As a further means of assessing effective BRUVS deployment periods, the mean times of first arrival (at those BRUVS on which the species were recorded) were determined for each species in the Cayman Islands and the Amirante Islands (Figure 7). For two species in the Cayman Islands, the mean first arrival time was close to 60 min, while the remaining four species took longer (mean of 71–76 min). In the Amirante Islands, the mean first arrival time was less than 60 min for four species and was about an hour for two other species, but in two species, *C. albimarginatus* and great hammerhead (*Sphyrna mokarran*) shark, the mean arrival time was greater than an hour. It was notable that some individual *C. amblyrhynchos* and *C. melanopterus* (in the Amirante Islands), and *G. cirratum* and *C. perezi* (in the Cayman Islands), took more than 2 h to arrive.

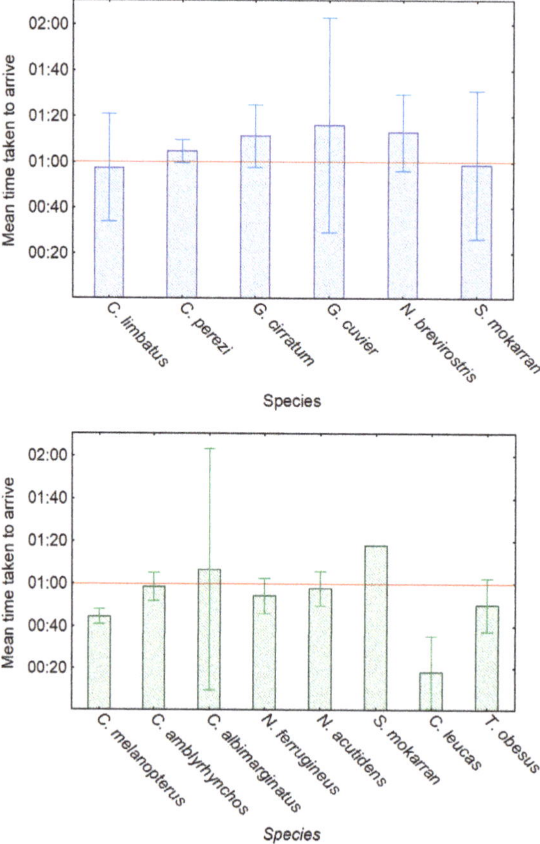

Figure 7. Mean and 95% confidence interval of time taken (hh:mm) by shark species to arrive at the BRUVS camera. Top: Cayman Islands, Bottom: Amirante Islands. The red line indicating a mean time taken to arrive of 1 h (60 min) is included for convenience.

In the Amirante Islands, the variation in first arrival time across species was statistically significant (ANOVA: $F_{(7,533)}$ = 5.05, p = 0.00001), with *C. leucas* when present arriving significantly earlier than *C. albimarginatus*, *C. amblyrhynchos*, *C. melanopterus*, *N. ferrugineus*, *N. acutidens* and *T. obesus* (Post hoc LSD tests: p < 0.006, respectively). *C. melanopterus*'s first arrival time was significantly shorter than those for *C. amblyrhynchos*, *N. ferrugineus* and *N. acutidens* (Post hoc LSD tests: p <0.04, respectively). There was no significant variation in first arrival time across species in the Cayman Islands (ANOVA: $F_{(5,411)}$ = 0.19, p = 0.9).

3.6. Re-Sighting Rates

The application of photo-identification methodology allowed many individuals to be recognised again if they were re-sighted. The proportions of the more commonly recorded species that were re-sighted during the study periods on a different BRUVS deployment ranged from 0% for *C. limbatus* to 10.1% for *G. cirratum* in the Cayman Islands, and from 4.5% for *N. ferrugineus* to 9.8% for *C. melanopterus* in the Amirante Islands (Table 5). In the Cayman Islands, a third of individual *G. cuvier* and *S. mokarran* were re-sighted, but only three individuals of either species were recorded in total. The mean of the re-sighting rates for each species across both locations was 12%; the combined re-sighting rate (from pooling all individuals irrespective of species) was 10.9%.

Table 5. The re-sightings rates on BRUVS for shark species (a) in the Cayman Islands and (b) in the Amirante Islands, both within years and between years. Only re-sights recorded on BRUVS videos are included here (as opposed to other re-sightings that were recorded during other related activities). Note that the total number of individuals re-sighted over the whole period is not necessarily the sum of the re-sights in each year, since some re-sights may be of individuals that have already been re-sighted.

Area	Species	Year	No. of Distinguishable Individuals	Re-Sights within Year	Re-Sights within Year as%	Re-Sights from Previous Years	Total Individuals Re-Sighted as%
	Carcharhinus limbatus	2015	1	0	0		
		2016	4	0	0	0	0
		2017	3	1	33.3	0	12.5
	Carcharhinus perezi	2015	44	2	4.5		
		2016	47	4	8.5	0	6.6
		2017	69	7	10.1	4	9.6
(a) Cayman Islands	*Galeocerdo cuvier*	2015	2	1	50		
		2016	1	0	0	0	33.3
		2017	1	0	0	1	50
	Ginglymostoma cirratum	2015	51	9	17.6		
		2016	68	2	2.9	1	10.1
		2017	79	13	16.5	21	28.8
	Negaprion brevirostris	2015	3	1	33.3		
		2016	6	0	0	0	11.1
		2017	4	1	25	1	9.1
	Sphyrna mokarran	2015	2	1	50		
		2016	1	0	0	0	33.3
		2017	1	0	0	0	33.3
	Carcharhinus albimarginatus	2015	2	0	0		
		2016	2	1	50	0	
		2017	0	0	0	0	25
	Carcharhinus amblyrhynchos	2015	23	1	4.3		
		2016	34	2	5.9	1	
		2017	34	0	0	2	6.8
(b) Amirante Islands	*Carcharhinus melanopterus*	2015	107	14	13.1		
		2016	79	7	8.9	2	
		2017	135	6	4.4	2	9.8
	Nebrius ferrugineus	2015	9	0	0		
		2016	17	1	5.9	0	
		2017	18	1	5.6	0	4.5
	Negaprion acutidens	2015	23	2	8.7		
		2016	21	2	9.5	0	
		2017	21	0	0	0	6.2
	Triaenodon obesus	2015	7	0	0		
		2016	8	1	12.5	0	
		2017	14	0	0	1	6.8

3.7. Movements of Individual Sharks

Re-sightings of individuals on a BRUVS different from those on which they were originally recorded provided information about shark movement patterns. In the Cayman Islands, individuals were re-sighted at distances of up to 10 km for *C. perezi* and 15 km for *G. cirratum* from the location of the original sighting (Table A5). In the Amirante Islands, individuals were re-sighted at distances of up to about 25 km for *C. melanopterus* shark, 33 km for *C. amblyrhynchos* shark, and about 27 km for *N. acutidens*, *N. ferrugineus* and *T. obesus* (Table A5), although both these last two species were also often re-sighted on BRUVS close to the stations at which they were originally recorded. While a majority of re-sights occurred within the same season, in the Amirante Islands some re-sights of *C. amblyrhynchos*, *C. melanopterus* and *N. acutidens* occurred after longer periods of time, the longest being of an individual *C. melanopterus* re-sighted after 814 days. In the Cayman Islands, some individuals were also detected after long periods of time, the longest being 429 days for *C. perezi* and 184 days for *G. cirratum*.

3.8. Population Estimates

For most species, the numbers of re-sightings were too low to permit the reliable estimation of population size, but using the Chapman derivative of the basic Lincoln–Peterson estimator successive estimates were possible for the numbers of *C. perezi* and *G. cirratum* present in the Cayman Islands study area. The mean estimated number of individually recognisable *C. perezi* over 2015–2017 was 76 ± 23 (S.E.), and of individually recognisable *G. cirratum* over the same period was 199 ± 42 (S.E.). Comparable estimates for *C. melanopterus* and *C. amblyrhynchos* in the Amirante Islands study area are in hand.

4. Discussion

Comparison of the present data with those obtained by other researchers in the same ocean regions is possible (Table 6 [10–12,42,46–48]), but confounded by differences in method, such as the time and depth ranges at which the BRUVS were deployed. In particular, while most researchers report MaxN for a deployment period of 60 min, relevant studies have used deployment periods ranging from 54 to 90 min, with one using 6 h [48]. Some authors (Brooks et al. [10]) have corrected for differences in deployment period by reporting numbers as catch per unit effort (CPUE) of one hour, on the assumption that MaxN increases linearly with time. However, in the present study, the values of both MaxN and IndN rise exponentially with time, hence longer deployment periods will appear to yield lower values of CPUE if presented this way. Nevertheless, some comment is possible. In the Cayman Islands, the abundance of two species of sharks was greater and of four species less than that reported in Eleuthera, the Bahamas. The abundances of *N. acutidens* appear markedly greater in the Amirante Islands than in the eastern Indian Ocean. Abundances of *C. melanopterus* and *N. ferrugineus* appear broadly comparable. More detailed analysis of the spatial and temporal patterns of abundance for the different species recorded in our study areas will be reported separately.

It is also important to note that when fish visible on BRUVS videos are counted (e.g., to determine MaxN), normally the results provide information only on relative abundance (that is, differences in abundance between locations or between occasions), but not on overall population size or density. Even though such population indices are widely used for wildlife monitoring and are generally assumed to be directly proportional to population density, in many field studies this assumption has been found to be invalid [49]. It is a weakness of BRUVS methodology as generally employed that the most commonly used measure, MaxN, does not increase linearly with actual abundance, but tends to saturate [22]. While this would be less of an issue for schooling or non-territorial species, it becomes more of a concern in species which are territorial or keep their distance from each other.

Table 6. A comparison of the values of MaxN reported by authors for BRUVS deployed (A) in the Caribbean Sea and tropical western mid-Atlantic Ocean, and (B) in the Indian Ocean. In (A), the area in the Bahamas was Eleuthera (1: Brooks et al. [10]) and in Belize it was Glover's Reef (6: Bond et al. [11]). (B) included an area in the BMR in the BIOT, Indian Ocean (3: Tickler et al. [12]), MOU74 and Rowley Shoals, E. Australia (4: Meekan et al. [42]), Raja Ampat in Indonesia (5: Jaiteh et al. [46]; 6: Beer [47]), and Houtman Abrolhos Islands, W. Australia (7: Santana et al. [48]). * denotes unfished or protected areas. Catch per unit effort (CPUE) values for the present study were calculated as the shark MaxN per hour on BRUVS.

Shark Species	Ocean	Literature Review							Current Study		
		CPUE	BRUVS N	Country	Years	Depth m	Reference	Area	CPUE	BRUVS N	Years
Carcharhinus limbatus	A	0.003	279	Bahamas	2008–2009	4–15	1	CI	0.01	451	2015–2017
		1.43	200	Belize	2009–2010	10–25	2		0.128	451	2015–2017
Carcharhinus perezi		0.28	200	Belize	2009–2010	10–25	2				
		0.081	279	Bahamas	2008–2009	4–15	1				
Galeocerdo cuvier		0.013	279	Bahamas	2008–2009	4–15	1		0.004	451	2015–2017
Ginglymostoma cirratum		0.215	279	Bahamas	2008–2009	4–15	1		0.178	451	2015–2017
Negaprion brevirostris		0.035	279	Bahamas	2008–2009	4–15	1		0.013	451	2015–2017
Sphyrna mokarran		0.012	279	Bahamas	2008–2009	4–15	1		0.004	451	2015–2017
Carcharhinus albimarginatus	B	0.19	138	Indian Ocean	2012	10–80	3	AI	0.01	210	2015–2017
		0	24–41	E Australia	2003–2004	5–71	4				
		0.306	80	E Australia *			4				
Carcharhinus amblyrhynchos		1.35	138	Indian Ocean	2012	10–80	3		0.168	210	2015–2017
		0.216	58	E Australia	2003–2004	5–71	4				
		0.398	24–41	E Australia *			4				
		0.1	80	Indonesia	2012	1–34	5				
		0.55	80	Indonesia *	2012	1–34	5				
Carcharhinus melanopterus		0.1	138	Indian Ocean	2012	10–80	3		0.328	210	2015–2017
		0.5	80	Indonesia *	2012	1–34	5				
		0.2	80	Indonesia	2012	2–85	5				
		0.3	328	Indonesia	2015	1–34	6				
Galeocerdo cuvier		0.01	138	Indian Ocean	2012	10–80	3		0	210	2015–17
		0	24–41	E Australia	2003–2004	5–71	4				
		0.056	80	E Australia *			4				
		0.007	31	W Australia	2012	35–106	7				
Nebrius ferrugineus		0.15	138	Indian Ocean	2012	10–80	3		0.098	210	2015–2017
		0	24–41	E Australia	2003–2004	5–71	4				
		0.019	80	E Australia *			4				
Negaprion acutidens		0.025	328	Indonesia	2015	2–85	6		0.125	210	2015–2017
Sphyrna mokarran		0.01	138	Indian Ocean	2012	10–80	3		0.003	210	2015–2017
		0.028	24–41	E Australia	2003–2004	5–71	4				
		0.018	80	E Australia *			4				
Triaenodon obesus		0.2	138	Indian Ocean	2012	10–80	3		0.065	210	2015–2017
		0.273	24–41	E Australia	2003–2004	5–71	4				
		0.139	80	E Australia *			4				
		0.025	328	Indonesia	2015	2–85	4				

BRUVS N = number of BRUVS deployed. Years indicate when the study was undertaken. AI = Amirante Islands, CI = Cayman Islands.

In the present study, we sought to address this issue by identifying sharks as individuals and determining the actual number of sharks visiting each BRUVS. As this study confirmed, most individual sharks carry distinctive marks or features allowing them to be identified as individuals. Some marks are natural while others are acquired, with older individuals tending to have more characteristics through injuries and scars than neonates and juveniles. In 12 of the 14 species recorded, at least two-thirds of the observed sharks could be recognised as individuals through such features. In most cases where it was not possible to discern any distinguishing features, this was because the shark's appearance on camera was too brief or too distant. Some species, notably *C. albimarginatus* and *C. leucas*, in which the percentages of distinguishable individuals were the lowest (9.1% and 16.7%, respectively), never closely approached the bait even though they may have been attracted to the vicinity of the BRUVS by olfactory cues. With other species the proportion of individually identified sharks was lower than it might otherwise have been because of high turbidity and poor visibility at some stations. This was especially the case in the sandy lagoon in St. Joseph Atoll, where only a minority of individuals could be identified. In particular, this affected the percentage of individually identifiable *C. melanopterus* (85.3%), which otherwise are readily identified as individuals because of the very variable shape of the black patches on their dorsal fins (see also [50]). In summary, while many sharks can be recognised as individuals with certainty, a proportion of them viewed on BRUVS cannot be distinguished. Thus, in many cases the difference between the true number of visiting sharks and MaxN will be even greater than between MaxN and IndN.

The species for which for most BRUVS IndN was greater than MaxN are those that are largely coastal by habitat. In contrast, more open-water species *C. albimarginatus*, *C. limbatus*, *G. cuvier* and *S. mokarran*, appeared rarely and only as single individuals. Mean IndN was significantly greater than MaxN for each deployment period overall (60, 120, 150 min). Further, in most species, including *C. limbatus*, *C. perezi* and *G. cirratum* in the Cayman Islands and *C. amblyrhynchos*, *C. melanopterus*, *N. ferrugineus* and *T. obesus* in the Amirante Islands, the mean ratio of IndN/MaxN increased with recording period. These findings indicate that in at least these two study areas, for most species an increased deployment period resulted in visits by significantly more individuals, not simply in return visits by the same individuals. Thus, IndN may at times be a more useful index than MaxN, especially for assessing ongoing population trends in the least abundant species, such as *S. mokarran*.

A comparison of the data obtained after different BRUVS deployment periods also showed that while most shark species present might be recorded during the course of survey work using 60 min deployments, nevertheless both the number of species recorded and the numbers of individuals of most species increased with longer BRUVS deployment periods. Interestingly, in a very recent study by Torres et al. [38], a deployment time of 24 h showed that while no additional species were observed after 60 min deployment, additional individuals continued to arrive for a long time. In the present study, the 120 and 150 min deployment periods showed a greater number of both individuals and species, with some individuals of four species taking over 2 h to arrive. An example of a non-shark elasmobranch arriving late during a recording session was a bowmouth guitarfish, *Rhina ancylostoma*, which arrived at a BRUVS only 2 h 17 min 7 s after deployment. Thus we concluded that extended recording periods may be advantageous, provided the fieldwork schedule permits this.

Besides deployment period, the number of sharks attracted to a BRUVS, and hence the precision of monitoring, is also influenced by other factors. Our experience over 10 years suggests that the quality, as well as the quantity, of bait can be important. The oil in bait appears to be a component particularly attractive to sharks, but the oil content can vary greatly not only between bait species, but with the season in which it is caught (relative to fish maturity and reproductive state) for fat content, where they are fished, and with catch storage conditions. For example, the fat content of fished tuna has been reported to vary from 0.4 to 13.5%, and of sardine from 1.2 to 18% [51–54]. The quantity of bait may in our experience be less critical. We used less bait than many researchers, but in a pilot study we found that using over five times the amount of bait used here (1550 g as opposed to 300 g) resulted in only a small, non-significant increase in the numbers of sharks arriving at the BRUVS.

The method of containment of bait on the BRUVS is also important. When, during earlier work, bait was enclosed only in a large mesh-size bag, most or all of it could be quickly removed by sharks, but also by scavenging fishes such as species of snapper (*Lutjanus*), emperor (*Lethrinus*), durgon (*Melichthys*) and octopus. In such circumstances, all of the bait could be consumed in well under an hour, in which case extended deployment would not result in further sharks being recorded. In both studies described here, double bags of fine mesh inside larger mesh were employed, so that the scent of the bait could still disperse readily, while the bait itself was difficult for fishes to extract. Other workers have used perforated cans as bait containers [23,32]. As in any monitoring program, the data obtained from BRUVS-based shark monitoring programs will be more reliable if such considerations are taken into account and all aspects of the method standardised as far as possible.

As well as providing more precise estimates of abundance, the application of photo-identification to BRUVS can provide information on the behaviour of individuals. Re-sightings showed that some individuals of most species moved distances of up to 33 km across the study area. *C. melanopterus*, *G. cirratum* and *N. ferrugineus*, however, were also often re-sighted close to the stations where they were originally recorded. Data from acoustic tagging work undertaken in both study areas and elsewhere have provided evidence that some of the species studied can move over even greater distances than evident from the BRUVS data. For example, one *C. perezi* was recorded moving 125 km from Grand Cayman to Little Cayman and the same distance back again to Grand Cayman [39]. Similarly, in the present study individual *C. amblyrhynchos* in the Amirante Islands were observed on different BRUVS up to 20 km apart. Both long- and short-range movements of this species have also been documented in New Caledonian waters by Bonnin et al. [55], who suggested that the long-range movement of adults was motivated by mating opportunity. In contrast, only a few of the >50 tagged *C. melanopterus* in the Amirante Islands have been recorded crossing the <1 km wide channel between D'Arros and St. Joseph Atoll [56], even though in Polynesia this species has been recorded crossing open water between islands up to 50 km apart [57].

If the low rates of re-sighting that were observed in the present study are typical of such photo-identification work, then the movement data generated will be more limited than can be obtained using acoustic tagging, which frequently records the locations of individuals if they are present within a study area, provided that a dense network of receivers is installed over the area of interest. Information derived from photo-identification work may nevertheless be of value where the considerable funding required to mount an effective acoustic tagging study is unavailable. Furthermore, where BRUVS are deployed on a regular basis within a specific area, photo-identification data can reveal whether individual sharks make frequent use of an area or are resident within it.

It was also anticipated that re-sightings data might allow the application of mark–recapture modelling to estimate the population of different species within the study areas. In practice, the numbers of individuals of even the more frequently recorded species re-sighted on a different BRUVS during the same sampling campaign were low (ranging from 2 to 9%), and the numbers re-sighted in the following year even lower (0.4–2.7%). Encouragingly, however, low rates of re-sighting suggest that known individuals must be mixing with a larger population of other individuals. Thus the model used indicated for the Cayman Islands a mean effective population sizes of 76 ± 23 (S.E.) *C. perezi* and 199 ± 42 (S.E.) *G. cirratum*. These estimates are of interest in light of shark conservation efforts in Cayman, where all shark species have been protected since April 2015 [39,40]. It will be important to determine whether estimated population sizes of these species increase with time. Local opinion surveys indicated that the shark protection legislation had overwhelming public support [39], being opposed by only a small minority of boat-based fishers who may be tempted to ignore it.

There are also two factors which confound the interpretation of the population estimates. Firstly, the calculated population sizes are only of individually identifiable sharks. Following Gore et al. [25], these numbers need to be corrected upwards to allow for the proportions of sharks (of the species concerned) that lacked distinguishing marks. However, a simple proportional correction based on the proportions of individually identifiable sharks may result in an over-estimate, because some of the sightings of non-identifiable sharks will also have been re-sightings. Furthermore, while some sharks could not be identified as individuals because they lacked distinguishing features, others could not be identified as individuals since they did not approach the BRUVS closely enough or because the water was too turbid, and so could in fact have been known individuals.

A second uncertainty relates to the size of area to which these population estimates apply. The three Cayman Islands have a combined terrestrial area of only 264 km^2 and an estimated length of coast of 160 km, most of which is fringed by coral reef. These reefs drop steeply into water >500 m deep within 2 km of the shore. However, *C. perezi*, for example, have been recorded at depths of 150 m or more [58–60], and in the Cayman Islands were recorded moving between Grand Cayman and Little Cayman, islands 125 km apart separated by water > 1000 m deep [39]. Further, it seems from the results of conventional and acoustic tagging studies of sharks on reefs that, while some individuals may be resident or semi-resident in an area, many are only occasionally present in the area or may never be recorded again [39,58,61,62]. Thus, some of the individuals observed in the present study may be foraging and mixing over a much larger region than the 160–320 km^2 shelf area immediately surrounding the individual Cayman Islands. Comparable analysis is underway of the Amirante Islands data, but is confounded by the facts that the study area covers only a third of the archipelago and that the area did not achieve marine protected area status until 2019.

5. Conclusions

The results of the present study demonstrate that the photo-identification of individual sharks can be applied to BRUVS videos and that in most species this approach can result in a greater ability to distinguish between different abundances. This was most evident for species in which there were usually too few sharks for more than one individual to appear on screen at a time. Extended recording periods enhanced this benefit, increasing both the number of shark species and the numbers of individual sharks detected per BRUVS, at least up to 150 min. The application of photo-identification to BRUVS can also provide information about the movement ecology of a species; this is helpful since it indicates for management the appropriate size of the protected area or the foraging range of specific individuals of interest or concern. Importantly, the method clarifies whether within a study area the sharks recorded on BRUVS are frequently the same individuals or usually different ones. Additionally, the method can generate indicative estimates of population size, which can be of value in provoking either concern or reassurance (depending on the numbers) in relation to local shark conservation programs. However, the process of matching images is very time consuming (in the present study the videos from 661 BRUVS were analysed, an undertaking which required about 1500 h of desk time), and the reliable detection of distinguishing features on BRUVS videos requires patience and experience. Sherman et al. [24] similarly noted that identifying individuals does provide better accuracy and additional information, but the determination of actual numbers takes time. MaxN is a simpler and quicker method, and is particularly appropriate for large-scale surveys or routine monitoring. On the other hand, the more sensitive estimates attained by determining IndN are valuable if attempting to monitor population trends of a scarce species or in a smaller area, such as an island or an MPA. IndN is sensitive to differences in abundance that MaxN would not detect. The balance of costs and benefits in applying this approach will depend on the aims and circumstances of any particular project.

6. Research Ethics

The work on endangered shark species was carried out in the Cayman Islands with permission from and on behalf of the Cayman Islands Department of Environment, and, in the Amirantes, under a letter of approval from the Republic of Seychelles Ministry of Environment, Energy and Climate Change.

Author Contributions: Conceptualisation, M.G. and R.O.; Data curation, M.G., J.K. and C.M.; Formal analysis, M.G., R.O., J.K. and C.M.; Funding acquisition, M.G., R.O. and C.C.; Investigation, M.G., R.O., C.C., J.K., C.M. and E.B.; Methodology, M.G., R.O. and E.B.; Project administration, M.G.; Resources, C.C.; Supervision, M.G. and R.O.; Validation, R.O.; Writing—original draft, M.G., R.O. and J.K.; Writing—review and editing, M.G., R.O., C.C., J.K., C.M. and E.B. All authors have read and agreed to the published version of the manuscript.

Funding: The work in the Cayman Islands was supported by a Darwin Initiative grant (DPLUS036) from the UK Department of Environment, Food & Rural Affairs, by the Department of Environment, Cayman Islands, and by the CayBrew Whitetip Shark Fund. Work in the Seychelles was supported by the Marine Research Facility, North Obhur, Jeddah, Saudi Arabia. RO was part-supported through a research chair at King Abdulaziz University, Jeddah, funded by HRH Prince Khalid bin Sultan.

Acknowledgments: We thank the Cayman Islands Department of Environment and all the staff on D'Arros Island, Seychelles, for their support and assistance. We also thank Peter Davies, Martin Eaton, Frida Lara and Robert Nestler for assistance with fieldwork and Tim Austin and James Lea for advice.

Conflicts of Interest: The authors declare no conflict of interest.

Appendix A

Table A1. Comparison between mean IndN and mean MaxN for shark species in (a) the Cayman Islands and (b) the Amirante Islands for 60, 120 and 150 min deployment periods, with differences tested using post hoc Wilcoxon matched pairs test (statistic Z), n = number of cases and p = probability values for Figures 5 and 6.

Area	Species	Deployment Period	n	t	p
	All species	60	451	3.96	<0.0001
		120	447	6.78	<0.0001
		150	228	4.89	<0.0001
	Carcharhinus limbatus	60	368		
		120	364		
		150	168		
	Carcharhinus perezi	60	368	2.01	0.05
		120	364	3.91	0.0001
		150	168	2.90	0.004
(a) Cayman Islands	*Galeocerdo cuvier*	60	368		
		120	363	1.00	0.32
		150	168		
	Ginglymostoma cirratum	60	368	2.86	0.005
		120	364	4.86	<0.00001
		150	168	3.75	0.0002
	Negaprion brevirostris	60	368		
		120	364	1.00	0.32
		150	168		
	Sphyrna mokarran	60	368		
		120	364		
		150	168	1.00	0.32
	All species	60	210	5.95	<0.0001
		120	210	7.48	<0.0001
		150	79	4.99	<0.0001
	Carcharhinus albimarginatus	60	210		
		120	210		
		150	94		
	Carcharhinus amblyrhynchos	60	210	2.35	0.02
		120	210	3.73	0.0002
		150	94	3.32	0.001
	Carcharhinus leucas	60	210	1.00	0.32
		120	210	1.00	0.32
		150	94		
(b) Amirante Islands	*Carcharhinus melanopterus*	60	210	5.31	<0.00001
		120	210	6.20	<0.00001
		150	94	4.94	<0.00001
	Nebrius ferrugineus	60	210	1.42	0.16
		120	210	2.26	0.02
		150	94	1.42	0.16
	Negaprion acutidens	60	210	2.68	0.01
		120	210	3.53	0.001
		150	94	2.16	0.03
	Sphyrna mokarran	60	210		
		120	210		
		150	94		
	Triaeonodon obesus	60	210		
		120	210	1.64	0.10
		150	94	1.65	0.10

Table A2. The ratio of IndN to MaxN for shark species for deployment periods of increasing length (60, 120 and 150 min) in the Cayman Islands and Amirante Islands. n = number of BRUVS included in analyses. IndN = number of individuals identified. MaxN = maximum number of individuals observed on screen at any one time. n decreases with increasing time periods, because not all recordings ran for 120 min, and a minority for 150 min.

Cayman Islands				Amirante Islands			
Species	**Period**	*n*	**Ratio IndN/MaxN**	**Species**	**Period**	*n*	**Ratio IndN/MaxN**
Carcharhinus limbatus	60	451	1	*Carcharhinus albimarginatus*	60	210	1
	120	445	1.11		120	210	1
	150	228	1.2		150	78	1
Carcharhinus perezi	60	451	1.12	*Carcharhinus amblyrhynchos*	60	210	1.19
	120	445	1.21		120	210	1.36
	150	228	1.26		150	78	1.45
Galeocerdo cuvier	60	451	1	*Carcharhinus leucas*	60	210	1.11
	120	445	1		120	210	1.08
	150	228	1		150	78	-
Ginglymostoma cirratum	60	451	1.16	*Carcharhinus melanopterus*	60	210	2.07
	120	445	1.27		120	210	2.45
	150	228	1.34		150	78	2.59
Negaprion brevirostris	60	451	1	*Nebrius ferrugineus*	60	210	1.08
	120	445	1.07		120	210	1.13
	150	228	1		150	78	1.15
Sphyrna mokarran	60	451	1	*Negaprion acutidens*	60	210	1.23
	120	445	1		120	210	1.3
	150	228	1		150	78	1.29
				Sphyrna mokarran	60	210	-
					120	210	1
					150	78	1
				Triaenodon obesus	60	210	1
					120	210	1.15
					150	78	1.27

Table A3. Statistical comparison of pairs of deployment periods of IndN and MaxN for all shark species combined in (a) Cayman Islands and (b) Amirante Islands, using post hoc Wilcoxon matched pairs tests (statistic Z), n = number of cases and p = probability values for Table 3.

Area	Period	IndN			MaxN		
		n	Z	*p*	*n*	Z	*p*
(a) Cayman Islands	60 vs. 120 min	70	7.3	0.00001	64	6.6	<0.00001
	120 vs. 150 min	14	3.3	0.001	14	2.9	0.004
(b) Amirante Islands	60 vs. 120 min	37	5.3	<0.00001	27	4.5	<0.00001
	120 vs. 150 min	5	2.0	0.04	2	1.3	0.2

Table A4. Statistical comparison for pairs of BRUVS deployment periods (60, 120, 150 min) of IndN and of MaxN for each shark species separately in (a) Cayman Islands and (b) Amirante Islands, using post hoc Wilcoxon matched pairs tests (statistic Z), n = number of cases and p = probability values for Table 4.

Area	Shark Species	Period	IndN			MaxN		
			n	Z	p	n	Z	p
(a) Cayman Islands	*Carcharhinus limbatus*	60 vs. 120	3	1.6	0.1			
	Carcharhinus perezi	60 vs. 120	27	4.5	<0.0001	25	4.4	0.00001
		120 vs. 150	6	2.2	0.028	6	2.2	0.03
	Ginglymostoma cirratum	60 vs. 120	39	5.4	<0.0001	33	5	0.00001
		120 vs. 150	8	2.5	0.011	8	1.9	0.06
	Negaprion brevirostris	60 vs. 120	4	1.83	0.07	4	1.8	0.07
(b) Amirante Islands	*Carcharhinus amblyrhynchos*	60 vs. 120	9	2.7	0.008	9	2.7	0.008
		120 vs. 150	2	1.3	0.2	2	1.3	0.2
	Carcharhinus melanopterus	60 vs. 120	7	2.4	0.018	7	2.4	0.018
	Nebrius ferrugineus	60 vs. 120	4	1.8	0.07	4	1.8	0.07
	Negaprion acutidens	60 vs. 120	7	2.4	0.018	7	2.4	0.018
	Triaenodon obesus	60 vs. 120	4	1.8	0.07	4	1.8	0.07

Table A5. The range of distances and the respective time periods taken for each shark species.

Species	Distance (km)	Time (d)
C. amblyrhynchos	2.9–33.0	<1–405
C. melanopterus	0.3–24.5	<1–814
C. perezi	1.2–9.7	181–429
G. cirratum	0.02–15.3	<1–185
G. cuvier	0.38	<1
N. accutidens	0.7–27.5	<1–111
N. brevirostris	1.03	<1
N. ferrugineus	0.5–27.0	<1–2
S. mokarran	2.23	<1
T. obesus	26.91	374

Data availability: 10.6084/m9.figshare.13061546.

References

1. Baum, J.K.; Myers, R.A.; Kehler, D.G.; Worm, B.; Harley, S.J.; Doherty, P.A. Collapse and conservation of shark populations in the Northwest Atlantic. *Science* **2003**, *299*, 389–392. [CrossRef] [PubMed]

2. Robbins, W.D.; Hisano, M.; Connolly, S.R.; Choat, J.H. Ongoing collapse of coral-reef shark populations. *Curr. Biol.* **2006**, *16*, 2314–2319. [CrossRef] [PubMed]

3. Ward-Paige, C.A.; Mora, C.; Lotze, H.K.; Pattengill-Semmens, C.; McClenachan, L.; Arias-Castro, E.; Myers, R.A. Large-scale absence of sharks on reefs in the greater-Caribbean: A footprint of human pressures. *PLoS ONE* **2010**, *5*, e11968. [CrossRef] [PubMed]

4. Clarke, S.C.; McAllister, M.K.; Milner-Gulland, E.J.; Kirkwood, G.P.; Michielsens, C.G.J.; Agnew, D.J.; Pikitch, E.K.; Nakano, H.; Shivji, M.S. Global estimates of shark catches using trade records from commercial markets. *Ecol. Lett.* **2006**, *9*, 1115–1126. [CrossRef] [PubMed]

5. Clarke, S.; Milner-Gulland, E.J.; Bjørndal, T. Social, economic and regulatory drivers of the shark fin trade. *Mar. Resour. Econ.* **2007**, *22*, 305–327. [CrossRef]

6. Dulvy, N.K.; Fowler, S.L.; Musick, J.A.; Cavanagh, R.D.; Kyne, P.M.; Harrison, L.R.; Carlson, J.K.; Davidson, L.N.K.; Fordham, S.V.; Francis, M.P.; et al. Extinction risk and conservation of the world's sharks and rays. *elife* **2014**, *3*, e00590. [CrossRef]

7. Dulvy, N.K.; Baum, J.K.; Clarke, S.; Compagno, L.J.V.; Cortés, E.; Domingo, A.; Fordham, S.; Fowler, S.; Francis, M.P.; Gibson, C.; et al. You can swim but you can't hide: The global status and conservation of oceanic pelagic sharks and rays. *Aquat. Cons.* **2008**, *18*, 459–482. [CrossRef]

8. Robinson, J.P.W.; Robinson, J.; Gerry, C.; Govinden, R.; Freshwater, C.; Graham, N.A.J. Diversification insulates fisher catch and revenue in heavily exploited tropical fisheries. *Sci. Adv.* **2020**, *6*, eaaz0587. [CrossRef]

9. Heupel, M.R.; Papastamatiou, Y.P.; Espinoza, M.; Green, M.E.; Simpfendorfer, C.A. Reef Shark Science—Key Questions and Future Directions. *Front. Mar. Sci.* **2019**, *6*, 12. [CrossRef]

10. Brooks, E.J.; Sloman, K.A.; Sims, D.W.; Danylchuk, A.J. Validating the use of baited remote underwater video surveys for assessing the diversity, distribution and abundance of sharks in the Bahamas. *End. Species Res.* **2011**, *13*, 231–243. [CrossRef]

11. Bond, M.E.; Babcock, E.A.; Pikitch, E.K.; Abercrombie, D.L.; Lamb, N.F.; Chapman, D.D. Abundance in Marine Reserves on the Mesoamerican Barrier Reef. *PLoS ONE* **2012**, *7*, e32983. [CrossRef]

12. Tickler, D.M.; Letessier, T.B.; Koldewey, H.J.; Meeuwig, J.J. Drivers of abundance and spatial distribution of reef-associated sharks in an isolated atoll reef system. *PLoS ONE* **2017**, *12*, e0177374.

13. Langlois, T.; Williams, J.; Monk, J.; Bouchet, P.; Currey, L.; Goetze, J.; Harasti, D.; Huveneers, C.; Ierodiaconou, D.; Malcolm, H.; et al. Marine sampling field manual for benthic stereo BRUVS (Baited Remote Underwater Videos). In *Field Manuals for Marine Sampling to Monitor Australian Waters*; Przeslawski, R., Foster, S., Eds.; National Environmental Science Programme (NESP): Hobart, Australia, 2018; pp. 82–104.

14. Cappo, M.; Stowar, M.; Syms, C.; Johansson, C.; Cooper, T. Fish-habitat associations in the region offshore from James price point—A rapid assessment using baited remote underwater video stations (BRUVS). *J. R. Soc. West. Aust.* **2011**, *94*, 303–321.

15. Harvey, E.S.; McLean, D.L.; Frusher, S.; Haywood, M.D.E.; Newman, S.J.; Williams, A. *The Use of Bruvs as a Tool for Assessing Marine Fisheries and Ecosystems: A Review of the Hurdles and Potential*; Fisheries Research and Development Corporation, and the University of Western Australia: Crawley, Australia, 2013; ISBN 978-1-74052-265-6.

16. Ashworth, J.S.; Ormond, R.F.G. Effects of fishing pressure and trophic group on abundance and spillover across boundaries of a no-take zone. *Biol. Conserv.* **2005**, *121*, 333–344. [CrossRef]

17. Hill, J.; Wilkinson, C. *Methods for Ecological Monitoring of Coral Reefs*; Australian Institute of Marine Science: Townsville, Australia, 2004; p. 117.

18. McCauley, D.J.; McLean, K.A.; Bauer, J.; Young, H.S.; Micheli, F. Evaluating the performance of methods for estimating the abundance of rapidly declining coastal shark populations. *Ecol. Appl.* **2012**, *22*, 385–392. [CrossRef]

19. Bassett, D.K.; Montgomery, J.C. Investigating nocturnal fish populations in situ using baited underwater video: With special reference to their olfactory capabilities. *J. Exp. Mar. Biol. Ecol.* **2011**, *409*, 194–199. [CrossRef]

20. Campbell, M.D.; Pollack, A.G.; Gledhill, C.T.; Switzer, T.S.; DeVries, D.A. Comparison of relative abundance indices calculated from two methods of generating video count data. *Fish. Res.* **2015**, *170*, 125–133. [CrossRef]

21. Schobernd, Z.H.; Bachele, N.M.; Conn, P.B. Examining the utility of alternative video monitoring metrics for indexing reef fish abundance. *Can. J. Fish. Aquat. Sci.* **2014**, *71*, 464–471. [CrossRef]

22. Stobart, B.; Diaz, D.; Alvarez, F.; Alonso, C.; Mallol, S.; Goni, R. Performance of Baited Underwater Video: Does It Underestimate Abundance at High Population Densities? *PLoS ONE* **2015**, *10*, e0127559. [CrossRef] [PubMed]

23. Irigoyen, A.J.; De Wysiecki, A.M.; Trobbiani, G.; Bovcon, N.; Awruch, C.A.; Argemi, F.; Jaureguizar, A.J. Habitat use, seasonality and demography of an apex predator: Sevengill shark *Notorynchus cepedianus* in northern Patagonia. *Mar. Ecol. Prog. Ser.* **2018**, *603*, 147–160. [CrossRef]

24. Sherman, C.S.; Chin, A.; Heupel, M.R.; Simpfendorfer, C.A. Are we underestimating elasmobranch abundances on baiter remote underwater video systems (BRUVS) using traditional metrics? *J. Exp. Mar. Biol. Ecol.* **2018**, *503*, 80–85. [CrossRef]

25. Gore, M.A.; Frey, P.H.; Ormond, R.F.; Allan, H.; Gilkes, G. Use of Photo-Identification and Mark-Recapture methodology to assess basking shark (*Cetorhinus maximus*) populations. *PLoS ONE* **2016**, *11*, e0150160. [CrossRef]

26. Anderson, S.D.; Goldman, K.J. Photographic evidence of white shark movements in California waters. *Calif. Fish Game* **1996**, *82*, 182–186.

27. Mourier, J.; Vercelloni, J.; Planes, S. Evidence of social communities in a spatially structured network of a free-ranging shark species. *Anim. Behav.* **2012**, *83*, 389–401. [CrossRef]

28. Devine, B.M.; Wheeland, L.J.; Fisher, J.A.D. First estimates of Greenland shark (*Somniosus microcephalus*) local abundances in Arctic waters. *Sci. Rep.* **2018**, *8*, 974. [CrossRef]

29. Castro, A.L.F.; Rosa, R.S. Use of natural marks on population estimates of the nurse shark, *Ginglymostoma cirratum*, at Atol das Rocas Biological Reserve, Brazil. *Environ. Biol. Fishes* **2005**, *72*, 213–221. [CrossRef]

30. Buray, N.; Mourier, J.; Planes, S.; Clua, E. Underwater photo-identification of sicklefin lemon sharks, *Negaprion acutidens*, at Moorea (French Polynesia). *Cybium* **2009**, *33*, 21–27.

31. Meekan, M.G.; Bradshaw, C.J.; Press, M.; McLean, C.; Richards, A.; Quasnichka, S.; Taylor, J.G. Population size and structure of whale sharks *Rhincodon typus* at Ningaloo Reef, Western Australia. *Mar. Ecol. Prog. Ser.* **2006**, *319*, 275–285. [CrossRef]

32. Rowat, D.; Speed, C.W.; Meekan, M.G.; Gore, M.A.; Bradshaw, C.J. Population abundance and apparent survival of the vulnerable whale shark *Rhincodon typus* in the Seychelles aggregation. *Oryx* **2009**, *43*, 591–598. [CrossRef]

33. Benjamins, S.; Dodd, J.; Thorburn, J.; Milway, V.A.; Campbell, R.; Bailey, D.M. Evaluating the potential of photo-identification as a monitoring tool for flapper skate (*Dipturus intermedius*). *Aquat. Cons.* **2018**, *28*, 1360–1373. [CrossRef]

34. Harasti, D.; Lee, K.A.; Laird, R.; Bradford, R.; Bruce, B. Use of stereo baited remote underwater video systems to estimate the presence and size of white sharks (*Carcharodon carcharias*). *Mar. Fresh. Res.* **2016**, *68*, 1391–1396. [CrossRef]

35. Langlois, T.J.; Harvey, E.S.; Fitzpatrick, B.; Meeuwig, J.J.; Shedrawi, G.; Watson, D.L. Cost-efficient sampling of fish assemblages: Comparison of baited video stations and diver video transects. *Aquat. Biol.* **2010**, *9*, 155–168. [CrossRef]

36. Boussarie, G.; Bakker, J.; Wangensteen, O.S.; Mariani, S.; Bonnin, L.; Juhel, J.-B.; Kiszka, J.J.; Kulbicki, M.; Manel, S.; Robbins, W.D.; et al. Environmental DNA illuminates the dark diversity of sharks. *Sci. Adv.* **2018**, *4*, eaap9661. [CrossRef]

37. Asher, J. A Deeper Look at Hawaiian Coral Reef Fish Assemblages: A Comparison of Survey Approaches and Assessments of Shallow to Mesophotic Communities. Ph.D. Thesis, Curtin University, Perth, Australia, April 2017.

38. Torres, A.; Abril, A.-M.; Clue, E. A time-extended (24 h) Baited Remote Underwater Video (BRUV) for monitoring pelagic and nocturnal marine species. *J. Mar. Sci. Eng.* **2020**, *8*, 208. [CrossRef]

39. Ormond, R.; Gore, M.; Bladon, A.; Dubock, O.; Kohler, J.; Millar, C. Protecting Cayman Island Sharks: Monitoring, Movement and Motive. In Proceedings of the 69th Gulf and Caribbean Fisheries Institute, Grand Cayman, Cayman Islands, 7–11 November 2016.

40. Cayman Islands Government. The National Conservation Law, 2013 (Law 24 of 2013). Supplement No. 1 published with Extraordinary Gazette No. 9, 5 February 2014. Available online: https://legislation.gov.ky/cms/images/LEGISLATION/PRINCIPAL/2013/2013-0024/NationalConservationLaw_Law%2024%20of%202013.pdf (accessed on 6 October 2020).

41. OpenStreetMap Foundation. OpenStreetMap. Available online: https://www.openstreetmap.org/copyright (accessed on 6 October 2020).

42. Meekan, M.; Cappo, M.; Carleton, J.; Marriott, R. *Surveys of Shark and Fin-Fish Abundance on Reefs within the Mou74 Box and Rowley Shoals Using Baited Remote Underwater Video Systems*; Australian Institute of Marine Science: Townsville QLD, Australia, 2006; Dewey Number 333.956311.

43. Seber, G.A.F. *The Estimation of Animal Abundance and Related Parameters*, 2nd ed.; Griffin: London, UK, 1982.

44. Cliff, G.; van der Elst, R.P.; Govender, A.; Witthuhn, T.K.; Bullen, E.M. First Estimates of Mortality and Population Size of White Sharks on the South African Coast. In *Great White Sharks: The Biology of Carcharodon Carcharias*; Klimley, A.P., Ainley, D.G., Eds.; Academic Press: San Diego, CA, USA, 1996; pp. 393–400.

45. Chapman, D.G. *Some Properties of the Hypergeometric Distribution with Applications to Zoological Censuses*; University of California Publications in Statistics: Dordrecht, The Netherlands, 1951.

46. Jaiteh, V.F.; Lindfield, S.J.; Mangubhai, S.; Warren, C.; Fitzpatrick, B.; Loneragan, N.R. Higher Abundance of Marine Predators and Changes in Fishers' Behavior Following Spatial Protection within the World's Biggest Shark Fishery. *Front. Mar. Sci.* **2016**, *3*, 43. [CrossRef]

47. Beer, A.J.E. Diversity and Abundance of Sharks in no-Take and Fished Sites in the Marine Protected aRea Network of Raja Ampat, West Papua, Indonesia, using Baited Remote Underwater Video (BRUVs). Master's Thesis, Royal Roads University, Victoria, BC, Canada, February 2015.

48. Santana-Garcon, J.; Braccini, M.; Langlois, T.J.; Newman, S.J.; McAuley, R.B.; Harvey, E.S. Calibration of pelagic stereo-BRUVs and scientific longline surveys for sampling sharks. *Methods Ecol. Evol.* **2014**, *5*, 824–833. [CrossRef]

49. Pollock, K.H.; Nichols, J.D.; Simons, T.R.; Farnsworth, G.L.; Bailey, L.L.; Sauer, J.R. Large scale wildlife monitoring studies: Statistical methods for design and analysis. *Environmetrics* **2002**, *13*, 105–119. [CrossRef]

50. Porcher, I.F. On the gestation period of the blackfin reef shark, *Carcharhinus melanopterus*, in waters off Moorea, French Polynesia. *Mar. Biol.* **2005**, *146*, 1207–1211. [CrossRef]

51. Balcik-Misir, G.; Tufan, B.; Köse, S. Monthly variation of total lipid and fatty acid contents of Atlantic bonito, Sarda sarda (Bloch,1793) of Black Sea. *Int. J. Food Sci. Technol.* **2014**, *49*, 2668–2677. [CrossRef]

52. Mahaliyana, A.; Jinadasa, B.; Liyanage, N.; Jayasinghe, G.; Jayamanne, S. Nutritional composition of skipjack tuna (Katsuwonus pelamis) caught from the oceanic waters around Sri Lankae. *Am. J. Food Nutr.* **2015**, *3*, 106–111.

53. Bagthasingh, C.; Aran, S.; Vetri, V.; Innocen, A.; Kannaiyan, S. Seasonal variation in the proximate composition of sardine (*Sardinella gibbosa*) from Thoothukudi coast. *J. Food Compos. Anal.* **2016**, *49*, 9–18.

54. Gámez-Meza, N.; Higuera-Ciapara, I.; Calderson de la Barca, A.; Vásquez-Moreno, L.; Noriega-Rodrìguez, J.; Angulo-Guerrero, O. Seasonal variation in the fatty acid composition and quality of sardine oil from *Sardinops sagax caeruleus* of the Gulf of California. *Lipids* **1999**, *34*, 639–642. [CrossRef]

55. Bonnin, L.; Robbins, W.D.; Boussarie, G.; Kiszka, J.J.; Dagorn, L.; Mouillot, D.; Vigliola, L. Repeated long-range migrations of adult males in a common Indo-Pacific reef shark. *Coral Reefs* **2019**, *38*, 1121–1132. [CrossRef]

56. Lea, J.S.E.; Humphries, N.E.; von Brandis, R.G.; Clarke, C.R.; Sims, D.W. Acoustic telemetry and network analysis reveal the space use of multiple reef predators and enhance marine protected area design. *Proc. R. Soc. B* **2016**, *283*, 20160717. [CrossRef]

57. Mourier, J.; Planes, S. Direct genetic evidence for reproductive philopatry and associated fine-scale migrations in female blacktip reef sharks (*Carcharhinus melanopterus*) in French Polynesia. *Mol. Ecol.* **2013**, *22*, 201–214. [CrossRef]

58. Chapman, D.D.; Pikitch, E.K.; Babcock, E.A.; Shivji, M.S. Marine reserve design and evaluation using automated acoustic telemetry: A case-study involving coral reef associated sharks in the Mesoamerican Caribbean. *Mar. Technol. Soc. J.* **2005**, *39*, 42–53. [CrossRef]

59. Chapman, D.D.; Pikitch, E.K.; Babcock, E.A.; Shivji, M.S. Deep-diving and diel changes in vertical habitat use by Caribbean reef sharks *Carcharhinus perezi*. *Mar. Ecol. Prog. Ser.* **2007**, *344*, 271–275. [CrossRef]

60. Clarke, C.; Lea, J.S.E.; Ormond, R.F.G. Reef-use and residency patterns of a baited population of silky sharks, *Carcharhinus falciformis*, in the Red Sea. *Mar. Fresh. Res.* **2011**, *62*, 668–675. [CrossRef]

61. Heupel, M.R.; Simpfendorfer, C.A.; Fitzpatrick, R. Large–scale movement and reef fidelity of grey reef sharks. *PLoS ONE* **2010**, *5*, e9650. [CrossRef]

62. Vianna, G.M.S.; Meekan, M.G.; Meeuwig, J.J.; Speed, C.W. Environmental Influences on Patterns of Vertical Movement and Site Fidelity of Grey Reef Sharks (*Carcharhinus amblyrhynchos*) at Aggregation Sites. *PLoS ONE* **2013**, *8*, e60331. [CrossRef]

Publisher's Note: MDPI stays neutral with regard to jurisdictional claims in published maps and institutional affiliations.

Article

Use of Polyphosphates and Soluble Pyrophosphatase Activity in the Seaweed *Ulva pseudorotundata*

Juan J. Vergara *, Patricia Herrera-Pérez, Fernando G. Brun and José Lucas Pérez-Lloréns

Departamento de Biología, Facultad de Ciencias del Mar y Ambientales, Universidad de Cádiz, E-11510 Puerto Real, Cádiz, Spain; patricia.herrera@uca.es (P.H.-P.); fernando.brun@uca.es (F.G.B.); joselucas.perez@uca.es (J.L.P.-L.)
* Correspondence: juanjose.vergara@uca.es

Received: 1 October 2020; Accepted: 14 December 2020; Published: 16 December 2020

Abstract: The hydrolytic activity of different types of polyphosphates, and the induction of soluble pyrophosphatase (sPPase; EC 3.6.1.1) activity have been assessed in cell extracts of nutrient limited green seaweed *Ulva pseudorotundata* Cormaci, Furnari & Alongi subjected to different phosphorus regimes. Following a long period of nutrient limitation, the addition of different types of (poly)phosphates to artificial seawater enhanced growth rates on fresh weight and area, but not on dry weight bases. Chlorophyll and internal P content were affected by P supply. In contrast, internal soluble reactive P was kept low and was little affected by P additions. Soluble protein content increased in all treatments, as ammonium was added to prevent N limitation. The C:N:P atomic ratio revealed great changes depending on the nutrient regime along the experiment. Cell extracts of *U. pseudorotundata* were capable of hydrolyzing polyphosphates of different chain lengths (pyro, tripoly, trimeta, and polyphosphates) at high rates. The sPPase activity was kept very low in P limited plants. Following N and different kind of P additions, sPPase activity was kept low in the control, but slightly stimulated after 3 days when expressed on a protein basis. The highest activities were found at the end of the experiment under pyro and polyphosphate additions (7 days). The importance of alternative P sources to phosphate and the potential role of internal soluble pyrophosphatases in macroalgae are discussed.

Keywords: phosphorus; polyphosphates; pyrophosphate; pyrophosphatase activity; seaweed; *Ulva*

---

1. Introduction

Although nitrogen (N) is considered to be the main limiting nutrient in the ocean [1], phosphorus (P) can also limit primary production of algae in some coastal areas [2]. The major P form for algae is inorganic ortho-phosphate (Pi), whose uptake in seaweeds occurs through specific transport systems [3–5]. Besides the main Pi source (i.e., from the water column and sediments), the hydrolysis of organic P monoesters by the periplasmic enzyme alkaline phosphatase also renders Pi [6]. However, although Pi is the main form of P used by algae, their growth can also be supported by more complex inorganic P compounds [7]. Accordingly, pyrophosphates (PPi) and polyphosphates (polyP) could be important sources of P [8]. However, these compounds are not molybdenum sensitive, not being quantified in standard analytical measurements of soluble reactive phosphorus (SRP) [9,10]. These Pi polymers, which can contribute to the P pool in the coastal ocean, are mostly derived from anthropogenic activities [9]. However, since P is taken up as Pi, enzyme cleavage of Pi polymers at the cell surface is needed. Data on the role of these alternative sources of P in seaweeds are scarce.

In contrast to P uptake kinetics and transmembrane transport mechanisms, little is known about P internal cellular metabolism. Marine algae, as opposed to most terrestrial plants, are able to store Pi polymers as polyP [2,8]. Studies of ^{31}P-NMR on the chlorophyte *Ulva lactuca*, identified polyP as the main intracellular storage for P [11]. This polyP pool was accumulated under external Pi availability, while it was mobilized under P limitation. The PolyP polymers have also been reported in other *Ulva* spp., in the Rhodophyta *Ceramium* sp. [11] and in *Porphyra purpurea* [12], among other species.

Soluble pyrophosphatase (sPPase) (EC 3.6.1.1) is a ubiquitous enzyme essential for cell anabolism [13]. It is located in cellular organelles (plastids and mitochondria) of a variety of photosynthetic autotrophs [14], where many anabolic and biosynthetic reactions occur [15]. This enzyme is responsible for PPi cleavage and Pi regeneration in vivo, thus allowing biosynthetic reactions to proceed [16]. Its function is also essential to replenish Pi for phosphorylation [17]. Despite the importance of this enzyme, the few data available are mostly from photosynthetic prokaryotes and some microalgal groups [14], whereas studies in seaweeds are much scarcer, e.g., [7].

Under this framework, the aim of this work was double: (1) to assess the capacity of cell extracts of the green macroalgae *U. pseudorotundata* to hydrolyze Pi polymers, and (2) to study the effect of different P sources (i.e., Pi, PPi and PolyP) on the response of nutrient-limited algae (growth, internal C:N:P composition, chlorophyll, protein content and internal sPPase activity). Results indicate the capacity of *U. pseudorotundata* to exploit alternative sources of P for growth, as well as the induction of internal sPPase activity in cell extracts of nutrient-limited specimens following P additions.

2. Materials and Methods

2.1. Plant Material and Preculture Conditions

Seaweed specimens, identified as *Ulva pseudorotundata* Cormaci, Furnari & Alongi, according to [18], were collected from earthern pools at an intensive fish farm (Acuinova S. L., San Fernando, Cádiz, southern Spain). Once in the laboratory, thalli were rinsed with seawater and cleaned of mud and epiphytes. Some of these field-collected specimens were used both, to optimize sPPase assays and to determine the hydrolytic activities of different kinds of polyphosphates in cell extracts (see below). Pretreatment: the remaining biomass was grown in aquaria with constant aeration and seawater pumped from a natural underground well (pH 7.3, salinity 39.6, 35 μM NH_4^+, 0.8 μM PO_4^{3-}, and negligible NO_3^- levels, mean values) for 70 days before starting the experiments (see below). Lighting (145 μmol photons m^{-2} s^{-1}, which is saturating for this species [19]) was provided with two fluorescent lamps (Phillips cool white) in a 12 h:12 h light-dark photoperiod. Mean pH and temperature were 8.4 and 25.1 °C, respectively, along this pre-treatment phase. Culture was kept at a high biomass density (16 g FW L^{-1}) in a total volume of 20 L, and seawater was renewed weekly. The aquarium was cleaned at every seawater change with distilled deionised water plus sodium hypochlorite. Germanium dioxide (2 mg L^{-1}, final concentration) was added to inhibit diatom growth [20]. The objective of this long preculture period at a high biomass density was to obtain nutrient-depleted thalli (i.e., with very low N and P quota), despite there not being a complete absence of PO_4^{3-} and NH_4^+ in seawater. In fact, the growth rate was very low along this period (data not shown) and N and P quotas approached the subsistence ones for this species (see initial conditions in results), with a N:P atomic ratio of about 15, close to the Redfield ratio, indicating an incipient N and P co-limitation for *Ulva* species [21].

2.2. Experimental Design

The N and P-limited thalli were grown under four different P treatments: (1) without P (control), (2) with orthophosphate (Pi, 5 μM final concentration), (3) with pyrophosphate (PPi, 5 μM final concentration) and (4) with polyphosphate (PolyP, 5 μM final concentration). Inorganic N (as NH_4^+, 5 μM final concentration) was also added to all treatments (including the control) to avoid a parallel N co-limitation. All chemical reagents for P analytics were from Sigma, and polyphosphate used is referred to as "phosphate glass", with a mean length chain of 13–18 Pi molecules. Thallus discs (2 cm diameter) were punched with a cork borer and grown in 1 L flasks (see below) within an incubation chamber (Koxka model EC−540-F) at 20 °C and at a photon flux of 200 μmol m^{-2} s^{-1} in a 12 h:12 h light-dark photoperiod provided by fluorescent tubes (cool white, Phillips). Each of the four treatments (*n* = 3 per treatment) were run for seven days in 1 L-aerated flasks (12 in total) at a biomass density about 1 g FW L^{-1} (c.a. 20 discs per flask) with synthetic seawater [22], which allowed us to control nutrient availability. During the experiment, seawater and *Ulva* samples were taken at days 0, 2, 4 and 7. Seawater was renewed at days 2 and 4.

2.3. Analyses

Three discs were randomly sampled in each of the 12 flasks, measured and weighted (fresh and dry, after drying in an oven at 60 °C for 48 h) to estimate growth rates assuming an exponential growth model. Fresh weight/area (g FW m^{-2}) and DW/FW ratios were also estimated. Total chlorophyll ($a + b$) concentration was determined according to [23], after extracting liposoluble pigments in N, N, dimethyl formamide for 24 h at 4 °C in darkness. Internal C and N concentrations were determined from dried biomass at the beginning and at the end of the experiment in a Perkin-Elmer 240-C elemental analyser. The time course for the C:N ratio was not monitored since there was not enough algae material to be collected. The SRP was determined according to [24]. Total internal P was also determined according to [24], after acid digestion in an autoclave (121 °C, one hour, Raypa Sterilmatic AE-150). In both cases, absorbance was measured at 885 nm (Hitachi, U-1100 spectrophotometer, Tokyo, Japan).

Cell extracts were prepared by grinding *Ulva* discs with liquid N$_2$ with a mortar and a pestle. Cell extract was resuspended in 50 mM Tris buffer pH 7.5 at an approximate ratio of 100 mg FW in 1.5 mL buffer. Tris buffer was supplemented with 1 mM DTT and 1 mM PMSF to protect active enzyme groups from oxidation and to inhibit proteases. Cell extracts were centrifuged at 15,000× g for 15 min at 4 °C. Cell debris and membrane fractions were discarded and soluble components were recovered, frozen in liquid N$_2$, and stored at −80 °C for enzyme assays. An aliquot of cell extracts was taken for soluble protein determination, to scale enzyme activities. Soluble protein content of cell extracts was determined according to [25].

The sPPase activity was measured in cell extracts according to [26], monitoring the Pi production at 25 °C for 10 min. Assay medium contained Tris buffer 50 mM pH 8.0, 20 mM of substrate (either PPi, tripolyP, trimetaP or polyP depending on the treatment), 20 mM Mg^{2+}, and an appropriate quantity of cell extract. The ion Mg^{2+} is an essential cofactor for sPPase activity for most of the assayed organisms [14]. In preliminary tests, a 200 μL extract was assayed in a total volume of 1 mL, and 500 μL of the assay medium was assayed for Pi production. Given the high activity of the enzyme, 50 μL cell extract was used instead of 200 μL, and 100 μL of the assay medium was diluted with 400 μL Tris buffer pH 8.0 before measuring Pi, to reduce the sensibility of the assay, yielding a dilution factor of 20× in comparison to preliminary assays. Immediately, an aliquot was taken and mixed with ammonium molybdate and ascorbic acid to measure Pi production.

A unit of sPPase activity was defined as the quantity of enzyme activity that yields 1 μmol Pi per minute in the assay conditions stated above. This activity was scaled either to proteins or DW. In all enzyme assays two blanks were run in parallel: (1) in absence of the substrate (PPi), which indicates the amount of internal SRP in the sample, and (2) in the absence of cell extract, to determine spontaneous PPi hydrolysis and/or SRP contamination of the substrate. These assays were done with 200 μL extract, and no further dilutions were performed for the measurements. Internal SRP values were quite low compared to the rate of Pi production by the enzyme activity, and the spontaneous hydrolysis was almost undetectable (data not shown).

2.4. Statistics

The effect of P source and time on thallus area, fresh weight:area, dry weight:fresh weight ratio, internal SRP, internal total phosphorus, soluble protein and sPPase activity were analyzed by a 2-way ANOVA (Table A1). Prior to any statistical analysis, data were checked for normality (Shapiro–Wilk normality test) and homocedasticity (Bartlett test of homogeneity of variances test). The effects of different P sources on enzymatic activity and different P sources on C:N:P tissue composition were analyzed by means of 1 way ANOVA (significant Tukey's test differences indicated in Tables 1 and 2).

Post hoc comparisons among means were tested by the Tukey test [27]. In all cases, the null hypothesis was rejected at the 5% significance level and data were presented as mean ± SE.

Table 1. *Ulva pseudorotundata*. Enzyme soluble hydrolytic activities for different polyphosphate substrates in cell extracts of field collected samples. Data are expressed as mean (n = 3–4) ± SE. Different letters indicate significant differences among treatments at α = 0.05 following Tukey's test and 1 way ANOVA.

Substrate	Enzyme Activity (U·mg prot^{-1})
Pyrophosphate	4.71 ± 0.10 [a]
Tripoliphosphate	4.87 ± 0.10 [a]
Trimetaphosphate	2.12 ± 0.01 [b]
Polyphosphate	16.94 ± 0.11 [c]

Table 2. *Ulva pseudorotundata*. Internal C:N:P composition at the beginning of the experiment and after 7 days under different phosphorus treatments. Data are expressed as mean (n = 3) ± SE. Different letters indicate significant differences among treatments at α = 0.05 following Tukey's test and 1 way ANOVA.

	Initial	Control	Phosphate	Pyrophosphate	Polyphosphate
Carbon (%DW)	22.7 ± 0.9 [a]	25.2 ± 0.4 [b]	27.1 ± 0.1 [b]	27.6 ± 0.3 [b]	27.3 ± 0.3 [b]
Nitrogen (%DW)	1.16 ± 0.03 [a]	3.65 ± 0.09 [b]	4.57 ± 0.10 [c]	4.81 ± 0.13 [c]	4.04 ± 0.12 [b]
Phosphorus (%DW)	0.17 ± 0.01 [a]	0.17 ± 0.01 [a]	0.32 ± 0.02 [b]	0.43 ± 0.01 [c]	0.40 ± 0.01 [c]
C:N (by atoms)	22.8 ± 0.5 [a]	8.08 ± 0.32 [b]	6.94 ± 0.13 [b]	6.68 ± 0.12 [b]	7.91 ± 0.13 [b]
C:P (by atoms)	349 ± 13 [a]	387 ± 9 [a]	220 ± 14 [b]	166 ± 7 [b]	176 ± 4 [b]
N:P (by atoms)	15.3 ± 0.5 [a]	48.0 ± 1.3 [b]	31.8 ± 2.2 [b]	25.0 ± 1.5 [bc]	22.2 ± 0.8 [c]

3. Results

The hydrolytic capacity of several polyphosphate compounds assayed with soluble cell extracts from *Ulva pseudorotundata*, as a first approach, is shown in Table 1. Hydrolytic activities were recorded for all the substrates tested. Both sPPase and tripolyphosphatase activities showed similar values, whereas the cyclic trimetaphosphate was hydrolysed at a lower rate. In contrast, polyphosphatase activity was 3–4 times higher than that of the sPPase.

Ulva pseudorotundata was maintained for a long period under nutrient limitation (preculture) and, subsequently, grown under different P sources, but with NH_4^+ addition to avoid N-limitation. Seaweed growth rates were affected by P treatments, but their values depended largely on the plant attributes used in the calculations (Figure 1a). Thus, normalization by fresh weight (FW) resulted in higher values, with those obtained under different P additions being higher than the controls. In contrast, growth rates were minimal when normalized by dry weight (DW), being unaffected by P treatments. On area basis, the growth rates were also lower than those estimated on a FW basis, but were affected by P additions. The time-course of *U. pseudorotundata* disc area grown under different P treatments is shown in Figure 1b. A positive thallus expansion was observed regardless of P treatment, with the highest values under PPi and polyP enrichment, and the lowest in the control. As a consequence of the different water gain and area expansion, the FW/area ratio increased along time in all treatments (Figure 2a). In contrast, the DW/FW ratio showed the opposite trend, decreasing along time in all the treatments (Figure 2b).

Total chlorophyll content was also affected by P sources (Figure 3), with the lowest values recorded in the controls. Chlorophylls *a* and *b* displayed the same trend, as the chlorophyll *a/b* ratio was constant along the experiment (data not shown).

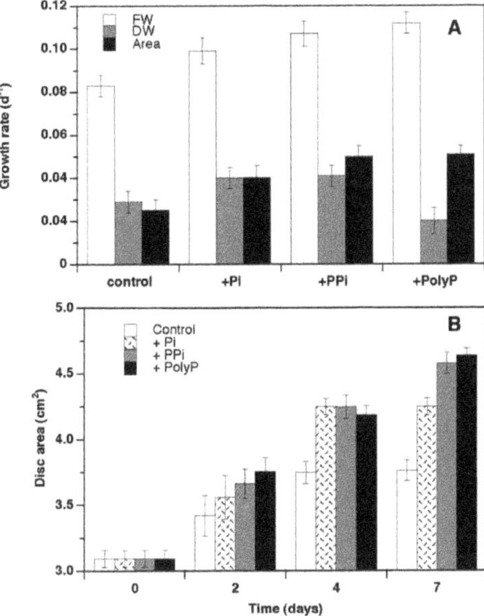

Figure 1. *Ulva pseudorotundata.* (**A**) Specific growth rates on fresh weight (FW), dry weight (DW) and area basis of *seaweed* discs cultured under different P regimes (control without P, Pi, PPi, and PolyP) for 7 days. Data are means ± SE (*n* = 3 independent cultures). (**B**) Disc area of *seaweed* discs along the experiment as a function of the different P regimes. Data are means ± SE (*n* = 6–9 discs).

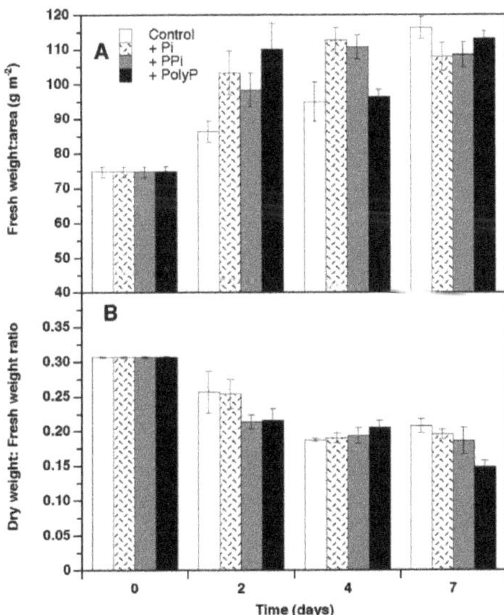

Figure 2. *Ulva pseudorotundata.* (**A**) Time course of fresh weight/area ratio and (**B**) dry weight/fresh weight ratio, as a function of different P regimes. Data are means ± SE (*n* = 6–9 discs).

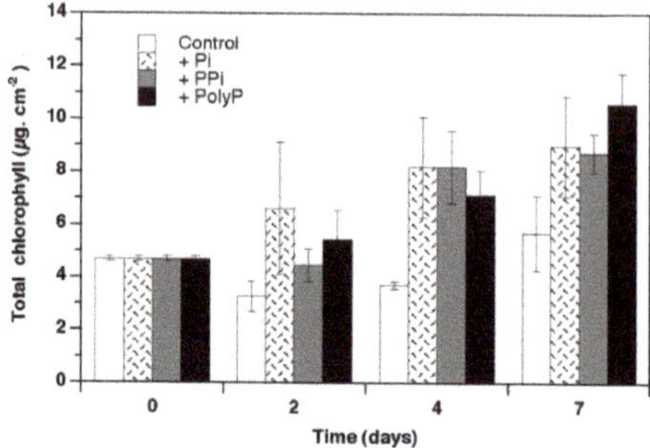

Figure 3. *Ulva pseudorotundata*. Time course of total ($a + b$) chlorophyll content as a function of different P regimes. Data are means $\pm$ SE ($n = 3$).

Whereas total internal P content was unaffected in the controls along time, it doubled the initial values under different P sources, especially for PPi and PolyP (Figure 4a). In contrast, internal SRP content was kept at quite low levels, (1–4 μmol Pi g^{-1} DW). Although a transient increase after 2 days was recorded for the PPi treatment, all values were fairly constant along the experiment in all treatments (Figure 4b).

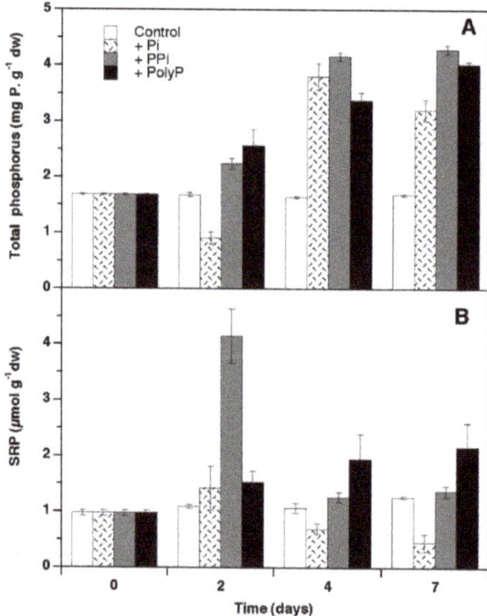

Figure 4. *Ulva pseudorotundata*. (**A**) Time course of total internal P and (**B**) internal soluble reactive P (SRP), as a function of different P regimes. Data are means $\pm$ SE ($n = 3$).

Seawater SRP was also monitored along the experiment (Figure 5). At the beginning, it was undetectable in both the control and the PPi treatment, but values of 7.5 μM and 0.5 μM were recorded for Pi treatment and PolyP treatments, respectively, indicating, in the latter case, the possibility of a certain degree of Pi contamination in the added substrate (added PolyP was about 5 μM). Significant Pi levels were detected in PolyP-enriched cultures, probably as a consequence of the partial hydrolysis of long chain Pi molecules. In contrast, Pi was not detected in either Pi or PPi treatments, except for a high value in the PPi treatment on the fourth day. To check for the spontaneous hydrolysis of PPi and PolyP in seawater, 1 L aerated flasks were maintained without algae for 3 days, and SRP from PPi and polyP hydrolysis was measured. Values ranged from 0.06 to 0.16 µM Pi (almost undetectable) for PPi and from 0.50 µM to 0.66 µM for polyP, indicating the lack of a significant polyP hydrolysis and suggesting Pi contamination in the polyP substrate used for enrichment.

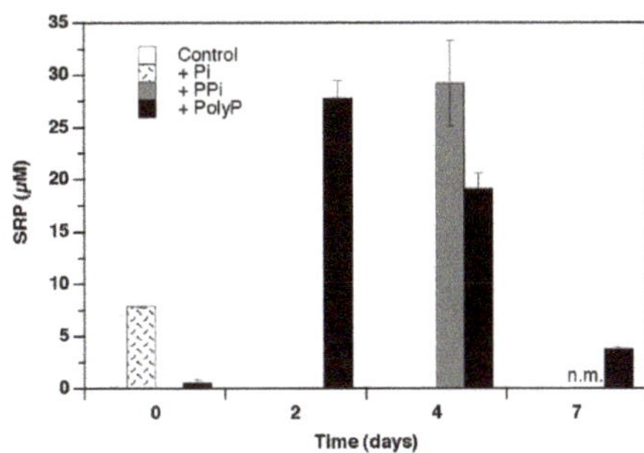

Figure 5. *Ulva pseudorotundata*. Time course of SRP in seawater, as a function of different P regimes. Data are means ± SE (*n* = 3).

In contrast to total internal P content, the soluble protein content increased along the experiment regardless of the P treatment as expected, since all cultures were supplemented with N (Figure 6). This enhancement was especially noticeable after 4 days.

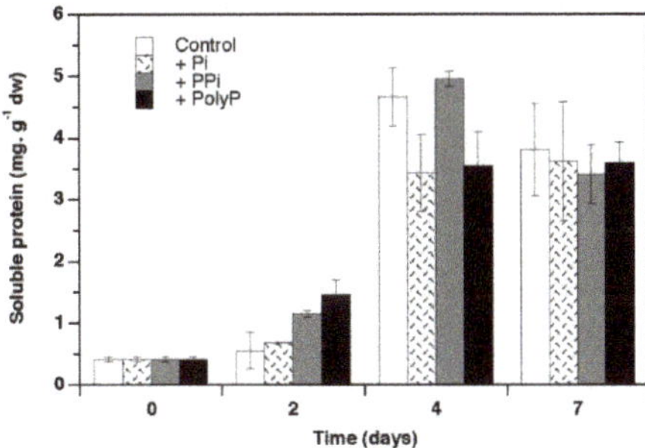

Figure 6. *Ulva pseudorotundata*. Time course of soluble protein content as a function of different P regimes. Data are means ± SE (*n* = 3).

Tissue C, N and P composition, as well as the C:N, C:P and N:P atomic ratios are shown in Table 2. These data showed great variations caused by different P supplies. In parallel with thallus expansion and growth, internal C content increased at the end of the incubations. Tissue N content increased by 3–4 times at the end of the experiment, with slightly higher values in the P enriched treatments than in the control. Tissue P content almost doubled the initial values in all P treatments, whereas no changes were observed in the control. As a consequence of the noticeable changes in N and P composition, and, to a lesser extent in C, C:N:P atomic ratios displayed great variations along the experiment. Thus, the C:N atomic ratio dropped from 22.8 down to about 8 in the control, and even more in the P enriched treatments. The C:P values were driven by P changes, being higher in the control than in P enriched treatments. As expected, the largest increase in the N:P atomic ratios was recorded for the control, since no P source was added. However, the N:P ratio also showed higher values following N and P additions than initial thalli, denoting a greater N enrichment than P in the cell composition.

The sPPase activity in cell extracts was assayed along the experiment and scaled on both a protein and DW basis (Figure 7). Remarkable low activities were recorded in the control at the beginning of the experiment (almost undetectable on a DW basis, Figure 7B). There was a transient increase of activity in the control after 2 days when expressed on a protein basis, then this activity was maintained at low levels. The maximum level for Pi treatment was recorded after 4 days, and the highest levels were recorded for PPi and specially polyP enrichment at the end of the experiment (7 days). These values for sPPase activity were at the same order of magnitude as those recorded in field-collected plants (Table 1).

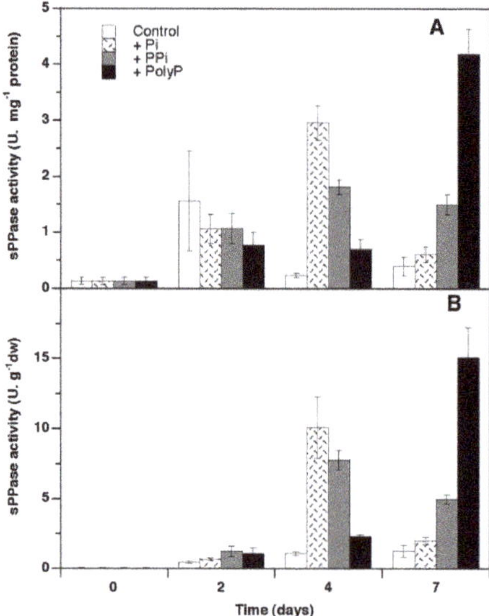

Figure 7. *Ulva pseudorotundata*. (**A**) Time course of sPPase activity on a protein basis and (**B**) on a dry weight, DW basis, as a function of different P regimes. Data are means ± SE (n = 3).

4. Discussion

4.1. Growth and Phosphorus Use

Ulva pseudorotundata grew and accumulated P regardless of the P source supplied in the incubations. The long preculture period resulted in nutrient limited thalli, with internal N and P cell quotas close to subsistence values for *Ulva* species [28] and a N:P ratio close to the Redfield ratio. Nutrients (N, P) resupply at the onset of the experimental phase (culture), resulted in thallus growth and cell C and N gain regardless of the P form supplied (treatment) even in the control without P supply, as N was also added. The highest growth rates were achieved under PPi and PolyP additions, which could be explained by the higher content of Pi per mol of substrate of both substrates (i.e., PPi and PolyP, 2 and 13–18 Pi molecules, respectively). The metabolic stimulation driven by N and/or P additions was also reflected by the increase of the FW/area and the drop of the DW/FW ratios. This could be a result of the increase in cell turgor (increased water content) and cell division. Therefore, in the short term, there was a stimulation of growth (on fresh weight and area basis), which can be indicative of cell division in this bi-stromatic sheet like species [19]. Initial tissue C content was lower than mean values for *Ulva* species (about 27 %DW, [29]), but increased in all treatments as a result of the metabolic stimulation caused by N and/or P supply. The growth rates recorded in this study were much lower than maximum ones measured in this species both in the laboratory and in the field (0.2 to 0.3 d⁻¹) [19,29]. These low values were probably because thalli were nutrient limited, and the immediate response was, once nutrients were supplied, to replenish cell nutrient quotas before achieving maximum growth rates [30]. In this way, cell N and P increased in all treatments (excepting P for the control, as expected).

Growth supported by different P sources has been previously reported in macroalgae [7]. Phosphorus is taken up through specific Pi transporters. Pi usually comes from SRP or from dissolved organic P molecules broken by alkaline phosphatase enzymes. But if the available P sources are pyro or polyposphates, an external hydrolytic activity is needed. This activity can be mediated through plasma membrane specific pyro or polyphosphatases, and/or from non-specific alkaline phosphatases located in the periplasmic space [31]. These non-specific phosphatases can even be released to culture medium, as shown by [7] in the green seaweed *Cladophora glomerata*. Although we used synthetic seawater, epiphytic bacteria occurring on *Ulva* thalli surfaces, can account for some of the PPi and polyP hydrolysis, also leaving some free Pi available for the macroalgae.

4.2. Soluble Pyrophosphatase Activity

Soluble cell extracts of field-collected *U. pseudorotundata* showed hydrolytic capacity for a variety of polyphosphates of different chain lengths. The activities measured in this study correspond to the enzymes located in the soluble fraction, as cell extracts were centrifuged and, thus, cell wall and membrane components were eliminated in the pellet. The high catalytic activity values, typical for this enzyme, were on the same order of magnitude or even higher than those recorded in several photosynthetic microorganisms [14]. Hydrolysis of PPi and tripolyposphate yielded similar rates. In contrast, lower values were obtained for trimetaphosphate, probably because of its cyclic structure. Polyphosphatase activity was higher than sPPase as a result of the breakdown of a polymeric substrate, causing a cascade of hydrolytic reactions, as it was observed in the cyanobacterium *Synechocystis* sp. strain PCC6803 [14].

As far as we are aware, there are few data on sPPase activity in marine macroalgae [7]. This sPPase activity has been successfully measured, displaying large changes depending on nutrient status and growth state. Thus, sPPase activity was almost undetectable in long-term nutrient-limited macroalgae. The transient increase in the control, at least on a protein basis, indicated that the response is not only dependent on P supply, but a more general response to metabolic stimulation caused by N addition. The sPPase induction has been observed in *Synechocystis* sp. strain PCC6803. This cyanobacterium showed a biphasic curve: a first sPPase maximum during the exponential growth period, and a second one, of lower magnitude, in the stationary phase [32]. As stated in the introduction, this is a key enzyme to maintain the biosynthetic reactions, and therefore, it was stimulated in response to nutrient supply in *U. pseudorotundata*. In contrast, quite low levels of sPPase activities were recorded in nutrient limited algae (initial values). It should be taken into account that preculture lasted 70 days, leaving sPPase activities at basal levels. After this long period of nutrient limitation, the maximum levels of sPPase activity registered for the Pi, PPi treatments after 4 days and polyP after 7 days showed the stimulation of sPPase activities in parallel to the stimulation of protein synthesis and growth after several days of supply of N and different P sources.

4.3. Ecological Implications

Some studies have revealed the importance of polyP molecules in contributing to the total available P pool of aquatic sediments [9,10]. However, the measurement of SRP misses these important Pi polymers than are non-molybdenum sensitive. The green macroalga *Ulva pseudorotundata* occurs in dense mats in the intertidal mudflats of the Palmones river estuary (Southern Spain) [29], reducing not only underneath light availability [33], but also (more than 40 times) the net phosphorus flux from sediment to the water column [34]. As cited above, this sediment may contain not only Pi, but also inorganic polymeric forms, that, as evidenced in our study, can support *U. pseudorotundata* growth. In fact, *U. pseudorotundata* eventually exhibits huge biomass accumulations, with canopies that can reach a thallus area index of 17 layers [29]. This is a great barrier for the flux of P between sediment and water. Therefore, it can be concluded that this overlooked P pool must be analysed and considered when studying biogeochemical cycles in coastal ecosystems and nutrient-mass balances, especially when green tides of macroalgae develop. Methods such as ^{31}P-NMR and/or PPi-dependent enzymatic assays to measure Pi polymers should be adopted in current protocols to understand the phosphorus cycle, especially in human impacted coastal areas.

Author Contributions: Conceptualization, J.J.V., F.G.B. and J.L.P.-L.; methodology, J.J.V., F.G.B. and P.H.-P.; lab research, J.J.V., F.G.B. and P.H.-P.; data curation, P.H.-P., J.J.V. and F.G.B.; writing—original draft preparation, J.J.V., F.G.B. and J.L.P.-L.; writing—review and editing J.J.V.; supervision, J.J.V.; funding acquisition, J.L.P.-L. and F.G.B. All authors have read and agreed to the published version of the manuscript.

Funding: This research has been supported by the project CTM2017-85365-R from the Spanish Ministry of Economy, Industry and Competitiveness.

Conflicts of Interest: The authors declare no conflict of interest.

Appendix A

Table A1. *Ulva pseudorotundata*. Statistic results of 2-way ANOVAs analyzing the effects of Pi treatments and time on the different variables analyzed. $* = p < 0.05$; $** = p < 0.01$; $*** = p < 0.001$. n.s. Non-significant. df. Degrees of freedom.

Variable	Effect	df Treatment/ df Error	F Value, p
Disc area (cm^2)	Time	3/32	123.6 ***
	Pi treatment	3/32	18.4 ***
	Interaction	9/32	4.8 ***
Fresh weight: area (g FW m^{-2})	Time	3/32	60.9 ***
	Pi treatment	3/32	3.6 *
	Interaction	9/32	2.6 *
Dry weight: fresh weight ratio	Time	3/32	60.2 ***
	Pi treatment	3/32	2.6 n.s.
	Interaction	9/32	2.2 n.s.
Total Chlorophyll ($\mu g\ cm^{-2}$)	Time	3/32	10.9 ***
	Pi treatment	3/32	7.0 ***
	Interaction	9/32	1.1 ns
Total cell P (mg g^{-1}DW)	Time	3/32	177.8 ***
	Pi treatment	3/32	96.8 ***
	Interaction	9/32	39.7 ***
Cell Phosphate ($\mu mol\ g^{-1}$DW)	Time	3/32	10.2 ***
	Pi treatment	3/32	19.9 ***
	Interaction	9/32	7.4 ***
Soluble protein (mg g^{-1}DW)	Time	3/32	1223 ***
	Pi treatment	3/32	1.8 n.s.
	Interaction	9/32	2.4 *
sPPase activity (U g^{-1}DW)	Time	3/32	53.1 ***
	Pi treatment	3/32	16.3 ***
	Interaction	9/32	24.1 ***
sPPase activity (U mg^{-1}protein)	Time	3/32	21.0 ***
	Pi treatment	3/32	5.0 **
	Interaction	9/32	14.6 ***

References

1. Falkowski, P.G.; Raven, J.A. *Aquatic Photosynthesis*; Blackwell Science: Maiden, MA, USA, 1997; 375p.
2. Lobban, C.S.; Harrison, P.G. *Seaweed Ecology and Physiology*; Cambridge University Press: New York, NY, USA, 1994; 366p. [CrossRef]
3. Hurd, C.L.; Dring, M.J. Phosphate uptake by intertidal fucoid algae in relation to zonation and season. *Mar. Biol.* **1990**, *107*, 281–290. [CrossRef]
4. Lavery, P.S.; McComb, A.J. The nutritional eco-physiology of *Chaetomorpha linum* and *Ulva rigida* in Peel Inlet, Western Australia. *Bot. Mar.* **1991**, *34*, 251–260. [CrossRef]
5. Schachtman, D.P.; Reid, R.J.; Ayling, S.M. Phosphorus uptake by plants: From soil to cell. *Plant Physiol.* **1998**, *116*, 447–453. [CrossRef] [PubMed]

6.   Hernández, I.; Niell, F.X.; Whitton, B.A. Phosphatase activity of benthic marine algae. An overview. *J. Appl. Phycol.* **2002**, *14*, 475–487. [CrossRef]

7.   Lin, C.K. Accumulation of water soluble phosphorus and hydrolysis of polyphosphates by *Cladophora glomerata* (Chlorophyceae). *J. Phycol.* **1977**, *13*, 46–51. [CrossRef]

8.   Sanz-Luque, E.; Bhaya, D.; Grossman, A.R. Polyphosphate: A Multifunctional Metabolite in Cyanobacteria and Algae. *Front. Plant Sci.* **2020**, *11*, 938. [CrossRef]

9.   Sundareshwar, P.V.; Morris, J.T.; Pellechia, P.J.; Cohen, H.J.; Porter, D.E.; Jones, B.C. Occurrence and ecological implications of pyrophosphate in estuaries. *Limnol. Oceanogr.* **2001**, *46*, 1570–1577. [CrossRef]

10. Ahlgreen, J.; Tranvik, L.; Gogoll, A.; Waldebäck, M.; Markides, K.; Rydin, E. Sediment depth attenuation of biogenic phosphorus compounds measured by ^{31}P NMR. *Environ. Sci. Technol.* **2005**, *39*, 867–872. [CrossRef]

11. Lundberg, P.I.; Weich, R.G.; Jensén, P.; Vogel, H.J. Phosphorus-31 and nitrogen-14 NMR studies of the uptake of phosphorus and nitrogen compounds in the marine macroalgae *Ulva lactuca*. *Plant Physiol.* **1989**, *89*, 1380–1387. [CrossRef]

12. Chopin, T.; Morais, T.; Belyea, E.; Belfry, S. Polyphosphate and siliceous granules in the macroscopic gametophyte of the red alga *Porphyra purpurea* (Bangiophyceae, Rhodophyta). *Bot. Mar.* **2004**, *47*, 272–280. [CrossRef]

13. Gómez-García, M.R.; Serrano, A. Expression studies of two paralogous ppa genes encoding distinct Family I pyrophosphatases in marine unicellular cyanobacteria reveal inactivation of the typical cyanobacterial gene. *Biochem. Biophys. Res. Commun.* **2002**, *295*, 890–897. [CrossRef]

14. Gómez-García, M.R. Caracterización Molecular de la Pirofosfatasa Inorgánica Soluble de Microorganismos Fotosintéticos y Plástidos. Ph.D. Thesis, University of Seville, Sevilla, Spain, 2001; p. 266.

15. Pérez-Castiñeira, J.R.; Gómez-García, R.; López-Marqués, L.; Losada, M.; Serrano, A. Enzymatic systems of inorganic pyrophosphate bioenergetics in photosynthetic and heterotrophic protists: Remnants or metabolic cornerstones? *Int. Microbiol.* **2001**, *4*, 135–142. [CrossRef]

16. Lathi, R.; Pitkäranta, T.; Valve, E.; Ilta, I.; Kukko-Kalse, E.; Heinonen, J. Cloning and characterization of the gene encoding inorganic pyrophosphatase of *Escherichia coli* K-12. *J. Bacteriol.* **1988**, *170*, 5901–5907. [CrossRef]

17. Baltscheffsky, M.; Nyrén, P. The Synthesis and Utilization of Inorganic Pyrophosphate. In *Molecular Mechanisms in Bioenergetics*; Ernster, L., Ed.; Elsevier: Amsterdam, The Netherlands, 1984; pp. 187–206.

18. Cormaci, M.; Furnari, G.; Alongi, G. Flora marina bentonica del Mediterraneo: Chlorophyta. *Boll. dell'Accademia Gioenia Sci. Nat. Catania* **2014**, *47*, 11–436.

19. Pérez-Lloréns, J.L.; Vergara, J.J.; Pino, R.R.; Hernández, I.; Peralta, G.; Niell, F.X. The effect of photoacclimation on the photosynthetic physiology of *Ulva curvata* and *U. rotundata* (Ulvales, Chlorophyta). *Eur. J. Phycol.* **1996**, *31*, 349–359. [CrossRef]

20. Lewin, J. Silicon metabolism in diatoms. V. Germanium dioxide, a specific inhibitor of diatom growth. *Phycologia* **1966**, *6*, 1–12. [CrossRef]

21. Wheeler, P.A., Djörnsater, R. Seasonal fluctuations in tissue nitrogen, phosphorus, and N:P for five macroalgal species common to the Pacific Northwest coast. *J. Phycol.* **1992**, *28*, 1–6. [CrossRef]

22. Woelkerling, W.J.; Spencer, K.G.; West, J.A. Studies on selected *Corallinaceae* (Rhodophyta) and other algae in a defined marine culture medium. *J. Exp. Mar. Biol. Ecol.* **1983**, *67*, 61–77. [CrossRef]

23. Porra, R.J.; Thompson, W.A.; Kriedemann, P.E. Determination of accurate extinction coefficients and simultaneous equations for assaying chlorophylls a and b extracted with four different solvents: Verification of the concentration of chlorophyll standards by atomic absorption spectroscopy. *Biochim. Biophys. Acta* **1989**, *975*, 384–394. [CrossRef]

24. Murphy, J.; Riley, J.P. A modified single solution method for the determination of phosphate in natural waters. *Anal. Chim. Acta* **1962**, *26*, 31–36. [CrossRef]

25. Bradford, M.M. A rapid and sensitive method for the quantification of microgram quantities of protein utilizing the principle of protein-dye binding. *Anal. Biochem.* **1976**, *72*, 248–254. [CrossRef]

26. Ames, B.N. Assay of inorganic phosphate, total phosphate and phosphatases. *Methods Enzymol.* **1966**, *8*, 115–118. [CrossRef]

27. Zar, J.H. *Biostatistical Analysis*, 2nd ed.; Prentice-Hall, Inc.: Upper Saddle River, NJ, USA, 1984; 718p.

28. Lyngby, J.E.; Mortensen, S.; Ahrensberg, N. Bioassessment techniques for monitoring of eutrophication and nutrient limitation in coastal ecosystems. *Mar. Poll. Bull.* **1999**, *39*, 212–223. [CrossRef]

29. Hernández, I.; Peralta, G.; Pérez-Lloréns, J.L.; Vergara, J.J.; Niell, F.X. Biomass and dynamic of growth of *Ulva* species in Palmones River estuary. *J. Phycol.* **1997**, *33*, 764–772. [CrossRef]
30. Droop, M.R. 25 years of algal growth kinetics. A personal view. *Bot. Mar.* **1983**, *26*, 99–112. [CrossRef]
31. McComb, R.B.; Bowers, G.N., Jr.; Posen, S. *Alkaline Phosphatase*; Plenum Press: New York, NY, USA, 1979; 986p.
32. Gómez-García, M.R.; Losada, M.; Serrano, A. Concurrent transcriptional activation of *ppa* and *ppx* genes by phosphate deprivation in the cyanobacterium *Synechocystis* sp. strain PCC 6803. *Biochem. Biophys. Res. Commun.* **2003**, *302*, 601–609. [CrossRef]
33. Vergara, J.J.; Pérez-Lloréns, H.J.L.; Peralta, G.; Hernández, I.; Niell, F.X. Photosynthetic performance and light attenuation in *Ulva* canopies from Palmones river estuary. *J. Phycol.* **1997**, *33*, 773–779. [CrossRef]
34. Palomo, L.; Clavero, V.; Izquierdo, J.J.; Avilés, A.; Becerra, J.; Niell, F.X. Influence of macrophytes on sediment phosphorus accumulation in a eutrophic estuary (Palmones river, Southern Spain). *Aquat. Bot.* **2004**, *80*, 103–113. [CrossRef]

Publisher's Note: MDPI stays neutral with regard to jurisdictional claims in published maps and institutional affiliations.

Article

Structure and Dynamics of the Ras al Hadd Oceanic Dipole in the Arabian Sea

Adam Ayouche [1], Charly De Marez [1], Mathieu Morvan [1], Pierre L'Hegaret [1], Xavier Carton [1,*], Briac Le Vu [2] and Alexandre Stegner [2]

[1] Laboratoire d'Océanographie Physique et Spatiale, Institut Universitaire Européen de la Mer, Universite de Bretagne Occidentale, 29200 Brest, France; adam.ayouche@ifremer.fr (A.A.); demarez@univ-brest.fr (C.D.M.); mathieu.morvan@shom.fr (M.M.); lhegaret@univ-brest.fr (P.L.)
[2] Le Laboratoire de météorologie dynamique (LMD), Ecole Polytechnique, 91120 Palaiseau, France; briac.le-vu@lmd.polytechnique.fr (B.L.V.); alexandre.stegner@lmd.polytechnique.fr (A.S.)
* Correspondence: xcarton@univ-brest.fr; Tel.: +33-2-90-91-55-09

Abstract: The Ras al Hadd oceanic dipole is a recurrent association of a cyclone (to the northeast) and of an anticyclone (to the southwest), which forms in summer and breaks up at the end of autumn. It lies near the Ras al Hadd cape, southeast of the Arabian peninsula. Its size is on the order of 100 km. Along the axis of this dipole flows an intense jet, the Ras al Had jet. Using altimetric data and an eddy detection and tracking algorithm (AMEDA: Angular Momentum Eddy Detection and tracking Algorithm), we describe the life cycle of this oceanic dipole over a year (2014–2015). We also use the results of a numerical model (HYCOM, the HYbrid Coordinate Ocean Model) simulation, and hydrological data from ARGO profilers, to characterize the vertical structure of the two eddies composing the dipole, and their variability over a 15 year period. We show that (1) before the dipole is formed, the two eddies that will compose it, come from different locations to join near Ras al Hadd, (2) the dipole remains near Ras al Hadd during summer and fall while the wind stress (due to the summer monsoon wind) intensifies the cyclone, (3) both the anticyclone and the cyclone reach the depth of the Persian Gulf Water outflow, and (4) their horizontal radial velocity profile is often close to Gaussian but it can vary as the dipole interacts with neighboring eddies. As a conclusion, further work with a process model is recommended to quantify the interaction of this dipole with surrounding eddies and with the atmosphere.

Keywords: ras al hadd oceanic dipole; arabian sea; cyclonic and anticyclonic eddies; altimetric data; angular momentum eddy detection and tracking algorithm (AMEDA); HYCOM model; ARGO floats

Citation: Ayouche, A.; De Marez, C.; Morvan, M.; L'Hegaret, P.; Carton, X.; Le Vu, B.; Stegner, A. Structure and Dynamics of the Ras al Hadd Oceanic Dipole in the Arabian Sea. *Oceans* **2021**, *2*, 105–125. https://doi.org/10.3390/oceans2010007

Academic Editor: Michael W. Lomas

Received: 1 December 2020
Accepted: 27 January 2021
Published: 4 February 2021

Publisher's Note: MDPI stays neutral with regard to jurisdictional claims in published maps and institutional affiliations.

1. Introduction

Over the last three decades, there has been growing interest in the Arabian Sea and in its marginal gulfs and seas, for geopolitical, economic, and scientific reasons. These seas form a complex region in terms of oceanographic variability as they are influenced by the North Equatorial Current and by the Southwest Monsoon Current. The monsoon winds, blowing over this region, are highly variable with the seasons, coming strong and northeastward from May to September, weaker and southwestward from December to February, and quite weak and variable in direction during the two intermonsoon periods (March–April and October–November). These winds drive coastal and regional ocean currents, upwellings, and downwellings near the coast. These winds also play a major role in the vertical mixing of the upper ocean water masses [1,2]. The alongshore currents are therefore seasonally variable too. During the summer (southwest) monsoon, the Oman Coastal Current (OCC) flows northeastward, southeast of the Arabian Peninsula, and the West India Coastal Current (WICC) flows southward. In winter, the WICC flows northward (see Figure 1a).

133

Satellite observations and high resolution, primitive equation model simulations reveal that a strong turbulent activity takes place near the ocean surface in the Arabian Sea and in its marginal gulfs. The mesoscale (the mesoscale in the ocean characterizes motions of horizontal size from 10 to 250 km and time scales of 2 to 30 days) surface eddies (eddies are oceanic vortices) can be generated by the instability of alongshore currents [3,4] and are influenced by the wind stress curl and by Rossby waves [5,6]. These Rossby waves are generated at the western coast of India [7] and propagate westward towards the south Arabian or Somali coasts. There, they can break and intensify pre-existing eddies. Since both these Rossby waves, and the coastal currents, are surface intensified, the eddies thus formed are also concentrated near the ocean surface. Some of these eddies have already been studied, some of them have received less attention.

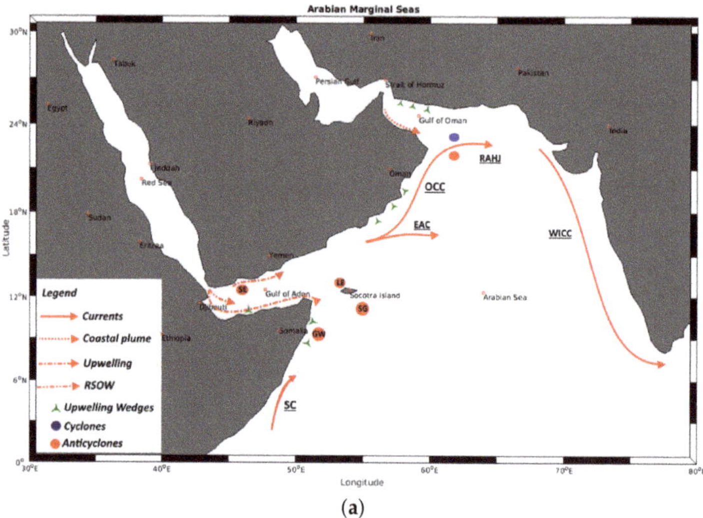

(a)

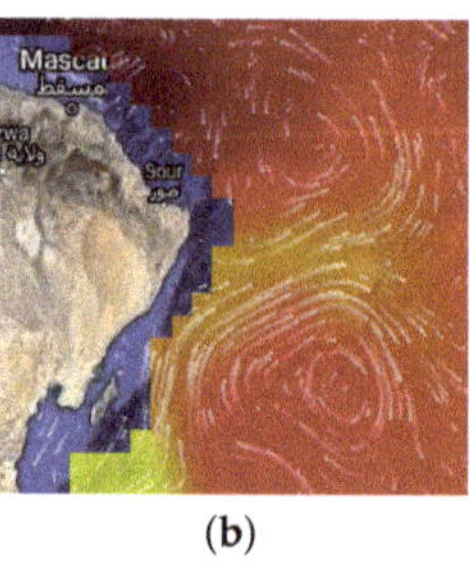

(b)

Figure 1. (a) Regional map of the Arabian Sea—in summer—displaying the main oceanographic phenomena: Currents, outflows, upwellings, and eddies and (b) snapshot of the Ras al Hadd dipole in summer 2020 with the cyclone to the northeast and the anticyclone to the southwest (from "Ocean Virtual laboratory web portal (ESA, OceanDataLab, https://ovl.oceandatalab.com (accessed on 1 December 2020)")).

Among the most often studied eddies in the region is the Great Whirl, a large surface eddy near the coast of Somalia. It is generated, and its intensity is modulated, by the summer monsoon wind stress curl, and by the incoming Rossby waves [8–10]. In the central Arabian Sea, Fisher et al. [11] observed several surface eddies via a moored currentmeter array. These eddies were generated during the summer monsoon wind stress. In the Gulf of Aden, the formation of eddies and their westward propagation are often due to wind stress during the summer monsoon [12] but current instabilities play an essential role in eddy dynamics during the fall intermonsoon and Rossby waves play a role during the winter monsoon. In November, eddies in the eastern Gulf of Aden are formed by a northward inflow of the Somali Current through the Socotra Passage [13]. In the Gulf of Aden, Bower and Furey [14] discovered two eddies, the Summer Eddy and the Lee Eddy, formed by the splitting of the Gulf of Aden Eddy at the end of the Spring intermonsoon. The Summer Eddy propagates westward into the Gulf of Aden, while the Lee Eddy remains fairly stationary north of Socotra Island and these two eddies are influenced by the wind stress curl.

The importance of these mesoscale eddies lies both in their interaction with the regional dynamics, or with the atmosphere, and in their effect on the marginal sea outflows. Indeed, the Red Sea and the Persian Gulf produce salty waters which are exported into the Arabian Sea via the straits of Bab el Mandeb and of Hormuz. These two water masses are the RSOW (Red Sea Outflow Water) also called RSW (Red Sea Water) and the PGW (Persian Gulf Water). After exiting the marginal seas, the RSW and PGW outflows mix with the surrounding water masses in the Gulf of Aden and in the Gulf of Oman, respectively, then, the RSW plume stabilizes at 600–1000 m and the PGW plume at a 250–350 m depth [15–17]. Due to the influence of mesoscale eddies or bottom topography [4,18–20], the PGW or RSW outflows are destabilized and they often shed smaller-scale eddies (of radii ranging between 5 and 20 km). These eddies, formed at a depth near the coast, drift away into the open ocean, carrying their warm and salty waters (see [3,21,22] for the small RSW eddies, and [23–25] for the PGW eddies). These small eddies contribute to the regional budgets of heat, salt, and momentum.

A less studied surface mesoscale eddy, or more precisely a pair of eddies, is the Ras al Hadd oceanic dipole. This association of a cyclone and of an anticyclone, exists from May to October, near Ras al Hadd, the southeastern-most cape of the Arabian Peninsula (see Figure 1b). An intense offshore jet, the Ras al Hadd jet, lies between the two eddies. It was studied by Böhm et al. [26] using high resolution radiometric data and Acoustic Doppler Current Profiler observations. Though Böhm et al. mentionned the presence of the two eddies near the jet, and noted their size of about 150 km, they mostly concentrated on the jet which advects cold water offshore during the summer monsoon. Indeed, this dipole marks the eastern edge of the Oman upwelling system which contains cold water. Apart from the study by Böhm et al., the Ras al Hadd dipole has not been studied in detail. Taking advantage of a new, high-resolution altimetric dataset, analyzed via vortex identification and tracking software (AMEDA: Angular Momentum Eddy Detection and tracking Algorithm), of ARGO profiling float data in the region (from the ARGO program), of a regional primitive equation model (HYCOM), the structure and the evolution of the Ras al Hadd dipole are examined here.

This paper is organized as follows: First, we present satellite and profiling float data, the AMEDA algorithm, and the HYCOM model simulations (Section 2). Then we analyze the Ras al Hadd dipole by studying its life cycle over a year (April 2014 to April 2015), and its horizontal and vertical structures using the altimetric and the float data, and the model results (Section 3), evaluating its inter-annual variability over a 15 year period. These results are discussed to present the impact of this dipole on its environment (Section 4). Finally, conclusions are drawn.

2. Materials and Methods

Time series of sea surface height data were obtained via a combination of up-to-date measurements by altimetric satellites (Topex/Poseidon, ERS-2, GFO, Jason-1, Envisat, Jason-2, Cryosat-2, Altika et HY2A). Using all available missions, an ADT (Absolute Dynamic Topography) product for the Arabian Sea was processed by CLS-Argos on a $\frac{1}{8}^{\circ}$ Mercator grid, with time intervals of 24 h, for the 2000–2015 period. Those products were processed by SSALTO/DUACS and distributed by AVISO+, https://www.aviso.altimetry. fr (accessed on 1 December 2020)) with support from CNES, Centre National d'Etudes

Spatiales, Toulouse, France The geostrophic velocity fields were then derived from absolute dynamic topography. The MDT used to calculate the ADT from SLA is described on the AVISO/CNES website https://www.aviso.altimetry.fr/en/data/products/auxiliary-products/mdt/mdt-description.html (accessed on 1 December 2020) and results from an average of sea surface height above the geoid over the 1993–2012 period. It takes into account the last geoid model GOCO05S (based on the complete GOCE mission and on 10.5 years of the GRACE mission) and 25 years of altimetric and in situ data (hydrological profiles, drifting buoys). The spatial resolution of this disseminated regional data set resolves the internal deformation radius, $R_d \approx$ 50–60 km, in the Arabian Sea. The spatio-temporal interpolation performed on the multiple altimetric tracks in order to build gridded products provides this $\frac{1}{8}^\circ$ resolution while the signal resolution of altimetry is about 100 km [27,28]. The sensitivity test performed on the AMEDA algorithm [29] shows that only eddies with a characteristic radius larger than 25 km could correctly be detected with the regional AVISO/CMEMS data set. In addition, a recent work [30] compared eddies identified in a $\frac{1}{60}^\circ$ resolution model with those found in the $\frac{1}{8}^\circ$ AVISO product used here and concluded that eddies with radii smaller than 25 km could not reliably be detected. The Ras al Hadd dipole eddies have radii larger than 45 km. Thus these two eddies can be detected with confidence in this AVISO/CMEMS product.

To relate vortex generation or intensification with the wind stress curl, the ECMWF (European Center of Medium-range Weather Forecast) meteorological model re-analyses (ERA-Interim) were used [31]. These re-analyses provided the regional wind every 6 h, with a $\frac{1}{4}^\circ$ resolution.

To support our analysis of the Ras al Hadd dipole, we also used HYCOM (Hybrid Coordinate Ocean Model) model simulations. These model simulations were performed by SHOM (Service Hydrographique et Oceanographique de la Marine, France). The model has a 5 km horizontal resolution, 40 vertical levels (the upper ones being sigma levels—or terrain following levels, and the lower ones, isopycnic levels—that is, isodensity levels). The atmospheric forcing for the ocean model was extracted from the Meteo-France atmospheric model results over the region with a $\frac{1}{4}^\circ$ horizontal resolution and a 6 h period for the year 2011. This year was chosen as typical of the meteorological variability in this region. The HYCOM model was initialized using a situation obtained from a global, lower resolution ocean model (MERCATOR PSY3, with $\frac{1}{4}^\circ$ resolution). The results of MERCATOR PSY3 were also used for the external boundary conditions of HYCOM (temperature, salinity, and currents). The geographical domain of the HYCOM model covers the Arabian Sea, the gulfs of Oman and of Aden, the Red Sea, and the Persian Gulf, north of 10° S, and west of 80° E. The tidal forcing of HYCOM was provided by TOPEX and was included in the model at the open boundaries. Vertical mixing was parameterized with the KPP (K-profile parameterization) scheme [32]. This numerical model has been described and validated in [18].

To identify vortices, and in particular the Ras al Hadd dipole, in the altimetric data, we used the AMEDA algorithm. The AMEDA (Angular Momentum Eddy Detection and tracking Algorithm) method is an algorithm that detects and tracks oceanic vortices in the horizontal (geostrophic) velocity fields derived from satellite measurements of the sea surface [29]. The AMEDA method is available and decribed on the website of the DYNED ATLAS project https://www.lmd.polytechnique.fr/dyned/methods (accessed on 1 December 2020) with its strengths and weaknesses, and in particular, a comparison of the geostrophic and cyclogeostrophic velocities. This comparison is also performed in [33]. The vortex centers are identified as extreme in local angular momentum, and each vortex boundary is determined via the connexity of the vortices thus found (the boundary must be a closed streamline). Tracking vortices in time relies on the spatial closeness and on the shape similarity between two vortices successively identified in time. Nevertheless, short gaps in the time series are allowed provided that the temporal continuity can be ensured (the distance between two successive positions of a given vortex must not exceed a threshold). The model also identifies merging (connection) and splitting of vortices, via the connection or separation of contours. This algorithm contains a minimal number of tunable parameters. It has been successfully tested on AVISO altimetry datasets, on ROMS model output, and on velocity fields issued from laboratory experiments. Here the AMEDA algorithm was applied to the high resolution altimetric maps.

Finally, ARGO profiling float data were extracted from the Coriolis database with only validated data with a high quality index that corresponded to the period of study, used. The ARGO profilers are deep reaching oceanic floats, equipped with temperature, conductivity (salinity) and possibly oxygen concentration sensors. These floats first dive to a given depth (the parking depth, usually between 500 and 1000 m), then they drift at this depth under the influence of the oceanic currents, for a prescribed duration (their cycle period, usually 5 or 10 days). Once every cycle, the ARGO floats move up to the sea surface, thus performing a vertical profile of temperature, salinity, and sometimes oxygen concentration, of the ocean. At the surface, the floats are positioned via GPS and they transmit their data back to the scientific institutes, via a satellite link. Then, they dive back to their parking depth for a new cycle. Therefore the vertical profile data are positioned but the exact trajectory of the ARGO floats at depth is unknown. The ARGO data used here (temperature and salinity only) were collected by floats with cycling periods of 5 or 10 days, and a parking depth of 1000 m. The Ras al Hadd AMEDA outputs for the studied period with their respective Argo profiles given in https://github.com/aayouche/RASALHADD (accessed on 1 December 2020).

3. Results

3.1. Seasonal Variability of the Ras al Hadd Dipole from Mid-April 2014 to Mid-April 2015

3.1.1. Origin and Evolution of the Two Eddies Composing the Dipole, Obtained from Surface Data Analysis

The AMEDA algorithm applied to the altimetric data (the satellite measurements of the sea surface elevation) allows the tracking in time of the eddies detected. This tracking can thus be reversed in time to determine the origin of the two eddies (the cyclone and the anticyclone) which form the Ras al Hadd dipole. These two eddies are detected in the altimetric data as early as April 2014. At that time, the two eddies lie far from each other, in the Arabian Sea. The cyclone is then at the mouth of the Gulf of Oman, near the Iranian/Pakistani coast. The anticyclone is then located east of Ras al Hadd. Altimetric maps show that the two eddies slowly migrate during the 2014 summer monsoon (see Figure 2). The cyclone (CE) drifts southeastward (early to mid-June), then it remains at the same location and elongates until mid-August. The anticyclone (AE) drifts southeastward early to mid June and then remains steady (at the same location). It intensifies in July due to the interaction with a few neighboring eddies and then weakens in mid-August.

During their migrations, the cyclone and anticyclone are altered by their mutual interaction, and by interactions with other eddies. For instance, in June 2014, the anticyclone becomes strongly elongated under the influence of a cyclone (see Figure 3a). The aspect ratio of the anticyclone (obtained from a best elliptical fit) increases beyond a value of 3, in early June 2014 (Figure 3d). Hydrodynamical theory states that an elliptical vortex with an aspect ratio larger than 3 is prone to breaking into two separate vortices [34]. And indeed, a temporary breaking of the anticyclone is observed in June.

In April 2015, the cyclone merges with another cyclone, which is then located to its east (see Figure 3b). Hydrodynamical theory states that two eddies can merge when the distance between them becomes smaller than 3 times their radii. In April 2015, the two cyclones are closer than this distance (see Figure 3c).

AMEDA also provides the time evolution of the cyclone and anticyclone radii, for a year (note that these two eddies may be associated in the dipole or separate during that period). Notable time variations of these radii reflect in part the successive interactions of the dipole with neighboring eddies (see Figure 4). In particular, we can see the growth of the cyclone by merger with a cyclone nearby, in August 2014 and in addition, what is noticeable is the growth of the cyclone radius from September to November 2014. A possible mechanism explaining this intensification is presented below (the effect of the wind). Between mid-January and early March 2015, the cyclone is elongated and loses mass due to the shear exerted by other eddies. In April 2015, it merges again with a cyclone nearby.

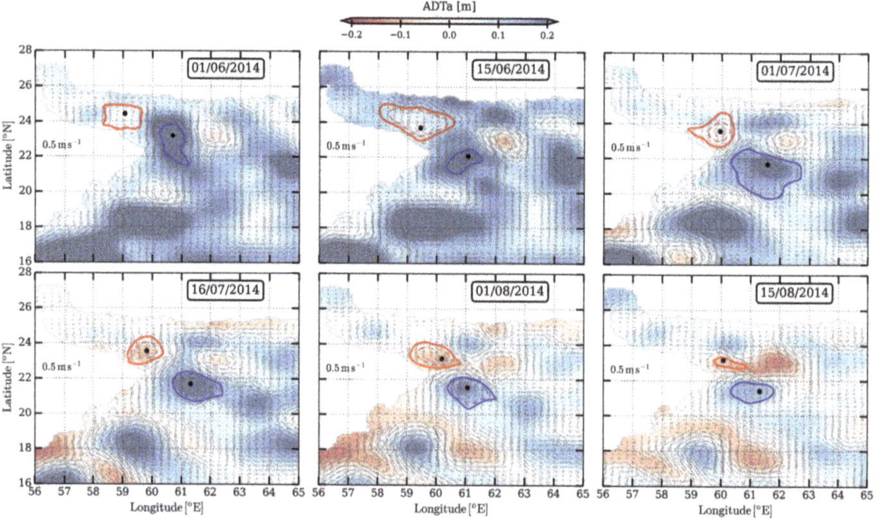

Figure 2. Snapshots of the Absolute Dynamic Topography (ADT) anomaly for the 2014 summer monsoon; the blue and red contours are for anticyclones and cyclones respectively; the dashed contours circle the eddy core; and solid lines indicate the outermost closed contours. The black dots represent the anticyclone and cyclone instantaneous center positions.

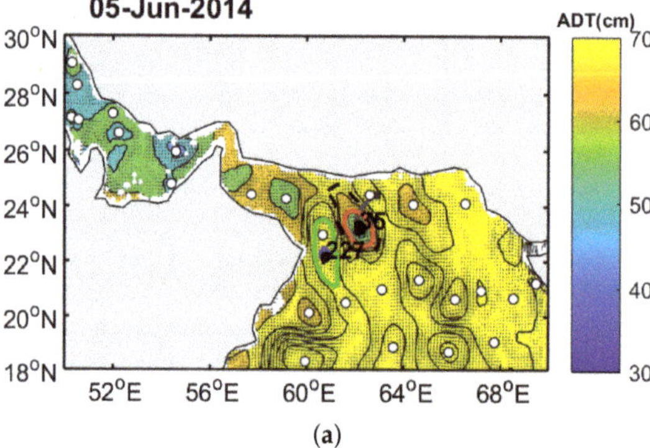

(a)

Figure 3. *Cont.*

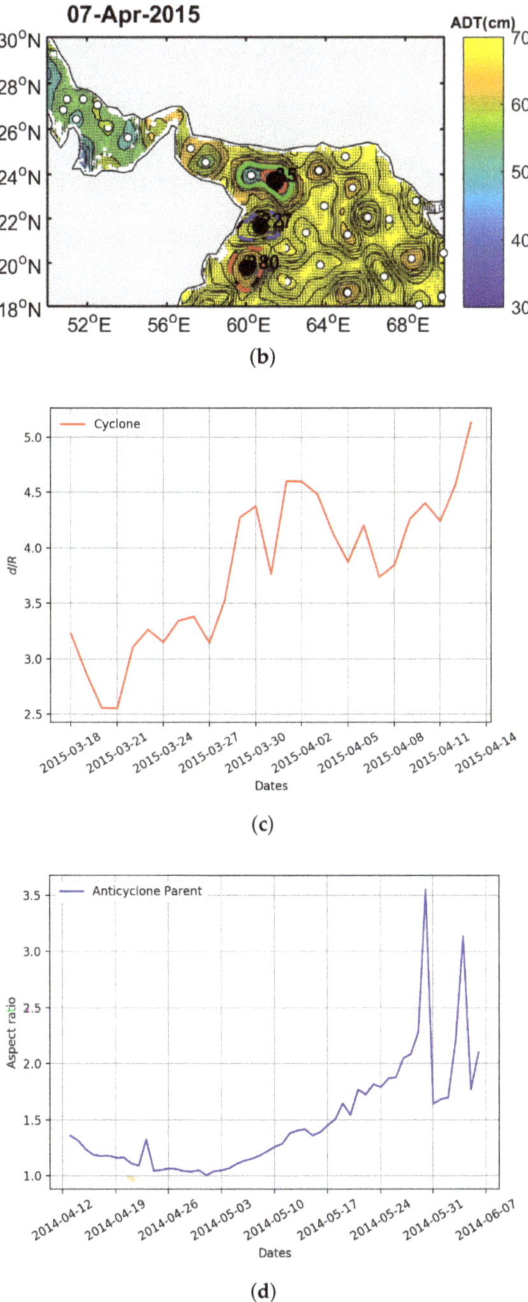

Figure 3. (**a**) Sea surface height map showing the elongation of the Ras al Hadd anticyclone (green contour, eddy 227) under the influence of cyclone 35, on 5 June 2014; (**b**) sea surface height map showing the merging of the Ras al Hadd cyclone with eddy 35 on 7 April 2015 (green contour); (**c**) ratio of mutual distance to the mean radius of the two cyclones for the merging event; and (**d**) aspect ratio of the anticyclone for the splitting event.

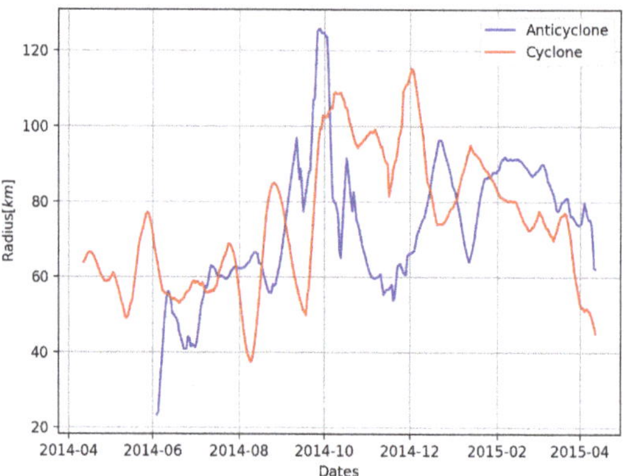

Figure 4. Time series of the Ras al Hadd cyclone and anticyclone radii from April 2014 to 2015.

The anticyclone of the dipole evolves as follows: In early October 2014, it grows in a favorable environment (with neighboring anticyclonic eddies) then, it splits and decreases in size. It intensifies again at the end of December 2014, early January 2015, in a favorable anticyclonic environment (via mass transfer between eddies). It decreases thereafter by elongation due to the shear of its cyclonic partner. Finally, it intensifies again via a merger with a neighboring anticyclone (in February–March 2015). These evolution of the two eddies of the dipole show the importance of vortex-vortex interactions in this region.

To explain another part of the eddy radii and intensity variations with time, we investigate the effect of the wind on the cyclone and on the anticyclone. We correlate the absolute dynamic topography anomaly (the Absolute Dynamic Topography (ADT) anomaly is the total ADT minus its instantaneous geographical average in a subdomain of 5×5 degrees around the eddy) and the wind stress curl for the Ras al Hadd dipole. In summer, the wind stress curl correlation was often positive with the cyclone and negative with the anticyclone (Figure 5). Indeed, during this period, the summer monsoon winds blow from Africa to Asia, and veer northward around Ras al Hadd. At this location, the wind stress curl is therefore positive in summer and early fall. During the winter monsoon and spring inter-monsoon, the wind direction is more variable and therefore the wind correlation with the eddies fluctuates more (see again Figure 4).

Now, we study the internal structure of each eddy. The internal velocity-pressure balance in a nearly circular eddy is the cyclogeostrophic balance:

$$V_\theta^2/r + f_0 V_\theta = (1/\rho)\, dp/dr$$

where V_θ is the azimuthal (tangential) velocity, r is the local radius of the point considered in the vortex, f_0 is the Coriolis parameter, ρ is seawater density, and dp/dr is the radial gradient of pressure. The first term in this equation is the centrifugal acceleration, the second is the Coriolis acceleration, and the right hand side in the pressure gradient. For large vortices, the balance lies essentially between the Coriolis acceleration and the pressure gradient and the equilibrium is geostrophic. For smaller and faster spinning eddies, the first term, the cyclostrophic correction to geostrophy, is not negligible. We calculate the ratio of the centrifugal to the Coriolis accelerations for the two eddies, when they are well developed (when their radius is close to 100 km):

$$Ce/Co = V_\theta/f_0 R$$

to estimate how nonlinear this balance is. This ratio is shown on Figure 6 with respect to the eddy radius, for each of them. The cyclostrophic correction reaches 20% in the eddy core (10 km from the eddy center) and decreases to about 11% for cyclones and 7%

for anticyclones beyond the radius of maximal velocity. We note that it never reaches a large value (e.g., 50%). Therefore, the geostrophic balance is a good approximation for the internal dynamics of these two eddies.

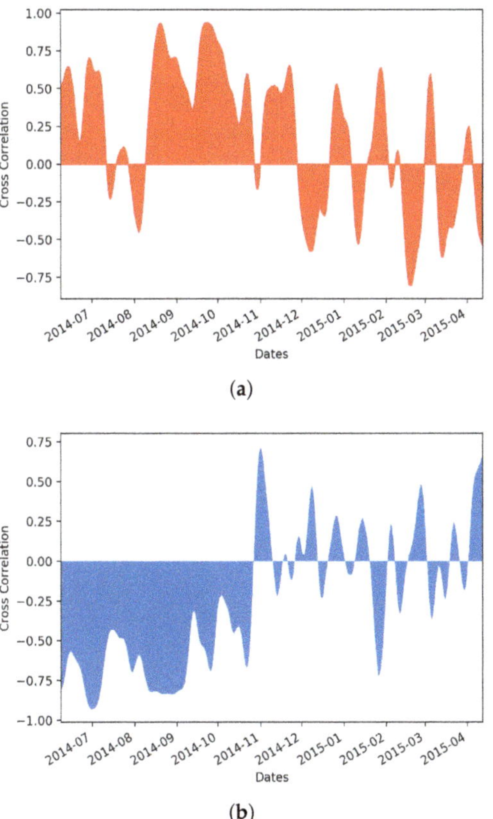

(**a**)

(**b**)

Figure 5. (**a**) Cross-correlation between the wind stress curl and the cyclone of the Ras al Hadd dipole with days starting in April 2014; (**b**) same as (**a**) for the correlation of the wind with the anticyclone of the dipole.

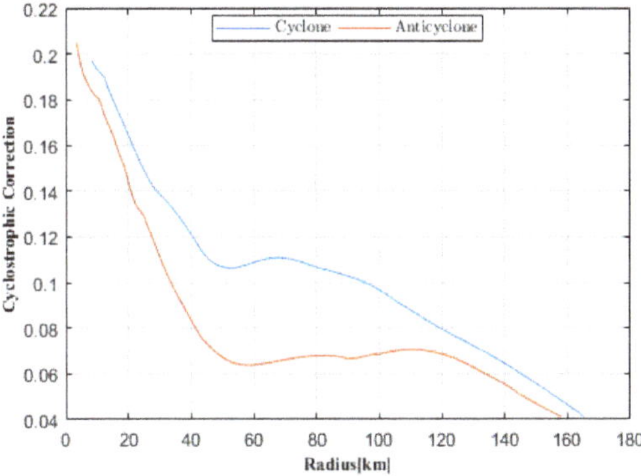

Figure 6. Ratio of the centrifugal to the Coriolis accelerations $V_\theta / f_0 R$ for the cyclone and the anticyclone of the Ras al Hadd dipole.

3.1.2. Vertical Structure of the Dipole from the HYCOM Model Results

To complement this surface analysis of the dipole, we present its vertical sections of temperature, salinity, density, and perpendicular current issued from the HYCOM model (see Figure 7). The HYCOM model reproduces well the location and intensity of the Ras al Hadd dipole during the summer monsoon period (see Figure A1). This numerical model has been validated with in-situ observations in previous studies and showed that the main water masses (e.g., the PGW) are well reproduced in the Gulf of Oman and in the northern Arabian Sea [24]. The sections presented in Figure 7a are snapshots extracted from the model. Firstly, the upwards or downwards curvature of the main pycnocline is clearly seen for each vortex, in particular, near a 200 m depth though a few small scale density features are observed near the surface (Figure 7a,b). The 200 m depth corresponds to a marked thermocline and halocline (associated with the presence of Persian Gulf Water), above which the Ras al Hadd eddy velocities are intensified and can reach 0.5 m/s. This magnitude compares well with that measured by ADCP [26]. The velocity field is slightly asymmetric with respect to each eddy vertical axis, but this is due to the deformation effect induced by the surrounding flow (Figure 7c). The vortex flow has a baroclinic component but no flow reversal occurs, at least down to 1500 m depth (not shown). The flow coherence above a 250 m depth indicates that the surface information (eddy drift, interaction or deformation) correctly characterizes the deeper structure of each eddy (Figure 7d).

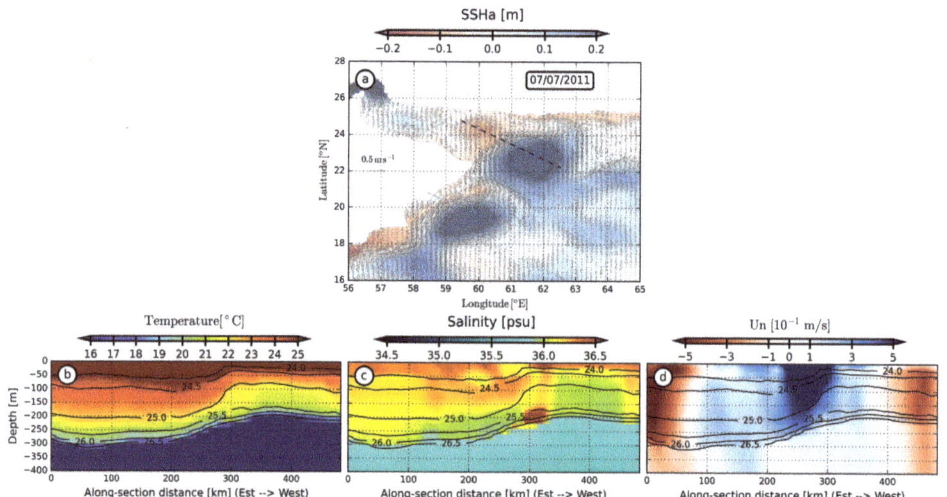

Figure 7. (**a**) Snapshot of the sea surface height (SSH) anomaly (HYCOM). The black dashed line represents the location of the vertical section. (**b–d**) Represent the temperature, salinity, and the cross-section velocity for the cyclone and anticyclone of the Ras al Hadd dipole, respectively. The black solid lines represent the isopycnals.

3.2. Interannual Evolution of the Dipole Over the 2000–2015 Period

Variations of the Surface Features of the Dipole

The AMEDA algorithm is now applied to altimetric data over the 2000–2015 period, to study the long-term variations of the Ras al Hadd dipole structure. During this period, years 2005 and 2009 are not presented because of the dipole position (southwest of the Ras al Hadd cape; see Figures A2 and A3). This dipole is long-living and similar in intensity (radius) to the the Ras al Hadd dipole observed during the other studied years. In this paper, we study the dipole only when it is located at the Ras al Hadd cape at the entrance of the Gulf of Oman (GO), which is not the case for these years. Another dipole is observed at the entrance of GO but is short lived (less than 3 months), therefore for these reasons the dipole is not presented for years 2005 and 2009. For both eddies composing the dipole, we show the time variation of: (1) The radius R_{max} of (2) the maximal azimuthal velocity V_{max} (defining each eddy core), (3) the steepness α of each eddy velocity profile fitted on:

$$V_\theta(r) = \frac{V_{max}r}{R_{max}} \exp(\frac{1}{\alpha}) \exp(-(r/R_{max})^\alpha/\alpha),$$

and (4) the dynamical Rossby number $Ro = \frac{\zeta_c}{2f}e^{-\frac{1}{\alpha}}$ where ζ_c is the maximal relative vorticity of each eddy and f is the Coriolis parameter. The calculation of the steepness α is accurate only for large eddies; this is the case of the two eddies composing the dipole. Figure 8 shows that, over 14 years, the anticyclone of the Ras al Hadd is slightly wider and spins often faster than the cyclone. A direct correlation with the interannual variability of the summer wind stress averaged over the Northwestern Arabian Sea (see Figure 7a of [35]) was not found for the dipole. This may be due to the fact that the dipole is both a local feature and that its intensity also varies by interaction with the surrounding eddies. For the anticyclone, the velocity profile is most often Gaussian ($\alpha \sim 2$), but for 2011 when this anticyclone was deformed, it created locally intense ($\alpha \geq 3$) and locally weaker ($\alpha \sim 1$) vorticity gradients. For the cyclone, the velocity profile was also often Gaussian. In 2004 and 2012, it was less steep ($\alpha \sim 1$), either because the cyclone was weaker (2012) or because it was not interacting much with neighboring eddies. The nonlinearity of oceanic eddies is also measured by the dynamical Rossby umber: During the summer monsoons of these 14 years, the values of Ro were below 0.3; note that the velocity deduced from altimetry is weaker than that measured at sea because of the low spatial resolution of altimetric data. Nevertheless, we can state that the geostrophic balance holds to a good accuracy for both eddies.

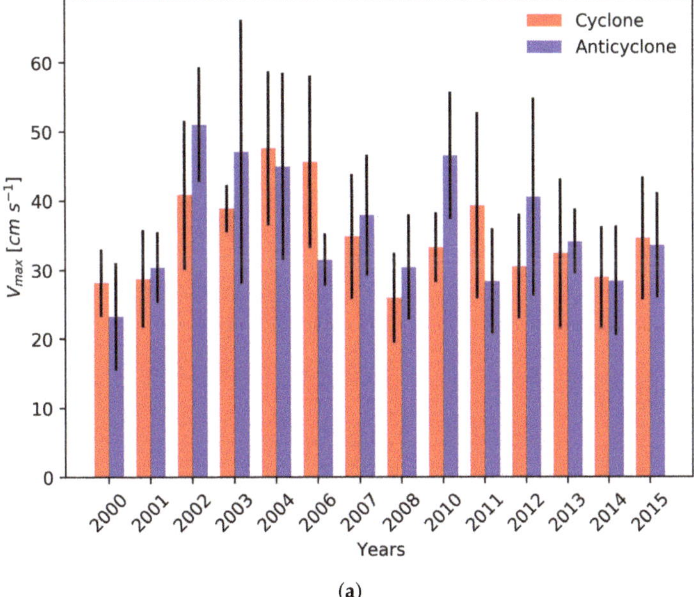

(a)

Figure 8. *Cont.*

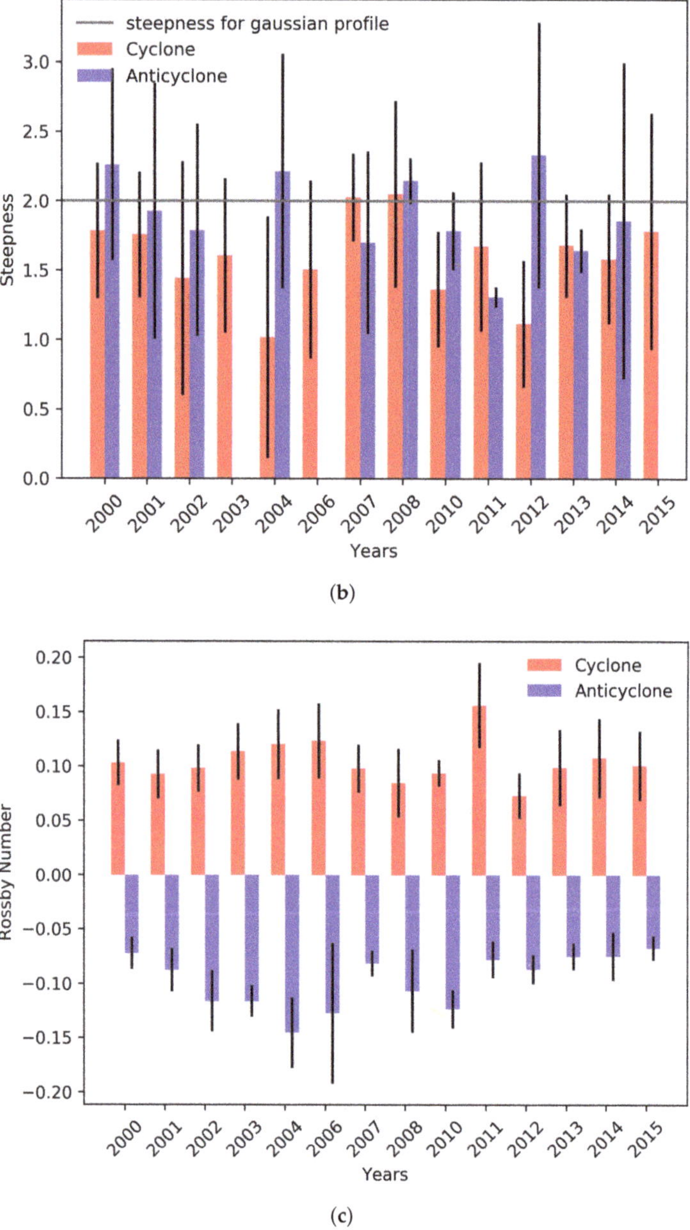

(**b**)

(**c**)

Figure 8. *Cont.*

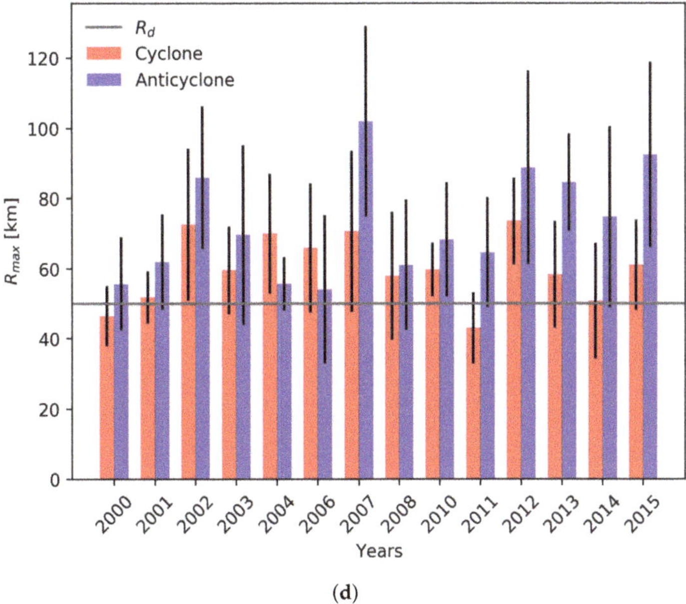

(**d**)

Figure 8. 2D (surface) statistics for both eddies of the Ras al Hadd dipole with error bars (vertical segments at the end of the colored lines): (**a**) Maximal azimuthal velocity V_{max}, (**b**) steepness of the radial profile of azimuthal velocity, (**c**) dynamical Rossby number Ro, and (**d**) radius of maximal velocity R_{max}. The horizontal gray line in the steepness plot indicates a Gaussian profile of velocity and the horizontal gray line in the R_{max} plot indicates the value of the internal Rossby radius of deformation.

3.3. Variations of the Vertical Structure of the Dipole

The vertical structure of the Ras al Hadd dipole is now evaluated using ARGO float co-localization in the two eddies. The co-localization is performed during the summer monsoon for 6 years (from 2010 to 2015). In the inner core of each eddy, the colocalized profiles give the structure of potential density and a few background profiles are obtained outside of the eddy (at 150 km from the eddy outermost closed contour). At each time step (when ARGO data are available), the inner core density anomaly is computed ($\rho_{eddy} - \rho_{background}$). The background state is calculated out of the outermost eddy closed contour (using an average over 5 points). Then a time average is calculated (over the summer monsoon) and over all the detected profiles. The number of ARGO floats co-located with the dipole varies from year to year. It is indicated on Figure 9b. The eddy depth is defined here as the depth of the maximum density anomaly. On Figure 9b, we can see that the anticyclone is globally deeper than the cyclone. The anticyclone reaches a 280 m depth for most of the 6 years while the cyclone depth is closer to 225 m. The year 2010 is of interest for the study; the number of co-localized profiles reaches 30 (Figure 9b) for both the cyclone and the anticyclone. For this year, the vertical distribution of the density anomaly (Figure 9a) shows a strong intensification near the surface (in the upper 50 m); these are mixed layer variations of the eddy density anomalies. They are negative for the anticyclone and positive for the cyclone. The density anomaly remains noticeable for the anticyclone and cyclone even below a 250 m depth. The thickness of each eddy can be computed from the vertical profiles of density anomalies. Choosing a 0.1 kg/m^3 value (an e-folding scale from the maximum) as a threshold for each eddy thickness, we obtain a thickness of 250 m for the anticyclone and of 300 m for the cyclone.

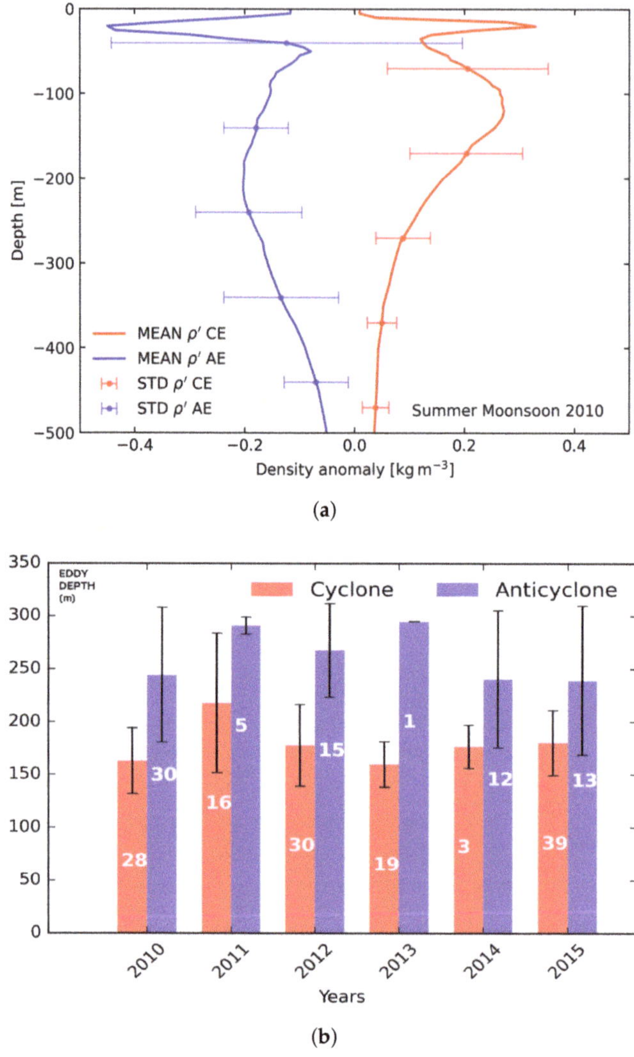

Figure 9. (**a**) Mean density anomaly (error bars for standard deviation (STD)) in the cyclone and anticyclone and (**b**) dipole depth as defined by the depth of the maximal density anomaly (the number of co-localized ARGO float profiles is indicated in white bold characters).

4. Discussion

The Ras al Hadd dipole is a recurrent feature composed of two eddies (a cyclone to the northeast and an anticyclone to the southwest) near the southeasternmost cape of the Arabian Peninsula, Ras al Hadd. It appears in the summer monsoon and breaks up in winter. Its position (at the cape or west of the cape) can vary from year to year. Its radius and azimuthal velocity also vary from year to year. We have found that, seasonally, the local wind stress curl intensifies the cyclone and weakens the anticyclone. It is likely that the anticyclone is strengthened both by the circulation at the upwelling front and by an upwelling Rossby wave (like the Great Whirl); a formal proof of this should be searched in forthcoming studies. Interannually, a correlation of the dipole intensity with the regionally averaged wind stress has not been evidenced. However, as Piontkovski et al. [35] indicate, the wind stress is spatially heterogeneous so that a regional average may not represent the local effect. The recurrence of this eddy dipole is also related to the Ras al Hadd front which separates the warm waters of the Gulf of Oman, from the colder water of the upwelling south of Oman. By inspecting the dipole life cycle, it was shown that each eddy evolved via strong interactions with neighboring eddies. The two eddies are fairly wide (100 km) and their Rossby number is small, so that the cyclostrophic correction to geostrophy is small. The radial profile (the velocity steepness profile) also evolves in time with the eddy interactions. It becomes steeper as the eddy periphery is eroded under the influence of neighboring eddies. It becomes smoother as the eddy undergoes few or no interaction and as dissipation slowly acts. On average over the years, the anticyclone is slightly steeper than the cyclone. In most cases, the two eddy velocity profiles are close to Gaussian (with a steepness in the range [2.25, 2.4]). The Burger Number of each eddy ($Bu = (\frac{R_d}{R_{max}})^2$) is of order unity, and the Rossby number is small, which shows that quasi-geostrophic dynamics should apply to model the evolution of these eddies, which are stable [34].

An order of magnitude of the kinetic energy and heat content stored in this dipole can be obtained. Assuming that the azimuthal velocity also decays vertically as a Gaussian, we obtain, from our eddy model, and with the maximal velocities observed, a kinetic energy on the order of:

$$K = \frac{1}{2} \rho \, \pi \int_0^\infty V_0^2 \, (r/R)^2 \, \exp[-2(r/R)^2] \, r \, dr \int_{-\infty}^0 \exp[-2(z/H)^2] dz$$

with $R = \sqrt{2}R_{max}$, $V_0 = \sqrt{2e}V_{max}$, $H = 250m$. For each eddy, this leads to:

$$K = \frac{\sqrt{\pi}}{8} K_0, \text{with } K_0 = \pi R^2 H \rho V_0^2 / 2.$$

Taking ρ = 1025 kg/m^3, $R = 5 \times 10^4$ m, H = 250 m, and V_0 = 1.15 m/s, we have $K \approx 10^{14}$ J for each eddy.

Via geostrophy and hydrostatics and assuming a vertical dependence of the azimuthal velocity as $\exp(-(z/H)^2)$, we obtain:

$$\rho'(r,z) = -Ro/Bu \, (z/H) \, \exp(-(r/R)^2) \, \exp(-(z/H)^2)$$

where ρ' is the density anomaly. Such a vertical dependence can indeed fit the vertical variation of the density anomaly shown in Figure 9a. Via a linear equation of state for the upper ocean, assuming that temperature variations follow density variations, we have:

$$T'(r,z) = T_0 \, (z/H) \, \exp(-(r/R)^2) \, \exp(-(z/H)^2)$$

where T' is the temperature anomaly associated with each eddy, and $T_0 = 2\,°C$ its maximum. The total heat (anomaly) content of each eddy is therefore:

$$Q = \rho C \int_0^\infty r \, dr \int_{-\infty}^0 dz T'$$

or $Q = \sqrt{\pi} M_e C T_0$ where M_e is the eddy mass and C is the heat capacity of seawater. This yields $Q = 2.85 \times 10^{19}$ J for each eddy.

Both values of eddy kinetic energy and of heat content are one order of magnitude smaller than those of the Gulf Stream rings. However the lifetime of the Ras al Hadd dipole (about 6 months) renders its contribution to the regional budgets of eddy kinetic energy and heat non negligible. Its heat fluxes towards the atmosphere should be evaluated in further work.

Furthermore, the Ras al Hadd eddies have a dynamical influence on deeper water masses, though the azimuthal velocity of these eddies is intensified in the upper 250 m of the water column. This surface intensification is due to at least two mechanisms: The stronger energy of the wind forced ocean above the thermocline, and the presence of a heterogeneous water mass at depth (PGW), which creates a strong pycnocline, attenuating the vertical penetration of the energy. The values of the azimuthal velocity for both eddies found in the present study are comparable with those measured at sea (with ADCP's) by [26]. The anticyclone spins slightly faster than the cyclone below a 250 m depth. However both eddies have a noticeable velocity at depth (about 0.25–0.3 m/s), which can influence the subsurface path of the PGW outflow in the region [25]. As noted by [25] and by [1], the strong surface-intensified eddies of the Gulf of Oman and of the Arabian Sea pull PGW fragments away from the coast. This has been modeled hydrodynamically by [19,20]. The surface eddies distort the alongshore outflow and also creates an unstable boundary layer vorticity sheet, which detaches offshore and breaks into subsurface submesoscale eddies containing PGW. These small eddies play a role in the transport of heat and salt from the Gulf of Oman to the Arabian Sea.

5. Conclusions

The objective of this paper was the description of the seasonal life cycle and of the inter-annual variability of the two eddies composing the Ras al Hadd oceanic dipole. Using altimetric data and eddy identification and tracking software (AMEDA), we were able to determine the characteristics of the two eddies composing it, for a year (2014–2015). We determined their origin and drift from their initial position to their final junction near Ras al Hadd. We showed the importance of eddy-eddy interactions during the lifetime of the dipole. The evolution of the dipole was also driven by the local wind stress curl, by the upwelling circulation, and by the incoming Rossby waves.

On a longer period (about 15 years), we noted that the anticyclone was larger than the cyclone of the Ras al Hadd dipole. Though the maximal internal velocities could reach 50 cm/s, in particular for the anticyclone, the cyclostrophic correction could be neglected as the dynamical Rossby number did not exceed 0.3. This is due to the large horizontal size of the eddies. On average, the two eddies have a Gaussian radial profile of velocity, but the steepness of this profile varies with the interactions undergone by each eddy. The vertical structure of both eddies is intensified above the layer of the Persian Gulf Water but influences it dynamically.

This leads us to propose process studies to deepen our understanding of this oceanic dipole. The quasi-geostrophic approximation holds for both eddies and this could be used first to perform a stability analysis of this dipole, with respect to velocity profile steepness, to the Burger number, assuming that the cyclone and anticyclone of the RAH dipole are isolated. Secondly, the interaction of this dipole with neighboring eddies and with the Rossby waves originating from the eastern boundary of the Arabian Sea deserves a specific analysis. Finally, the dipole interaction with the atmosphere should be assessed in detail, in particular for the heat and freshwater exchanges.

Author Contributions: Conceptualization, X.C., C.D.M., A.S. and A.A.; methodology, B.L.V., P.L. and A.S.; software and validation, B.L.V. and M.M.; analysis and investigation, A.A., C.D.M., M.M., P.L. and X.C.; writing—original draft preparation, A.A.; writing—review and editing, X.C. All authors have read and agreed to the published version of the manuscript.

Funding: This research was funded by SHOM contract HYCOM to UBO.

Data Availability Statement: The data sources are specified in the text.

Acknowledgments: This work is a contribution to the PHYSINDIEN research program.

Conflicts of Interest: The authors declare no conflict of interest.

Appendix A. Representation of the Ras al Hadd Dipole in the HYCOM Model

To validate the ability of the HYCOM model to represent the Ras al Hadd dipole, we present here snapshots of the sea surface height anomaly in the model compared with the AVISO MADT anomaly. The dipole location (near Ras al Hadd) was well represented in

HYCOM model. A small difference ($\sim$5–10 cm) in the amplitudes of each eddy of this dipole is observed. The HYCOM model represents a more intense dipole compared to AVISO data. Differences may be due to the different horizontal resolution of HYCOM and AVISO, to the HYCOM model atmospheric forcings, or to the parameterization of momentum or heat diffusion in the model. Still, HYCOM reproduces well the known location and radius of the cyclone and anticyclone of the Ras al Hadd dipole. It also provides a proper positioning and extent of the upwelling. Finally, the surrounding eddies in the model and in the data have similar sizes and intensities. At deeper levels (in particular for the Persian Gulf outflow), the HYCOM model has been validated by [24]. Thus we can properly use the HYCOM model results in this region.

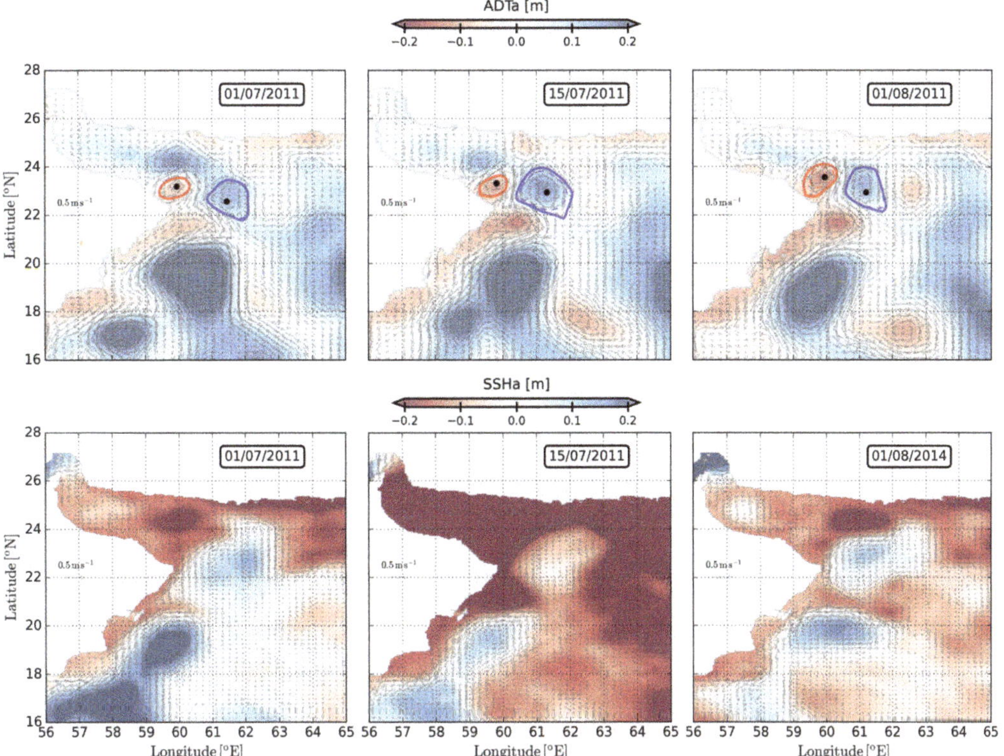

Figure A1. Snapshots of the ADT anomaly (AVISO) (**top row**). Black red and blue solid contours represent the external shape (closed contour) of the cyclone and anticyclone, respectively. The dashed contours characterize the inner core of these eddies. The black dot represents their centers. Snapshots of the SSH anomaly from HYCOM model (**bottom row**).

Appendix B. Details for Specific Years 2005 and 2009

In this appendix, the Ras al Hadd dipole locations and intensities are represented for the two specific years 2005 and 2009.

In our study, this dipole is considered to be long-living, intensified in summer monsoon and located near Ras al Hadd. Another important criterion for this dipole is the existence of the jet of Ras al Hadd between the cyclone (north) and anticyclone (south). This jet is also considered intense in summer monsoon season.

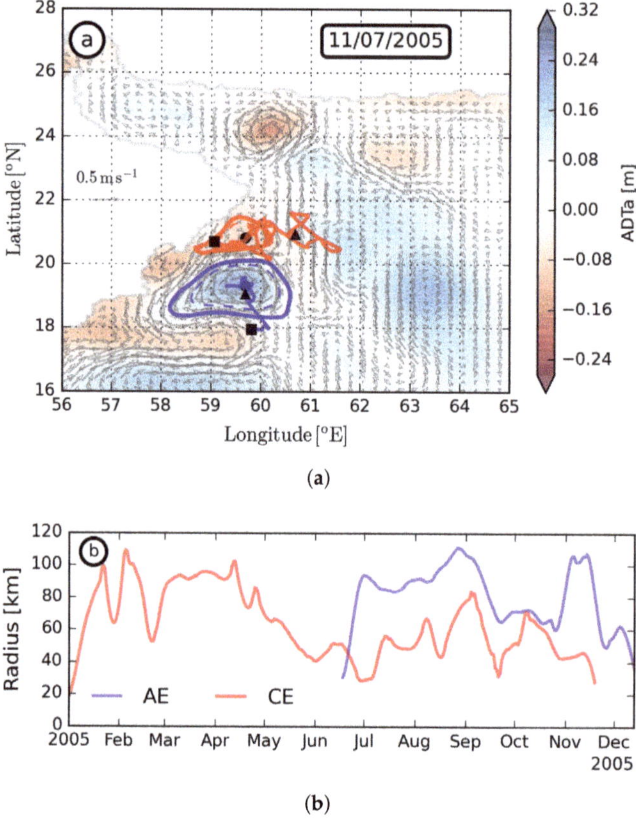

Figure A2. (**a**) Absolute Dynamic Topography anomaly for the 2005 year (snapshot); blue and red contours are for anticyclones and cyclones respectively, dashed contours are for eddy core, and solid contours for the last closed contour. The black dot represents the anticyclone and cyclone instantaneous center position. The solid red and blue trajectories are for cyclone and anticyclone, respectively during their lifetime. The black triangle and square represent the initial and final position of each eddy, respectively; (**b**) time series of the cyclone and anticyclone radii for 2005.

As can be seen in the following two figures, in 2005 and 2009, only a weak (and intermittent) dipole is observed near Ras al Hadd. The largest and most stable dipole lies southwest of Ras al Hadd. Thus, these two years are considered as "anomalous".

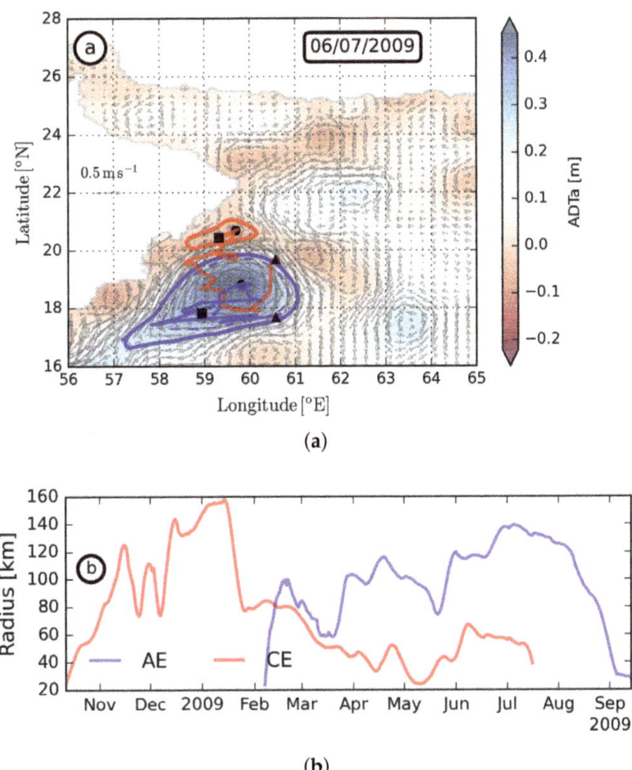

(a)

(b)

Figure A3. (**a**) Absolute Dynamic Topography anomaly for the 2009 year (snapshot); blue and red contours are for anticyclones and cyclones respectively, dashed contours are for eddy core, and solid contours for the last closed contour. The black dot represents the anticyclone and cyclone instantaneous center position. The solid red and blue trajectories are for cyclone and anticyclone, respectively during their lifetime. The black triangle and square represent the initial and final position of each eddy, respectively and (**b**) time series of the cyclone and anticyclone radii for year 2009.

We can suggest several reasons for these anomalous years which would have to be confirmed: A weaker summer monsoon wind would lead to a smaller extent of the upwelling region. Then, the eastern edge of this upwelling, which corresponds to the position of the anticyclone, would be west of its usual position near Ras al Hadd. Another possibility is that the eddy field in this region has interacted with the cyclone and anticyclone to the west of their usual positions and progressively intensified them there. Finally it is possible that the Oman Coastal Current is weaker these years and thus is unable to advect the dipole towards the cape.

References

1. Flagg, C.N.; Kim, H.S. Upper ocean currents in the northern arabian sea from shipboard ADCP measurements collected during the 1994–1996 US JGOFS and ONR programs. *Deep Sea Res. Part II* **1998**, *45*, 1917–1959. [CrossRef]
2. Lee, C.M.; Jones, B.H.; Brink, K.H.; Fisher, A.S. The upper-ocean respecton to monsoonal forcing in the Arabian Sea: Seasonal and spatial variability. *Deep Sea Res. II* **2000**, *47*, 1177–1226. [CrossRef]
3. Carton, X.; L'Hegaret, P.; Baraille, R. Mesoscale variability of water masses in the Arabian Sea as revealed by ARGO floats. *Ocean Sci.* **2012**, *8*, 227–248. [CrossRef]

4. Morvan, M.; L'Hegaret, P.; de Marez, C.; Carton, X.; Corréard, S.; Baraille, R. Life cycle of mesoscale eddies in the Gulf of Aden. *Geophys. Astrophys. Fluid Dyn.* **2020**, *114*, 631–649. [CrossRef]

5. Wang, Z.; DiMarco, S.F.; Jochens, A.E.; Ingle, S. High salinity events in the northern Arabian Sea and Sea of Oman. *Deep Sea Res. I* **2013**, *74*, 14–24. [CrossRef]

6. L'Hegaret, P.; Lacour, L.; Carton, X.; Baraille, R.; Correard, S. A seasonal dipolar eddy near Ras Al Hamra (Sea of Oman). *Ocean Dyn.* **2013**, *63*, 633–659. [CrossRef]

7. Brandt, P.; Stramma, L.; Schott, F.; Fischer, J.; Dengler, M.; Quadfasel, D. Annual Rossby waves in the Arabian Sea from TOPOX/POSEIDON altimeter and in situ data. *Deep Sea Res. II* **2002**, *49*, 1197–1210. [CrossRef]

8. Beal, L.M.; Hormann, V.; Lumpkin, R.; Folz, G.R. The response of the surface circulation of the Arabian Sea to Monsoonal forcing. *J. Phys. Oceanogr.* **2013**, *43*, 2008–2022. [CrossRef]

9. Beal, L.M.; Donohue, K.A. The Great Whirl: Observations of its seasonal development and interannual variability. *J. Geophys. Res. C* **2013**, *18*, 1—13. [CrossRef]

10. Vic, C.; Roullet, G.; Carton, X.; Capet, X. Mesoscale dynamics in the Arabian Sea and a focus on the Great Whirl lifecycle: A numerical investigation using ROMS. *J. Geophys. Res.* **2014**, *119*, 6422–6443. [CrossRef]

11. Fischer, A.S.; Weller, R.A.; Rudnick, D.L.; Eriksen, C.C.; Lee, C.M.; Brink, K.H.; Fox, C.A.; Leben, R.R. Mesoscale eddies, coastal upwelling and the upper ocean heat budget in the Arabian Sea. *Deep Sea Res. II* **2002**, *49*, 2231–2264. [CrossRef]

12. Al Saafani, M.A.; Shenoi, S.S.C.; Shankar, D.; Aparna, M.; Kurian, J.; Durand, F.; Vinayachandran, P.N. Westward motion of eddies into the Gulf of Aden from the Arabian Sea. *J. Geophys. Res. C* **2007**, *112*, 11004. [CrossRef]

13. Fratantoni, D.M.; Bower, A.S.; Johns, W.E.; Peters, H. Somali Current rings in the eastern Gulf of Aden. *J. Geophys. Res. C* **2006**, *111*, 09039. [CrossRef]

14. Bower, A.S.; Furey, H.H. Mesoscale eddies in the Gulf of Aden and their impact on the spreading of Red Sea Outflow Water. *Prog. Oceanogr.* **2012**, *96*, 14–39. [CrossRef]

15. Bower, A.S.; Hunt, H.D.; Price, J.F. Character and dynamics of the Red Sea and Persian Gulf outflows. *J. Geophys. Res. C* **2000**, *105*, 6387–6414. [CrossRef]

16. Pous, S.P.; Carton, X.; Lazure, P. Hydrology and circulation in the Strait of Hormuz and the Gulf of Oman; results from the GOGP99 Experiment: Part I. Strait of Hormuz. *J. Geophys. Res.* **2004**, *109*, 1–15. [CrossRef]

17. Pous, S.P.; Carton, X.; Lazure, P. Hydrology and circulation in the Strait of Hormuz and the Gulf of Oman; results from the GOGP99 Experiment: Part II. Gulf of Oman. *J. Geophys. Res.* **2004**, *109*, 1–26. [CrossRef]

18. Morvan, M.; Carton, X.; Gula, J.; L'Hegaret, P.; Vic, C.; Sokolovskiy, M.; Koshel, K. The life cycle of submesoscale eddies generated by topographic interactions. *Ocean Sci.* **2019**, *15*, 1531–1543. [CrossRef]

19. Morvan, M.; Carton, X.; Corréard, S.; Baraille, R. Submesoscale dynamics in the Gulf of Aden and the Gulf of Oman. *Fluids* **2020**, *5*, 146. [CrossRef]

20. Morvan, M.; de Marez, C.; L'Hegaret, P.; Carton, X. On the dynamics of an idealized bottom density current overflowing in a semi-enclosed basin: Mesoscale and submesoscale eddy generation. *Geophys. Astrophys. Fluid Dyn.* **2020**, *114*, 607–630. [CrossRef]

21. Meschanov, S.L.; Shapiro, G.I. A young lens of Red Sea Water in the Arabian Sea. *Deep Sea Res.* **1998**, *45*, 1–13. [CrossRef]

22. De Marez, C.; Carton, X.; Corréard, S.; L'Hegaret, P.; Morvan, M. Observation of a deep submesoscale cyclonic vortex in the Arabian Sea. *Geophys. Res. Lett.* **2020**, *47*, e2020GL087881. [CrossRef]

23. Senjyu, T.; Ishimaru, T.; Matsuyama, M.; Koike, Y. High salinity lens from the Strait of Hormuz. In *Offshore Environment of the ROPME Sea Area after the War-Related Oil Spill, Results of the 1993–1994 Umitaka-Maru Cruises*; Otsuki, A., Abdulraheem, M.Y., Reynolds, R.M., Eds.; Terra Scientific Publishing Company: Tokyo, Japan, 1998; pp. 35–48.

24. L'Hegaret, P.; Duarte, R.; Carton, X.; Vic, C.; Ciani, D.; Baraille, R.; Correard, S. Seasonal mesoscale variability in the Arabian Sea from HYCOM model and observations: Impact on the Persian Gulf Water path. *Ocean Sci.* **2015**, *11*, 667–693. [CrossRef]

25. L'Hegaret, P.; Carton, X.; Louazel, S.; Boutin, G. A submesoscale lens of Persian Gulf off the Omani coast in Spring 2011. *Ocean Sci.* **2016**, *12*, 687–701. [CrossRef]

26. Böhm, E.; Morrison, J.M.; Manghnani, V.; Kim, H.S.; Flagg, C.N. The Ras al Hadd Jet: Remotely sensed and acoustic Doppler current profiler observations in 1994–1995. *Deep Sea Res. Part II* **1999**, *46*, 1531–1549. [CrossRef]

27. Ballarotta, M.; Ubelmann, C.; Pujol, M.-I.; Taburet, G.; Fournier, F.; Legeais, J.-F.; Faugère, Y.; Delepoulle, A.; Chelton, D.; Dibarboure, G.; et al. On the resolutions of ocean altimetry maps. *Ocean Sci.* **2019**, *15*, 1091–1109. [CrossRef]

28. Quilfen, Y.; Chapron, B. On denoising satellite altimeter measurements for high-resolution geophysical signal analysis. *Adv. Space Res.* **2020**, in press. [CrossRef]

29. Le Vu, B.; Stegner, A.; Arsouze, T. Angular Momentum Eddy Detection and tracking Algorithm (AMEDA) and its application to coastal eddy formation. *J. Atmos. Ocean. Technol.* **2017**, *35*, 739–762. [CrossRef]

30. Amores, A.; Jorda, G.; Arsouze, T.; Sommer, J.L. Up to What Extent Can We Characterize Ocean Eddies Using Present-Day Gridded Altimetric Products? *J. Geophys. Res. Oceans* **2018**, *123*, 7220–7236. [CrossRef]

31. Dee, D.P.; Uppala, S.M.; Simmons, A.J.; Berrisford, P.; Poli, P.; Kobayashi, S.; Andrae, U.; Balmaseda, M.A.; Balsamo, G.; Bauer, D.P.; Bechtold, P.; et al. The ERA-Interim reanalysis: configuration and performance of the data assimilation system. *Q. J. R. Met. Soc.* **2011**, *137*, 553–597. [CrossRef]

32. Large, W.G.; McWilliams, J.C.; Doney, S.C. Oceanic vertical mixing: A review and a model with a nonlocal boundary layer parameterization. *Rev. Geophys.* **1994**, *32*, 363–403. [CrossRef]

33. Ioannou, A.; Stegner, A.; Tuel, A.; LeVu, B.; Dumas, F.; Speich, S. Cyclostrophic corrections of AVISO/DUACS surface velocities and its application to mesoscale eddies in the Mediterranean Sea. *J. Geophys. Res. Oceans* **2019**, *124*, 8913–8932. [CrossRef]
34. Carton X. Hydrodynamical modeling of oceanic vortices. *Surv. Geophys.* **2001**, *22*, 179–263. [CrossRef]
35. Piontkovski, S.A.; Al-Tarshi, M.H.; Al-Ismaili, S.M.; Al-Jardani, S.S.; Al-Alawi, Y.H. Inter-annual variability of mesoscale eddy occurrence in the western Arabian Sea. *Int. J. Oceans Oceanogr.* **2019**, *13*, 1–23.

Article

Ocean Circulation Drives the Variability of the Carbon System in the Eastern Tropical Atlantic

Nathalie Lefèvre [1,*], **Carlos Mejia** [1], **Dmitry Khvorostyanov** [1], **Laurence Beaumont** [2] **and Urbain Koffi** [3]

[1] Institut de Recherche Pour le Développement, Laboratoire d'Océanographie et du Climat Expérimentations et Approches Numériques (LOCEAN), Sorbonne Université, Centre National de la Recherche Scientifique, Muséum National d'Histoire Naturel, 4 Place Jussieu, 75005 Paris, France; carlos.mejia@locean.ipsl.fr (C.M.); dimitry.khvorostyanov@locean.ipsl.fr (D.K.)

[2] Division Technique de l'Institut des Sciences de l'Univers, 1 Place Aristide Briand, 92195 Meudon, France; laurence.beaumont@cnrs.fr

[3] Laboratoire des Sciences Physiques, Fondamentales et Appliquées de l'École Normale Supérieure, Abidjan 08 BP 10, Côte d'Ivoire; urban.koffi@ensabj.ci

* Correspondence: nathalie.lefevre@locean.ipsl.fr; Tel.: +33-1-4427-2857

Abstract: The carbon system in the eastern tropical Atlantic remains poorly known. The variability and drivers of the carbon system are assessed using surface dissolved inorganic carbon (DIC), alkalinity (TA) and fugacity of CO_2 (fCO_2) measured in the 12° N–12° S, 12° W–12° E region from 2005 to 2019. A relationship linking DIC to temperature, salinity and year has been determined, with salinity being the strongest predictor. The seasonal variations of DIC, ranging from 80 to 120 μmol kg^{-1}, are more important than the year-to-year variability that is less than 50 μmol kg^{-1} over the 2010–2019 period. DIC and TA concentrations are lower in the northern part of the basin where surface waters are fresher and warmer. Carbon supply dominates over biological carbon uptake during the productive upwelling period from July to September. The lowest DIC and TA are located in the Congo plume. The influence of the Congo is still observed at the mooring at 6° S, 8° E as shown by large salinity and chlorophyll variations. Nevertheless, this site is a source of CO_2 emissions into the atmosphere.

Keywords: carbon cycle; tropical Atlantic; dissolved inorganic carbon; alkalinity

Citation: Lefèvre, N.; Mejia, C.; Khvorostyanov, D.; Beaumont, L.; Koffi, U. Ocean Circulation Drives the Variability of the Carbon System in the Eastern Tropical Atlantic. *Oceans* **2021**, *2*, 126–148. https://doi.org/10.3390/oceans2010008

Academic Editor: Michael W. Lomas

Received: 14 December 2020
Accepted: 3 February 2021
Published: 8 February 2021

Publisher's Note: MDPI stays neutral with regard to jurisdictional claims in published maps and institutional affiliations.

1. Introduction

Tropical regions are strong sources of CO_2 to the atmosphere due to the equatorial upwellings, bringing CO_2-rich waters to the surface, and to high sea surface temperatures. Among all tropical regions, the Pacific Ocean is the most studied as El Niño events develop there and they are the dominant process governing the interannual variability of the air–sea CO_2 flux (e.g., [1]). The Atlantic Ocean has been less studied with relatively fewer observations and modelling analyses [2]. The Western Tropical Atlantic (WTA) and the Eastern Tropical Atlantic (ETA) receive the discharge of the two largest rivers of the world with the Amazon, near the equator in the west, and the Congo near 6° S in the east. Numerous studies have highlighted the impact of the Amazon outflow on productivity and carbon uptake in the WTA [3–10]. The dynamics of the Congo plume and its influence on the coastal ocean of the ETA are less documented than the Amazon plume [11]. The Congo plume, like large river plumes, supplies inorganic nutrients, which stimulates biological production. There is evidence of high chlorophyll concentrations (>10 mg m^{-3}) near its delta [12]. A few studies reported carbon uptake in the outer plume of the Congo river [13,14]. However, the extent of its influence on carbon properties in the ETA is poorly known.

Here, we focus on the 12° N–12° S, 12° W–12° E region defined as the ETA. This region is dominated by the seasonal variation of equatorial and coastal upwellings. The

Atlantic cold tongue, which spreads between the African coast and 20° W south of the equator, usually forms in June and lasts nearly 5 months [15]. Its formation coincides with the intensification of the southeasterly trade winds. Its spatial extent is maximum in July–September as well as its cooling [15]. The coastal upwelling along the coast between Cape Palmas (Ivory Coast) and Cotonou (Benin) is also seasonal with the main cooling period occurring at the same time, from July to September and a minor upwelling occurring from January to February [16,17].

Earlier studies showed that the tropical Atlantic Ocean has a low productivity except in areas where nutrients are brought to the surface such as the equatorial region and the coastal upwelling [18]. In the Gulf of Guinea, along the northern African coast, the highest chlorophyll concentrations are observed in July–September and coincide with the presence of the upwelling, which leads to large fish catches [19,20]. This productive region is part of the Guinea Current Large Marine Ecosystem extending from the Guinea-Bissau to the Cape Lopez in Gabon [21].

The equatorial upwelling is also a productive region with a main phytoplankton bloom in July–August, when the Atlantic cold tongue is most developed, and a secondary bloom in December as observed by chlorophyll concentrations from ocean color satellite [22]. The surface productivity is explained by the vertical supply of nutrients that fuel the phytoplankton development [23]. When the thermocline is deeper, the vertical supply is more limited and the surface productivity remains low [24]. Chlorophyll and nitrate are the biochemical variables for which extensive data exist and they were used for modelling studies in this region [25].

Carbon observations collected during oceanographic cruises in the 1980s enabled estimates of the source of CO_2 in this region (e.g., [26,27]). The seasonality of the flux and the CO_2 outgassing have been further documented by hourly monitoring of the fugacity of CO_2 (fCO_2) at the time-series station at 6° S, 10° W, e.g., [28]. More recently, a coupled physical-biogeochemical model was used to compare the air–sea CO_2 flux between the equatorial Atlantic and the equatorial Pacific [2]. One of the main result was that the variability of the sea surface partial pressure of CO_2 (pCO_2) appears to be driven by sea surface temperature (SST) in the equatorial Atlantic whereas it is driven by dissolved inorganic carbon (DIC) in the equatorial Pacific. However, the authors pointed out that the driving parameter in the equatorial Atlantic was subject to high uncertainty due to the lack of observations. They recommended regional analyses of key variables that regulate oceanic pCO_2 such as DIC.

The objective of this paper is to analyze the seasonal and interannual variability of the carbon system in the Eastern Tropical Atlantic and the impact of the Congo outflow, using surface carbon measurements from 2005 to 2019 and satellite data. We determine the driving factors of DIC and alkalinity (TA) spatial variations and provide empirical relationships for the ETA.

2. Materials and Methods

As part of the US–France–Brazil prediction and research moored array in the tropical Atlantic (PIRATA) program, five moorings have been deployed to monitor temperature, salinity, wind speed and precipitation in the eastern Tropical Atlantic [29]. The moorings are serviced and replaced every year during the French cruises. Two of these moorings have been equipped with a CO_2 CARIOCA sensor to monitor seawater fCO_2 at about 1.5 m depth. The fCO_2 has been measured hourly by spectrophotometry at 6° S, 10° W since 2006 [30] and at 6° S, 8° E from 2017 to 2019. The accuracy of the CARIOCA sensor is estimated at ±3 µatm. The fCO_2 data are archived by the Surface Ocean CO_2 Atlas project (SOCAT, www.socat.info (accessed on 31 October 2020)). The buoy at 6° S, 8° E drifted in 2018 and again in 2019; therefore, this site has been abandoned. During the cruises for servicing the PIRATA moorings, seawater samples were taken for inorganic carbon and alkalinity (National Center for Environmental Information (NCEI), Ocean Carbon Data System (OCADS), https://www.ncei.noaa.gov/access/ (accessed on 31 October 2020), accession numbers 0108086–0108091, 0171189–0171191, 0171193–0171197). The samples were analyzed by potentiometric titration with a closed-cell using the method of Edmond et al. [31]. The system was calibrated by Certified Reference Materials (CRMs) provided by Prof. A. Dickson (Scripps Institution of Oceanography, San Diego, CA, USA). The accuracy of TA and DIC is estimated at ± 3 μmol kg^{-1} and ± 5 μmol kg^{-1}, respectively. In June 2006 and March 2019, seawater and atmospheric fCO$_2$ were measured underway during the PIRATA cruises (EGEE 3 and PIRATA FR-29), using a CO$_2$ system including an infrared LiCor 7000 analyzer [14]. The accuracy of fCO$_2$ is estimated at ± 2 μatm. The PIRATA cruises have provided a good coverage of carbon measurements from ship and moorings in the $10°$ S–$6°$ N, $12°$ W–$12°$ E region (Figure 1).

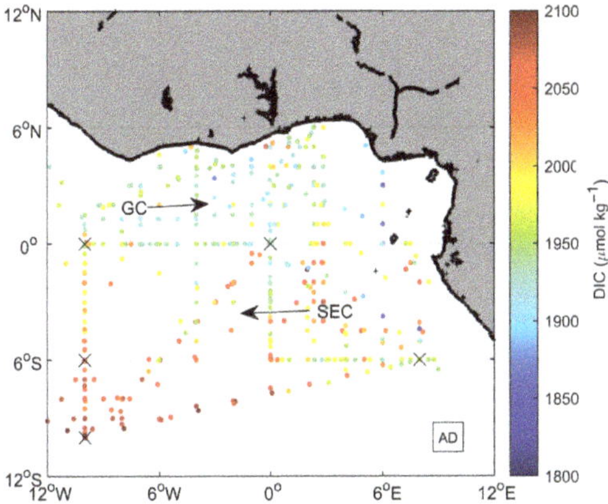

Figure 1. Distribution of DIC observations (2005–2019) and location of the five PIRATA moorings ($\times$). The main zonal surface currents are the eastward Guinea Current (GC) and the westward South Equatorial Current (SEC). The location of the Angola Dome (AD) is indicated.

The sampling covers the two main surface currents: the eastward Guinea Current (GC) in the north and the westward South Equatorial Current (SEC) in the south. The boundary between the two currents is located approximately along $3°$ N [32]. The carbon system includes four parameters: fCO$_2$, DIC, TA and pH. The knowledge of two parameters allows the calculation of the other two using the equations of the carbon system. At the deployment of the moorings and during the EGEE 3 and PIRATA FR-29 cruises, the three parameters fCO$_2$, DIC and TA were measured so that the carbon system was overdetermined. Two parameters are used to calculate the remaining one that is subsequently compared with the measurement. The calculation is performed with the CO2SYS software for Matlab [33] using the dissociation constants of Mehrbach et al. [34] refitted by Dickson and Millero [35].

A number of 119 concomitant measurements of DIC, TA and fCO$_2$ are used to check the consistency of the carbon system (Table 1).

Table 1. Statistics of the calculation of a carbon parameter from a given pair of parameters given by the Root Mean Squared Error (RMSE) and the correlation coefficient (r) between the calculated and the 119 measured values.

Pair	Calculated Parameter	RMSE	r
TA–DIC	fCO$_2$	11.7 μatm	0.90
DIC–fCO$_2$	TA	7.4 μmol kg^{-1}	0.99
fCO$_2$–TA	DIC	5.9 μmol kg^{-1}	0.99

From the three possible pairs of parameters, the fCO$_2$–TA pair gives a relatively low uncertainty to calculate DIC. The pair DIC–TA gives a high uncertainty to calculate fCO$_2$, which is explained by the strong co-variation between DIC and TA ($r = 0.96$). This is in agreement with previous works estimating combined uncertainties in calculated parameters from different measurement pairs and reporting DIC–TA as the worst pair of parameters to calculate fCO$_2$ [36,37].

The air–sea flux of CO$_2$, FCO$_2$, is calculated at the PIRATA moorings as follows:

$$\text{FCO}_2 = k * sol * \Delta\text{fCO}_2 \tag{1}$$

where k is the gas transfer velocity depending on the wind speed [38], *sol* is the solubility of seawater CO$_2$ [39] and ΔfCO$_2$ is the difference of fCO$_2$ between the ocean and the atmosphere. A positive flux means a source of CO$_2$ to the atmosphere. The PIRATA sites are equipped with an anemometer but, due to the failure of the sensor, there are some data gaps. Therefore, we use the wind speeds from the Cross-Calibrated Multiplatform (CCMP) product version 2 [40]. They are usually in good agreement with the PIRATA winds.

Empirical relationships are a useful means to estimate data from more accessible parameters, such as temperature and salinity, when no observation is available. In order to calculate DIC and pH from fCO$_2$ at the moorings, we use the following empirical TA–sea surface salinity (SSS) relationship determined by Koffi et al. [41] with 190 observations collected from 2005 to 2007:

$$\text{TA} (\pm 7.2) = 65.52 (\pm 0.77) * \text{SSS} + 2.50 (\pm 27.22),\ r^2 = 0.97, \tag{2}$$

Using 638 observations of TA from 2005 to 2019, the RMSE is 8.5 µmol kg^{-1} and r is over 0.99, which confirms the robustness of this TA–SSS relationship for the eastern tropical Atlantic. The salinity-normalized TA given by TA*35/SSS gives an average of 2298.1 $\pm$ 8.1 µmol kg^{-1} and is close to the value of 2291 $\pm$ 4 µmol kg^{-1} given by [42] for the Atlantic between 30° S and 30° N for SST > 20 °C.

Relationships between DIC and physical parameters are investigated through various techniques such as MLR (multiple linear regressions), decision trees, random forest and feed forward neural networks (Appendix A).

Statistical tests are used to compare different groups of data. When the parameters do not follow a normal distribution, we use the Wilcoxon rank sum test, equivalent to the Mann–Whitney U test. In particular, we tested whether the data located north of the equator were significantly different from the data located south of it.

We examine the link between the year-to-year variability and the tropical southern Atlantic (TSA) index. The TSA index is an indicator of the SST in the eastern tropical South Atlantic. It is the average of SST in the box 30° W–10° E, 20° S–0° [43].

Satellite images of SST, SSS, chlorophyll and precipitation provide the spatial environmental conditions in the ETA, and their seasonal variability in the region. The chlorophyll concentrations are monthly composites from MODIS (Moderate-Resolution Imaging Spectroradiometer) Aqua on a 4 km grid. The monthly composites of SST are on a 9 km grid. MODIS data are available from July 2002 to January 2020. The sea surface salinity fields are 18-day Gaussian means from SMOS (Soil Moisture and Ocean Salinity Satellite) at a resolution of 25 km (CATDS CEC LOCEAN debias V4.0, doi:10.17882/52804). The data from SMOS cover the period from January 2010 to September 2019. A monthly climatology of SSS is calculated over this period. The precipitations are from the Tropical Rainfall Measuring Mission (TRMM) and are available monthly, on a 0.25° grid. The monthly climatology of rain rate is calculated for the 2005–2019 period.

3. Results

3.1. Environmental Setting

In the 12° S–12° N, 12° W–12° E region, the MODIS climatological SST varies from 24.55 °C in July to 27.01 °C in March. During the warm season, warm water spreads over most of the basin as shown for the month of March (Figure 2a). From June to September, the surface water is cooler (<25.5 °C) as the equatorial and coastal upwellings are well developed. The months of March and August are chosen to illustrate the contrasts between the two seasons (Figure 2a,b). During the cold season, a central West African upwelling occurs in the northern part of the Gulf of Guinea between approximately 8° W and 4° E, north of 2° N, from June to October [17]. The cooling is visible in the SST map of August as a small band of cooler SST along the coast (Figure 2b). Along the coast of Gabon, around 10° E near the equator, the coastal upwelling merges with the equatorial upwelling forming the Atlantic cold tongue (ACT). In the southernmost part of the basin, austral winter cooling leads to the coldest SST.

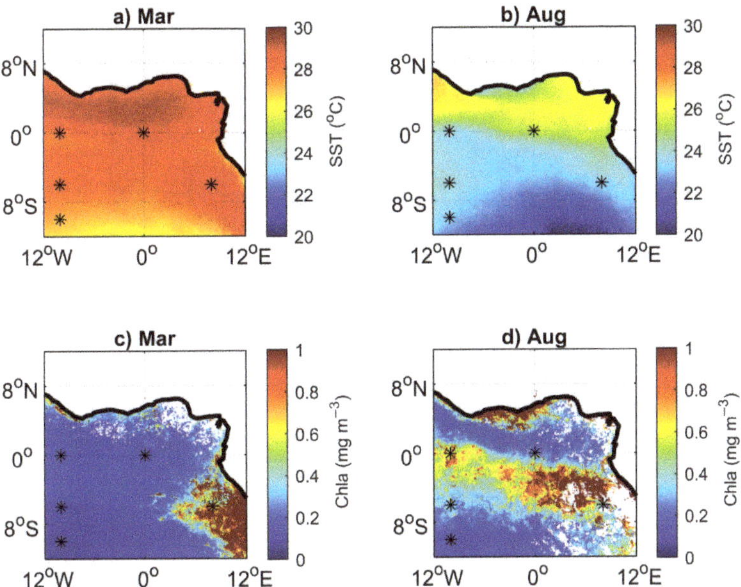

Figure 2. Distribution of the climatological sea surface temperature (SST) for (**a**) March and (**b**) August and chlorophyll a (Chla) for (**c**) March and (**d**) August from Moderate-Resolution Imaging Spectroradiometer (MODIS). The five PIRATA moorings are indicated by the black stars (*).

During the warm season, the productivity is low and the chlorophyll concentrations remain close to 0.2 mg m^{-3} except along the coastline, and in the Congo plume near 6° S, 10° E (Figure 2c). During the cold season, the chlorophyll concentrations increase along the coast, especially in the coastal upwelling of the Gulf of Guinea, and in the equatorial upwelling (Figure 2d). In the south, near 10° S, 9° E, the Angola Dome is maintained by the dynamic uplift of the thermocline, and it is also a productive region [44]. Signorini et al. [12] explain the surface distribution of the chlorophyll in the tropical Atlantic by the upwelling, defined as large vertical excursions of the thermocline, and by the river plumes. Both processes supply nutrients that lead to biological activity. Productive waters are advected by surface currents and arrive at the PIRATA moorings with the exception of the mooring at 10° S, 10° W that remains in very low chlorophyll waters throughout the year (Figure 2c,d). From 2010 to 2019, the MODIS chlorophyll concentrations during the July–September months are significantly higher than during the remaining months of the period (p-value < 0.0001), in agreement with previous studies [22]. This confirms that the cold season corresponds to the productive season.

The region is also affected by freshwater input from precipitation and river runoff. The maximum precipitation is an indicator of the position of the Intertropical Convergence Zone (ITCZ). The ITCZ migrates seasonally and is located at its northernmost position in July and at its southernmost position in March. In the ETA, the ITCZ is over the ocean during the warm season from approximately December to June. In March, it is located slightly north of the equator (Figure 3a) whereas in August it is over the land (Figure 3b). The precipitation feeds the numerous rivers that discharge into the Atlantic, which affects the distribution of SSS in the basin. The northern part of the basin is fresher than the southern part because of the presence of the ITCZ, and of the runoff of the rivers (e.g., Niger, Volta) located in the eastern Gulf of Guinea. North of about 2° N, the eastward Guinea Current, an extension of the North Equatorial Counter Current (NECC), is relatively fresh as it receives large precipitation due to the presence of the Intertropical Convergence Zone (ITCZ).

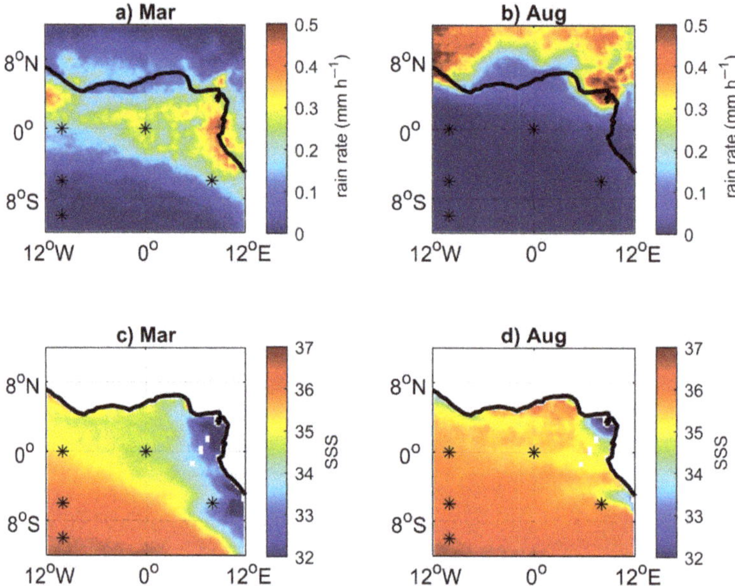

Figure 3. Distribution of the Tropical Rainfall Measuring Mission (TRMM) rain rate for (**a**) March and (**b**) August and Soil Moisture and Ocean Salinity Satellite (SMOS) SSS for (**c**) March and (**d**) August DIC. The PIRATA moorings are indicated by the black stars (∗).

South of the equator, the areas of freshwater are located along the coastline, associated to high precipitation, and near the Congo mouth at 6° S (Figure 3). The Congo River has a bimodal hydrological cycle with its maximum river discharge in December and a secondary peak of discharge in May, whereas the minimum discharge is in August and March [45]. In March, the SSS distribution follows the precipitations with low SSS (<35) associated to high precipitation in the northern and southeastern parts of the basin (Figure 3a,c). In August, the salinity is higher and low salinity (<34) is encountered near the rivers mouths of the Volta and Niger in the north and of the Congo in the south (Figure 3d). During this month of minimum discharge, the extension of low SSS water near the Congo mouth is limited.

The physical conditions in this region show a north–south difference between warmer and fresher waters in the north and colder and saltier waters in the south because of the precipitations, river outflows and the coastal and equatorial upwellings. In the northern part of the basin, the climatological SST ranges between 25.26 °C in July and 28.82 °C in February and the SSS is below 35.2, reached in August. In the southern part, the SST varies from 22.41 °C in July to 27.41 °C in February, and the SSS varies from 35.2 in April to 35.88 in July. The north–south gradient is visible on the satellite SST throughout the year. Even during the warm season, as illustrated in March (Figure 2a), the surface water is warmer north of the equator than south of it.

3.2. Variability and Drivers of the Carbon System

The DIC concentrations of the samples collected from 2005 to 2019, west of $10°$ E, range from 1618 to 2133 μmol kg^{-1} with the lowest concentrations ($<$1800 μmol kg^{-1}) close to the Congo outflow and the highest ($>$2100 μmol kg^{-1}) in the southwestern part of the region. The distribution of the DIC samples highlights the lower concentrations north of the equator compared to the south (Figure 1). As TA is strongly correlated to SSS, TA exhibits also a north–south difference. In the north, TA is lower with values ranging from 2114 to 2369 μmol kg^{-1} and a mean of 2278 $\pm$ 38 μmol kg^{-1}. In the south, TA has a mean value of 2324 $\pm$ 66 μmol kg^{-1} with a maximum of 2433 μmol kg^{-1} reached south of $8°$ S, near $4°$ W where evaporation dominates. The lowest TA value of 1832 μmol kg^{-1} is found in the Congo plume. Warm and fresh waters are located in the GC system and exhibit lower TA and DIC concentrations whereas colder and saltier water mass transported by the SEC have higher alkalinity and DIC content. The distributions of SST, SSS, DIC and TA north and south of the equator are significantly different ($p < 0.001$).

DIC concentrations vary according to the main system of currents and depend on the water masses present in the region, which are characterized by their SST and SSS. This led Koffi et al. [41] to propose an empirical relationship using data over 2005–2007. However, due to the atmospheric CO_2 increase and air–sea exchange, CO_2 increases in the ocean over time and the relationship presents a bias. The difference between the observations and the regression increases over time, which confirms that the relationship is no longer valid for recent years. According to the atmospheric CO_2 measurements during EGEE 3 and PIRATA FR-29, the atmospheric fCO_2 has increased from 367.6 $\pm$ 1.7 μatm in June 2006 to 392.2 $\pm$ 2.4 μatm in March 2019, which corresponds to a rate of 1.9 μatm yr^{-1}.

In order to take into account the CO_2 increase over time, we develop a new multiple linear regression (MLR) by adding a time variable:

$$\text{DIC} = -4585.8 \, (\pm \, 371.16) - 12.46 \, (\pm \, 0.53) * \text{SST} + 54.36 \, (\pm \, 0.93) * \text{SSS} + 2.49 \, (\pm \, 0.19) * \text{Year} \tag{3}$$

This relationship is based on 637 samples collected in the $12°$ W–$10°$ E, $10°$ S–$10°$ N region from 2005 to 2019. The RMSE is 15.3 μmol kg^{-1} and $r^2 = 0.93$. The negative sign of the SST coefficient in Equation (3) reflects the contribution of the upwelling with cold SST leading to higher concentrations of DIC. The same regression is performed after normalization of the data (by subtracting the mean and dividing by the standard deviation) in order to determine the weight of each predictor. The coefficients for SST, SSS and Year are -0.42, 0.75 and 0.20, respectively. The stronger predictor is SSS, which is the dominant factor driving the DIC variations. However, without taking into account the SST, the variance explained would drop from 93% to 87% and the RMSE would increase to 21 μmol kg^{-1}.

Equation (3) is determined using discrete samples of DIC and needs to be validated with independent data. Surface fCO_2 has been measured underway during two cruises (EGEE 3 and PIRATA FR-29) and at two moorings ($6°$ S, $10°$ W and $6°$ S, $8°$ E) on an hourly basis. In order to check the robustness of Equation (3), DIC is calculated from the alkalinity–SSS relationship and the fCO_2 measurements. The calculated DIC is then compared with the predicted DIC (Table 2). DIC given by Equation (3) is in good agreement with DIC calculated from underway fCO_2. The correlation coefficient is always over 0.96 and the RMSE is within the uncertainty given by Equation (3). The highest RMSE is found at $6°$ S, $8°$ E.

Using monthly MODIS SST and SMOS SSS fields on a $0.25°$ grid, the monthly regional distribution of DIC can be estimated using Equation (3) to map its spatial variations (Figure 4). The monthly climatology of DIC is built for the January 2010 to September 2019 period. Overall, the monthly maps of DIC are similar to the SSS maps with lower values in the northern part of the basin and close to the river mouths.

Table 2. Comparison between the DIC predicted by Equation (3) and DIC calculated from measured fCO$_2$ and alkalinity from Equation (2). The statistics include the Root Mean Squared Error (RMSE), the correlation coefficient (r) and the number of data (N). The data of the moorings correspond to daily means.

Mooring or Cruise	RMSE *	r	N	Time Period
6° S, 10° W	9.7 µmol kg^{-1}	0.96	6611	2006–2017
6° S, 8° E	14.4 µmol kg^{-1}	0.99	239	2017–2019
EGEE 3	9.8 µmol kg^{-1}	0.98	6895	2006
PIRATA FR-29	10.0 µmol kg^{-1}	0.99	4462	2019

* RMSE using the coefficients of Equation (3) with all decimal places are calculated in Appendix A.

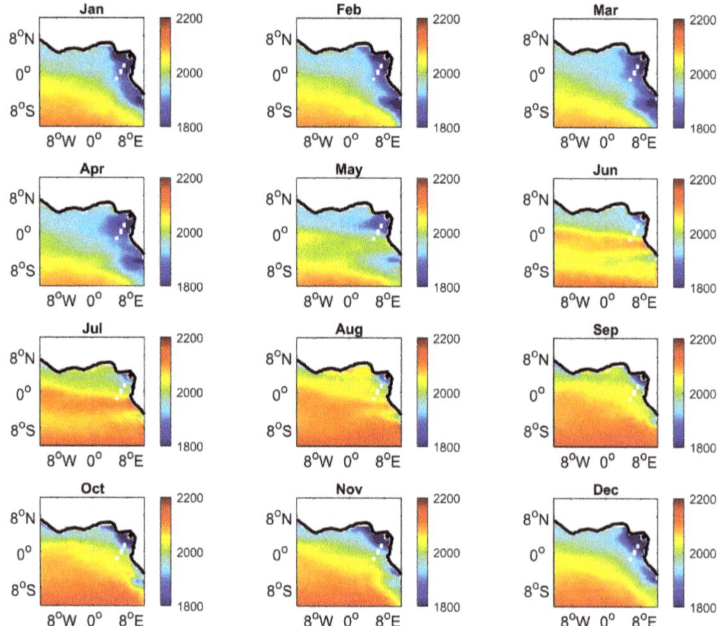

Figure 4. Climatology of DIC (in µmol kg^{-1}) obtained from Equation (3) with the monthly fields of MODIS SST and SMOS SSS. The DIC climatology is calculated over the January 2010 to September 2019 period.

Lower DIC concentrations are associated with the northern water mass and range from a mean value of 1924 µmol kg^{-1} in April to 2027 µmol kg^{-1} in August. Higher DIC concentrations are found in the southern water mass with a mean value of 2001 µmol kg^{-1} in April to 2096 µmol kg^{-1} in August. The monthly north–south DIC concentrations are significantly different throughout the year (p-value < 0.0001). During the upwelling period (July–September), the DIC concentrations are significantly higher than during the warm season (p-value < 0.0001). Thus, for the upwelling period, the mean DIC is 2070 ± 70 µmol kg^{-1}, whereas it is 2007 ± 28 µmol kg^{-1} during the warm season over the 2010–2019 period. In the upwelling season, the north–south contrast in DIC corresponds to a north–south SST difference. For example, in August, warm surface waters (>26 °C) are separated from cold waters (<24 °C) with the frontal zone located around the equator (Figure 2b). This front is also observed on the DIC maps from June to September (Figure 4).

3.3. Impact of the Congo Plume

The Congo has by far the strongest river discharge in this region. Near the Congo mouth, the carbon system is affected by the Congo discharge and is examined from nine observations taken east of 10° E, around 6° S, over the period 2013–2018 from March to June. The lowest salinity (27.3 psu) was observed in March 2018, at the time of the secondary minimum of the Congo discharge. There was no sampling during the period of high Congo discharge, in December and May. Using the nine samples near the Congo outflow, the mixing between the river and oceanic waters gives the following equation:

$$\text{DIC} = 50.60 \, (\pm 1.97) * \text{SSS} + 231.73 \, (\pm 62.19) \tag{4}$$

with a RMSE of 16.0 µmol kg^{-1} and $r^2 = 0.99$. The relationship covers the region close to the Congo mouth from 6° S to 5° S, east of 10° E. Further west, carbon data are available at the mooring at 6° S, 8° E. Hourly fCO$_2$ has been monitored at 6° S, 8° E from 2017 to 2019 with a CARIOCA sensor. This dataset includes observations from March to August. In addition, DIC samples have been taken at this site. These DIC samples and the DIC calculated from daily averages of fCO$_2$ and TA–SSS tend to depart from the mixing line (Figure 5). The RMSE is 32 µmol kg^{-1} (N = 358) and 24 µmol kg^{-1} (N = 11) for the calculated DIC and for the DIC samples at 6° S, 8° E, respectively.

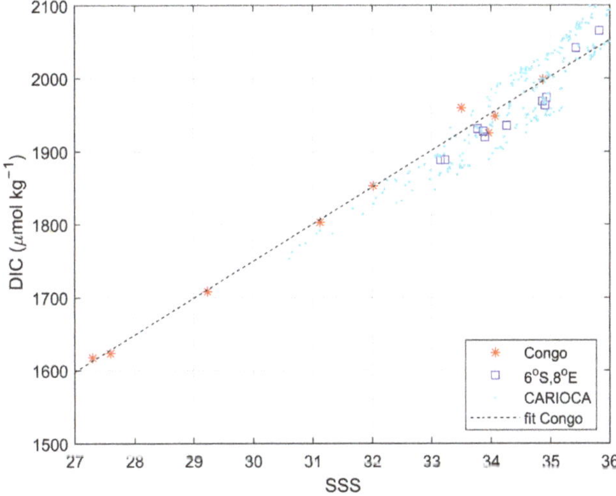

Figure 5. DIC–SSS relationship obtained with the samples near the Congo outflow (dashed line) with DIC samples used for the regression (red stars), DIC samples at 6° S, 8° E (blue squares) and DIC calculated (cyan dots) from daily fCO$_2$ measured at 6° S, 8° E and TA calculated from SSS at 6° S, 8° E.

The DIC data at 6° S, 8° E are better estimated by Equation (3), giving a RMSE of 14 µmol kg^{-1} for both the 11 samples taken at the site and the DIC calculated from the daily fCO$_2$. This suggests that the conservative mixing applies near the River mouth but it is no longer valid further west, at 8° E. Nevertheless, the influence of the Congo at the mooring is noticeable. Using satellite based data products, Hopkins et al. [46] show that the Congo plume extends northwest along the coastline or west-southwest into the Atlantic transported by the SEC. From MODIS chlorophyll images (Figure 2c,d), high chlorophyll concentrations are observed at the mooring at 6° S, 8° E in both seasons and originate from the coast near the Congo mouth.

Using all the available data measured at 6° S, 8° E over the 2006–2019 period, the SSS are grouped by month (Figure 6a). A data gap occurred from 2007 to 2013 as no mooring was deployed. This site exhibits large salinity variations. The lowest salinity of 30.22 psu, occurs in May 2018 during the secondary maximum discharge of the Congo. The numerous outliers usually point towards lower salinity than the median. In December, during the maximum Congo discharge, the salinity of the outliers ranges from 32 to 29 psu. The large range of salinity of the outliers (>1 psu) is usually observed around the maximum discharge periods, December and May. The median salinity of December is over 35 psu, which indicates that the plume has not reached the mooring yet, except for some years as in 2006 when a salinity close to 29 psu was observed. The lowest SSS median is in January, which suggests that the Congo plume, transported by the westward SEC, reaches the mooring. Based on satellite images, Hopkins et al. [46] found that the maximum offshore extent of the Congo plume occurs between December and April with a January peak. The freshwater supplied by the secondary peak of discharge, in May, tend to move towards the Gulf of Guinea [47]. The minimum Congo discharge occurs in August, which corresponds to the highest salinity at 6° S, 8° E. The measured values of SSS and DIC, calculated from fCO_2 at the mooring in 2017, 2018 and 2019, and averaged for each month, are also indicated (Figure 6).

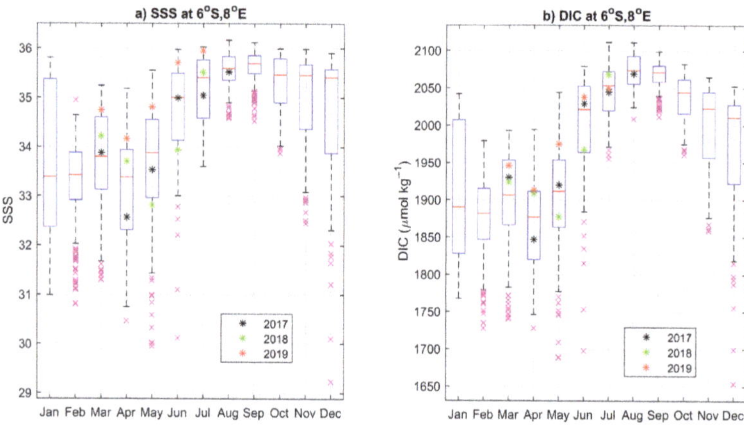

Figure 6. (**a**) Distribution of SSS at 6° S, 8° E grouped by month using SSS observations from 2006 to 2019 (the data at 6° S, 8° E are for the periods 28 June 2006 to 10 June 2007 and 5 July 2013 to 8 August 2019). The central mark of the box corresponds to the median with the bottom and top edges of the box indicating the 25th and 75th percentiles. The whiskers extend to the extreme data points and the crosses (×) are considered as outliers. (**b**) Same for the distribution of DIC at 6° S, 8° E calculated with Equation (3) using the SSS and SST recorded at the PIRATA mooring. The monthly measurements of SSS and DIC calculated from fCO_2 at the PIRATA mooring are indicated for 2017 (black stars), 2018 (green stars) and 2019 (red stars).

The DIC variations follow closely the SSS variations with the highest salinity corresponding to the highest DIC in August, the largest variations in January and the lowest DIC concentrations occurring during the January–May period (Figure 6b). The year-to-year variability is very pronounced as shown by the samples taken at this site. In April 2017, the salinity is lower than 1 psu and DIC is higher than 50 μmol kg^{-1} compared to April 2018 and 2019. Likewise, in June 2018 a low salinity (34 psu) corresponds to a low DIC (1968 μmol kg^{-1}) whereas in June 2017 and 2019 the salinity is over 35 psu and the DIC concentrations over 2029 μmol kg^{-1}.

The distribution of the chlorophyll from MODIS is examined at the 6° S, 8° E mooring (Figure 7). The chlorophyll concentrations at this site are relatively high, which is in contrast with the other four PIRATA moorings where concentrations are always lower than 1.5 mg m^{-3} at 6° S, 10° W and lower than 3 mg m^{-3} at the two equatorial moorings. The lowest chlorophyll concentrations (<0.25 mg m^{-3}) are encountered at 10° S, 10° W as this site is outside the area of influence of both the Atlantic cold tongue and the Congo plume.

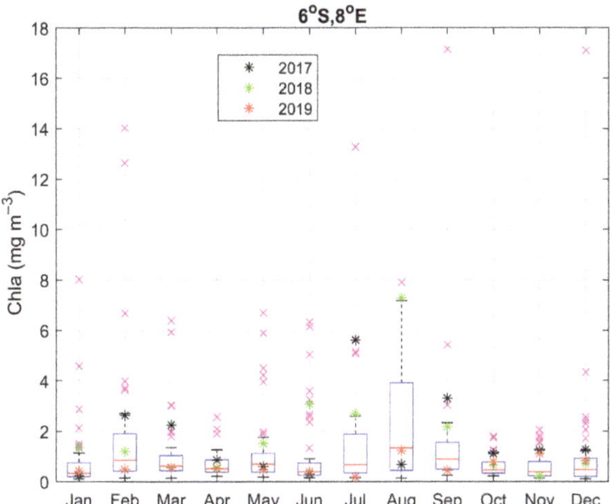

Figure 7. Distribution of MODIS chlorophyll at 6° S, 8° E grouped by month using data from 2002 to 2019. The central mark of the box corresponds to the median with the bottom and top edges of the box indicating the 25th and 75th percentiles. The whiskers extend to the extreme data points and the crosses are considered as outliers.

At 6° S, 8° E, the highest concentrations of MODIS chlorophyll *a* occur in July–September and, to a lesser extent, in January–February, which is in agreement with the results of Signorini et al. (1999) using SeaWiFS (Sea-viewing Wide Field-of-view Sensor) chlorophyll *a* from 1997 to 1998. The concentrations of the years 2017, 2018 and 2019 are indicated (Figure 7) to compare with the DIC variations (Figure 6b). The chlorophyll *a* from MODIS is highly variable at 6° S, 8° E with the highest concentrations in August, as shown by the whisker reaching over 7 mg m^{-3} (Figure 7), when the concentrations of DIC are the highest. There is no clear link between DIC and chlorophyll concentrations as the highest chlorophyll are not associated with low DIC that would occur if strong carbon uptake were taking place. In July–September, both chlorophyll and DIC concentrations are high. This suggests that biological activity is not the dominant driver of carbon variations at this site.

3.4. Year-to-Year Variability of the Carbon Parameters

In the ETA, the longest time-series of carbon parameters is the monitoring of hourly fCO$_2$ at 6° S, 10° W that started in 2006. At 6° S, 8° E, the monitoring of fCO$_2$ is rather limited with observations from 2017 to 2019 and a large data gap in 2018–2019 due to the drift of the buoy. The daily DIC at these two sites, calculated from daily fCO$_2$, is compared with the DIC from the MLR, and with the samples taken during the PIRATA cruises at both locations (Figure 8a,b). The MLR reproduces rather well the variations of DIC at 6° S, 10° W but overestimates the DIC concentrations in some years, such as in July–September 2011 where the difference is over 35 µmol kg^{-1}. The timing of the decreases and increases in DIC is in good agreement between the DIC calculated by the MLR and the DIC calculated from underway fCO$_2$. At 6° S, 10° W, DIC values are higher in July–November and lower in April–June each year. At 6° S, 8° E the daily variations of DIC tend to show the lowest concentrations in April and the highest in August but with large variations. For example, DIC concentrations vary by more than 150 µmol kg^{-1} within days in April 2017 (Figure 8b). In April 2019, the lowest DIC is much higher than in April 2017 with a difference reaching 125 µmol kg^{-1}. Nevertheless, the MLR captures well the high frequency variations of DIC.

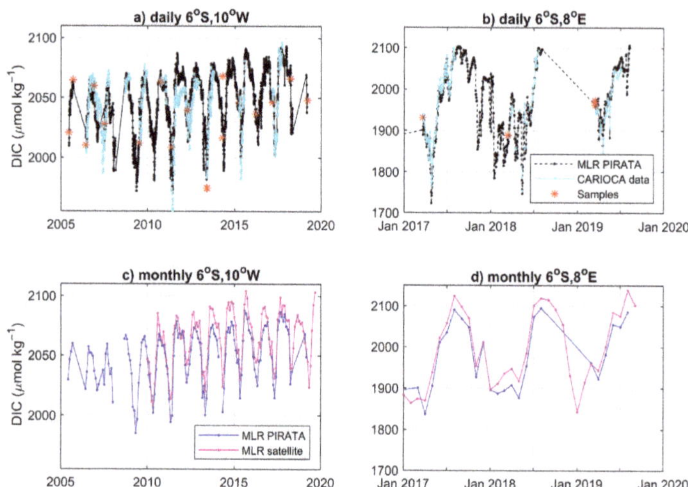

Figure 8. (**a**) Daily DIC determined from the multiple linear regressions (MLR) using SST and SSS recorded at the 6° S, 10° W mooring since 2006 (in black) and calculated from measured seawater fCO_2 and TA–SSS (in cyan) at 6° S, 10° W The DIC samples are in red. (**b**) as in (**a**) for the mooring at 6° S, 8° E from 2017 to 2019. (**c**) Monthly DIC calculated with the MLR using SST and SSS recorded at the mooring (in black) and using satellite SST and SSS (in blue) collocated at 6° S, 10° W. (**d**) same as in (**c**) for the mooring at 6° S, 8° E from 2017 to 2019.

As expected, given the discrepancy observed in the daily variations, monthly DIC derived from satellite data tend to give higher concentrations than DIC calculated with the PIRATA data but reproduce the seasonal variations of DIC (Figure 8c,d). The monthly DIC smooth significantly the variations. It is especially visible at 6° S, 8° E on the MLR with a reduced range of variations of DIC, decreasing from 460 to 320 µmol kg^{-1}. The seasonal cycle of DIC at 6° S, 8° E is more distinct than on the daily variations. Given that the satellite pixels correspond to a much larger area than the observations at the mooring, there is expected discrepancy between the two products but overall, the MLR reproduces well the seasonal cycle and the year-to-year variations of DIC.

After verifying the DIC variations over time at the two sites where observations are available, we examine the monthly DIC variations given by the MLR using satellite data at regional scale. The DIC concentrations are lower in the north than in the south with the maximum values occurring from July to September each year in both regions (Figure 9). The monthly means of DIC over the entire region are closer to the DIC variations observed in the south in Figure 9 because the region north of the equator includes a large section of land and, hence, the weight of this region is much smaller in the global average. The lowest concentrations of DIC occur in March–April in the south, whereas, north of the equator, low DIC concentrations (<1950 µmol kg^{-1}) occur over a much longer period, usually from November to May. The difference between the highest and the lowest DIC concentrations range from about 80 to 125 µmol kg^{-1} in both regions. This range is much higher than the year-to-year variability of DIC that is less than 50 µmol kg^{-1} over the 2010–2019 period. On annual average, over the whole region, the DIC variations are less than 35 µmol kg^{-1}. In both the northern and southern regions, the seasonal variability is responsible for the large variations of DIC (Figure 9).

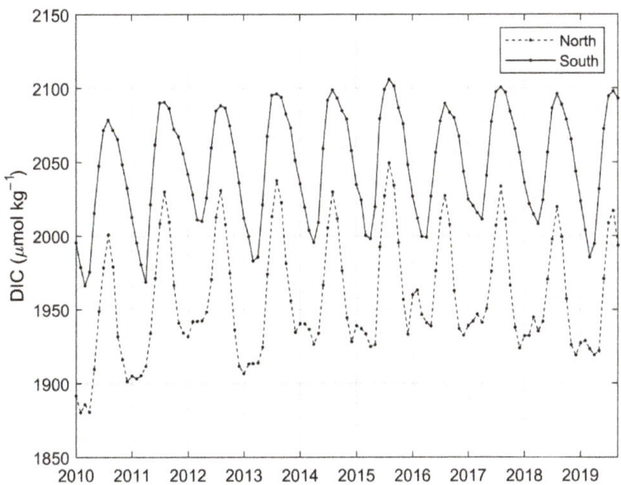

Figure 9. Monthly DIC determined from the MLR using satellite SST and SSS for the region north and south of the equator between January 2010 and September 2019.

The year 2010 exhibits the lowest annual average of DIC in the north (p-value = 0.006) and in the south but, in the southern region, it is not significantly different from the other years. The salinity data retrieved from SMOS are of lower quality from January to March 2010, at the beginning of the series. If we remove these 3 months in the time series, the annual mean of DIC in 2010 is no longer significantly different from the other years in both regions. The highest DIC annual average occurs in 2015 north of the equator and in 2017 south of it, but these years are not significantly different from the other years. Overall, the year-to-year variability of DIC does not highlight any anomalous year over the 2010–2019 period.

4. Discussion

4.1. Main Features of the Carbon Parameters

Both DIC and TA are strongly correlated to physical parameters. The TA–SSS relationship is robust and does not vary much with new observations. The DIC can be estimated by taking into account the water masses, characterized by their temperature and salinity, and an increase in oceanic surface CO_2 over time due to the atmospheric CO_2 increase. In stratified regions, like the tropics, the oceanic CO_2 tends to follow the rate of the atmospheric increase. The MLR used here shows that SSS, SST and year describe well the DIC variations in this region. Other methods, taking into account non-linear processes, do not perform significantly better than the MLR (Appendix A), which implies that the DIC dependency on the SSS, SST, and the year is essentially linear. However, feed forward neural networks show slightly better results with values of the first and last quantiles of DIC, suggesting that a non-linear dependency exists between DIC and input variables (Appendix A). Low concentrations of DIC are mostly encountered in coastal area in river plumes. As most samples are in the open ocean region, the MLR is adequate here. However, a non-linear method would be required when more samples in the coastal zone become available.

The DIC distribution in the 12° W–10° E, 12° S–12° N region (Figure 4) shows high DIC concentrations in most part of the ETA during the upwelling season. Overall, the DIC values range from 1945 to 2090 μmol kg^{-1} with an average of 2024 $\pm$ 37 μmol kg^{-1} from 2010 to 2019. The highest DIC values occur during the cold and productive season, in July–September each year (Figure 9). In this season, the MODIS chlorophyll is significantly higher. The vertical supply of subsurface waters is responsible for bringing CO_2-rich waters to the surface that spread over the basin. It is the dominant process and the biological activity is not strong enough to counterbalance the carbon supply.

The DIC concentrations in the ETA are similar to DIC observations collected further west. In the Western Tropical Atlantic, in the SEC, DIC concentrations average 2045 ± 40 µmol kg^{-1} in April–September and 2058 ± 22 µmol kg^{-1} in October–March from analyses made over the period 1989–2014 [48]. In the ETA, south of the equator, the DIC concentrations, between 2010 and 2019, average 2059 ± 39 µmol kg^{-1} in April–September and 2036 ± 32 µmol kg^{-1} in October–March.

The lowest DIC concentrations in the ETA are observed east of 10° E near the Congo mouth. In this area, DIC is better estimated with the conservative mixing Equation (4) than with the MLR. The slope of Equation (4) is different from the ones given by published relationships (Table 3). As previously noticed [49], the value of the slope is sensitive to the salinity range. Using all the data or data with salinity higher than 33 changes the slope from 46.5 µmol kg^{-1}/psu to 50.6 µmol kg^{-1}/psu [49]. It also changes the end-member (DIC at SSS = 0) for the Congo from 355 ± 48 µmol kg^{-1} to 221 ± 97 µmol kg^{-1}. As no sample was taken during the highest Congo discharge, our lowest salinity is above 27. Nevertheless, the end-member for the Congo River of 231.7 ± 62.2 µmol kg^{-1} is in good agreement with the average riverine DIC concentrations of 258 ± 29 µmol kg^{-1} measured at Brazzaville–Kinshasa by [45].

Table 3. Mixing equations, including Equation (4), between the Congo River and oceanic waters.

Mixing Equation	Salinity Range	Reference
DIC = 50.6 ($\pm$2.0) $*$ SSS + 231.7 ($\pm$62.2)	S > 27	This work
DIC = 54.0 $*$ S + 109	S > 33	[13]
DIC = 46.5 ($\pm$1) $*$ SSS + 355 ($\pm$48)	S > 22	[49]

TA values are also lower near the Congo mouth and are better estimated by the following equation than by Equation (2):

$$TA = 61.89 \, (\pm \, 0.70) * SSS + 137.51 \, (\pm \, 22.08) \tag{5}$$

which gives a TA end-member of 137.51 ± 22.08 µmol kg^{-1}, in agreement with the riverine TA values ranging from 85 to 235 µmol kg^{-1}, the variations being strongly dependent on the Congo discharge unlike the DIC end-member that is relatively constant [45].

The chlorophyll images show the high biological activity near the Congo mouth, which decreases DIC concentrations and leads to carbon uptake as previously proved by the estimates of the CO_2 flux in this region [13,14]. The mouth of the Congo is located at 6° S, 12.3° E and its plume can be traced 400 to 1000 km from the mouth. The influence of the Congo plume is evidenced at 6° S, 8° E by the large salinity (Figure 6a) and chlorophyll variations (Figure 7), as well as the spatial distributions of chlorophyll (Figure 2c,d) and salinity (Figure 3c,d). In order to relate the DIC distribution (Figure 4) to the chlorophyll climatology, the chlorophyll fields were re-gridded on a 0.25-degree map and the correlation between DIC and chlorophyll concentrations was calculated at each grid point from January to December (Figure 10).

The blue regions correspond to negative correlations and coincide with high chlorophyll concentrations and low DIC, which are caused by carbon uptake. The positive correlations (in red) correspond to regions where chlorophyll and DIC follow the same seasonal cycle. In the equatorial upwelling region, high DIC and high chlorophyll concentrations are observed during the upwelling season (July–September) and lower concentrations occur during the rest of the year, which explains the positive correlation between DIC and Chla. Outside the upwelling region, low chlorophyll concentrations are observed but follow the seasonal cycle and the DIC–Chla correlation is also positive. Near river outflows, high chlorophyll concentrations are usually associated with low SSS and low DIC concentrations leading to negative DIC–Chla correlations. The spatial extension of the negative correlations indicates the areas of influence of the river discharge. As SSS is a strong driver of DIC variations in the ETA, the negative DIC–Chla correlations correspond to areas with low SSS and high chlorophyll concentrations.

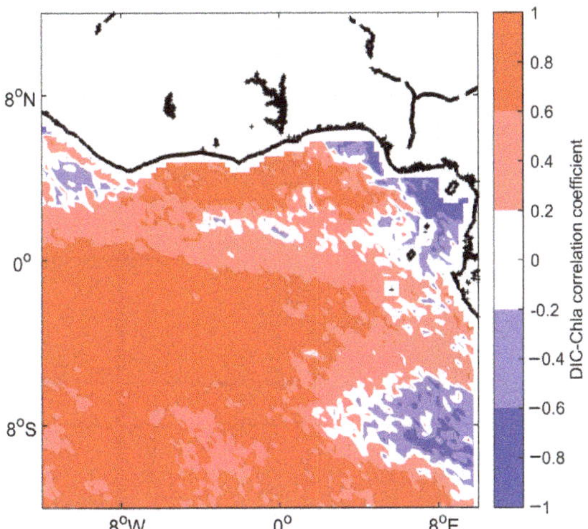

Figure 10. Map of correlations between chlorophyll and DIC concentrations calculated over 12 climatological months.

At 6° S, 8° E, low DIC and TA are observed especially in March–April during the period of maximum offshore extension of the Congo plume [46]. Nevertheless, the DIC observations at this site are better reproduced by the MLR, which suggests rapid mixing of the plume with oceanic waters during its westward propagation.

Observations made at 6° S, 8° E are compared with those at 6° S, 10° W. Unfortunately, measurements of fCO_2 are available at both sites for 6 months only, from March to August 2017. The CO_2 flux is much lower at 6° S, 8° E than at 6° S, 10° W (Table 4). However, the difference of fCO_2 between the ocean and the atmosphere, ΔfCO_2, is similar at both sites. DIC and SSS, and hence TA, are significantly lower at 6° S, 8° E. An increase in DIC would increase fCO_2 but the increase in TA would decrease fCO_2. The increase in DIC from 6° S, 8° E to 6° S, 10° W is compensated by the increase in alkalinity, so that overall, the mean fCO_2 (and ΔfCO_2) remains similar to the mean fCO_2 at 6° S, 8° E.

Table 4. Mean and standard deviation of the carbon parameters, salinity and temperature at 6° S, 8° E and 6° S, 10° W from March to August 2017 from observations at the PIRATA moorings. The winds are from the Cross-Calibrated Multiplatform (CCMP).

Site	CO_2 Flux (mmol m^{-2}d^{-1})	ΔfCO_2 (µatm)	DIC (µmol kg^{-1})	SSS	SST (°C)	Wind (m s^{-1})
6° S, 8° E	3.06 ± 1.74	59 ± 28	1973 ± 87	34.3 ± 1.1	26.6 ± 2.8	5.0 ± 0.6
6° S, 10° W	5.61 ± 1.49	55 ± 16	2047 ± 27	35.9 ± 0.2	26.9 ± 1.7	7.1 ± 0.7

The flux is significantly lower at 6° S, 8° E because the wind speed is much weaker close to the coast than offshore. However, the CO_2 flux remains positive with only slight negative values, or close to zero, that may occur in March–April in some years (Figure 10). The flux and the ΔfCO_2 are strongly correlated at this site ($r = 0.95$). During this time, the offshore spatial extension of the Congo plume is maximum [46] and the SSS at 6° S, 8° E is low (Figure 6a). The surface waters coming at the mooring are from the Congo plume and have low CO_2 content because of the biological activity near the Congo mouth and the low CO_2 content of the Congo waters. Although DIC–TA is the worst pair to calculate fCO_2, the flux has been reconstructed at 6° S, 8° E to examine its annual variations (Figure 11, MLR sat). Some differences can reach more than 2 mmol m^{-2} d^{-1} as in June 2017 but the reconstruction suggests that the site is a source of CO_2 with some possible CO_2 uptake occurring during the first months of the year.

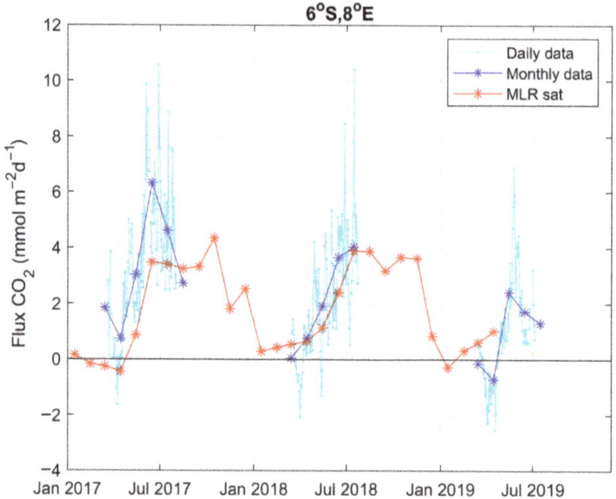

Figure 11. CO_2 flux from daily fCO_2 observations at 6° S, 8° E (in cyan), averaged monthly (in blue) and using the MLR with satellite SST and SSS (in red).

4.2. Year-to-Year Variability

Using the MLR and the SST and SSS satellite fields from 2010 to 2019 at regional scale, the anomalies of DIC are calculated and compared with the anomalies of the physical parameters (Figure 12). As salinity is a strong predictor of DIC, the monthly variations of DIC anomalies from 2010 to 2019 are correlated with SSSA ($r = 0.87$). The anomalies of DIC are negatively correlated with SSTA and the correlation is much weaker ($r = -0.38$). The anomalies of SST in the ETA are related to the TSA index that cover a much larger area of the South Atlantic (30° W–10° E, 20° S–0°), with a correlation coefficient between SSTA and TSA of 0.77.

In the tropical region, the year 2010 is characterized by much warmer temperatures whereas 2012 is a cold year [50]. The positive SSTA in 2010 are observed in the northern Tropical Atlantic and affect the air–sea CO_2 flux [51]. In the ETA, SSTA are close to zero suggesting that the anomalous conditions of 2010 are limited to the Northern Tropical Atlantic. However, the ETA presents negative SSS anomalies throughout the year 2010 that result in negative DIC anomalies (Figure 12a). There was no anomaly of precipitation in the ETA in 2010 but, during the first months of 2010, the position of the ITCZ in the tropical Atlantic remained north of the equator instead of migrating south of it. As the ITCZ is associated with high precipitations, lower-than-usual salinities advected by the NECC might explain the negative SSSA in the ETA. When comparing the SSS anomalies in 2010 north and south of the equator, the SSSA are significantly lower north of the equator (mean of −0.36) than south of it (mean of −0.16, *p*-value = 0.0024). Nevertheless, the impact of the SSSA anomalies on DIC is moderate as it does not exceed 40 µmol kg^{-1}, which is much less than the seasonal DIC variations in this region, as shown, for example, in Figure 9.

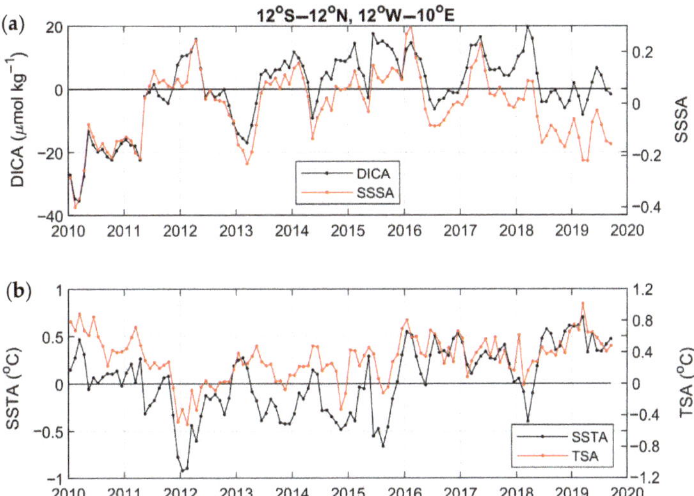

Figure 12. (**a**) Monthly DIC and SSS anomalies (DICA and SSSA), (**b**) SST anomalies (SSTA) and Tropical Southern Atlantic (TSA) index from January 2010 to September 2019.

The other strong anomaly in the tropical Atlantic is the cold event of 2012 mainly observed in the northeastern Atlantic and the southern hemisphere [50]. From November 2011 to early 2012, a strong cooling anomaly (SSTA < 0) occurs in the region and is associated with a negative TSA index (Figure 12b). The distribution of fCO_2 at 6° S, 10° W was affected by the cooling with a decrease in fCO_2 associated to a decrease in SST during January–March 2012 compared to other years [52]. However, the SST variations have a moderate effect on DIC and this cooling anomaly does not affect much the DIC concentrations (Figure 12a).

According to [2], interannual fCO_2 is mainly driven by SST in the equatorial Atlantic extending from 50° W to 5° E and by DIC in the equatorial Pacific. The fCO_2 monitoring is still sparse in the tropical Atlantic to conclude. Nevertheless, the interannual events of 2010 and 2012 affected fCO_2 and suggest an important role of SST anomalies whereas in the ETA, DIC is mainly driven by SSS variations. The use of DIC and TA to calculate fCO_2 does not reproduce well the fCO_2 variations observed at 6° S, 10° W, which might be explained by the strong co-variation of fCO_2 and SST. Using pH that has a strong dependency with SST with either DIC or TA would better reproduce observed fCO_2. Monitoring fCO_2 and calculating TA, with the robust TA–SSS relationship, allow the calculation of pH. This would be a substitute for direct pH measurements. Given the strong seasonality of carbon parameters in this region, long term monitoring is necessary to be able to quantify the acidification rate in this region.

5. Conclusions

The DIC distribution has been examined in the ETA (12° W–12° E, 12° S–12° N) as well as the factors controlling its distribution. DIC strongly depends on the water masses and increases over time due to the atmospheric CO_2 increase. In the northern part of the basin, relatively fresh and warm waters are associated with lower DIC concentrations compared to the southern part where colder and saltier waters are enriched in CO_2. A multiple linear regression between DIC and SST, SSS and year has been determined with SSS as the main predictor of DIC. Other regression techniques, such as decision tree, random forest and feed forward neural networks, do not significantly improve the DIC estimation except for low DIC concentrations corresponding to coastal data. However, a neural network would be required when more coastal data become available as non-linearity occurs at low values.

East of 10° E, the strong influence of the Congo plume is evidenced with a different relationship which corresponds to conservative mixing between the river and oceanic waters. The end-members of 231.7 ± 62.2 µmol kg^{-1} for DIC and 137.5 ± 22.1 µmol kg^{-1} for TA are in good agreement with published riverine measurements. The Congo plume reaches its more offshore spatial extension during December–April and affects the distribution of DIC at the 6° S, 8° E mooring. Nevertheless, the conservative mixing is no longer valid as the plume rapidly mixes with oceanic waters. Consequently, the carbon uptake is restricted to the region close to the Congo mouth and the carbon supplied by the upwelling season is the dominant process in the ETA. The seasonal cycle dominates the variations of DIC in the region, whereas the year-to-year variability of DIC is less than 40 µmol kg^{-1} over the 2010–2019 period. SSS is the main driver of DIC in this region and the cold event of 2012 does not lead to significant DIC anomalies.

Although no pH measurements are made, the observations of fCO$_2$ and the calculation of TA from SSS allow the calculation of pH. Pursuing the monitoring of fCO$_2$ at time-series stations will contribute to the quantification of the rate of ocean acidification in this region.

Author Contributions: Conceptualization, N.L.; methodology and validation, N.L., C.M. and D.K.; writing—original draft preparation, N.L.; writing—review and editing, N.L., C.M., D.K., L.B. and U.K. All authors have read and agreed to the published version of the manuscript.

Funding: This research was funded by the Institut de Recherche pour le Développement (IRD), the European integrated project AtlantOS (grant agreement 633211), and the French Integrated Carbon Observation System (ICOS-France).

Data Availability Statement: The data presented in this study are openly available and the references are given in the paper.

Acknowledgments: We are very grateful to the volunteers who collected the samples during the PIRATA cruises and to the US IMAGO from the IRD, especially F. Roubaud, J. Grelet and P. Rousselot for the installation and deployments of the CO$_2$ sensors. DIC and TA have been analyzed at the service national d'analyses des paramètres océaniques du carbone in LOCEAN. The L3_DEBIAS_LOCEAN_v4 Sea Surface Salinity maps have been produced by LOCEAN/Institut Pierre Simon Laplace (UMR CNRS/UPMC/IRD/MNHN) laboratory and ACRI-st company that participate in the Ocean Salinity Expertise Center (CECOS) of Centre Aval de Traitement des Donnees SMOS (CATDS). This product is distributed by the Ocean Salinity Expertise Center (CECOS) of the CNES-IFREMER Centre Aval de Traitement des Donnees SMOS (CATDS), at IFREMER, Plouzane (France). We thank two anonymous reviewers for their constructive comments.

Conflicts of Interest: The authors declare no conflict of interest.

Appendix A. Regression Methods and Results

Four data-driven regression methods (Multiple Linear Regression, Decision Tree, Random Forest and Feed Forward Neural Networks) have been applied to estimate DIC in the Eastern Tropical Atlantic. Figure A1 presents the dataset with the relationships between DIC and the predictors (SST, SSS and Year) and the distribution divided by DIC quantiles (in color).

Decision Trees (DT) are a non-parametric supervised learning method used for classification and regression. The goal is to create a model that predicts the value of a target variable by learning simple decision rules inferred from the data features. During the tree construction, the dataset is iteratively split by the parameters maximizing the variance of the predicted variable.

Random Forest (RF) algorithms are ensemble methods using so called perturb-and-combine techniques [53] designed for decision trees. This means a diverse set of regressors is created by introducing randomness in the model construction. The prediction of the ensemble is given as the averaged prediction of the individual regressors. Each tree in the ensemble is built from a sample drawn with replacement (i.e., a bootstrap sample) from the training set. In addition, when splitting each node during the construction of a tree, the best split is found either from all input features or a random subset of a predefined size.

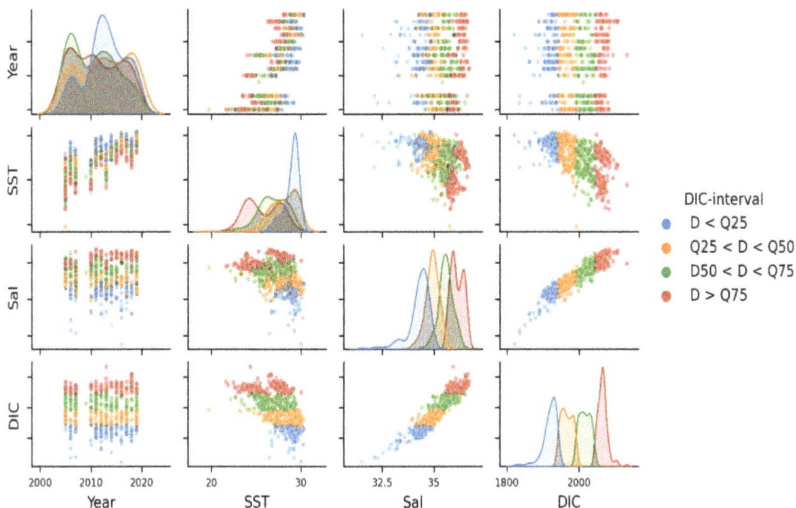

Figure A1. Scatter plots between variables and density plots for each variable (on the main diagonal). Data are colored according to the intervals of the quantiles of DIC (25%, 50%, 75%).

Feed forward neutral networks (NN) are part of artificial neural networks. NN are non-linear functions able to uniformly approximate a wide class of functions encountered in physics [54], which include continuous and derivable functions with respect to each variable. The minimization used allows a global optimization of the set of parameters. Feed forward neural networks are a broadly applied technique in classification and regression tasks. Table A1 summarizes the characteristics of the feed forward Neural model. We use keras API [55] in Python using TensorFlow [56] as back-end.

Table A1. Characteristics of the neutral networks (NN).

Meta-Parameters	Specifications
Input variables	Year, SST, SSS, SinDoY, CosDoY
Output variables	DIC
Train/Val	DIC 4 quantiles (1/6 for Val from each quant.)
Normalization	Center-reduction (both Input and Output)
Architecture	Layers size: Input: 5, Hidden: 15, Output: 1

The relationship between the DIC and its potential predictors, such as the sea surface salinity (SSS), sea surface temperature (SST), the interannual trend (expressed by the year variable) have been explored using four regression models and four independent observation datasets.

For the NN, the intra-annual variations were taken into account by adding two variables: sinDoY and cosDoY, representing the sine and cosine of the day of the year, respectively, normalized between 0 and 2π. Having a validation dataset is fundamental to avoid overfitting on trained data. From the data available for training, i.e., the DIC (2005–2019), two datasets were created for training and validation processes. We applied the following method to get a homogeneous representation of data in the validation dataset: the 637 measurements of DIC (2005–2019) were separated into 4 groups according to the 25%, 50%, 75% quantiles of DIC (this corresponds to the 1941.1 µmol kg^{-1}, 1992.2 µmol kg^{-1} and 2044.6 µmol kg^{-1} DIC values). Finally, data in each group were sorted by SST and a sample was chosen for validation, leaving the rest for training. Best results were obtained by selecting one data out of six. The sizes of these datasets are given in Table A2.

Table A2. Data distribution for the NN showing the total number of data (N), the number of data for training (N_{Train}) and for validation (N_{Val}) according to the 25%, 50%, 75% quantiles (Q25, Q50, Q75) of the DIC. The corresponding values of DIC are indicated in $\mu mol\ kg^{-1}$.

	Condition	N	N_{Train}	N_{Val}
DIC < Q25	DIC < 1941.1	159	132	27
Q25 < DIC < Q50	$1941.1 \leq$ DIC < 1992.2	159	132	27
Q50 < DIC < Q75	$1992.2 \leq$ DIC < 2044.6	158	132	26
DIC > Q75	$DIC \geq 2044.6$	161	135	27

The NN is then compared with the MLR using the same groups of data (Table A3). The two methods give similar RMSE but the NN tends to perform better than the MLR for low DIC values.

Table A3. RMSE for each DIC interval for both MLR and NN methods tested on the two time-series stations and the two cruises.

Mooring or Cruise	Regression Method	DIC < Q25 RMSE [$\mu mol\ kg^{-1}$]	Q25 < DIC < Q50 RMSE [$\mu mol\ kg^{-1}$]	Q50 < DIC < Q75 RMSE [$\mu mol\ kg^{-1}$]	DIC > Q75 RMSE [$\mu mol\ kg^{-1}$]
6° S, 10° W	MLR	NaN	13.8	7.2	7.8
	NN	NaN	13.4	7.3	7.0
6° S, 8° E	MLR	18.0	15.0	12.7	9.4
	NN	14.0	15.0	11.5	8.9
EGEE 3	MLR	7.1	8.3	12.5	18.1
	NN	4.8	7.5	9.9	14.1
PIRATA-FR-29	MLR	11.2	8.1	4.8	6.9
	NN	8.1	11.1	5.3	9.2

The different methods, MLR, NN, DT and RF are compared without separating the data into groups (Table A4). The NN usually performs slightly better than the other methods but, as it does not lead to significant improvement, we use the MLR, which is the simplest method. This also means that the dependencies of the DIC on its main predictors, SSS, SST, and year, are mostly linear. Among these features, the SSS has the largest impact. For random forests and decision trees, the relative importance of SSS is more than 90% (up to 95% for 6° S, 10° W), while that of the SST is about 5% (calculated as total variance reduction brought by the corresponding variable).

Although not very different, the DT is slightly less efficient than the other methods. The DT tend to group the predicted values into clusters, due to relatively small tree depths and consequently small numbers of leaves (i.e., discrete cases with observations attributed to them). DTs typically exhibit high variance and tend to lead to overfitting. RFs decrease the variance of the ensemble estimator. By taking an average of tree predictions, with injected randomness, some errors can cancel out. In our examples RF yield slightly better results that DT for all stations, in most cases being close to the MLR model. Nevertheless, all methods used here give similar results.

Table A4. Same as Table 2, but including a comparison between the multiple linear regression (MLR), the decision tree (DT), the random forest (RF) and the feed forward neural networks (NN) methods.

Mooring or Cruise	Regression Method	RMSE (μmol kg^{-1})	*r*	N	Time Period
6° S, 10° W	MLR	7.7	0.96	6611	2006–2017
	DT	14.1	0.87		
	RF	9.8	0.94		
	NN	7.3	0.97		
6° S, 8° E	MLR	14.8	0.98	239	2017–2019
	DT	25	0.95		
	RF	24	0.95		
	NN	12.8	0.99		
EGEE 3	MLR	11.8	0.97	6895	2006
	DT	14	0.96		
	RF	10	0.98		
	NN	9.3	0.98		
PIRATA FR-29	MLR	8.1	0.99	4462	2019
	DT	11	0.98		
	RF	9	0.98		
	NN	9.4	0.99		

References

1. Le Quéré, C.; Orr, J.C.; Monfray, P.; Aumont, O.; Madec, G. Interannual variability of the oceanic sink of CO_2 from 1979 through 1997. *Glob. Biogeochem. Cycles* **2000**, *14*, 1247–1266. [CrossRef]
2. Wang, X.; Murtugudde, R.; Hackert, E.; Wang, J.; Beauchamp, J. Seasonal to decadal variations of sea surface pCO_2 and sea-air CO_2 flux in the equatorial oceans over 1984–2013: A basin-scale comparison of the Pacific and Atlantic Oceans. *Glob. Biogeochem. Cycles* **2015**, *29*, 597–609. [CrossRef]
3. Cooley, S.R.; Coles, V.J.; Subramaniam, A.; Yager, P.L. Seasonal variations in the Amazon plume-related atmospheric carbon sink. *Glob. Biogeochem. Cycles* **2007**, *21*. [CrossRef]
4. Cooley, S.R.; Yager, P.L. Physical and biological contributions to the western tropical North Atlantic Ocean carbon sink formed by the Amazon River plume. *J. Geophys. Res.* **2006**, *111*. [CrossRef]
5. Subramaniam, A.; Yager, P.L.; Carpenter, E.J.; Mahaffey, C.; Björkman, K.; Cooley, S.; Kustka, A.B.; Montoya, J.P.; Sañudo-Wilhelmy, S.A.; Shipe, R.; et al. Amazon River enhances diazotrophy and carbon sequestration in the tropical North Atlantic Ocean. *Proc. Natl. Acad. Sci. USA* **2008**, *105*, 10460–10465. [CrossRef] [PubMed]
6. Ternon, J.F.; Oudot, C.; Dessier, A.; Diverrès, D. A seasonal tropical sink for atmospheric CO_2 in the Atlantic Ocean: The role of the Amazon River discharge. *Mar. Chem.* **2000**, *68*, 183–201. [CrossRef]
7. Körtzinger, A. A significant sink of CO_2 in the tropical Atlantic Ocean associated with the Amazon River plume. *Geophys. Res. Lett.* **2003**, *30*, 2287. [CrossRef]
8. Lefèvre, N.; Flores, M.M.; Gaspar, F.L.; Rocha, C.; Jiang, S.; Araujo, M.; Ibánhez, J.S.P. Net heterotrophy in the Amazon continental shelf changes rapidly to a sink of CO_2 in the outer Amazon plume. *Front. Mar. Sci.* **2017**, *4*, 278. [CrossRef]
9. Ibánhez, J.S.P.; Araujo, M.; Lefèvre, N. The overlooked tropical oceanic CO_2 sink. *Geophys. Res. Lett.* **2016**, *43*, 3804–3812. [CrossRef]
10. Ibánhez, J.S.P.; Diverrès, D.; Araujo, M.; Lefèvre, N. Seasonal and interannual variability of sea-air CO_2 fluxes in the tropical Atlantic affected by the Amazon River plume. *Glob. Biogeochem. Cycles* **2015**, *29*, 1640–1655. [CrossRef]
11. Chao, Y.; Farrara, J.D.; Schumann, G.; Andreadis, K.M.; Moller, D. Sea surface salinity variability in response to the Congo river discharge. *Cont. Shelf Res.* **2015**, *99*, 35–45. [CrossRef]
12. Signorini, S.R.; Murtugudde, R.G.; McClain, C.R.; Christian, J.R.; Picaut, J.; Busalacchi, A.J. Biological and physical signatures in the tropical and subtropical Atlantic. *J. Geophys. Res. Ocean.* **1999**, *104*, 18367–18382. [CrossRef]
13. Bakker, D.C.E.; De Baar, H.J.W.; De Jong, E. Dissolved carbon dioxide in tropical East Atlantic surface waters. *Phys. Chem. Earth* **1999**, *24*, 399–404. [CrossRef]
14. Lefèvre, N. Low CO_2 concentrations in the Gulf of Guinea during the upwelling season in 2006. *Mar. Chem.* **2009**, *113*, 93–101. [CrossRef]
15. Caniaux, G.; Giordani, H.; Redelsperger, J.-L.; Guichard, F.; Key, E.; Wade, M. Coupling between the Atlantic cold tongue and the West African monsoon in boreal spring and summer. *J. Geophys. Res.* **2011**, *116*. [CrossRef]
16. Hisard, P. Variations saisonnières à l'équateur dans le golfe de Guinée. *Cah. ORSTOM Sér. Océanograph.* **1973**, *11*, 349–358.
17. Hardman-Mountford, N.J.; McGlade, J.M. Seasonal and interannual variability of oceanographic processes in the Gulf of Guinea: An investigation using AVHRR sea surface temperature data. *Int. J. Remote Sens.* **2003**, *24*, 3247–3268. [CrossRef]

18. Herbland, A.; Le Borgne, R.; Le Bouteiller, A.; Voituriez, B. Structure hydrologique et production primaire dans l'Atlantique tropical oriental (Hydrological structure and primary production in the eastern tropical Atlantic Ocean). *Océanograph. Trop.* **1983**, *18*, 249–293.

19. Djagoua, É.V.; Affian, K.; Larouche, P.; Saley, B. Variabilité saisonnière et interannuelle de la chlorophylle en surface de la mer sur le plateau continental de la Côte d'Ivoire à l'aide des images de SeaWiFS de 1997 à 2004. *Télédétection* **2006**, *6*, 143–151.

20. Nieto, K.; Mélin, F. Variability of chlorophyll-a concentration in the Gulf of Guinea and its relation to physical oceanographic variables. *Prog. Oceanogr.* **2017**, *151*, 97–115. [CrossRef]

21. Abe, J.; Brown, B.; Ajao, E.A.; Donkor, S. Local to regional polycentric levels of governance of the Guinea Current Large Marine Ecosystem. *Environ. Dev.* **2016**, *17*, 287–295. [CrossRef]

22. Radenac, M.H.; Jouanno, J.; Tchamabi, C.C.; Awo, M.; Bourlès, B.; Arnault, S.; Aumont, O. Physical drivers of the nitrate seasonal variability in the Atlantic cold tongue. *Biogeosciences* **2020**, *17*, 529–545. [CrossRef]

23. Jouanno, J.; Marin, F.; Du Penhoat, Y.; Molines, J.M.; Sheinbaum, J. Seasonal Modes of Surface Cooling in the Gulf of Guinea. *J. Phys. Oceanogr.* **2011**, *41*, 1408–1416. [CrossRef]

24. Voituriez, B.; Herbland, A. Etude de la production pélagique de la zone équatoriale de l'Atlantique à 4oW: I—Relations entre la structure hydrologique et la production primaire. *Cah. ORSTOM. Sér. Océanograph.* **1977**, *15*, 313–331.

25. Christian, J.R.; Murtugudde, R. Tropical Atlantic variability in a coupled physical–biogeochemical ocean model. *Deep Sea Res. Part II Top. Stud. Oceanogr.* **2003**, *50*, 2947–2969. [CrossRef]

26. Andrié, C.; Oudot, C.; Genthon, C.; Merlivat, L. CO_2 fluxes in the tropical Atlantic during FOCAL cruises. *J. Geophys. Res.* **1986**, *91*, 11741–11755. [CrossRef]

27. Oudot, C.; Ternon, J.F.; Lecomte, J. Measurements of atmospheric and oceanic CO_2 in the tropical Atlantic: 10 years after the 1982–1984 FOCAL cruises. *Tellus* **1995**, *47*, 70–85. [CrossRef]

28. Parard, G.; Lefèvre, N.; Boutin, J. Sea water fugacity of CO_2 at the PIRATA mooring at 6oS, 10oW. *Tellus B* **2010**, *62*, 636–648. [CrossRef]

29. Lefèvre, N.; Veleda, D.; Araujo, M.; Caniaux, G. Variability and trends of carbon parameters at a time series in the eastern tropical Atlantic. *Tellus B* **2016**, *68*, 1147. [CrossRef]

30. Bourlès, B.; Araujo, M.; McPhaden, M.J.; Brandt, P.; Foltz, G.R.; Lumpkin, R.; Giordani, H.; Hernandez, F.; Lefèvre, N.; Nobre, P.; et al. PIRATA: A Sustained Observing System for Tropical Atlantic Climate Research and Forecasting. *Earth Space Sci.* **2019**, *6*, 577–616. [CrossRef]

31. Lefèvre, N.; Guillot, A.; Beaumont, L.; Danguy, T. Variability of fCO_2 in the Eastern Tropical Atlantic from a moored buoy. *J. Geophys. Res.* **2008**, *113*, C01015. [CrossRef]

32. Edmond, J.M. High precision determination of titration alkalinity and total carbon dioxide content of seawater by potentiometric titration. *Deep Sea Res.* **1970**, *17*, 737–750.

33. Stramma, L.; Schott, F. The mean flow field of the tropical Atlantic Ocean. *Deep Sea Res. Part II* **1999**, *46*, 279–303. [CrossRef]

34. Van Heuven, S.; Pierrot, D.; Rae, J.W.B.; Lewis, E.; Wallace, D.W.R. *MATLAB Program Developed for CO_2 System Calculations*; Carbon Dioxide Information Analysis Center: Oak Ridge, TN, USA, 2011.

35. Mehrbach, C.; Culberson, C.H.; Hawley, J.E.; Pytkowicz, R.M. Measurement of the apparent dissociation constants of carbonic acid in seawater at atmospheric pressure. *Limnol. Oceanogr.* **1973**, *18*, 897–907. [CrossRef]

36. Dickson, A.G.; Millero, F.J. A comparison of the equilibrium constants for the dissociation of carbonic acid in seawater media. *Deep Sea Res.* **1987**, *34*, 1733–1743. [CrossRef]

37. McLaughlin, K.; Weisberg, S.B.; Dickson, A.G.; Hofmann, G.E.; Newton, J.A.; Aseltine-Neilson, D.; Barton, A.; Cudd, S.; Feely, R.A.; Jefferds, I.W.; et al. Core Principles of the California Current Acidification Network: Linking Chemistry, Physics, and Ecological Effects. *Oceanography* **2015**, *28*, 160–169. [CrossRef]

38. Millero, F.J.; Byrne, R.H.; Wanninkhof, R.; Feely, R.A.; Clayton, T.; Murphy, P.; Lamb, M.F. The internal consistency of CO_2 measurements in the equatorial Pacific. *Mar. Chem.* **1993**, *44*, 269–280. [CrossRef]

39. Sweeney, C.; Gloor, E.; Jacobson, A.R.; Key, R.M.; McKinley, G.; Sarmiento, J.L.; Wanninkhof, R. Constraining global air-sea gas exchange for CO_2 with recent bomb14C measurements. *Glob. Biogeochem. Cycles* **2007**, *21*. [CrossRef]

40. Weiss, R.F. CO_2 in water and seawater: The solubility of a non-ideal gas. *Mar. Chem.* **1974**, *2*, 203–215. [CrossRef]

41. Wentz, F.J.J.; Scott, R.; Hoffman, M.; Leidner, R.; Atlas, J.A. *Remote Sensing Systems Cross-Calibrated Multi-Platform (CCMP) 6-Hourly Ocean Vector Wind Analysis Product on 0.25 deg Grid*; Version 2.0.; Remote Sensing Systems: Santa Rosa, CA, USA, 2015. Available online: www.remss.com/measurements/ccmp (accessed on 30 September 2020).

42. Koffi, U.; Lefèvre, N.; Kouadio, G.; Boutin, J. Surface CO_2 parameters and air-sea CO_2 flux distribution in the eastern equatorial Atlantic Ocean. *J. Mar. Syst.* **2010**, *82*, 135–144. [CrossRef]

43. Millero, F.J.; Lee, K.; Roche, M. Distribution of alkalinity in the surface waters of the major oceans. *Mar. Chem.* **1998**, *60*, 111–130. [CrossRef]

44. Enfield, D.B.; Mestas, A.M.; Mayer, D.A.; Cid-Serrano, L. How ubiquitous is the dipole relationship in tropical Atlantic sea surface temperatures? *J. Geophys. Res.* **1999**, *104*, 7841–7848. [CrossRef]

45. Gallardo, Y.; Dandonneau, Y.; Voituriez, B. *Variabilité, Circulation et Chlorophylle Dans la Région du Dôme d'Angola en Février-Mars 1971*; Documents Scientifiques; Centre de Recherches Océanographiques: Abidjan, Cote d'Ivoire, 1974; Volume 5, pp. 1–51.

46. Wang, Z.A.; Bienvenu, D.J.; Mann, P.J.; Hoering, K.A.; Poulsen, J.R.; Spencer, R.G.M.; Holmes, R.M. Inorganic carbon speciation and fluxes in the Congo River. *Geophys. Res. Lett.* **2013**, *40*, 511–516. [CrossRef]
47. Hopkins, J.; Lucas, M.; Dufau, C.; Sutton, M.; Stum, J.; Lauret, O.; Channelliere, C. Detection and variability of the Congo River plume from satellite derived sea surface temperature, salinity, ocean colour and sea level. *Remote Sens. Environ.* **2013**, *139*, 365–385. [CrossRef]
48. Materia, S.; Gualdi, S.; Navarra, A.; Terray, L. The effect of Congo River freshwater discharge on Eastern Equatorial Atlantic climate variability. *Clim. Dyn.* **2012**, *39*, 2109–2125. [CrossRef]
49. Bonou, F.K.; Noriega, C.; Lefèvre, N.; Araujo, M. Distribution of CO_2 parameters in the Western Tropical Atlantic Ocean. *Dyn. Atmos. Ocean.* **2016**, *73*, 47–60. [CrossRef]
50. Vangriesheim, A.; Pierre, C.; Aminot, A.; Metzl, N.; Baurand, F.; Caprais, J.-C. The influence of Congo River discharges in the surface and deep layers of the Gulf of Guinea. *Deep Sea Res. Part II Top. Stud. Oceanogr.* **2009**, *56*, 2183–2196. [CrossRef]
51. Neto, A.V.N.; Giordani, H.; Caniaux, G.; Araujo, M. Seasonal and Interannual Mixed-Layer Heat Budget Variability in the Western Tropical Atlantic From Argo Floats (2007–2012). *J. Geophys. Res. Ocean.* **2018**, *123*, 5298–5322. [CrossRef]
52. Ibánhez, J.S.P.; Flores, M.; Lefèvre, N. Collapse of the tropical and subtropical North Atlantic CO_2 sink in boreal spring of 2010. *Sci. Rep.* **2017**, *7*, 41694. [CrossRef]
53. Breiman, L. Random Forests. *Marchine Learn.* **2001**, *45*, 5–32. [CrossRef]
54. Bishop, C.M. *Networks for Pattern Recognition*; Oxford University Press: Oxford, UK, 1995.
55. Chollet, F. Keras. 2015. Available online: https://github.com/keras-team/keras (accessed on 31 October 2020).
56. Dean, J.; Monga, R. TensorFlow: Large-Scale Machine Learning on Heterogeneous Systems. 2015. Available online: https://www.tensorflow.org/ (accessed on 31 October 2020).

Article

Millennial-Scale Environmental Variability in Late Quaternary Deep-Sea Sediments from the Demerara Rise, NE Coast of South America

Steve Lund * and Ellen Platzman

Department of Earth Sciences, University of Southern California, Los Angeles, CA 90007, USA; platzman@usc.edu
* Correspondence: slund@usc.edu

Abstract: We carried out a rock magnetic study of two deep-sea gravity cores from the Demerara Rise, NE South America. Our previous studies provided radiocarbon and paleomagnetic chronologies for these cores. This study presents detailed rock magnetic measurements on these cores in order to characterize the rock magnetic mineralogy and grain size as indicators of the overall clastic fraction. We measured the magnetic susceptibility, anhysteretic remanence, and isothermal remanence and demagnetized the remanences at several alternating field demagnetization levels. The magnetic intensities estimate the magnetic material concentration (and indirectly the overall clastic fraction) in the cores. Ratios of rock magnetic parameters indicate the relative grain size of the magnetic material (and indirectly the overall clastic grain size). Rock magnetic intensity parameters and rock magnetic ratios both vary systematically and synchronously over the last 30,000 years in both cores. There is a multi-millennial-scale cyclicity, with intervals of high magnetic intensity (high magnetic and clastic content) with low magnetic ratios (coarser magnetic and clastic grain size), alternating in sequence with intervals of low magnetic intensity with high magnetic ratios (finer grain size). There is also a higher-frequency millennial-scale variability in intensity superposed on the multi-millennial-scale variability. There are nine (A–I) multi-millennial-scale intervals in the cores. Intervals A, C, E, G, and I have high magnetic and clastic content with coarser overall magnetic and clastic grain size and are likely intervals of enhanced rainfall and runoff from the NE South American margin to the coastal ocean. In contrast, intervals B, D, F, and H have lower clastic flux with finer overall grain size, probably indicating lower continental rainfall and runoff. During the Holocene, high rainfall and runoff intervals can be related to cooler times and low rainfall and runoff to warmer times. The opposite pattern existed during the Pleistocene, with higher rainfall and runoff during interstadial conditions and lower rainfall and runoff during stadial conditions. We noted a similar pattern of Pleistocene multi-millennial-scale variability in a transect of deep-sea sediment cores along the NE Brazilian margin, from the Cariaco Basin (~10 N) to the NE Brazilian margin (~1° N–4° S). However the NW part of this transect (Cariaco Basin, Demerara Rise, Amazon Fan) has an out-of-phase relationship with the SE part of the transect (NE Brazilian margin) between warm–cold and wet–dry conditions. One possible cause of the high–low rainfall and runoff patterns might be oscillation of the Intertropical Convergence Zone (ITCZ), with higher rainfall and runoff associated with a more southerly average position of the ITCZ and lower rainfall and runoff associated with a more northerly average position of the ITCZ.

Keywords: Demerara Rise; millennial-scale variability; Late Quaternary paleoclimate

Citation: Lund, S.; Platzman, E. Millennial-Scale Environmental Variability in Late Quaternary Deep-Sea Sediments from the Demerara Rise, NE Coast of South America. *Oceans* **2021**, *2*, 246–265. https://doi.org/10.3390/oceans2010015

Academic Editor: Michael W. Lomas

Received: 30 December 2020
Accepted: 22 February 2021
Published: 5 March 2021

Publisher's Note: MDPI stays neutral with regard to jurisdictional claims in published maps and institutional affiliations.

1. Introduction

The circulation pattern within the western Equatorial Atlantic Ocean and the large sediment supply from the Amazon River create a complex sedimentation system off the coast of NE South America. The Amazon continental shelf accumulates about five hundred million tons of sediment annually (Kuehl et al., 1986) [1], carried by about 20% of the world's

179

freshwater entering into the ocean from the Amazon River (Franzinelli and Potter, 1983) [2]. Over 80% of the sediment discharged by the Amazon River plume originates from the Andes (Milliman and Meade, 1993; Meade, 1994) [3,4]. This large flux of sediment creates the Amazon Fan, the third largest modern deep-sea fan, an expanded continental shelf found off the coast of NE South America, as well as a distal sediment blanket on the adjacent Demerara Rise and Abyssal Plain. These sediments and their stratigraphic variability provide a unique record of Equatorial Atlantic ocean circulation and Amazon River runoff and their climate and environmental variability (e.g., Damuth and Fairidge, 1970; Damuth and Kumar, 1975; Damuth and Flood, 1985; Flood et al., 1995; Maslin et al., 2000; Maslin and Burns, 2000) [5–10].

Cruise KNR197-3 (chief scientists D. Oppo and W. Curry) collected multicores, gravity cores, and long piston cores along a transect of the Demerara Rise perpendicular to the local coastline (Figure 1). The cores were collected in varying water depths from ~500 m to 3200 m. The sediment cores were taken to better understand the Late Quaternary sequestration of carbon and related physical sedimentation. This paper examines the Late Quaternary pattern of physical sedimentation in the uppermost portion of the Demerara Rise (core depths of less than 1100 m) based on rock magnetic measurements. This part of the Demerara Rise transect has the highest sediment accumulation rates throughout the Late Quaternary. We have previously published a paleomagnetic secular variation record for some of these cores (Lund et al., 2017) [11], along with radiocarbon dating (Huang et al., 2014) [12], which provide a chronostratigraphic framework for our study. We think this study of Late Quaternary physical sediment variability on the Demerara Rise can be used to estimate the environmental pattern of Amazon River plume sedimentation into the deep ocean adjacent to NE South America for the last ~30 ka.

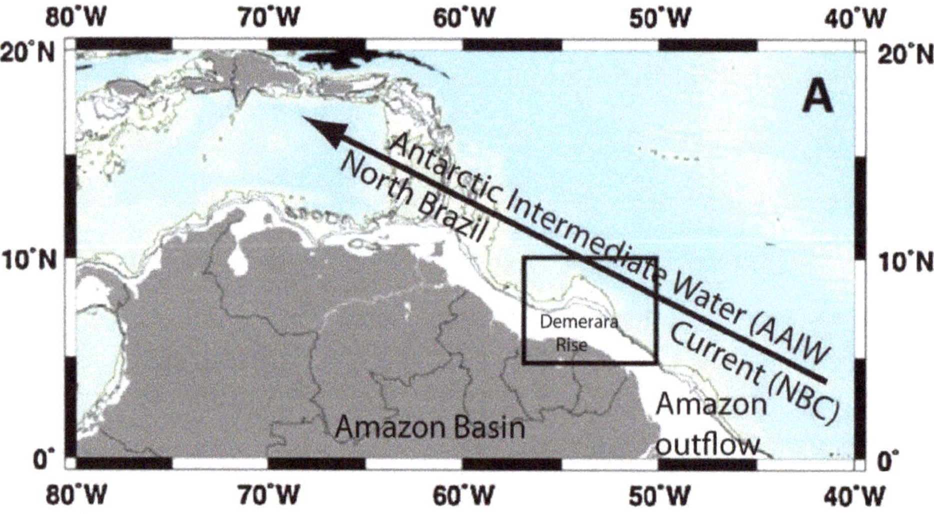

Figure 1. *Cont.*

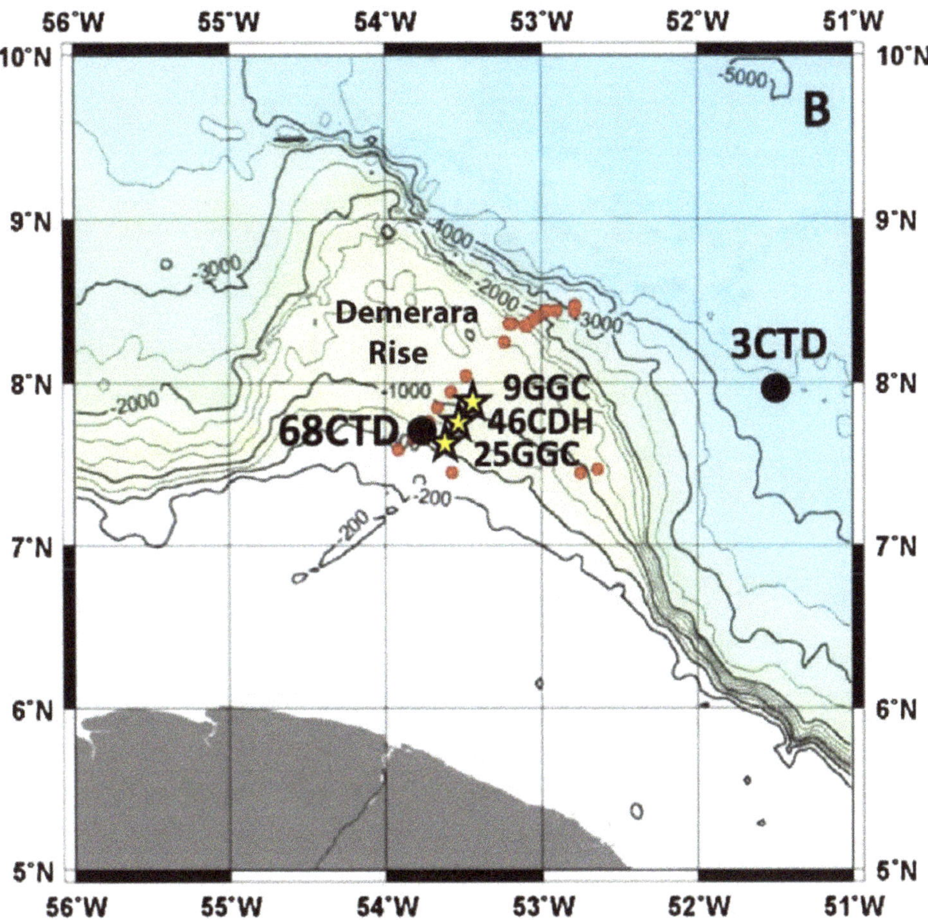

Figure 1. (**A**) Map of the northeastern continental margin of South America showing the Demerara Rise and general flow of the surface North Brazil Current. (**B**) Demerara Rise bathymetry and locations of our studied cores (stars).

2. Regional Ocean Circulation

The ocean circulation pattern in the western Equatorial Atlantic Ocean is dominated by the Atlantic Meridional Overturning Circulation (AMOC) (Berger and Wefer, 1996; Schmitz and McCartney, 1993) [13,14]. This system brings warm water into the North Atlantic Ocean, where it is transformed into North Atlantic Deep Water (NADW), which then migrates to the South Atlantic Ocean (Metcalf and Stalcup, 1967; Kirchner et al., 2009) [15,16]. The principal surface ocean current along the northeastern margin of South America is the North Brazil Current (NBC) (Figure 1). The NBC forms off the easternmost margin of Brazil, where the westward-flowing South Equatorial Current (SEC) bifurcates to form the northward-flowing NBC and the southward-flowing Brazil Current (Stramma et al., 1995) [17]. The NBC crosses the equator as a warm-water current, where it connects to the North Equatorial Current (NEC), Caribbean Current, and Gulf Stream (Schmitz and McCartney, 1993) [14]. The NBC plays an important role in the cross-equatorial transfer of heat to the North Atlantic Ocean (Metcalf and Stalcup, 1967) [15]. The NBC also distributes the sediment plume associated with Amazon River discharge northward along the continental shelf and Demerara Rise–Abyssal Plain (Ruhlemann et al., 2001) [18].

Three deep-ocean currents flow along the NE margin of South America below the NBC (e.g., Huang et al., 2014) [12]. Antarctic Intermediate Water (AAIW) flows northward at depths of ~500–1500 m, directly below the NBC, and is an important source of nutrients and carbon to the North Atlantic Ocean. NADW flows southward at depths of ~2000–4000 m below AAIW. The deepest parts of the Demerara Abyssal Plain may see northward flow of Antarctic Bottom Water (AABW) at depths below 4000 m.

3. Regional Deep-Sea Sedimentation

All of the currents carry varying amounts of suspended sediment, depending partly on flow velocities and partly on advection of sediment at the surface by aeolian processes or coastal runoff. The most significant source of advected sediment is the Amazon River discharge. This large flux of sediment creates the Amazon Fan, the third largest modern deep-sea fan, and an expanded continental shelf. The Amazon delta and adjacent shelf currently trap perhaps two-thirds of the discharged sediment (Figueiredo and Nittrouer, 1995; Nittrouer et al., 1995) [19,20], however the finest sediment fraction is probably dispersed beyond the shelf by the NBC. This flux of Amazon River water and sediment beyond the shelf environment brings vast amounts of nutrients to the surface waters of the western Equatorial Atlantic Ocean, causing large blooms of phytoplankton (Smith and DeMaster, 1996) [21].

The western Equatorial Atlantic has two major modes of sedimentation caused by sea-level changes. Higher sea levels associated with interglacial periods foster the deposition of most sediment from the Amazon River onto the north Brazil continental shelf (Nittrouer and DeMaster, 1986; Nittrouer et al., 1995) [20,22]. A lesser portion of the sediment is transported as a suspended load either by longshore currents toward the northwest, where it is deposited along the continental shelf as far as French Guinea, or by the NBC, where it is deposited onto the Amazon Fan and adjacent Demerara Rise–Abyssal Plain (Lund et al., 2018) [23]. Lower sea levels associated with glacial periods cause sediment to bypass the exposed continental shelf and be deposited directly into the NBC and head of the Amazon Submarine Canyon (Damuth and Fairbridge, 1970) [5] and onto the fan.

The "on–off" supply of sediment to the Amazon Fan results in distinctive sediments for glacial versus interglacial intervals. Glacial sediments on the abyssal plain of the most recent glacial period consist of mottled and dark grey hemipelagic clays, a low concentration of carbonate and foraminifera, and occasional redeposited terrigenous sand and silt layers. Above the glacial sediments, Holocene pelagic sediments consist of tan foraminiferal marls or oozes with diverse foraminifera assemblages (Damuth, 1977; Damuth and Flood, 1985) [7,24]. Holocene and interglacial sediments in the western Equatorial Atlantic are characterized by lower concentrations of magnetic material (mostly magnetite) and by finer-grained magnetic material (Meinert and Bloemendal, 1989) [25]. In contrast, the glacial sediments have higher concentrations of magnetic material (mostly magnetite) that is coarser-grained.

Climate also plays a major role in the amount and distribution of sediment deposited on the Amazon Fan, adjacent continental shelf, and more distal Demerara Rise–Abyssal Plain at shorter timescales. The seasonal migration of the Intertropical Convergence Zone (ITCZ) creates distinctive wet and dry seasons and has a large influence on environmental patterns of the coastal region (e.g., Wainer and Soares, 1997) [26]. Boreal summer placement of the ITCZ near 10° N causes dryness in the Amazon Basin and strong precipitation further to the north, including the Cariaco Basin (Haug et al., 2001) [27]. During the boreal winter, the ITCZ moves over the Amazon Basin (0°–10° S), creating a pronounced wet season (austral summer monsoon from December to February).

There is evidence at millennial timescales that the average annual placement of the ITCZ may be biased in its position. During the Early–Middle Holocene (approximately 8000–4000 years BP), there is widespread evidence that the climatic conditions in the tropical Andes were significantly drier than present (e.g., Mayle et al., 2004; Mayle and Power, 2008) [28,29]. More southerly areas experienced similar increased aridity in the Early Holocene. For example, the driest climatic conditions at Lake Junin (11° S), Lake Titicaca (14–17° S), and Sajama Mountain (18° S) were centered at ca. 10,000, 5500, and 4000 years BP, respectively [30–33] (Thompson et al. 1998; Seltzer et al., 2000; Abbott et al., 2003; Bush et al., 2005). In contrast, more northerly sites in Venezuela and Columbia (including Cariaco Basin) showed significant increases in rainfall in the Early Holocene, as the ITCZ migrated even more northerly during summer months (Haug et al., 2001) [27]. Overall, the sediments of the northern South American margin and their stratigraphic variability provide a correlative record of Amazon Basin climate and environmental variability (e.g., Maslin et al., 2000, Lund et al., 2018) [9,23].

4. Core Descriptions

Cruise KNR197-3 collected a sequence of cores at ~500 m to ~3000 m water depth in a transect outward from the South American margin (Figure 1) as part of a paleoceanographic study of Atlantic Meridional Overturning Circulation (AMOC) in the Atlantic Ocean. Typically, multicores (MC), gravity cores (GGC), and long piston cores (CDH) were collected to provide a complete, composite sedimentologic record along the transect. The focus of our rock magnetic study is 9GGC (405 cm) and 25GGC (352 cm) gravity cores with companion 10CMC (31 cm) and 25DMC (32 cm) multicores (Figure 1, Table 1). We also make use of a 46CDH piston core because it has a large number of radiocarbon dates and XRF measurements that can be transferred into the gravity cores based on magnetic susceptibility (CHI) profile correlations. The radiocarbon dates (Huang et al., 2014) [12] indicate that our paleomagnetic or rock magnetic study covers the last ~19 ka in core 25GGC and ~31 ka in core 9GGC.

Table 1. Site locations and descriptions of Demerara Rise cores discussed in this paper.

Core	Latitude	Longitude	Water Depth (m)	Length (cm)	C14 Dates
10CMC	7.93° N	306.11° E	1100	31	0
9GGC	7.93° N	306.11° E	1100	405	6
26DMC	7.70° N	306.21° E	671	32	0
25GGC	7.70° N	306.21° E	671	352	8
46CDH	7.84° N	306.34° E	947	3150	13

5. Rock Magnetic Measurements

U-channel samples were collected from both gravity cores for our rock magnetic studies. The multicores were sampled discretely with contiguous 2 × 2 × 2 cm cubes because the cores were too short (<50 cm) for good-quality u-channel measurements. Paleomagnetic measurements of these cores have been previously described by Lund et al. (2017) [11]. All cores were subjected to whole-core magnetic susceptibility measurements while onboard the ship. The u-channeled sediments were given a series of rock magnetic measurements after completion of the paleomagnetic studies. All samples were given an artificial anhysteretic remanent magnetization (ARM) sequence with a bias field of 0.05 milliTesla (mT) and an alternating field of 100 mT. After sample measurement, all samples were subjected to a sequence of ARM af demagnetization and measurements at 10 mT, 20 mT, 40 mT, and 60 mT. Finally, all samples were given another artificial isothermal remanent magnetization (IRM) sequence with a pulsed 1 T field. (This 1-T artificial remanence is usually referred to as a saturation IRM (SIRM), because all magnetite and titanomagnetite grains, which hold almost all of the natural remanence (NRM), are saturated.) After sample measurement, all samples were subjected to a sequence of SIRM af demagnetization and measurements at 10 mT, 20 mT, 40 mT, and 60 mT. One u-channel from each of the gravity cores also had their ARMs and SIRMs subjected to 80 mT and 100 mT af demagnetization. Lund et al. (2017, 2018) [11,23] summarized the AF coercivity patterns of the NRM, ARM, and SIRM. Those papers also considered variations in the ARM/Chi ratio, which showed no evidence of fine-grained (bacterial) magnetite. These studies both suggest that detrital magnetite is the dominant magnetic mineral in these sediments. The resulting CHI, ARM, and SIRM intensities for both gravity cores are shown in Figures 2 and 3. The magnetic intensity patterns in the two multicores and neighboring multicores (Lund et al., 2018) [2] are consistent with the pattern in the uppermost 50 cm of the gravity cores. This provides corroborating evidence that the gravity cores did not lose significant surface sediment in the coring process.

The magnetic intensity variations reflect changes in the percentage of detrital magnetic material (and presumably total clastic material) in the sediment. Magnetite is the most common magnetic mineral noted in several other studies of deep-sea sediments from the Equatorial Atlantic–Amazon Fan region (Meinert and Bloemendal, 1989; Maslin et al., 2000; Franke et al., 2007; Bendle et al., 2010) [9,25,34,35]. Our previous rock magnetic studies of deep-sea sediments from the neighboring Demerara Abyssal Plain also support this view (Lund et al., 2018) [23]. These studies also indicate that ferric oxides (mostly hematite and goethite) are present in varying small amounts in the sediments. The magnetic intensity variations may be due to varying clastic flux (presumably mostly from the Amazon drainage) or varying dilution by biogenic components due to variable upwelling and biological productivity.

Our ARM and Chi measurements and ratios of ARM or SIRM under different levels of demagnetization (SIRM20/SIRM0; SIRM40/SIRM20, ARM20/ARM0, ARM40/ARM20) estimate the grain size distribution of the magnetic mineral fraction (e.g., Stacy and Banerjee, 1974; Bloemendal et al., 1988) [36,37]. For all rock magnetic ratios, higher (lower) values indicate a finer-grained (coarser-grained) set of magnetic grains. Variations in magnetic grain size may be due simply to a finer-grained or coarser-grained source of overall clastic material; they might also reflect changes in the distribution of magnetic minerals if more than one is significant in concentration.

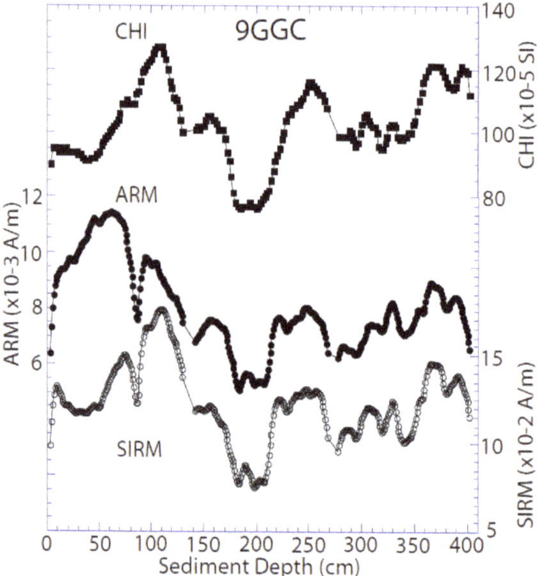

Figure 2. Rock magnetic intensity variability of gravity core 9GGC. Individual rock magnetic intensities are shown as Chi, anhysteretic remanence (ARM), and saturation isothermal remanence (SIRM) values. See text for further discussion.

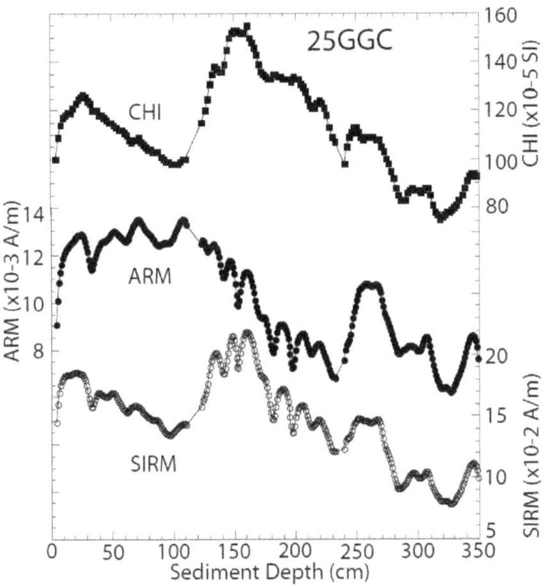

Figure 3. Rock magnetic intensity variability of gravity core 25GGC. Individual rock magnetic intensities are shown as Chi, anhysteretic remanence (ARM), and saturation isothermal remanence (SIRM) values. See text for further discussion.

Selected samples from gravity core 9GGC were also given an extra series of magnetic hysteresis measurements. The samples were cycled through a MicroMag up to 1 Tesla (T). Values of remanent magnetization (Mr), saturation magnetization (Ms), coercivity field (Hc), coercivity of remanence field (Hcr), and field intensity (H) were calculated to estimate the grain size distribution and percentage of ferric magnetic minerals (goethite, hematite) versus titanomagnetite minerals (primarily magnetite). H represents the amount of magnetic material that acquires a remanence between 0.3 and 1 T (ferric magnetic minerals, primarily hematite). The primary purpose of these measurements was to assess the percentage of ferric magnetic minerals that primarily develop in Amazon Basin soils and reflect the relative importance of erosion from the Andes (primarily magnetite and titanomagnetite) versus the Amazon Basin itself (a more significant portion of goethite and hematite) on total Amazon drainage to the coastal ocean.

6. Development of a Revised Magnetic Chronostratigraphy Technique

Lund et al. (2017) [11] developed chronostratigraphies for cores 9GGC and 25GGC by first correlating the two cores based on environmental magnetic and paleomagnetic variability and then using that correlation to assess the radiocarbon dates in cores 9GGC and 25GGC. (Selected radiocarbon dates from 46CDH were also tied into the two gravity cores using magnetic susceptibility correlations.) The final chronostratigraphies for 9GGC and 25GGC were then derived from a selection of radiocarbon dates tied to each core individually and connected with linear interpolation. We then identified 52 paleomagnetic and rock magnetic features that both cores have in common. We estimated the ages of each feature based on the chronologies outlined by Lund et al. (2017) [11]. We found age differences between the same correlatable features in both cores of as much as 1000 years because of differences in linear interpolation between radiocarbon dates in the two cores.

We developed a revised magnetic chronostratigraphy technique for these two cores that provides more consistent ages for environmental and paleomagnetic features. The starting point was to date the 52 environmental magnetic or paleomagnetic correlation features in the two cores (summarized in Supplementary Table S1) based on the original chronostratigraphies developed by Lund et al. (2017) [11]. The 52 feature ages in 9GGC and 25GGC were then averaged. Those feature ages were then used to rebuild chronologies for both cores. The results are shown in Figure 4. Figure 4 shows the original Lund et al. (2017) [11] chronologies based on calibrated radiocarbon dates (large open circles in Figure 4), with linear interpolation and the revised chronology based on the 52 feature ages summarized in Supplementary Table S1. This procedure makes PSV and environmental variability more synchronous between the two cores. It does not, however, improve the overall chronology of the sediments. It simply makes the two records chronologically consistent. There are two underlying assumptions in this procedure that we think are warranted. First, we presume that the sediment remanence acquisition process is comparable in both cores. Second, we presume that the sedimentation process near our study cores is not time transgressive. Both assumptions seem reasonable based on the strong (±2 cm) correlation between both PSV and rock magnetic variability among the two cores over their entire lengths.

The revised chronostratigraphies were used to estimate sedimentation rate variability in both cores. The results are shown in Figure 5. There is a notable overall pulse of high sedimentation around the Pleistocene–Holocene transition (~10–16 ka). Superposed on this are multi-millennial variations in sedimentation rates that we can tie to our rock magnetic variability described below. In particular, note that colder intervals of the Late Pleistocene (Younger Dryas (YD) and Heinrich events H1 and H2) have lower sedimentation rates than the neighboring warmer intervals (Bolling–Allerod (B/A) and Dansgard–Oeschger (D/O) interstadial events 2–4).

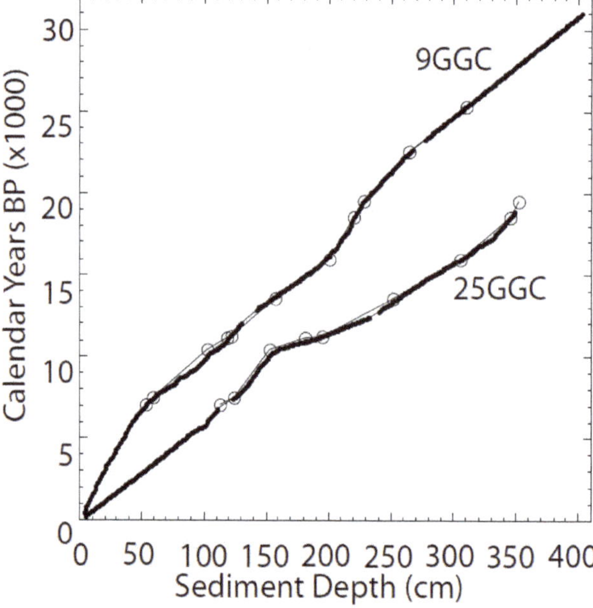

Figure 4. Time–depth plots for cores 9GGC and 25GGC. The open circles are calibrated radiocarbon dates used by Lund et al. (2017) [11] with linear interpolation to dates the cores. We built a revised chronology, as summarized in Supplementary Table S1, to create the more detailed chronologies shown here (dark lines).

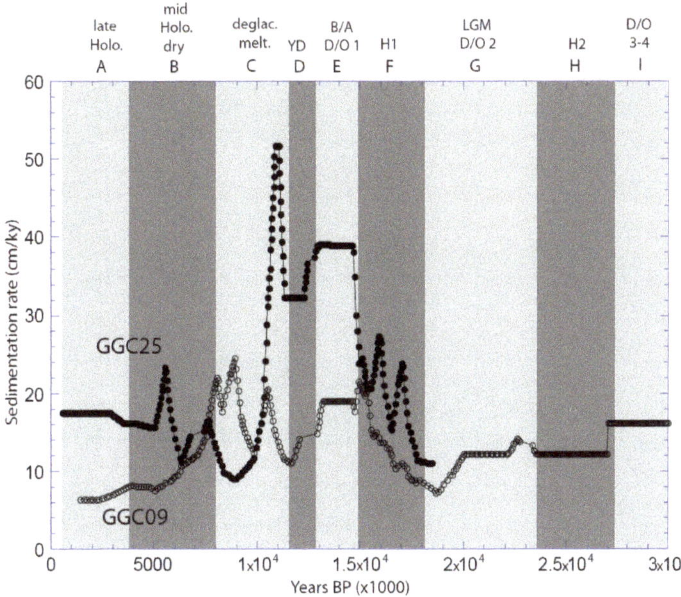

Figure 5. Sedimentation rates for GGC9 and GGC25. There is a multi-millennial-scale alternation in sedimentation rates (faster–slower) that is indicated by grey versus white zones labeled A–I. These correspond to labeled zones in Figures 6–9.

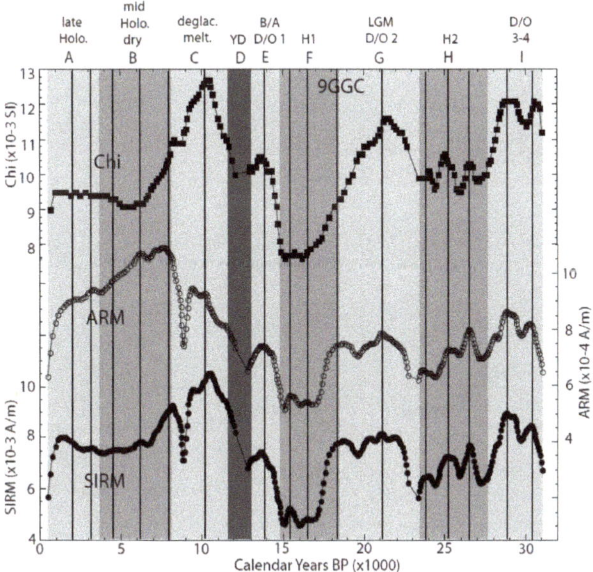

Figure 6. Rock magnetic intensity variability for core 9GGC versus time. Alternating light and dark grey zones are multi-millennial-scale oscillations in magnetic intensity (light = higher intensity, dark = lower intensity) labeled A-I. Vertical lines estimate finer-scale millennial intensity peaks that are superposed on the longer-term variability.

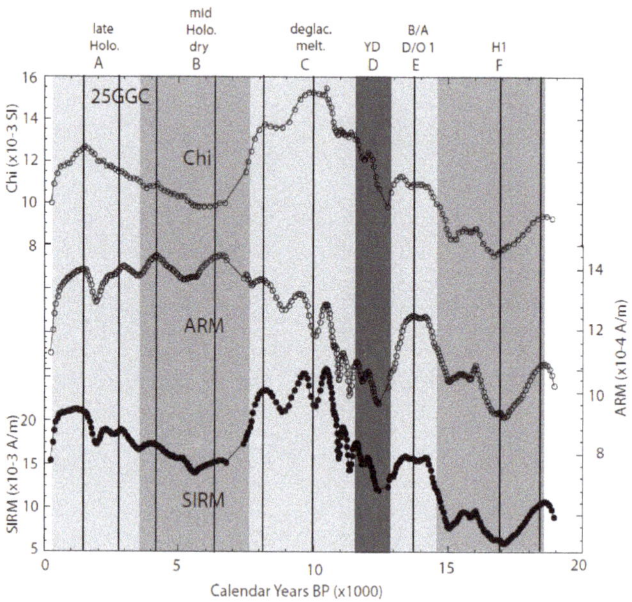

Figure 7. Rock magnetic intensity variability for core 25GGC versus time. Alternating light and dark grey zones are multi-millennial-scale oscillations in magnetic intensity (light = higher intensity; dark = lower intensity) labeled A–I. Vertical lines estimate finer-scale millennial intensity peaks that are superposed on the longer-term variability.

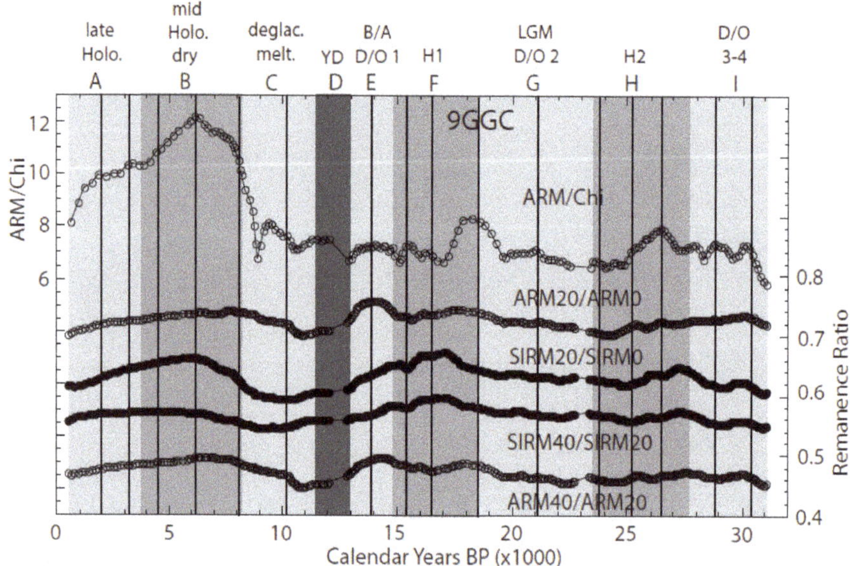

Figure 8. Rock magnetic coercivity variability for core 9GGC versus time. Alternating light and dark grey zones are multi-millennial-scale oscillations in magnetic coercivity (light = lower coercivity or coarser-grained; dark = higher coercivity or finer-grained) labeled A–I. Vertical lines indicate finer-scale millennial intensity peaks noted in Figures 6 and 7. There is no significant evidence for coercivity variations associated with those intensity variations.

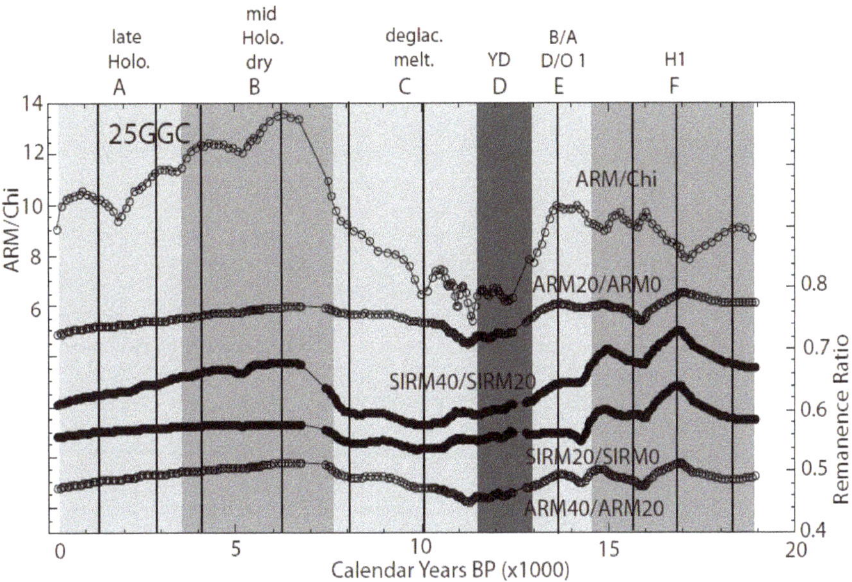

Figure 9. Rock magnetic coercivity variability for core 25GGC versus time. Alternating light and dark grey zones are multi-millennial-scale oscillations in magnetic coercivity (light = lower coercivity or coarser-grained; dark = higher coercivity or finer-grained) labeled A–I. Vertical lines indicate finer-scale millennial intensity peaks in Figures 6 and 7. There is no significant evidence for coercivity variations associated with those intensity variations.

7. Environmental Variability

Figure 6 (9GGC) and Figure 7 (25GGC) display the pattern of magnetic intensity variation as a function of time over the last 30 ka. The two gravity cores have the same pattern of intensity variation, with six multi-millennial-scale intervals in common (A–F in Figures 6 and 7). Core 25GGC also displays three more intervals (G-I) that are older than the data in 9GGC. Each of these intervals lasts ~1500–4000 years. The intervals oscillate between high and low intensity. The intensity variations could be associated with variations in the rate of clastic flux to the coastal ocean or with variations in biological productivity causing biogenic carbonate or silica dilution.

Figure 8 (9GGC) and Figure 9 (25GGC) display the pattern of magnetic coercivity variations over the last 30 ka. The coercivity varies between high (finer-grained clastics) and low (coarser-grained clastics) values and is the same pattern in both cores. The timing of coercivity variations is also consistent with the pattern of intensity variations. High intensities generally go with lower coercivities. This relationship between intensity and coercivity strongly suggests that clastic flux variations in amount and grain size are the primary cause of the observed multi-millennial-scale variability. Higher rates of clastic flux probably indicate stronger outflow from the Amazon delta; this should also yield overall coarser clastic flux. This pattern is consistent with the sedimentation rate variations noted in Figure 5. High intensity–coarser clastic flux intervals that suggest stronger Amazon outflow are also intervals of higher than average sedimentation rate. Additionally, the lower intensity–finer clastic intervals that suggest lower Amazon outflow are also intervals of lower than average sedimentation rate.

Several other deep-sea sediment studies discussed below use Ti/Ca or Fe/Ca from XRF studies to assess clastic flux variability. Delia Oppo (unpublished) [38] supplied us with an XRF and magnetic susceptibility record for core 46CDH (Figure 1, Table 1). The results are plotted in Figure 10. We correlated our multi-millennial-scale variability (intervals A–I) to the Ti/Ca results using the magnetic susceptibility records. The high (low)

Ti/Ca intervals (relatively high clastic flux) are consistent with our high-intensity–coarser (low intensity–finer) rock magnetic intervals.

Intervals A–I can be easily associated with known regional to global climate variations over the last 30 ka. Intervals A–C can be associated with Late Holocene (A) and Middle Holocene drought (B) in the Amazon Basin (Maslin and Burns, 2000; Lund et al., 2018) [10,23] and final deglaciation (C) with high Amazon outflow (Maslin et al., 2000; Maslin and Burns, 2000) [9,10]. These three intervals represent Holocene variations in clastic flux, most likely due to variations in multi-millennial-scale rainfall or glacial melting. The low intensity (and modestly higher coercivity, finer-grained) of intervals D, F, and H can be associated with cold intervals (stadials) of the Younger Dryas (YD) and Heinrich events 1 (F) and 2 (H). This observation is consistent with the results of Maslin et al. (2011) [39] for the YD and Heinrich event 1. These cold intervals can also be interpreted as relatively dry intervals. The intervening higher flux (coarser -rained) intervals are associated with the warmer (interstadial) Bolling–Allerod (D/O) cycle 1 (E), D/O cycle 2 (G), and D/O cycles 3–4 (I). These intervals probably represent less cold and wetter intervals within the Amazon Basin. These are all labeled on Figures 6–9. There is one complication to this analysis. It is likely that the colder Pleistocene intervals (YD, Heinrich events) are slower North Brazil Current intervals. This change in flow speed might have contributed to the variations in sedimentation rate. However, a number of published studies (discussed below) that show this same multi-millennial-scale pattern of sediment variability all associate the variation predominantly with variable rainfall or runoff from the Amazon Basin to the NE South American margin.

There is also a higher-frequency scale of variability in these records that is more subdued. This pattern is indicated by the solid lines in Figures 6–9. The lines indicate local intensity peaks. The relationship with coercivity is more complicated and subdued at best or perhaps there is no relationship at all. These cycles occur in all of the longer-duration intervals noted above. They suggest a higher-frequency oscillation in Amazon outflow, with the black lines indicating times of higher flow (higher clastic flux). These more subtle oscillations in intensity are consistent between cores 9GGC and 25GGC. However, the lack of any notable coercivity variability makes their interpretation more questionable. Table 2 presents the intensity peak ages and the durations between successive peaks. The average duration is ~1750 ± 450 years.

We finally assess the evidence for variations in ferric (primarily hematite)/ferrous (magnetite and titanomagnetite) ratios in the clastic sediments. These ratios indicate the relative importance of ferrous versus ferric components to the overall sediment magnetization. Hysteresis measurements of selected samples from 9GGC are shown in Figure 11. The rock magnetic data suggest that the magnetization is primarily magnetite–titanomagnetite in the size range of ~5–15 microns. This distribution could have varying percentages of coarser (MD) or finer (SD) grains and could still be within the pseudo-single domain (PSD) range for magnetite (Figure 11A). The hysteresis data let us estimate the amount of magnetization that is acquired with increasing magnetic fields; magnetic contributions between 0.3 and 1.0 T are usually referred to as hard components (H), which is associated with ferric minerals (primarily hematite). The ratio of total magnetic remanence to H (H/SIRM; Figure 11B) estimates the relative portion of ferric magnetization in any sediment sample. The time interval 6–9 ka has the lowest ratio of the last 25 ka. Bendle et al. (2010) [35] and Lund et al. (2018) [23] have both documented low H/SIRM ratios in the Early–Middle Holocene (6–9 ka) of deep-sea sediments from the Demerara Abyssal Plain; our results corroborate this. They all indicate that the Amazon Basin itself was relatively dry during the Early–Middle Holocene, with lower than average erosion of Amazon Basin soils (relative to contributions from the Andes).

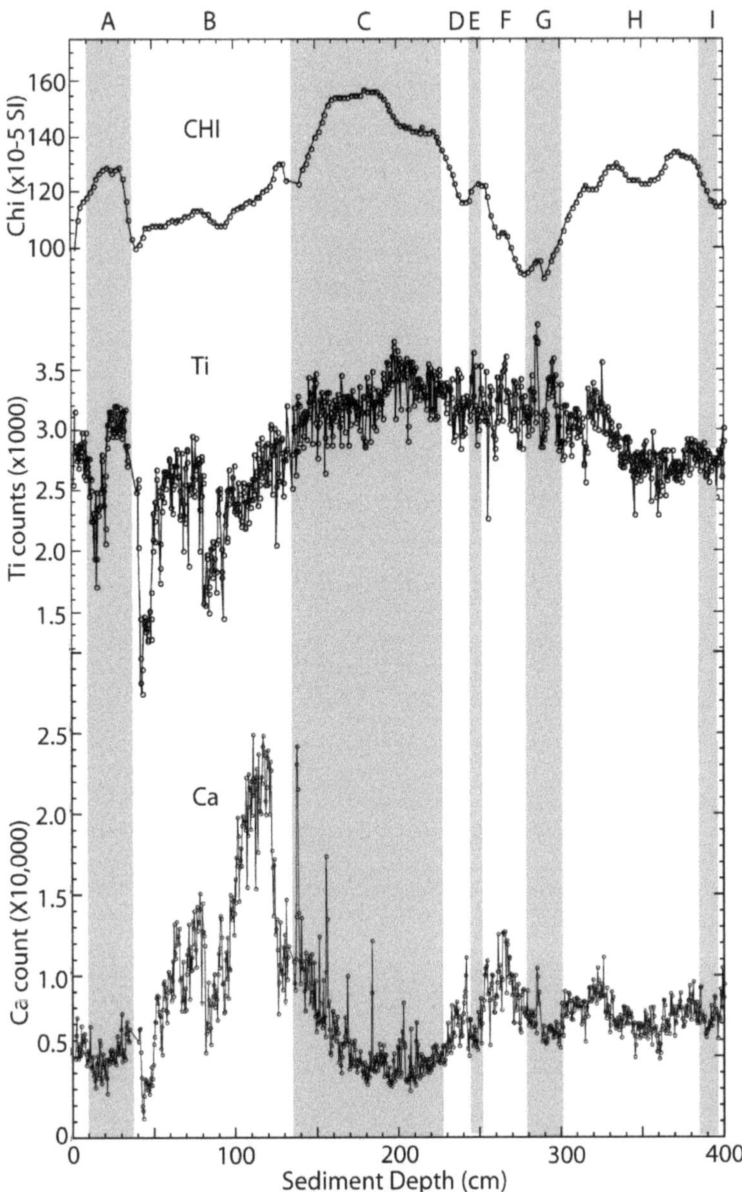

Figure 10. Proxy data for core 46CDH (Figure 1, Table 1), situated directly between gravity cores 9GGC and 25GGC. Top—magnetic susceptibility record; the variability can be directly correlated to data from 9GGC (Figure 2) and 25GGC (Figure 3). Below are Ti and Ca counts from unpublished XRF data on 46CDH (Delia Oppo, personal communication). The rock magnetic intensity zonation is consistent with the XRF data. High chi goes with high Ti (a clastic indicator; low chi goes with high Ca (a biologic indicator).

Table 2. Timing of millennial-scale intensity peaks and durations between successive intensity peaks.

9GGC	Duration	25GGC	Duration	Average	Duration
YBP	**YRS**	**YBP**	**YRS**	**YBP**	**YRS**
1900		1500		1700	
3200	1300	2700	1200	2950	1250
4600	1400	4200	1500	4400	1450
6100	1500	6300	2100	6200	1800
8000	1900	8100	1800	8050	1850
10,200	2200	10,000	1900	10,100	2050
11,800	1600	11,800	1800	11,800	1700
13,900	2100	13,700	1900	13,800	2000
15,400	1500	15,800	2100	15,600	1800
16,500	1100	16,900	1100	16,700	1100
18,400	1900	18,500	1600	18,450	1750
21,200	2800			21,200	2750
23,800	2600			23,800	2600
25,200	1400			25,200	1400
26,400	1200			26,400	1200
28,800	2400			28,800	2400
30,400	1600			30,400	1600
Ave Dur.	1781		1700		1794
SD	513		346		482

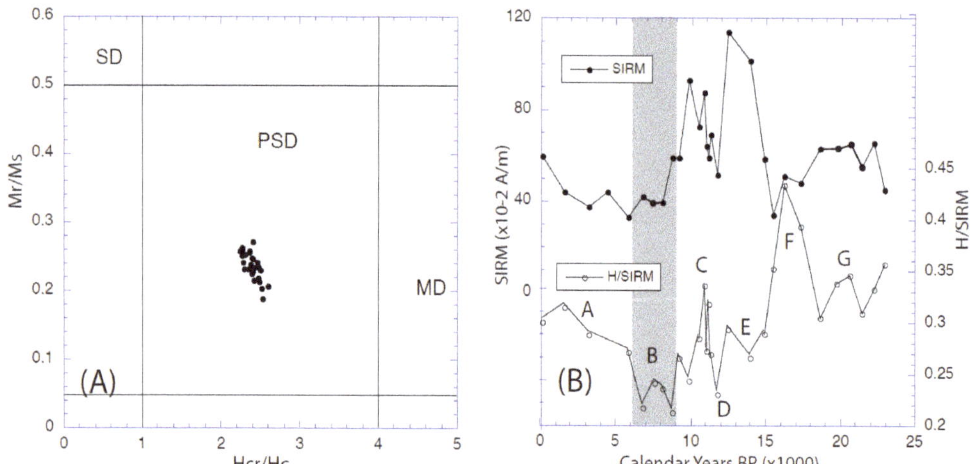

Figure 11. Magnetic hysteresis measurements of selected sediment horizons in core 25GGC. (**A**) Plot of remanent magnetization (Mr)/saturation magnetization (Ms) versus coercivity of remanence (Hcr)/coercivity (Hc). All samples fall within the pseudo-single domain (PSD) range. (**B**) Saturation isothermal remanence (SIRM) and field (H)/SIRM plotted versus time in 25GGC. See text for further discussion.

8. Comparison with Other Records

The first evidence for millennial-scale variability in the tropical Atlantic region was noted by Grimm et al. (1993) [40] in Lake Tulane (27.6° N) in Florida and Curry and Oppo (1997) [41] in the Ceara Rise (5° N), also on the NE South American continental margin. The Lake Tulane study documented enhanced rainfall during Pleistocene North Atlantic cold (Heinrich) events. The Ceara Rise study linked coherent multi-millennial-scale tropical sea-surface temperature and high-latitude North Atlantic temperature variations. Locally, we can compare our deep-sea sediment records with others from the NE margin of South America.

Lund et al. (2018) [23] studied rock magnetic variability of Holocene deep-sea sediments from the Amazon Fan–Demerara Abyssal Plain. Our results from this study are the same. Our Pleistocene results (~7° N) can be compared with three other studies along the NE South American margin: the Cariaco Basin (~10° N), Demerara Rise–Amazon Fan (~4° N–7° N), and outboard of the NE Brazilian margin (~1° N–3° S). All of these studies show strong evidence of multi-millennial-scale variability from 12–30 ka, dominated by colder phases associated with the Younger Dryas and Heinrich events 1 and 2 versus warmer phases associated with the Bolling–Allerod and DO cycles 1, 2, and 3.

Our results are consistent with results from the Cariaco Basin (Haug et al. 2001; Deplazes et al., 2013) [27,42] showing lower sedimentation rates and clastic flux during the Younger Dryas, H1, and H2 associated with dryer conditions along NE South America, while the intervening intervals have higher clastic flux associated with heavier rainfall and runoff. These variations are associated with shifts in the long-term position of the Intertropical Convergence Zone (ITCZ).

Haggi et al. (2017) [43] studied the isotopic composition of plant waxes in a deep-sea sediment core from the same region (7° N) as our cores. They suggested that the Amazon rainforest was affected by intrusions of savannah or open framework vegetation types during H1 and H2 associated with drier conditions. The intervening intervals indicated higher Amazon rainfall. This pattern is completely consistent with our Pleistocene multi-millennial-scale variability. They attributed these long-term variations to changes in atmospheric circulation.

Crivellari et al. (2018) [44] studied the H1 interval in more detail in the same deep-sea core as Haggi et al. (2017) [43]. They identified an early part of the H1 interval (HS1a: 17–18 ka) that had higher influx of sediment to the site, however the core of HS1 (HS1b: 15–17 ka) showed a decrease in terrestrial flux. Our results do not show evidence of this two-part structure, however our study and the study by Haggi et al. (2017) [43] do show a broadly drier response to H1.

A group of four deep-sea sediment studies further to the south (1° N–4° S) along the NE Brazilian margin also show the same pattern of multi-millennial-scale environmental variability but with mostly an opposite phasing: colder intervals had higher clastic flux and warmer intervals had lower clastic flux. However, there is some disagreement among these four studies that could be due to chronology issues. Arz et al. (1998) [45] used Ti/Ca XRF ratios to document multi-millennial-scale pulses of significant clastic sedimentation into the Atlantic slope along the NE corner of Brazil (3.5° S). One pulse correlates to H1, but two pulses correlate to warm Bolling–Allerod and DO interstadial 2 phases; the intervening Younger Dryas and H2 cold intervals had low clastic flux. Nace et al. (2014) [46] studied a marine core along the NE Brazilian margin at 0.3° N. They also identified multi-millennial-scale pulses of high clastic flux associated with Ti/Ca XRF data, which they associated with increased precipitation or weathering on the adjacent continent. They correlated these events to multiple (cold) Heinrich events. However, their data actually showed high clastic pulses associated with H1 (~16 ka) and (warm) D/O cycles 2–3 (~30 ka) and low clastic flux at ~26 ka (cold H2). This pattern is consistent with the data from the study by Arz et al. (1998) [45]. Zhang et al. (2015) [47] studied Fe/Ca XRF data from a deep-sea core at 1.6° S and noted higher clastic flux during all three late Pleistocene cold intervals, namely the Younger Dryas, H1, and H2 events. Mulitza et al. (2017) [48] studied a deep-sea sediment core (2° S) in the same region and concluded that both the Younger Dryas and H1 cold intervals were intervals of increased rainfall and higher clastic flux based on Fe/Ca XRF data. These results are opposite to the Younger Dryas and H2 results of Arz et al. (1998) [45] from the same region. Overall, these records do generally support an opposite phase of high clastic flux in cold intervals than we noted for the Demerara Rise.

Speleothem records from the Amazon Basin display similar complexity in Late Pleistocene rainfall variability. Cheng et al. (2013) [49] suggested that broadly the eastern Amazon region had lower rainfall during the late Pleistocene versus the western Amazon region. This might support our observation that NW marine records (Cariaco Basin, Demerara Rise) had an out-of-phase relationship for rainfall and runoff with SE marine records (NE Brazilian margin). Cheng et al (2013) [49] also noted that both regions had multi-millennial-scale variability on top of longer-duration orbital-scale variability with wetter intervals in NE Brazil (Rio Grande do Norte record) during the H1 and H2 stadials. This is consistent with the NE Brazilian margin marine records noted above. Wang et al. (2017) [50] also noted a similar pattern of multi-millenial-scale rainfall variability in their Paraiso speleothem record (~3° S) of NE Brazil.

9. Discussion

Cores 9GGC and 25GGC are dominated by millennial- to multi-millennial-scale variability throughout their entire extent. It is worthwhile noting that the timescales of variability do not significantly change between the Holocene and the Pleistocene. The dominant variability seems to follow a simple oscillatory multi-millennial pattern of wet versus dry conditions in the Amazon Basin (stronger and faster Amazon outflow with coarser clastic grains during wetter conditions versus lower and weaker Amazon outflow with finer clastic grains during drier conditions) in NE South America. Intervals A, C, E, G, and I are associated with relatively wet conditions (and strong Amazon flow), while intervals B, D, F, and H are associated with relatively dry conditions (and weaker Amazon flow). The sediments reach the Demerara Rise from the Amazon outflow under control of the North Brazil Current after passing the local shelf. Coarser (finer) clastic grains are more (or less) likely to reach the North Brazil current under conditions of higher (or lower) Amazon flow. The deep-sea sediment records south of the Amazon Fan (NE Brazilian margin, 1° N–4° S) suggest an out-of-phase relationship with our Demerara Rise results and suggest that NE Brazil to the east of the Amazon River delta receives most rainfall and runoff when the Demerara Rise and Cariaco Basin are drier. This also suggests that when we consider Amazon River outflow versus NE Brazil outflow, we are talking about western versus eastern Amazon Basin patterns as a whole.

Multi-millennial-scale intervals A–C are of the Holocene–Glacial transition in age. This interval seems to associate low rainfall and Amazon flow to warmer conditions (Middle Holocene warmth—B) and higher rainfall and Amazon flow to cooler conditions (Late Holocene—A; glacial transition—C). On the other hand, the Pleistocene intervals (e.g., Stuiver and Grootes, 2000 [51]; Cronin, 2009 [52]) D–I have an opposite relationship between Amazon flow and rainfall and temperature. Intervals D (Younger Dryas), F (H1), and H (H2) are all times of low rainfall and Amazon flow with colder conditions, while intervals E (Bolling–Allerod), G (D/O interstadial 2), and I (D/O interstadials 3–4) are times of strong rainfall and Amazon flow and warmer conditions. On this basis, we argue that multi-millennial-scale cyclicity in the Holocene is out of phase with multi-millennial-scale variability in the Pleistocene.

The higher-frequency magnetic intensity variability indicated by vertical black lines in Figures 6–9 suggests another level of wet–dry oscillation associated with Amazon Basin hydrology and Amazon outflow. However, the data do not show a consistent associated variation in coercivity, and so the interpretation of this more subtle higher-frequency variability is more questionable. Even so, there is a great deal of evidence for millennial-scale environmental variability (e.g., Bond et al., 1997 [53]), most dominantly with ~1500 year cyclicity. The average duration of our successive millennial-scale high intensity peaks is 1794 ± 482 years (Table 2). All of our cycles that are common to both 9GGC and 25GGC have an average of ~1600 years. The pre 20 ka interval in 9GGC alone shows three intervals with ~2600 year cyclicity (underlined in Table 2). As noted previously, our millennial-scale intensity cycles are not as notable as our longer-term cycles. If we are simply missing three cycles in the pre 20ka interval of 9GGC then it seems that our millennial-scale intensity variability is consistent with previous estimates of ~1500 year cyclicity.

Several of the published studies discussed above have commented on possible mechanisms for tropical environmental variability. One process that fits the overall data is multi-millennial-sale variation in the placement of the Intertropical Convergence Zone (ITCZ). More southerly average placement will enhance NE South America and Amazon Basin rainfall. More northerly placement will lead to drier conditions. Haug et al. (2001) [27] suggested this mechanism caused a wet early Holocene wet interval in the Cariaco Basin (10.5° N) further to the northwest along the NE South American margin. This occurred at the same time as our interval B (early Holocene dry conditions), which was also noted in other nearby cores of the Demerara Rise by Lund et al. (2018) [23]. This pattern is also consistent with Lake Tulane wet intervals during Pleistocene Heinrich (cold) events versus our dry events in cold intervals D, F, and H.

The dominance of millennial- to multi-millennial-scale environmental variability in the tropical North Atlantic Ocean is also interesting to note given the recent study by Obrochta et al. (2012) [54] on variability of a similar scale in the high-latitude North Atlantic Ocean (ODP Site 609). They focused on the Late Quaternary ice-rafting evidence for a "1500-year cycle" in climate variability over the last glacial cycle. They suggested that much of that cyclicity might be better interpreted as an admixture of ~1000 and ~2000 year cycles (Obrochta et al., 2012) [54]. It seems clear that the entire Atlantic region has been subject to continuing millennial- to multi-millennial-scale variability over the Late Quaternary, however the cause(s) of that regional climate pattern and the relationship between tropical and high-latitude North Atlantic variability are still not certain.

10. Conclusions

We have carried out a rock magnetic study of two deep-sea gravity cores (9GGC and 25GGC) from the Demerara Rise, NE South America. Previous studies (Huang et al., 2014 [12]; Lund et al., 2017 [11]) provided radiocarbon and paleomagnetic chronologies for these cores. In this study, we performed detailed rock magnetic measurements on these cores to characterize the rock magnetic mineralogy, magnetic concentration, and magnetic grain size as an indicator of the overall clastic fraction of the cores. We measured the magnetic susceptibility (chi), anhysteretic remanence (ARM), and isothermal remanence (SIRM) and demagnetized the remanences at several levels of af demagnetization. The magnetic intensities estimate the proportion of magnetic material (and indirectly overall clastic fraction) in the cores. The ratios of rock magnetic parameters (ARM/chi, $ARM_{10}/ARM0$, $SIRM_{10}/SIRM0$) indicate the relative grain size of the magnetic material (and indirectly the overall clastic grain size). Selected hysteresis measurements also gave us a sense of the variation in ferric (goethite, hematite)/ferrous (magnetite, titanomagnetite) ratios for the magnetic minerals.

The rock magnetic intensity parameters (chi, ARM, SIRM) and the rock magnetic ratios (ARM/chi, $ARM_{10}/ARM0$, $SIRM_{10}/SIRM0$) both vary systematically and synchronously over the last 30,000 years in both cores. There is clear evidence of a multi-millennial-scale pattern of cyclicity, with intervals of high magnetic intensity (high magnetic and clastic content) and low magnetic ratios (coarser magnetic and clastic grain size) alternating in sequence. There is also evidence of finer millennial-scale variability in intensity superposed on the multi-millennial-scale variability, with an average cycle duration of ~1800 ± 500 years. There are nine (A–I) multi-millennial-scale intervals in the cores. Intervals A, C, E, G, and I have high magnetic and clastic content with coarser overall magnetic and clastic grain size and are likely intervals of enhanced rainfall and runoff from the NE South American margin to the coastal ocean. Alternatively, intervals B, D, F, and H represent periods of lower clastic flux with finer overall grain size, probably indicating lower rainfall and runoff from the continental margin. During the Holocene, high rainfall and runoff intervals can be related to cooler times and low rainfall and runoff to warmer times. The opposite pattern is true during the Pleistocene, with higher rainfall and runoff occurring during interstadial conditions and lower rainfall and runoff occurring during stadial conditions. The highest ferric/ferrous ratios occur during Holocene interval B (low rainfall and runoff) during the Early Holocene warm interval ~5–9 ka. We associated that interval with drought conditions in the Amazon region, consistent with previous estimates of Amazon flow in this interval (Maslin and Burns, 2000 [10]; Bendle et al., 2010 [35]; Lund et al., 2018 [23]).

We noted a similar pattern of Pleistocene multi-millennial-scale variability in published marine deep-sea core studies from Cariaco Basin (~10° N) to the NE Brazilian margin (~1° N–4° S). These studies documented clastic flux variability using XRF measurements of Ti and Fe. However, it seems likely that the NW part of this transect (Cariaco Basin, Demerara Rise, Amazon fan) has out-of-phase warm–cold wet–dry relationships that support the notion of the eastern Amazon Basin and western Amazon being alternately wetter or drier. One possible cause of the high–low rainfall–runoff patterns might be oscillation of the Intertropical Convergence Zone (ITCZ), with higher rainfall and runoff associated with a more southerly average position of the ITCZ and lower rainfall–runoff associated with a more northerly average position of the ITCZ.

Supplementary Materials: Supplementary materials can be found at https://www.mdpi.com/2673-1924/2/1/15/s1, Supplementary Table S1: Correlations for revised chronostratigraphy.

Author Contributions: Writing—S.L., Data acquisition—S.L. and E.P., data analysis—S.L. and E.P. All authors have read and agreed to the published version of the manuscript.

Funding: NSF grant EAR1547605.

Conflicts of Interest: The authors declare no conflict of interest.

References

1. Kuehl, S.A.; DeMaster, D.J.; Nittroucr, C.A. Nature of sediment accumulation on the Amazon continental shelf. *Cont. Shelf Res.* **1986**, *6*, 209–225. [CrossRef]
2. Franzinelli, E.; Potter, P. Petrology, chemistry and texture of Amazon River sands, Amazon River system. *J. Geol.* **1983**, *91*, 23–39. [CrossRef]
3. Milliman, J.; Meade, R. World-wide delivery of river sediments to the oceans. *J. Geol.* **1983**, *91*, 1–21. [CrossRef]
4. Meade, R. Suspended sediment of the modern Amazon and Orinoco Rivers. *Quat. Int.* **1994**, *21*, 29–39. [CrossRef]
5. Damuth, J.E.; Fairbridge, R.W. Equatorial Atlantic deep-sea arkosic sands and ice-age aridity in tropical South America. *Geol. Soc. Am. Bull.* **1970**, *81*, 189–206. [CrossRef]
6. Damuth, J.; Kumar, N. Amazon cone, morphology, sediments, age, and growth pattern. *Geol. Soc. Am. Bull.* **1975**, *86*, 863–878. [CrossRef]
7. Damuth, J.E.; Flood, R.D. Amazon Fan, Atlantic Ocean. In *Submarine Fans and Related Turbidite Systems*; Bouma, A.H., Barnes, N.E., Nomwk, W.R., Eds.; Springer: New York, NY, USA, 1985; pp. 91–106.
8. Flood, R.D.; Piper, D.J.W. Shipboard Scientific Party. In *Proceedings of the Ocean Drilling Program, Initial Reports 155*; Texas A&M University Digital Library: Killeen, TX, USA, 1995. [CrossRef]
9. Maslin, M.; Durham, E.; Burns, S.; Platzman, E.; Grootes, P.; Greig, S.; Nadau, M.; Schleicher, M.; Pflauman, U.; Lomax, B.; et al. Paleoreconstruction of the Amazon River freshwater and sediment discharge using sediments recovered from Site 942 on the Amazon Fan. *J. Quat. Sci.* **2000**, *15*, 419–434. [CrossRef]
10. Maslin, M.; Burns, S. Reconstruction of the Amazon Basin effective moisture availability over the past 14,000 years. *Science* **2000**, *290*, 2285–2289.
11. Lund, S.; Oppo, D.; Curry, W. Late quaternary paleomagnetic secular variation recorded in deep-sea sediments from the Demerara Rise, Equatorial west Atlantic Ocean. *Phys. Earth Planet. Int.* **2017**, *272*, 17–26. [CrossRef]
12. Huang, K.; Oppo, D.; Curry, W. Decreased influence of Antarctic Intermediate Water in the tropical Atlantic during North atlantic cold events. *Earth Planet. Sci. Lett.* **2014**, *389*, 200–208. [CrossRef]
13. Berger, W.H.; Wefer, G. Central themes of South Atlantic circulation. In *The South Atlantic: Present and Past Circulation*; Wefer, G., Berger, W.H., Siedler, G., Webb, D.J., Eds.; Springer: Heidelberg, Germany, 1996; pp. 1–11.
14. Schmitz, W.; McCartney, W. On the North Alatntic circulation. *Rev. Geophys.* **1993**, *31*, 29–49. [CrossRef]
15. Metcalf, W.G.; Stalcup, M.C. Origin of the Atlantic Equatorial Undercurrent. *J. Geophys. Res.* **1967**, *72*, 4959–4975. [CrossRef]
16. Kirchner, K.; Rhein, M.; Huttl-Kabus, S.; Boning, C. On the spreading of South Atlantic water into the Northern Hemisphere. *J. Geophys. Res.* **2009**, *114*, C05019. [CrossRef]
17. Stramma, L.; Fischer, J.; Reppin, J. The North Brazil Undercurrent. *Deep Sea Res. Part I* **1995**, *4*, 773–795. [CrossRef]
18. Ruhlemann, C.; Diekmann, B.; Mulitza, S.; Frank, M. Late Quaternary changes of western equatorial Atlantic surface circulation and Amazon lowland climate recorded in Ceara Rise deep-sea sediments. *Paleoceanography* **2001**, *6*, 293–305. [CrossRef]
19. Figueiredo, A.; Nittrouer, C. New insights to high-resolution stratigraphy on the Amazon continental shelf. *Mar. Geol.* **1995**, *125*, 393–399. [CrossRef]

20. Nittrouer, C.; Kuehl, S.; Sternberg, R.; Figuerido, A.; Faria, L. An introduction to the geological significance of sediment transport and accumulation on the Amazon continental shelf. *Mar. Geol.* **1995**, *125*, 177–192. [CrossRef]

21. Smith, W.O.; Demaster, D.J. Phytoplankton biomass and productivity in the Amazon River plume: Correlation with seasonal river discharge. *Cont. Shelf Res.* **1996**, *16*, 291–319. [CrossRef]

22. Nittrouer, C.A.; DeMaster, D.J. Sedimentary processes on the Amazon continental shelf: Past, present and future research. *Cont. Shelf Res.* **1986**, *6*, 5–30. [CrossRef]

23. Lund, S.; Mortazavi, E.; Chong, L.; Platzman, E.; Berelson, W. Holocene sedimentation on the distal Amazon Fan/Demerara Abyssal Plain. *Mar. Geol.* **2018**, *404*, 147–157. [CrossRef]

24. Damuth, J. Late Quaternary sedimentation in the western equatorial Atlantic. *Geol. Soc. Am. Bull.* **1977**, *88*, 695–710. [CrossRef]

25. Meinert, J.; Bloemedal, J. A comparison of acoustic and rock-magnetic properties of equatorial Atalantic deep-sea sediments: Paleooceanographic implications. *Earth Planet. Sci. Lett.* **1989**, *94*, 291–300. [CrossRef]

26. Wainer, I.; Soares, J. Northeast Brazil and its decadal-scale relationship to wind stress and sea surface temperature. *Geophys. Res. Lett.* **1997**, *24*, 277–280. [CrossRef]

27. Haug, G.; Hughen, K.; Sigman, D.; Peterson, L.; Rohl, L. Southward migration of the intertropical convergence zone through the Holocene. *Science* **2001**, *293*, 1304–1308. [CrossRef] [PubMed]

28. Mayle, F.; Beerling, D.; Gosling, W.; Bush, M. Response of Amazonian ecosystems to climatic and atmosphereic carbon dioxide change since the last glacial maximum. *Philos. Trans. R. Soc. Lond. Ser. B Biol. Sci.* **2004**, *359*, 499–514. [CrossRef] [PubMed]

29. Mayle, F.; Power, M. Impact of a drier Early-Mid Holocene climate upon Amazonian forests. *Philos. Trans. R. Soc. Lond. Ser. B Biol. Sci.* **2008**, *363*, 1829–1838. [CrossRef] [PubMed]

30. Thompson, L.G.; Davis, M.E.; Mosley-Thompson, E.; Sowers, T.A.; Henderson, K.A.; Zagorodnov, V.S.; Lin, P.-N.; Mikhalenko, V.N.; Campen, R.K.; Bolzan, J.F.; et al. A 25,000-year tropical climate history from Bolivian ice cores. *Science* **1998**, *282*, 1858–1864. [CrossRef]

31. Seltzer, G.; Rodbell, D.; Burns, S. Isotopic evidence for late Quaternary climatic change in tropical South America. *Geology* **2000**, *28*, 35–38. [CrossRef]

32. Abbott, M.B.; Wolfe, B.B.; Wolfe, A.P.; Seltzer, G.O.; Aravena, R.; Mark, B.G.; Polissar, P.J.; Rodbell, D.T.; Rowe, H.D.; Vuille, M. Holocene paleohydrology and glacial history of the central Andes using multiproxy lake sediment studies. *Palaeogeogr. Palaeoclim. Palaeoecol.* **2003**, *194*, 123–138. [CrossRef]

33. Bush, M.B.; Hansen, B.C.S.; Rodbell, D.T.; Seltzer, G.O.; Young, K.R.; Leon, B.; Abbott, M.B.; Silman, M.R.; Gosling, W.D. A 17,000-year history of Andean climate and vegetation change from Laguna de Chochos, Peru. *J. Quat. Sci.* **2005**, *20*, 703–714. [CrossRef]

34. Franke, C.; von Dobeneck, T.; Drury, M.; Meeldjik, J.; Dekkers, M. Magnetic petrology of equatorial Atlntic sediments: Electron microscopy results and their implications for environmental magnetic interpretation. *Paleoceanography* **2007**, *22*. [CrossRef]

35. Bendle, J.A.; Weijers, J.W.H.; Maslin, M.A.; Sinninghe Damsté, J.S.; Schouten, S.; Hopmans, E.C.; Boot, C.S.; Pancost, R.D. Major changes in glacial and Holocene terrestrial temperatures and sources of organic carbon recorded in the Amazon fan by tetraether lipids. *Geochem. Geophys. Geosyst.* **2010**, *11*, 21–31. [CrossRef]

36. Stacey, F.D.; Banerjee, S.K. *The Physical Principles of Rock Magnetism*; Elsevier: Amsterdam, The Netherlands, 1974.

37. Bloemendal, J.; Lamb, B.; King, J. Paleoenvironmental implications of rock-magnetic properties of late Quaternary sediment cores from the eastern equatorial Atlantic. *Paleoceanography* **1988**, *3*, 61–87. [CrossRef]

38. Oppo, D. Xray Fluorescence Studies of CDH 46. Unpublished work. 2021.

39. Maslin, M.; Ettwein, V.; Wilson, K.; Guilderson, T.; Burns, S.; Leng, M. Dynamic boundary-monsoon intensity hypothesis: Evidence from the deglacial Amazon River discharge record. *Quat. Sci. Rev.* **2011**, *30*, 3823–3833. [CrossRef]

40. Grimm, E.; Jacobson, G.; Watts, W.; Nansen, B.; Maasch, K. A 50,000 year record of climate oscillations from Florida and its temporal correlation with the Heinrich Events. *Science* **1993**, *261*, 198–200. [CrossRef] [PubMed]

41. Curry, W.; Oppo, D. Synchronous high-frequency oscillations I ntropical sea surface temperatures and North Atlantic deep water production during the last glacial cycle. *Paleoceanography* **1997**, *12*, 1–14. [CrossRef]

42. Deplazes, G.; Lukge, A.; Peterson, L.; Timmerman, A.; Hamann, Y.; Hughen, K.; Rohl, U.; Laj, C.; Cane, M.; Sigman, D. Links between tropical rainfall and North Atlantic climate during the last glacial period. *Nat. Geosci.* **2013**, *6*, 213–217. [CrossRef]

43. Haggi, C.; Chiessi, C.; Merkel, U.; Mulitza, S.; Prange, M.; Schultz, M.; Schfus, E. Reponse of the Amazon rainforest to late Pleistocene climate variability. *Earth Planet. Sci. Lett.* **2017**, *479*, 50–59. [CrossRef]

44. Crivellari, S.; Chiessi, C.M.; Kuhnert, H.; Häggi, C.; Portilho-Ramos, R.D.C.; Zeng, J.-Y.; Zhang, Y.; Schefuß, E.; Mollenhauer, G.; Hefter, J.; et al. Increased Amazon freshwater discharge during late Heinrich Stadial 1. *Quat. Sci. Rev.* **2018**, *181*, 144–155. [CrossRef]

45. Arz, H.W.; Patzold, J.; Wefer, G. Correlated millennial-scale changes in surface hydrography and terrigenous sediment yields inferred from last-glacial marine deposits off northeastern Brazil. *Quat. Res.* **1998**, *50*, 157–166. [CrossRef]

46. Nace, T.E.; Baker, P.A.; Dwyer, G.S.; Silva, C.G.; Rigsby, C.A.; Burns, S.J.; Giosan, L.; Otto-Bliesner, B.; Liu, Z.; Zhu, J. The role of North Brazil current in the paleoclimate of the Brazilian NE margin and paleoceanography of the west tropical Atlantic during the late Quaternary. *Palaeogeogr. Palaeoclim. Palaeoecol.* **2014**, *415*, 3–13. [CrossRef]

47. Zhang, Y.; Chiessi, C.; Mulitza, S.; Zabel, M.; Trindade, R.; Hollanda, M.; Dantas, E.; Govin, A.; Tiedemann, R.; Wefer, G. Origin of increased terrestrial supply to the NE South American contintal margin during HS1 and the Younger Dryas. *Earth Planet. Sci. Lett.* **2015**, *432*, 493–500. [CrossRef]

48. Mulitza, S.; Chiessi, C.M.; Schefus, E.; Lippold, J.; Wichmann, D.; Antz, B.; Mackenson, A.; Paul, A.; Prange, M.; Rehfeld, K.; et al. Synchronous and proportional deglacia lchanges in Atlantic meridional overturning and NE Brazilian precipitation. *Paleoceanography* **2017**, *32*, 622–633. [CrossRef]
49. Cheng, H.; Sinha, A.; Cruz, F.; Wang, X.; Edwards, L.; D'Horta, F.; Ribas, C.; Vuille, M.; Stott, L.; Auler, A. Climate change patterns in Amazonia and biodiversity. *Nat. Commun.* **2013**, *4*, 1411. [CrossRef]
50. Wang, X.; Edwards, L.; Auler, A.; Chang, H.; Kong, X.; Wang, Y.; Cruz, F.; Dorale, J.; Chiang, H. Hydroclimate changes across the Amazon lowlands over the last 45,000 years. *Nature* **2017**, *541*, 204–207. [CrossRef] [PubMed]
51. Stuiver, M.; Grootes, P. GISP2 oxygen isotope ratios. *Quat. Res.* **2000**, *53*, 277–284. [CrossRef]
52. Cronin, T. *Paleoclimates: Understanding Climate Change Past and Present*; Columbia University Press: New York, NY, USA, 2009.
53. Bond, G.; Showers, W.; Cheseby, M.; Lotti, R.; Almasi, P.; deMenocal, P.; Priore, P.; Cullen, H.; Hajdas, I.; Bonani, G. A pervasive millenial-scale cycle in North Atlantic Holocene and glacial climates. *Science* **1997**, *278*, 1257–1266. [CrossRef]
54. Obrochta, S.; Miyahara, H.; Yokoyama, Y.; Crowley, T. A re-examination of evidence for the North Atlantic 1500-year cycle at Site 609. *Quat. Sci. Rev.* **2012**, *55*, 23–33. [CrossRef]

Article

The Interactive Role of Hydrocarbon Seeps, Hydrothermal Vents and Intermediate Antarctic/Mediterranean Water Masses on the Distribution of Some Vulnerable Deep-Sea Habitats in Mid Latitude NE Atlantic Ocean

Luis Somoza [1,*], José L. Rueda [2], Olga Sánchez-Guillamón [2], Teresa Medialdea [1], Blanca Rincón-Tomás [3], Francisco J. González [1], Desirée Palomino [2], Pedro Madureira [4], Enrique López-Pamo [1], Luis M. Fernández-Salas [5], Esther Santofimia [1], Ricardo León [1], Egidio Marino [1], María del Carmen Fernández-Puga [6] and Juan T. Vázquez [2]

[1] Marine Geology Division, Geological Survey of Spain (IGME), Ríos Rosas 23, 28003 Madrid, Spain; t.medialdea@igme.es (T.M.); fj.gonzalez@igme.es (F.J.G.); e.lopez@igme.es (E.L.-P.); E.santofimia@igme.es (E.S.); r.leon@igme.es (R.L.); e.marino@igme.es (E.M.)

[2] Instituto Español de Oceanografía (IEO), Centro Oceanográfico de Málaga, 29640 Málaga, Spain; jose.rueda@ieo.es (J.L.R.); olga.sanchez@ieo.es (O.S.-G.); desiree.palomino@ieo.es (D.P.); juantomas.vazquez@ieo.es (J.T.V.)

[3] Department of General Microbiology, Institute of Microbiology and Genetics, Georg-August University Göttingen, Wilhelmsplatz 1, 37073 Göttingen, Germany; b.rincontomas@gmail.com

[4] Estrutura de Missão para a Extensão da Plataforma Continental (EMEPC), Rua Costa Pinto 165, 2770-047 Paço de Arcos, Portugal; pedro.madureira@emepc.mm.gov.pt

[5] Instituto Español de Oceanografía (IEO), Centro Oceanográfico de Cádiz, Muelle de Levante (Puerto Pesquero), 11006 Cádiz, Spain; luismi.fernandez@ieo.es

[6] Departamento de Ciencias de la Tierra, Facultad de Ciencias del Mar y Ambientales/INMAR, Universidad de Cádiz (UCA), Av. República Saharaui s/n, 11510 Cádiz, Spain; mcarmen.fernandez@uca.es

* Correspondence: l.somoza@igme.es

Citation: Somoza, L.; Rueda, J.L.; Sánchez-Guillamón, O.; Medialdea, T.; Rincón-Tomás, B.; González, F.J.; Palomino, D.; Madureira, P.; López-Pamo, E.; Fernández-Salas, L.M.; et al. The Interactive Role of Hydrocarbon Seeps, Hydrothermal Vents and Intermediate Antarctic/Mediterranean Water Masses on the Distribution of Some Vulnerable Deep-Sea Habitats in Mid Latitude NE Atlantic Ocean. *Oceans* **2021**, *2*, 351–385. https://doi.org/10.3390/oceans2020021

Academic Editor: Antonio Bode

Received: 13 December 2020
Accepted: 16 April 2021
Published: 26 April 2021

Publisher's Note: MDPI stays neutral with regard to jurisdictional claims in published maps and institutional affiliations.

Abstract: In this work, we integrate five case studies harboring vulnerable deep-sea benthic habitats in different geological settings from mid latitude NE Atlantic Ocean (24–42° N). Data and images of specific deep-sea habitats were acquired with Remoted Operated Vehicle (ROV) sensors (temperature, salinity, potential density, O_2, CO_2, and CH_4). Besides documenting some key vulnerable deep-sea habitats, this study shows that the distribution of some deep-sea coral aggregations (including scleractinians, gorgonians, and antipatharians), deep-sea sponge aggregations and other deep-sea habitats are influenced by water masses' properties. Our data support that the distribution of scleractinian reefs and aggregations of other deep-sea corals, from subtropical to north Atlantic could be dependent of the latitudinal extents of the Antarctic Intermediate Waters (AAIW) and the Mediterranean Outflow Waters (MOW). Otherwise, the distribution of some vulnerable deep-sea habitats is influenced, at the local scale, by active hydrocarbon seeps (Gulf of Cádiz) and hydrothermal vents (El Hierro, Canary Island). The co-occurrence of deep-sea corals and chemosynthesis-based communities has been identified in methane seeps of the Gulf of Cádiz. Extensive beds of living deep-sea mussels (*Bathymodiolus mauritanicus*) and other chemosymbiotic bivalves occur closely to deep-sea coral aggregations (e.g., gorgonians, black corals) that colonize methane-derived authigenic carbonates.

Keywords: seafloor mapping; vulnerable deep-sea habitats; deep-sea corals; chemosynthesis-based communities; vulnerable marine ecosystem; Atlantic Ocean

1. Introduction

Maintaining the sustainable functioning of the global biosphere requires protection of deep-sea ecosystems, particularly because they face major changes related to human and climate-induced impacts [1]. For an effective protection, an improvement in knowledge and research is essential, because the deep sea is truly the Earth's last frontier. Although oceans cover more than 70% of the Earth's surface, less than 0.0001% of the deep sea has been explored to date at a high resolution [2]. Seafloor bathymetry and habitat mapping are basic tools used to improve management in some areas of conservation importance and scientific interest [1]. High-resolution seabed mapping is essential for improving the current knowledge and understanding of deep-seafloor morphology, key geological processes (including sedimentary depositional and erosional processes), habitat distribution, and typology as well as availability of mineral deposits and energy resources, among other aspects. Seabed mapping is also valuable for more applied purposes such as the deployment of submarine cables and pipelines, detection of marine geohazards, development of early warning systems for volcanic eruptions, tsunamis, and earthquakes, prevention of pollution, management of fisheries, analysis of potential environmental impacts associated with deep-sea mining and for safer shipping. In summary, high-resolution seabed mapping studies provide data and information for researchers, contractors, industry, mining companies, regulators and Non-Governmental Organizations to manage and mitigate impacts on marine environments.

The main aim of the present study is to integrate bathymetry and seafloor geomorphology, water masses properties, and the distribution of vulnerable deep-sea habitats (and their associated biota) along the southernmost NE Atlantic Ocean. These vulnerable deep-sea habitats include those conformed by deep-sea corals (in this study including deep-sea scleractinians, stony octocorals, bamboo corals, soft corals, gorgonians. and antipatharians), sea-pens, sponges and chemosynthesis-based communities. This integration has been applied to five case studies located between 24° and 42° N at 90–2500 m below sea level (mbsl) including: (1) Deep-sea scleractinian reefs (dominated by *Desmophyllum pertusum*) still thriving on the Galicia Bank; (2) Chemosynthesis-based communities of hydrocarbon seeps in the Gulf of Cádiz colonized by sponges, gorgonians, scleractinians, and black corals; (3) Deep-sea stony octocoral aggregations occurring on volcanic submarine ridges located along the Passage of Lanzarote; (4) New habitats formed after the Tagoro submarine eruption in 2011–2012; and (5) Aggregations of scleractinians, antipatharians, gorgonians, bamboo corals, and sponges on ferromanganese crusts of seamounts of the subtropical Canary Island Seamount Province (CISP). These five case study areas have been mapped in detail with MBES and ground truthed by means of ROV surveys and/or coring/dredging samples. The water mass properties for each specific habitat were obtained with in-situ ROV-mounted CTD (Conductivity–Temperature–Depth sound) and Niskin bottles. The comparison between vulnerable deep-sea habitats of these case studies allows us to analyze: (i) the potential relationship between water mass properties (salinity, oxygen, potential density, oxygen and methane concentration) for each specific habitat as a function of water depth and latitude, (ii) the influence of methane seeps on the distribution of some of these deep-sea habitats; and (iii) the influence of abrupt submarine eruptions and low-temperature hydrothermal vents in the Macaronesia volcanic archipelagos.

2. Geological and Oceanographic Settings

The five case-study areas include important types of world-wide recognized geomorphological seabed features such as banks, mud volcanoes (MVs), coral mounds and reefs, oceanic passages, contourite drifts, volcanic ridges, salt domes, seamounts, and submarine volcanoes [3,4].

2.1. The Galicia Bank

The first case study concerns banks and reefs of colonial scleractinians (mainly *D. pertusum*) located at 700–1800 mbsl on the Galicia Bank, NE Atlantic (Figures 1 and 2A). The Galicia Bank, with an area of 2117 km^2 (Figure 3A), constitutes the main structural high located on the NW Iberian continental margin, 200 km west off the Galician coast. The summit of the bank yields a contourite depositional system associated with the Mediterranean Outflow Water (MOW) and consists of erosive furrows and depositional bottom-current features such as drift and sediment wave fields [5,6]. Large pavement-like plates composed of phosphorite slabs and ferromanganese crusts dominate the top of the Galicia Bank [7]. Several water masses have been recognized to influence key processes of the Galicia Bank. Water masses above 500 mbsl are dominated by the Eastern North Atlantic Central Water current (ENACW). Below the minimum salinity located near 500 mbsl, the ENACW gradually mixes with the MOW reaching a peak in salinity values (36.1–36.5 psu) at approximately 1000 mbsl. The MOW, in this area, is characterized by an increase of the salinity values at 800 and 1200 mbsl and acts as a contour current above and around the Galicia Bank reaching velocities of 5–10 cm s^{-1} [8].

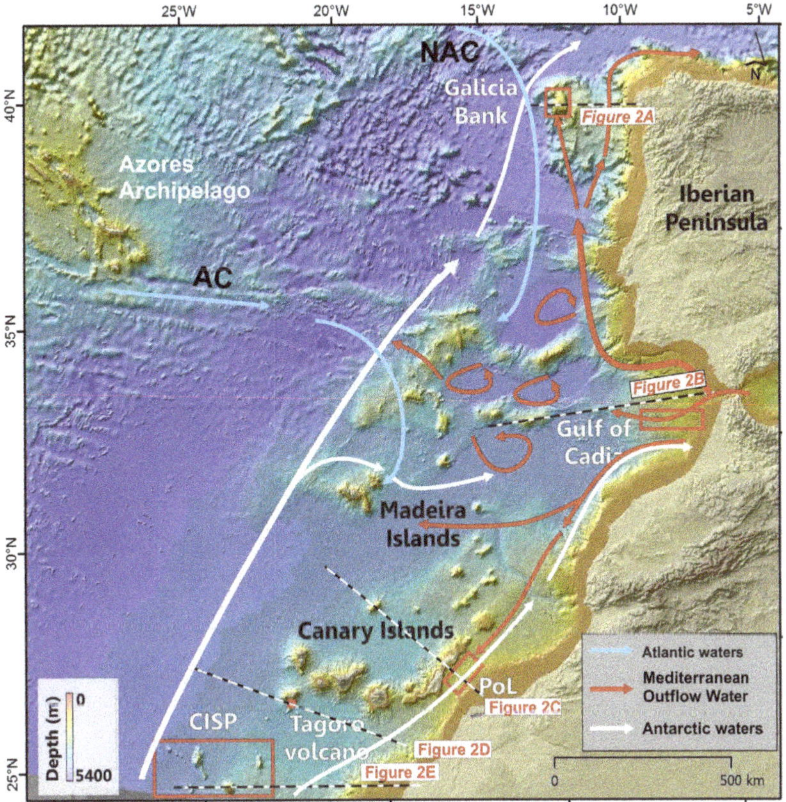

Figure 1. Location of case studies areas along the North East and Central Atlantic Ocean and schematic oceanographic circulation affecting the cases studies. CISP: Canary Islands Seamount Province; PoL: Passage of Lanzarote. Main currents from [8–12] are labelled as: North Atlantic Current (NAC), Azores Current (AC). Regional bathymetric map extracted from the EMODnet Project (http://www.emodnet.eu/bathymetry, accessed on 1 December 2020).

2.2. *Gulf of Cádiz and Moroccan Atlantic Margin*

The second case study refers to chemosynthesis-based habitats and other vulnerable deep-sea habitats (aggregations of sponges, gorgonians, antipatharians, among others) in mud volcanoes (MVs) located in the Moroccan Atlantic margin of the Gulf of Cádiz (GoC) between 34° and 35° N (Figures 1 and 2B). In the GoC, extensive geophysical research and seabed sampling have resulted in the discovery of 84 MVs and MVs-mud diapir complexes since 2000 [13–21]. Mud volcanism in the GoC is primarily associated with high concentrations of thermogenically-derived methane [22]. Different sources have been invoked for mud volcanism in the GoC: Pliocene and Late Miocene shales, the so-called Allochthonous

Unit of the GoC and even deeper sources of Mesozoic age [23,24]. Carbonates and ferro-manganese rocks such as methane-derived authigenic carbonates (MDACs) [25,26] formed by microbial-mediated anaerobic oxidation of methane (AOM) [27] and Fe-rich nodules formed by later exhumation, oxidation and diagenesis of buried AOM products [28]. Three types of intermediate water masses are currently considered in the case study of the GoC (Figure 1): the Eastern North Atlantic Central Water current (ENACW), the Mediterranean Outflow Water (MOW), the North Atlantic Deep Water current (NADW), and the Subarctic Intermediate Water current (SAIW).

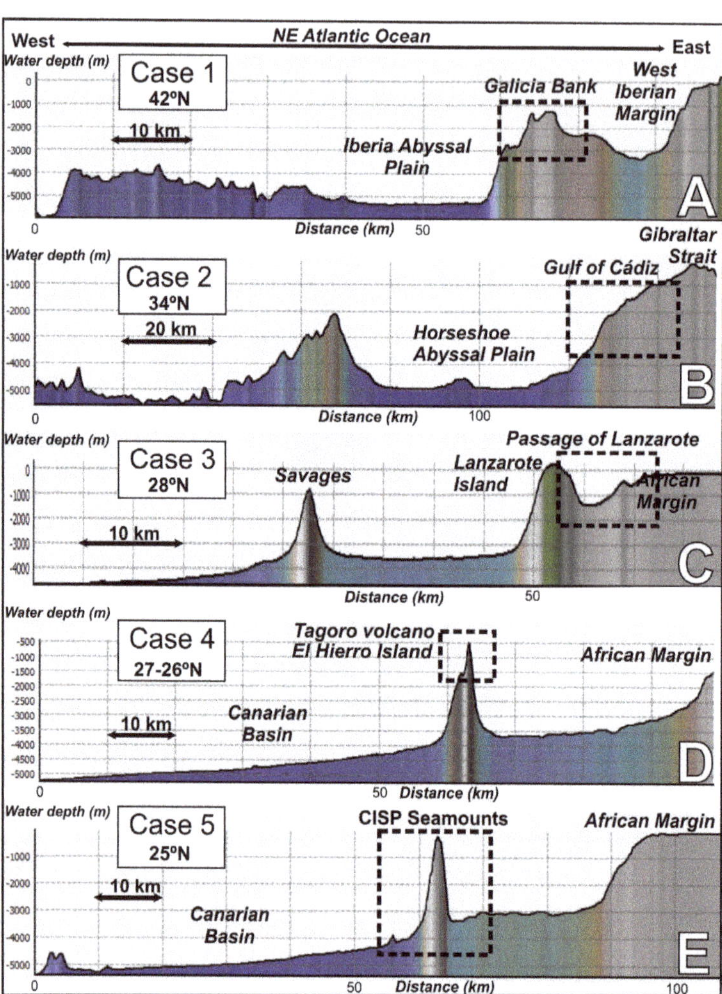

Figure 2. E–W regional bathymetric profiles along the North East and Central Atlantic Ocean showing the morpho-structural context and water depth distribution of the studied areas. (**A**) Galician margin; (**B**) Gulf of Cádiz; (**C**) Passage of Lanzarote (PoL), (**D**) Tagoro volcano, El Hierro Island, and (**E**) Canary Island Seamount Province (CISP). Profiles extracted from bathymetry of the Global Multi-Resolution Topography (GMRT) synthesis data set [29]. See Figure 1 for location.

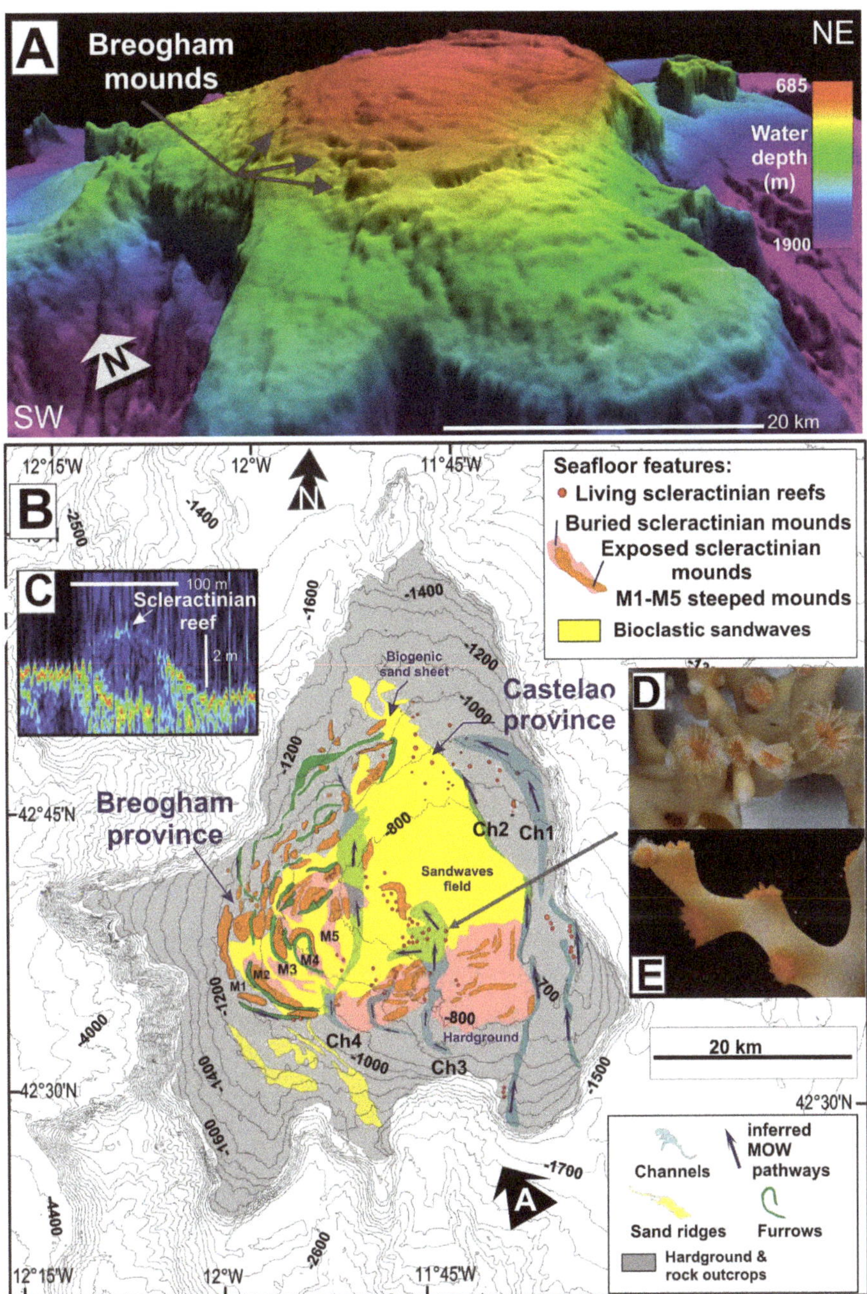

Figure 3. Seabed distribution of scleractinian reefs and mounds (M) in the Galicia Bank: (**A**) 3D shaded relief model using multibeam bathymetry data; Sunlight from the NW; (**B**) Morphological map including morpho-sedimentary features and scleractinian reefs identified with multibeam and backscatter data; (**C**) Ultra high-resolution sub-bottom profile crossing scleractinian reefs were (**D**,**E**) living *Desmophyllum pertusum* and *Madrepora oculata* appear.

2.3. The Passage of Lanzarote-Canary Islands-NW African Margin

The third case study focuses on the Passage of Lanzarote (PoL), between the Canary Islands (Fuerteventura-Lanzarote Islands) and the West African Continental Margin (Figure 1). The bathymetric profile transverse to the PoL is asymmetrical with the passage narrowing towards the south, showing a steeper western flank (Figure 2C). The PoL has maximum water depths at its center, ranging between 1240 and 1460 mbsl (Figures 1 and 2C). Two groups of seafloor morphologies stand out in the submarine relief of this passage [30,31]: (a) seamounts and hills related to volcanic ridges and salt domes and (b) bottom current-related features. Three main types of water masses have been identified circulating through the PoL (Figure 1) [10,32,33]: The upper thermocline North Atlantic Central Water current (NACW), which spans from the surface to the neutral density (Υn) value of 27.3 kg/m^3 (approximately 600 mbsl); and the Antarctic Intermediate Water current (AAIW) reflected by a clear S minimum and found below the NACW (roughly 700–1600 mbsl).The AAIW is well identified along the African continental margin with a predominantly northward flow which often exceed 0.02 m s^{-1}, appearing particularly intense (up to 0.06 m s^{-1} near the bottom) during the summer seasons [33]; the third, the Mediterranean Modified Water current (MW), a modified branch of the MOW identified in individual profiles by high-salinity peaks flowing southwards between 900 mbsl to the bottom with a mean southward flow of −0.05 Sv [33]. In winter and spring, the formation of active meddies (Mediterranean Outflow Water eddies) have been reported at the north of the PoL between 12° W and 15° W [33]. The occurrence of submarine mounds along this deep-water passage promotes the intensification of the turbulence at the interface between the AAIW and MW bottom currents [31,32].

2.4. El Hierro Island, Canary Islands

The fourth case study highlights the natural successional changes in newly formed volcanic substrates and habitats related to recent submarine eruptions that occurred in 2011–2012, south of El Hierro Island, Canary Islands (Figures 1 and 2D). Dead fish and patches of pale colored water at the sea surface indicated the onset of a submarine eruption on 10 October 2011 at 5 km distance from the town of La Restinga. The El Hierro eruption continued during late February, when seismicity decreased drastically until 6 March 2012. As a result of this shallow submarine eruption, a new submarine volcano grew from 375 to 89 mbsl [34]. Since 2016, the new volcano appears on the official hydrographic charts as "Tagoro" (meaning "stone circle for meeting place" in the aboriginal language of Canary Islands). The eruption consisted of two main phases of edifice construction intercalated with collapse events. After the cessation of the volcanic eruption, emissions of large quantities of iron and carbon dioxide from submarine hydrothermal vents were reported [35,36]. Several oceanographic expeditions were carried out during and after the eruption of the Tagoro volcano (Supplementary Table S1). These expeditions have allowed the identification of the development of new habitats after the eruption, associated with the emission of large quantities of iron and carbon dioxide from hydrothermal submarine vents.

2.5. The Canary Island Seamounts (23°48′N–26°30′N)

The fifth case study comprises the southern seamounts of the Canary Island Seamount Province (CISP), located southwest of the Canary Islands and 100 km west off the African coast (Figures 1 and 2E). The CISP is composed of more than 100 seamounts and submarine hills rising from ~5000 to 200 mbsl [37]. The CISP extends from the Lars-Essauira seamount, approximately 400 km from the north of Lanzarote (32.7° N, 13.2° W) to the Tropic seamount, 500 km at the southwest of the archipelago (23.8° N, 20.7° W). This case study focuses on the southern CISP area which is characterized by the presence of several large seamounts, including The Paps, Echo, Drago and Tropic, rising from ~5000 m to 200 mbsl (Figures 1 and 2E). These are the oldest seamounts of the CISP, dated between 91 and 119 Ma [38]. The seabed of these seamounts is mainly composed of ferromanganese crusts and nodules growing on basalt-sedimentary rock substrates containing high average amounts of Fe (23.5 wt%), Mn (16.1 wt%), and trace elements like Co (4700 µg/g), Ni (2800 µg/g), V (2400 µg/g), and Pb (1600 µg/g) [39–41]. The oldest age of initiation of the ferromanganese crusts growth was estimated at 90 Ma [40]. The CISP seamounts are located beneath the northern end of the Oxygen Minimum Zone (OMZ), which extends from 100 to 700 mbsl, from near the equator to 25° N [42]. The core of the OMZ is located at 400–500 mbsl, reaching very low oxygen values (<50 µmol/kg). The dissolved oxygen is also depleted by the high biological productivity in proximity to the Canary Islands due to upwelling and nutrient-rich currents [42]. At intermediate depths, the low oxygen waters are ventilated by the ENACW and by the South Atlantic Central Water current (SACW) above ~700 mbsl and by the AAIW at ~1000 mbsl. The North Atlantic Deep-Water current (NADW) flows at depths below 1500 mbsl and the Antarctic Bottom Water current (AABW) flows below ~4000 mbsl [33] (Figure 1). These deep bottom currents accelerate around the seamounts, increasing the supply of nutrients related to high inputs of Fe and P sourced from the Sahara dust [39].

3. Materials and Methods

Mapping the seafloor and its specific habitats (and Vulnerable Marine Ecosystems, VMEs) at regional scale requires interdisciplinary research teams using a wide suite of surveying techniques. Data of this work were collected mainly during the SUBVENT-2 cruise aboard RV *Sarmiento de Gamboa* [43]. Other data included in this work were collected during a set of extensive scientific expeditions listed in the Supplementary Table S1. A summary of the cruise details for each of the five case studies including data sets collected, MBES systems and ground truthing data sets are listed in Supplementary Table S1.

The multibeam bathymetry echosounder (MBES) Atlas DS 1 × 1 on board the RV *Sarmiento de Gamboa* has been used for seabed mapping of the distinct habitats and to plan ROV tracks (see Supplementary Table S1 for further details on MBES systems). MBES data were processed using CARIS HIPS&SIPS™ version 9 software (Teledyne CARIS, Frederiction, NB, Canada). The MBES data were used to generate Digital Terrain Models (DTM) at spatial resolutions of 30–50 m (depending on water depth). The multi-resolution DTM was used to produce regional sun-shaded image renders, perspective views and to extract margin-wide bathymetric profiles using Fledermaus™ version 7 software (QPS, Porstmouth, NH, USA) to interpret the submarine landscapes. It was also used to generate derivative products such as slope angle maps by means of ArcGIS™ desktop v. 12.4.

The ATLAS P-35 used aboard the RV *Sarmiento de Gamboa* is a parasound echosounder with two frequencies: a Primary High Frequency (PHF) at 20 kHz and a Secondary Low Frequency (SLF) at 4.5 kHz. PHF was used to record anomalies within the water column, whereas SLF was used as a sub-bottom profiler to record sediment/rock features (Supplementary Table S1).

Submersible observations and sampling of the seafloor were carried out with the ROV *Luso* (from Estructura de Missão para a Extensão da Plataforma Continental, Portugal), equipped with a high definition video camera Argus HD-SDI Camera (ARGUS Remote Systems AS, Laksevag, Norway), two robotic manipulators to recover biological and geological samples, a CTD (conductivity, temperature, and depth measurements) with fluorescence, turbidity and CO_2-, CH_4- and O_2- sensors, and four Niskin bottles for water sampling (Supplementary Table S1). Water samples (two replicates of 20 mL) were taken for methane determination using a gas chromatograph in cold seep areas. Rock-mineralization samples and associated biota were taken by means of conventional rectangular benthic dredges (0.8 m in width by 0.6 m in height) towed on the seafloor during 10 to 20 min at 1 knot speed. ROV tracks were planned after analysis and interpretation of previous MBES mapping (DTM of the bathymetry) and backscatter models. Gravity cores with a maximum length of 300 cm were taken on the summits and/or slopes of the mud volcanoes. The cores were photographed and a visual lithostratigraphic description was made in each case.

Regional oceanographic data were extracted from Conductivity-Temperature-Depth (CTD) of the World Ocean Database 2018 [44] of the National Centers for Environmental

Information (NOAA) (http://www.nodc.noaa.gov/OC5/indprod.html, accessed on 1 December 2020). We use Ocean Data View 4.0 software (Alfred-Wegener-Institut Helmholtz-Zentrum für Polar- und Meeresforschung, Bremerhaven, Germany) [45] for displaying regional oceanographic variables.

4. Results

4.1. Description of Vulnerable Deep-Sea Habitats of Each Case Study

In this section, the morphology and the type of habitats and communities in the five case studies from the north Atlantic (42° N) to subtropical Atlantic (23° N) have been described. Full details of substrate types, water depth ranges, geomorphology, habitat types, and the associated biota for each specific site are listed in the Supplementary Table S2.

4.1.1. Case Study 1: Scleractinian Reefs in the Galicia Bank (42°15′ N–42°55′ N)

A large number of live scleractinian colonies of *Desmophyllum pertusum* (previously named as *Lophelia pertusa*) and *Madrepora oculata* have been identified conforming reefs on the summit and flanks of the Galicia Bank at water depths between 620 and 1175 mbsl (Figure 3). The distribution of these scleractinian reefs has been mapped by means of MBES backscatter mosaic images, ultra-high resolution and high-resolution multichannel seismic reflection data and seabed sampling (Figure 3). A complete sequence of stepped elongated reefs (M1 to M5 in Figure 3), known as Breogham mounds and characterized by high backscatter strengths with a seabed expression, were identified on multibeam bathymetry along the western flank of the Galicia Bank (Figure 3A,B).

These semi-buried reefs display heights up to 70 m and widths of 450 m and are lined up in along-slope trending ridges from 826 to 1126 mbsl. Furthermore, on the top and at the eastern flank of the bank, a series of mini-reefs, 2–4 high and 80–100 m wide appear on a flat erosion surface at 780–750 mbsl. Mini reefs with living scleractinian colonies appear as acoustically transparent domes in very high-resolution sub-bottom profiles (Figure 3C). We interpret their low acoustic response ($\sim$1520 ms^{-1}) close to values for the water column, as due to the high contents of seawater within their skeletons, as well as the open space created by the skeletal growth patterns. Samples of these still thriving scleractinian reefs yielded living *D. pertusum* and *M. oculata* (Figure 3D,E). In contrast, older and semi-buried scleractinian reefs were partially cemented forming elongated mounds along the flanks of the bank, being characterized by higher backscatter responses [46]. In some cases, scleractinians, mollusc monoplacophora (*Laevipilina rolani*), and bryozoans [47] occurred on encrusted phosphorites and ferromanganese nodules substrates at 750–1400 mbsl in the southeast Galicia Bank.

4.1.2. Case Study 2: Chemosynthesis-Based Communities and Other Vulnerable Deep-Sea Habitats on Hydrocarbon Seeps of the Gulf of Cádiz (34° N–35° N)

Six MVs located between 350 and 3000 mbsl (Mercator, Algacel, Yuma, Madrid, Las Negras, and Bonjardim), were explored with the ROV "Luso" during the oceanographic expedition SUBVENT-2 [43] (Location shown in Figure 4A). Chemosynthesis-based communities and high methane concentrations were detected in these explored MVs. Reduced grey mud breccias were recovered by gravity cores in all cases, when testing the nature of extrusion of these cones.

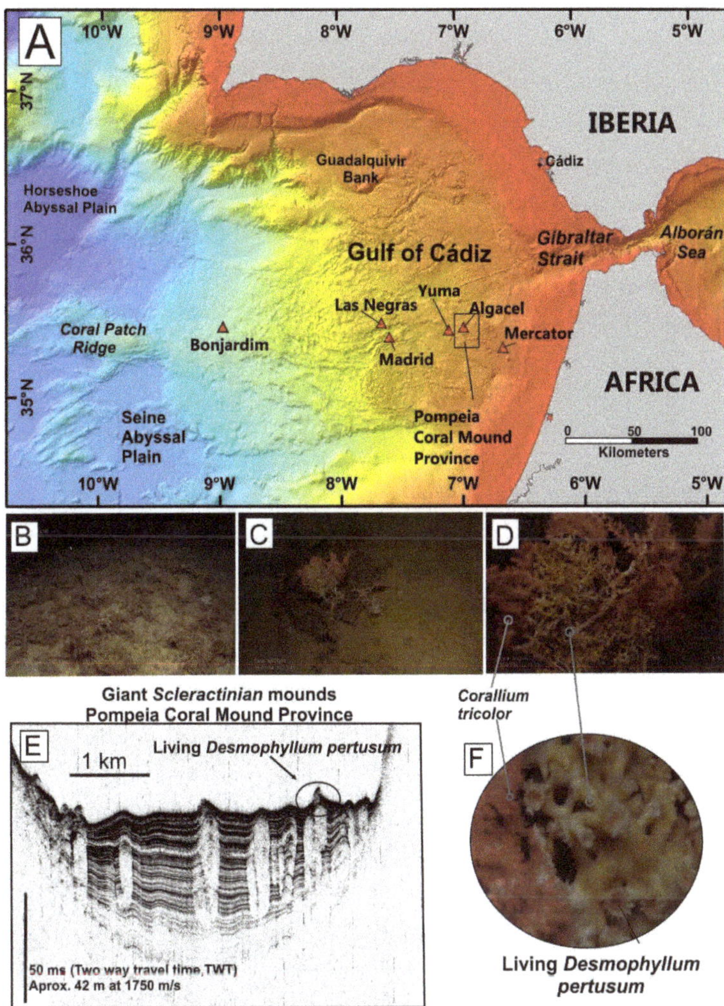

Figure 4. (**A**) Location of mud volcanoes explored in the present study in the Gulf of Cádiz along 35.5° N and the Pompeia Coral Mound Province. (**B**) Dead scleractinian skeletons at the top of the mounds. (**C,D,F**) Stony octocorals (*Corallium tricolor*) overlying live and dead colonies of *Desmophyllum pertusum* on mounds at the Pompeia Coral Province; (**E**) Ultrahigh-resolution sub-bottom profile showing giant coral mounds (transparent bodies) up to 42 m in height beneath the dead scleractinian reefs at the Pompeia Coral Mound Province.

The type of habitats identified in these MVs can be resumed into seven types (See details in Supplementary Table S2):

(1) Active methane seeps with chemosynthesis-based communities.
(2) Non-active pockmarks colonized by Cerianthids
(3) Aggregations of sponges, gorgonians, black corals, soft corals and bamboo corals colonizing methane-derived authigenic carbonates (MDACs)
(4) Hexactinellid sponge aggregations on muddy sediments and coral graveyards.
(5) Desmosponges aggregations on muddy sediments.
(6) Sea-pen (pennatulaceans) communities on micro-mounds muddy sediments.

(7) Graveyards of scleractinians colonized by stony octocorals.

Chemosynthesis-based communities were mainly detected on the summits and sometimes on the flanks (e.g., Algacel MV), of the surveyed MVs (Figure 5) and contained different types of taxa and rates of methane flux. Regarding this, in MVs with high methane fluxes such as Algacel MV (methane concentrations up to 97.60 nM) (Figure 5A–C), harbored living deep-sea mussel *Bathymodiolus mauritanicus* beds, containing both small- and large-size specimens were found. These beds formed large linear and circular mussel clumps (up to 10 m in diameter) surrounding emissions of intermittent gas bubbling with methane concentrations of 97.60 nM (Figure 6A–E).

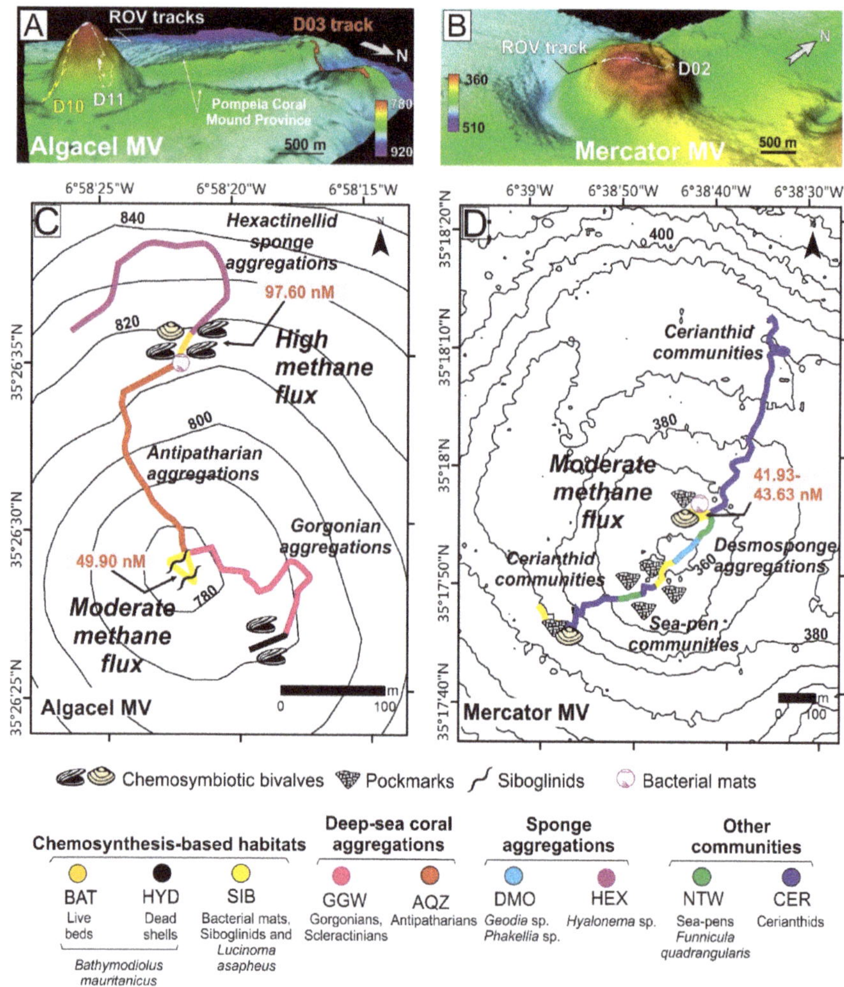

Figure 5. Vulnerable deep-sea habitats in mud volcanoes (MVs) of the Moroccan margin (Gulf of Cádiz). (**A**,**B**) 3D models of the Algacel and Mercator MVs as well as the Northern Pompeia Coral Ridge at the Pompeia Coral Mound Province, where ROV dives were carried out. Distribution of deep-sea habitats and organisms across (**C**) Algacel and (**D**) Mercator MVs. Methane concentrations are displayed in red letters. Location of MVs in Figure 4A. See further explanation in the text.

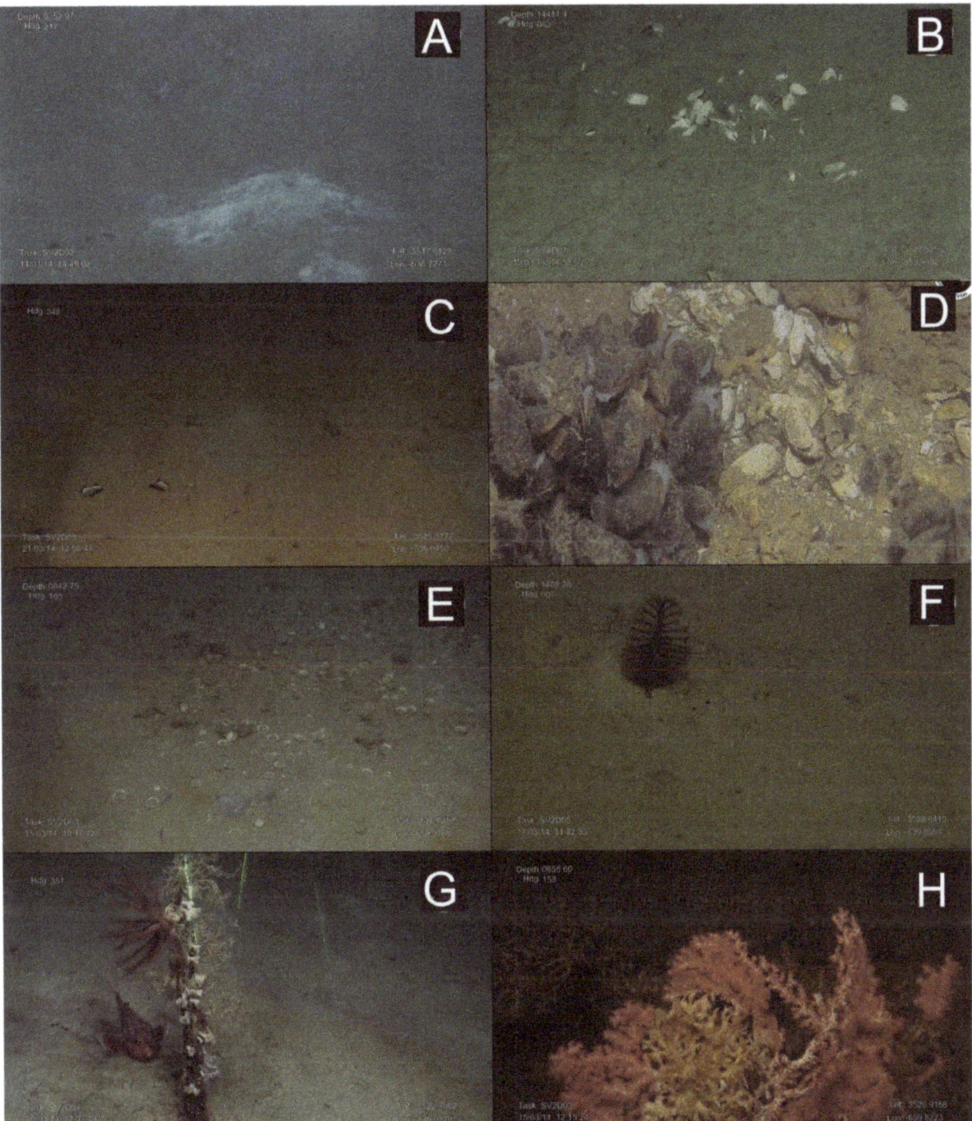

Figure 6. Underwater images obtained with the ROV "Luso"during the SUBVENT-2 expedition in mud volcanoes (MVs) of the Moroccan margin. (**A**): *Beggiatoa*-like bacterial mats and two cerianthids at the summit of the Mercator MV; (**B**): Accumulation of shells of the chemosymbiotic bivalve *Isorropodon megadesmus* on muddy substrates at the summit of the Bonjardim MV; (**C**): Shells of the chemosymbiotic bivalve *Acharax gadirae* at the summit of the Yuma MV; (**D**): Beds of the living chemosymbiotic *Bathymodiolus mauritanicus* in the Algacel MV; (**E**): Extensive graveyards of the chemosymbiotic bivalve *Lucinoma asapheus* at the Northern Pompeia Coral Ridge (Pompeia Coral Mound Province); (**F**): Sea-pen (*Pennatula* sp.) on muddy sediments at the Las Negras MV; (**G**): An iron bar of anthropogenic origin colonized by small gorgonians, solitary corals (*Desmophyllum dianthus*) and crinoids at the Yuma MV; (**H**): Living scleractinians(*D. pertusum*) and stony octocorals (*Corallium*) colonizing dead scleractinian skeletons in the Pompeia Coral Mound Province.

In MVs with moderate methane fluxes (30–50 nM), including the Mercator, Las Negras, and Madrid MVs, living chemosymbiotic bivalves such as *A. gadirae* and *Solemy aelarraichen-*

sis occurred with high densities of shells of *Thyasira vulcolutre, S. elarraichensis, Isorropodon megadesmus* and *B. mauritanicus*. Dense populations of frenulate polychaetes (*Siboglinum* sp.) also occured in some of these MVs with moderate methane fluxes (Figure 5B–D). Sulfur-oxidizing bacterial (SOB) mats forming macroscopically visible cohesive white patches of 10–30 cm in diameter were also detected (Figure 6A).

In MVs with low methane fluxes such as the Bonjardim MV (20.18–28.12 nM), the past occurrence of chemosynthesis-based communities is revealed by abundant shells of the chemosymbiotic bivalve *I. megadesmus* (Figure 6B).

Crater-like pockmarks identified on the MVs seafloor, 1–3 m in width, provide evidence of former blow-outs that are nowadays colonized with non-chemosynthetic organisms such as sea-pens (*Funiculina quadrangularis*), cerianthids, and annelids (*Hyalinoecia tubicola*).

Aggregations of gorgonians (*Swiftia*), antipatharians (*Bathypathes, Leiopathes, Stichopathes*), bamboo corals (*Chelidonisis*), and scleractinians (*D. pertusum*) colonize MDACs blocks (up to 1 m) formed by anaerobic oxidation of methane (AOM) on the summit and flanks of the MVs with high rates of diffuse methane release such as the Algacel MV (Figure 5C). Otherwise, large desmosponges aggregations (*Geodia* sp., *Phakellia* sp.) colonized blocks of MDACs formed on the summit and flanks of MVs with lower methane fluxes (Figure 5D). Along the flanks of some MVs, muddy sediments covered with scattered patches of dead scleractinians (coral rubble) and small MDAC gravels are colonized by hexactinellid sponges (*Pheronema, Hyalonema*) and gorgonians (*Radicipes* on soft substrates and *Swiftia* on MDAC gravels).

Other non-chemosymbiotic deep-sea species such as sea-pens (*Kophobelemnon* sp., *Pennatula, Anthoptilum*), cerianthids, holothurians, and decapod species of commercial interest (*Aristeomorpha foliacea, Plesiopenaeus edwardsianus*) were also detected on the soft substrates of MVs (Figure 6F).

Ridges and mounds with abundant scleractinian skeletons (mainly *D. pertusum* and *M. oculata*), but with scarce living *D. pertusum* colonies and stony octocoral colonies (e.g., *C. tricolor*) were detected westwards of the Algacel MV in the Pompeia Coral Mound Province (Figure 4). Patches of chemosymbiotic bivalve shells (mainly *L. asapheus* with some *Thyasiravulcolutre*), and scattered bacterial mats (*Beggiatoa*-like sulfur oxidizers) were also detected (Figure 6E,H). Ultra-high-resolution profiles below the scleractinian skeletons show that they represent the top of giant buried mounds that reach up to 42 m in height (equivalent to 50 ms two-way travel time, TWT, assuming an average sonic velocity of 1750 m s^{-1} for recent sediments) (Figure 4E). The potential causes of the regression of these giant coral mounds in the GoC will be considered in the Section 5. Methane concentrations on the seafloor of the ridge at the Pompeia Coral Mound Province range from 41.93 to 43.24 nM, indicating the occurrence of moderate seepage. The E-W orientation of these ridges at the westward side of the Algacel MV (Figure 5A) points to the influence of strong bottom currents causing strong turbulences on their lee face.

4.1.3. Case Study 3: Stony Octocorals, Gorgonians, Soft-Corals and Sponge Aggregations along the Passage of Lanzarote (28°30′ N–29° N)

The third case study focuses on the hard-substrate and sedimentary habitats of the Passage of Lanzarote (PoL) located between the Canary Islands (Fuerteventura-Lanzarote ridge) and the West African Continental margin (Figure 1). Three main types of morphologies can be described based on the bathymetric map along the PoL: Elongated mounds (M), volcanic ridges, volcanic cones and sub-circular sea-floor depressions (Figure 7)

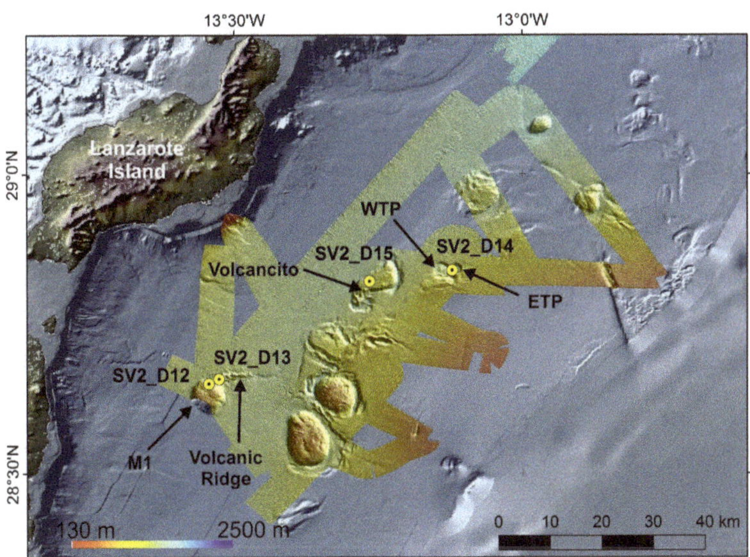

Figure 7. Bathymetric map obtained along the SUBVENT-2 expedition on board RV *Sarmiento de Gamboa*. The main characteristics of the seafloor such as sub-circular domes, volcanic ridges, marginal channels and circular to elongated depressions are highlighted. Locations of dives (from SV2_D12 to SV2_D15) are also shown as circles. M1: Dome; WTP: Western Twin Pool; ETP: Eastern Twin Pool.

Mounds are cylindrical or sub-rounded highs, 2–10 km in diameter with flat summits (e.g., M1 in Figure 7). The base of the mounds along the PoL ranges between 1225 and 1530 mbsl and their summits between 828 and 1336 mbsl. The volcanic ridges are 6 km long, 150 m high, E-W trending linear ridges detached from the mounds (e.g., the *Volcanic Ridge* in Figure 7). Volcanic cones are single structures 15–70 m height and 350–700 m in diameter crossing the flat summits of the mounds (e.g., the *Volcancito* in Figure 7). Six circular or slightly elliptical depressions have also been identified in the continental slope of the West African continental margin at the PoL [48]. They are located between 750 and 1415 mbsl, and their reliefs vary between 40 and 240 m. The length and width of their axes range from 2 and 4.5 km to 2 and 3.3 km, respectively. The surveyed deep depressions have named as the Twin Pools [48] (Western and Eastern Twin Pool, WTP and ETP respectively in Figure 7). Otherwise, the basin morphology of the PoL is dominated by channels and moats related to the interaction of the strong bottom currents with the seabed (Figure 7). Therefore, a NE-SW central channel 100 km in length and 1–5 km in width is located at 1290 mbsl in the central sector and deepens toward the NE (1460 mbsl) and towards the SW (1320 mbsl). Rimmed depressions around the mounds classified as moats are 0.5–2 km widths incising to 180 m depth. Moats are better developed along the east and south sides of the mounds reaching up to 1270–1530 mbsl at their bases.

The vulnerable deep-sea habitats identified within the PoL (Figure 8) can be resumed into the next types (see details in the Supplementary Table S2):

(1) Deep-Sea hexactinellid sponge aggregations intermixed with Actiniarian communities covering the soft, muddy sea floor of the summit of some mounds (e.g., *M1 mound*) at 830–850 mbsl.

(2) Sea-pens communities and aggregations of bamboo corals covering soft muddy bottoms with some coral rubble along the flanks of some mounds (e.g., *M1 mound*) at 1020–1100 mbsl.

(3) Aggregations of stony octocorals and of soft corals colonizing volcanic rock ridges (*Volcanic Ridge*) and single volcanic cones (e.g., *Volcancito Hill*) at 1020–1265 mbsl.

(4) Deep-Sea Desmosponges aggregations on deep giant circular depressions (*Western and Eastern Twin Pools*) covered by thin layers of fine sediments at 1200–1300 mbsl.

Figure 8. Underwater images obtained with the ROV "Luso"during the SUBVENT- 2 expedition along the PoL. (**A**): The hexactinellid sponge *Pheronema carpenteri* on soft substrates of the northern flank of M1 mound; (**B**): A Venus fly-trap anemone on soft substrates of the northern flank of M1 mound; (**C**): Brissingid starfishes colonizing small volcanic rocks at the summit of M1 mound; (**D**): The starfish *Nymphaster arenatus* on volcanic rocks at the summit of M1 mound; (**E**): Large colonies of the stony octocorals *Corallium* spp. (*C. niobe* and *C. tricolor*) on basaltic rocks of the volcanic ridge located at the northeastern part of M1 mound; (**F**): Basaltic rock colonized by different glass sponges, octocorals (*Corallium, Swiftia*) and hydrozoans at the Volcancito Hill. See Figure 7 for location.

Mounds along the PoL harbor a mixture of fauna from typical sedimentary and rocky habitats, including the hexactinellid sponge (*Pheronema carpenteri*), venus flytrap anemones, isolated stony octocorals (*Corallium*) and bamboo corals (*Acanella arbuscula*), decapod crustaceans (*A. foliacea, P. edwardsianus*), starfishes (*Hymenodiscus, Nymphaster*), holothurians (*Mesothuria*), and leather sea urchins (*Araeosoma*).

The volcanic ridges and volcanic cones are colonized by a wide variety of suspension-feeding species, such as stony octocorals (mainly *C. tricolor* and *C. niobe*), black corals (*Leiopathes glaberrima*), bamboo corals (*A. arbuscula*), and small and large gorgonians (*Swiftia*, Plexaurid gorgonians), together with crinoids and hexactinellid and lithistid sponges, among a wide variety of benthic species. Some areas covered by abundant coral rubble were colonized by isolated bamboo corals (*A. arbuscula*). The occurrence of strong bottom currents supports the fact that tree-like branches of some octocorals are oriented along an E-W direction (Figure 8E)

Giant circular depressions, such as WTP and ETP (Figure 7), contain a mixed seabed with soft and carbonated flagstones. The sandy and muddy substrates are colonized by typical sedimentary bathyal species such as sponges (*Thenea*), cerianthids, sea-pens (*Pennatula*), echinoderms (*Ceramaster, Mesothuria, Araeosoma*), decapods (*A. foliacea, P. edwardsianus, Plesionika* spp.) and large scalpellid barnacles. The rocky substrates are covered by a layer of sediment that is poorly colonized by fauna, including some massive demosponges (Pachastrellid-like sponges), small encrusting sponges, scattered hydroids and solitary scleractinians (*Caryophyllia*).

4.1.4. Case Study 4: Low-Temperature Hydrothermal Habitats at Tagoro Volcano, El Hierro Island (27°35′ N)

Recent submarine eruptions in El Hierro Island from 2011 to 2012 have provided the opportunity to map the development of a new submarine volcano, named as Tagoro, and the identification of habitats and communities associated with emissions of large quantities of iron and carbon dioxide from hydrothermal submarine vents.

The El Hierro eruption started on 10 October 2011, within a pre-existing submarine valley at 250–350 mbsl (Figure 9A). Initial bathymetric data collected on 25 October 2011, [49] identified a single cone around 205 mbsl (Figure 9B). MBES data from 29 November 2011, indicated that this volcano developed four vents with summits at 220, 195, 180 and 165 mbsl [34]. On 22 December, new bathymetric data showed a drastic collapse of the main edifice that was not present in previously acquired bathymetry. On 6 March 2012, the extrusive magmatic activity ceased 137 days after the onset of the eruption, leaving the summit of the new submarine volcano at 89 mbsl. Hydrothermal activity with intensive bubbling released from active vents continued until at least June 2012.

ROV observations in 2014 showed that the base of the Tagoro volcano is dominated by an area with extensive accumulations of scoriaceous bombs at 280 to 380 mbsl formed from basanitic lava emissions (Figure 9C,E,F). When observed closely, these bombs appeared to be either whole intact lava balloons or broken fragments of larger balloons. These bombs correspond to lava balloons floating on the sea surface during the eruption (Figure 9E). Echosounder images of the water column taken during the eruption overlie on the 3D multibeam bathymetry model showed high-reflective bright spots interpreted as these floating bombs sank along the flanks of the volcano (Figure 9A).

Two main habitats were recognized after two years of the cessation of the eruption (see details in Supplementary Table S2):

(1) Chemosynthesis—based habitats composed by a great proliferation of orange-brown Fe-oxidising bacteria draped the whole seafloor of the summit.
(2) Volcanic caves communities along the flanks of the volcano composed by small oysters and serpulids, shrimps and eels.

ROV images of the summit of the volcano show 5 m-high chimneys or hornitos with numerous degassing conduits punctuating their walls and marked by yellow mats, linked to sulfur-related bacteria (Figure 9D). Temperatures measured during the 2014 survey (Supplementary Table S1) showed abrupt increases in temperature up to ~2.69 °C, related to active hydrothermal chimneys or hornitos located at the volcano summit [50]. Peaks of CO_2 emissions were detected with the ROV sensors around these hornitos [50]. On the contrary, the flanks of the volcano showed recolonization of fauna such as small oysters (*Neopycnodonte cochlear*) and serpulids, shrimps (mainly *Plesionika*), and eels (*Conger conger, Gymnothorax*) living on caves and crevices generated by the cooling of the plastic bulbous lavas flowing downslope at 250–300 mbsl.

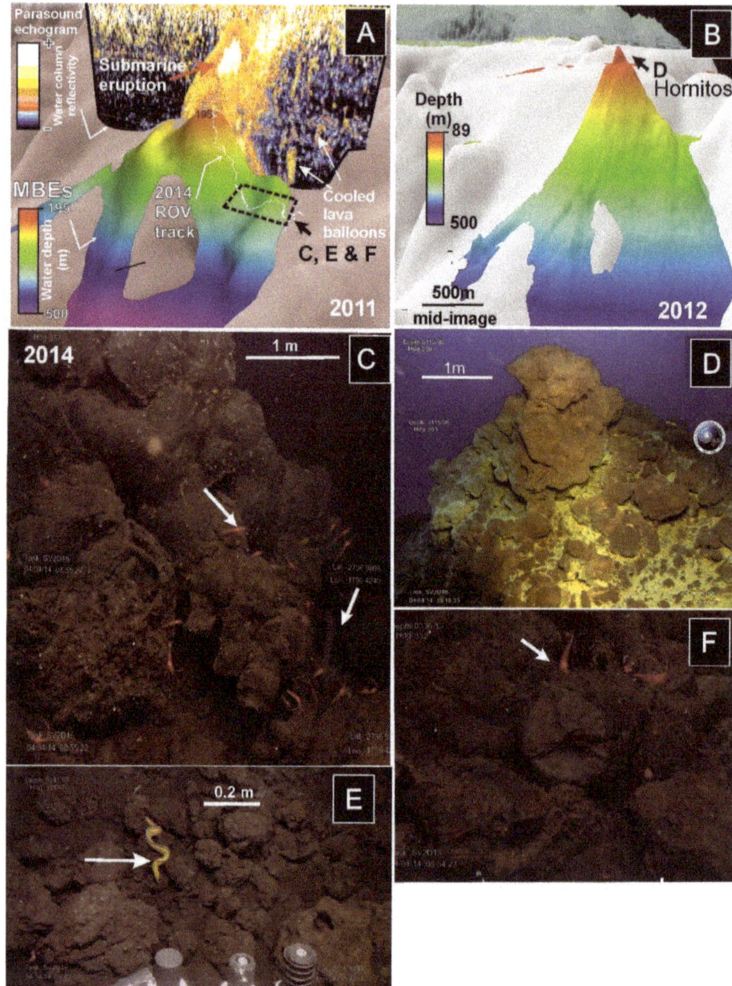

Figure 9. Sequence of MBES during and after El Hierro submarine eruption superimposed on previous bathymetry (in white colour): (**A**) Bathymetry from November 2011 during the eruption when the summit was located at 220 mbsl. Overlay on the bathymetry, a 2D image of parasound echosounder showing the volcanic eruption rising throughout the water column; (**B**) Bathymetry after the eruption in June 2012 (summit located at 89 mbsl). ROV images taken in April 2014 showing distinct habitats developed after the eruption: (**C**) Pillow-lavas flowing downslope with abundant shrimps (mainly *Plesionika*), serpulid worms and small oysters (*Neopycnodonte cochlear*, some of the small white dots on the rocks); (**D**) Hornitos (hydrothermal vents) developed at the summit of the volcano draped by yellow and orange iron-sulfide bacterial mats; (**E**) Lava balloons along the slope of the recent volcano showing demersal fauna such as moray eels (*Gymnothorax*); and (**F**) Accumulations of lava bombs colonized by shrimps (*Plesionika*).

4.1.5. Case Study 5: Ferromanganese Crust-Bearing Seamounts of Southern Canary Islands (23°48′ N–26°30′ N)

In this fifth case study, we present the data from four ferromanganese crust-bearing seamounts (Echo, The Paps, Drago, and Tropic) located in the west-southwest region of the CISP at water depths ranging from 300 to 4300 (Figure 10). The Echo Seamount is a sub-circular seamount of 10 km in diameter with the flat shallow summit located at only 300 mbsl and the base at −3700 mbsl (Figure 10A). The Paps seamount has the shape of a NW-SE ridge, 40 km length in the SE direction with the summit at 1600 mbsl and the base at 4300 mbsl (Figure 10B). The Tropic seamount is a star-shaped guyot, with its base located at 4200 mbsl and its summit at about 1000 mbsl (Figure 10C). The Drago Seamount is elliptical in shape with major axis NW-SE oriented. The summit is situated at 2200, whereas the base at 3000 mbsl (Figure 10D).

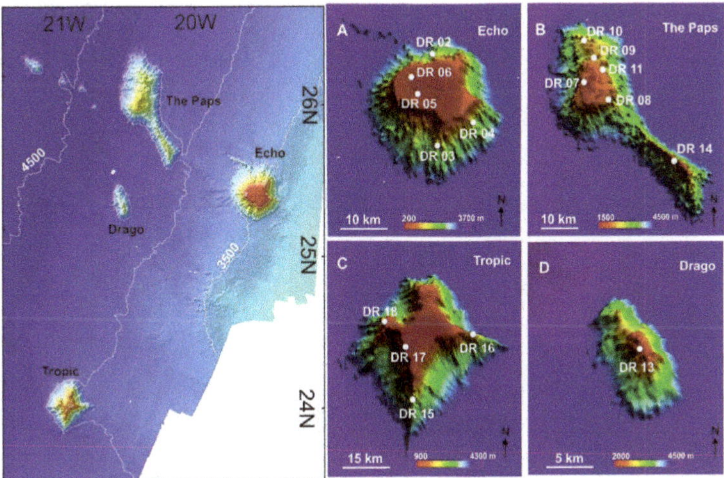

Figure 10. Location of the studied seamounts in the southwestern part of the Canary Islands Seamount Province (CISP) including the positions of the benthic dredge samples (DR) in (**A**): Echo Seamount; (**B**): The Paps Seamount; (**C**): Tropic Seamount; (**D**): Drago Seamount. See Supplementary Table S2.

The full list with the ROV data including substrate, morphology, water depth, oceanographic values, and associated taxa for each of the seamount surveyed can be found in the Supplementary Table S2. The habitats identified within the surveyed seamounts (Figure 11) can be resumed into the next types (Supplementary Table S3):

(1) Aggregations of scleractinians, sponges and antipatharians on the summits of shallow seamounts ranging from 300 to 1000 mbsl covered by ferromanganese crusts.
(2) Aggregations of gorgonians, antipatharians and bamboo corals along the flanks of shallow seamounts (>1000 mbsl).
(3) Aggregations of gorgonians, stony corals and deep-sea hexactinellid sponges on the summits and flanks of the deep seamounts (1600–2200 mbsl) covered by ferromanganese crusts.

The summits of shallow guyot-like seamounts (e.g., Echo Seamount) are colonized by habitat-forming species from hard substrates such as large sponges (*Pachastrella*, *Poecillastra*), scleractinians (mainly *Dendrophyllia cornigera*) and antipatharians (*Stichopathes*) (Figure 11A,B). Other dominant fauna includes molluscs (*Asperarca nodulosa*, *Clelandella*, different muricids), small hydroids and echinoderms (mainly cidarids and the crinoid *Thalassometra* cf. *lusitanica*). The southern flank of this shallow seamount (DR03, 1757–1949 mbsl, in Figure 10A) showed a higher abundance of sessile species that included large gorgonians (*Metallogorgia melanotrichos*), small encrusting sponges as well as antipatharians (*Bathypathes*) and different serpulids (Figure 11C,D). Solitary and colonial scleractinians (*Desmophyllum*-like corals, *Solenosmilia variabilis*), bamboo corals, and large brachiopods were also collected in that area (Figure 11E).

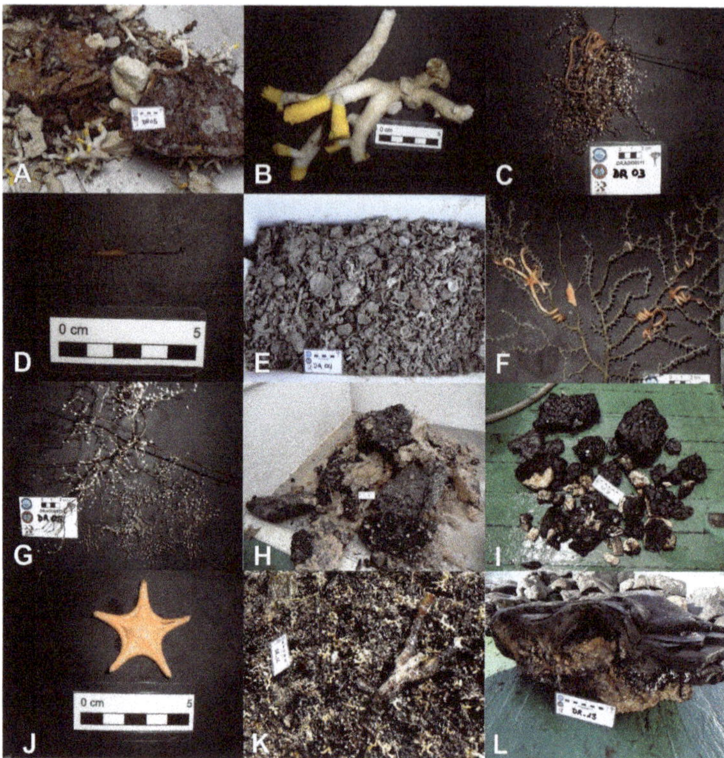

Figure 11. Images of some samples obtained during the DRAGO0511 expedition [48] in the Echo (**A–E**), The Paps (**F–J**), Tropic (**K**) and Drago (**L**) seamounts at the Canary Islands Seamount Province. (**A**): Rocks, sponges and corals; (**B**): A living colony of the scleractinian *Dendrophyllia cornigera* collected at the summit; (**C**): A gorgonian (*Metallogorgia melanotrichos*) harboring an euryalid ophiuroid collected at the flank; (**D**): A small antipatharian (*Bathypathes*) collected at the flank; (**E**): Coral rubble retrieved from the SE flank; (**F**): Holaxonian gorgonian with abundant epibionts captured at the summit; (**G**): *Metallogorgia* gorgonian collected at the summit; (**H**): Fe-Mn crusts and hexactinellid sponges retrieved from the summit; (**I**): Fe-Mn crust samples obtained at the flank; (**J**): Goniasterid starfish collected at the flank; (**K**): Coral rubble of scleractinians (mainly *Solenosmilia variabilis* and *Desmophyllum*-like cup corals) collected at the eastern; (**L**): Thick Fe-Mn crust on volcanoclastic rock substrate retrieved from the summit.

Ferromanganese crusts of The Paps Seamount (Figure 10B) are colonized by large and fragile deep-sea gorgonians (*Metallogorgia melanotrichos*, *Chrysogorgia*, *Paramuricea*) (Figure 11F–H), bamboo corals, and small soft corals (*Anthomasthus*), hexactinellid sponges (*Aphrocallistes*), together with crustaceans (*Munidopsis*) and echinoderms that are generally attached to the gorgonians (mainly the ophiuroids *Ophiocreas* and *Asteroschema*, as well as Antedonidae crinoids). The ferromanganese crust substrate of the northern flank of this seamount (DR10, 2839–3010 mbsl in Figure 10B) is colonized by hexactinellid sponges, gorgonians (*M. melanotrichos*) and echinoderms (brisingid and goniasterid starfishes, Figure 11I,J, ophiuroids on the gorgonians, e.g., *Ophiocreas* and *Asteroschema*).

The ferromanganese crusts and carbonate-phosphorite pavements of the guyot-like Tropic Seamount (Figure 10C) are colonized by scleractinians (mainly *Solenosmilia variabilis* and *Desmophyllum*-like solitary corals) (Figure 11K), large colonies of bamboo corals (*Acanella*, *Keratoisis* up to 60 cm long), hexactinellid sponges (mainly *Poliopogon amadou*), polychaetes (mainly eunicids), and stalked crinoids (*Endoxocrinus wyvillethomsoni*). The associated fauna included hexactinellid sponges, antipatharians (*Bathypathes*), ophiuroids, polychaetes and crustaceans (*Munidopsis*, balanids, pandalid shrimps).

The deepest seamount, the Drago Seamount (Figure 10D), harbored colonies of stony octocorals (*Corallium*) and the gorgonian *M. melanotrichos*, together with hexactinellid sponges (*Aphrocallistes, Regadrella*) and small unidentified gorgonians at its summit (DR13, 2290–2426 mbsl, Figure 10D) colonizing thick ferromanganese crusts (Figure 11L).

5. Discussion

5.1. Potential Drivers of Deep-Sea Habitat Distribution from Subtropical North Atlantic

This section discusses the potential drivers affecting the distribution and biodiversity of vulnerable deep-sea habitats, including both chemosynthesis and non-chemosynthesis-based habitats. Some of these drivers are: (a) the seafloor water mass properties as temperature, salinity, dissolved oxygen and potential density [12]; and (b) the active geological processes affecting the seafloor such as (i) methane seeps [51–55] and (ii) submarine volcanic eruptions followed by low-T hydrothermal degasification [50]. Furthermore, an overview of the effects of methane seeps and low-T hydrothermal vents following recent submarine eruptions on the distribution of deep-sea habitats at regional scale along the northeast Atlantic Ocean from 24° N to 42° N is provided.

5.2. Influence of Water Mass Properties on Vulnerable Deep-Sea Habitats

The relationships between water depths and water mass properties (temperature, salinity, dissolved oxygen, and potential density) for specific vulnerable deep-sea habitats highlight important considerations on suitable environmental conditions for them. The distribution of undercurrents along the NE Atlantic from the NW African Margin and Canary Islands to the north Iberian Margin is mainly driven by the outflow of the MOW at the Gibraltar Strait and by the intrusions of the Antarctic-derived AAIW currents at intermediate waters (Figure 12). The ROV-mounted CTDs data are in-situ measurements of the benthic layer, where the deep-sea habitats allow us to compare the regional distribution of water masses (Figure 12) with the local variables affecting the specific habitats (Figures 13–15). A full list of oceanographic measures in-situ for each site surveyed with the ROV-mounted CTD is shown in Supplementary Table S3.

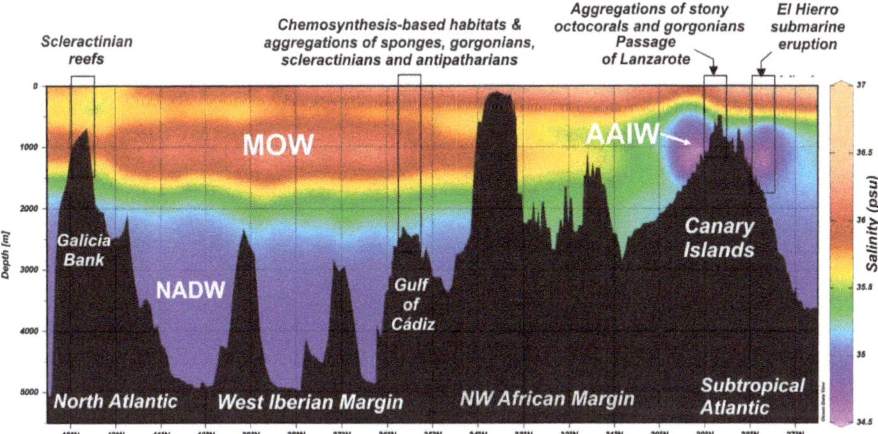

Figure 12. Distribution of water masses from the Subtropical Atlantic to the North East Atlantic based on regional salinity values. Mediterranean Outflow Waters are mainly flowing northwards from the Gibraltar Strait affecting the Gulf of Cádiz and west Iberian Margin. Otherwise, Antarctic Intermediate Waters are flowing northwards along the NW African Margin throughout the Passage of Lanzarote. Data from World Ocean Database 2018 [44].

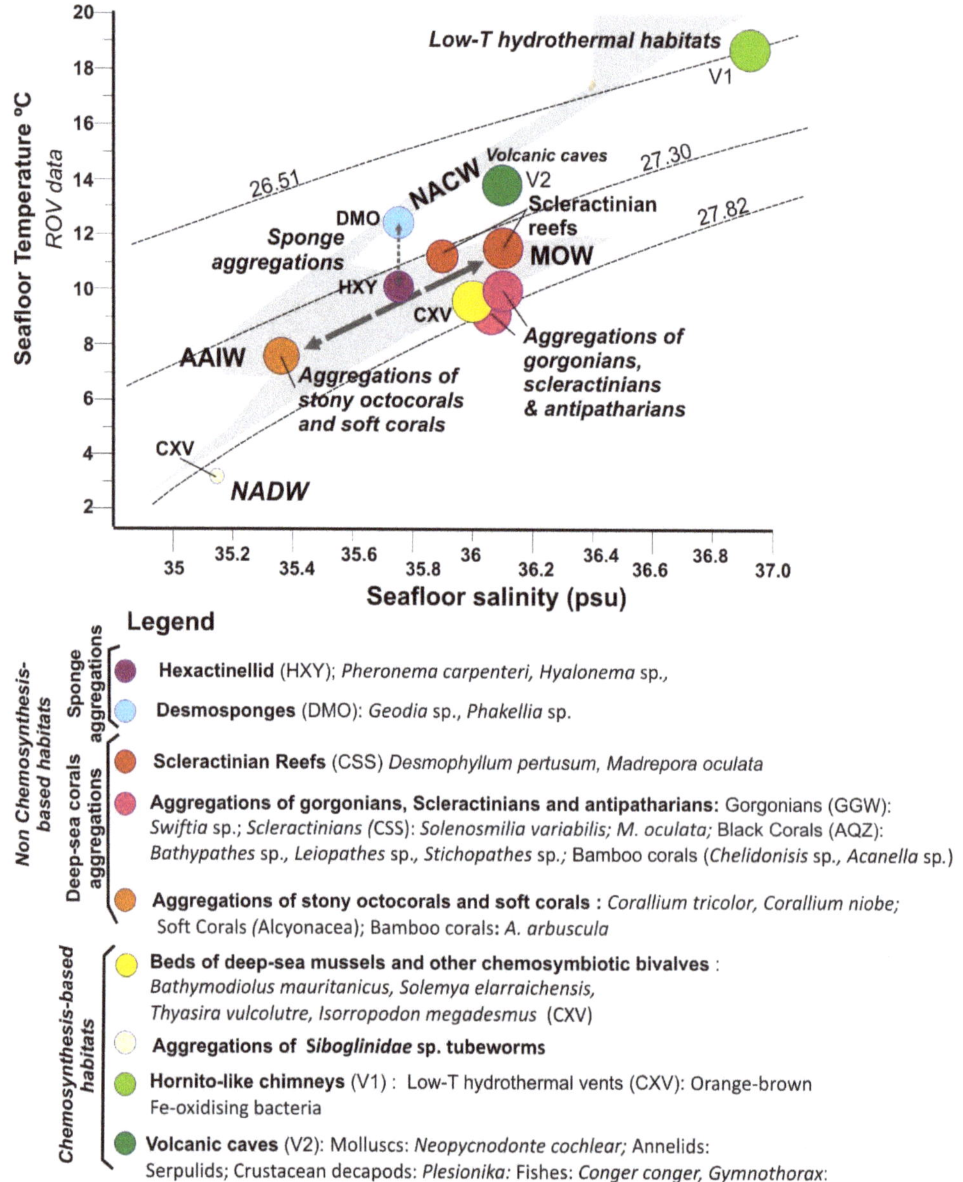

Figure 13. T-S diagram for the in-situ water mass conditions measured with the ROV for each of the type of habitats (full data listed in Supplementary Table S3). Background grey pattern corresponds to CTD data and dashed lines (from top to bottom) to constant potential density (isopycnal) contours = 26.51, 27.30 and 27.82 dividing four regions corresponding to the main different water masses defined by [33]: NACW, AAIW, MOW and NADW defined by [33]. Legend: FAO Taxa Classification and Codes for Vulnerable Marine Ecosystems (VMEs) (http://www.fao.org/in-action/vulnerable-marine-ecosystems/vme-database, accessed 1 December 2020).

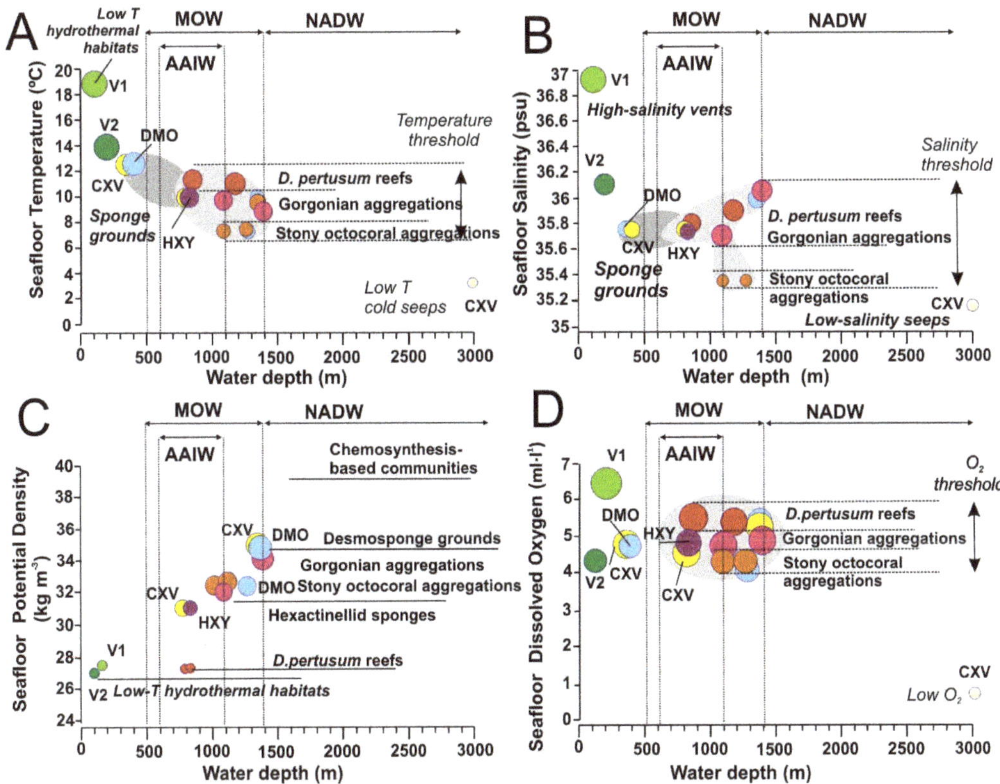

Figure 14. Seafloor water mass parameters vs. water depths for each of the identified vulnerable deep-sea habitats and their exposure to the main bottom currents: (**A**) Temperature; (**B**) Salinity; (**C**) Potential Density, and (**D**) Dissolved Oxygen. Color legend for deep-sea habitat types is the same as Figure 13. AAIW: Antarctic Intermediate Water current; MOW: Mediterranean Outflow Water; NADW: North Atlantic Deep-Water current. Data are full listed in Supplementary Table S3. Further explanation in the text.

Figure 13 shows the T-S diagram for the in-situ water mass conditions measured for each type of habitat (data can be found in Supplementary Table S3). In this T-S diagram, most of vulnerable deep-sea habitats are located within the field of the intermediate waters: AAIW and MOW bounded between 27.30 and 27.82 isopycnals.

The three types of deep-sea coral habitats are located between these two intermediate water masses. Therefore, the aggregations of stony octocorals and bamboo corals along the PoL are bathed by AAIW, which is reflected by a clear S minimum (S = 35.4). Otherwise, aggregations of gorgonians, scleractinians, and antipatharians of the GoC as well as scleractinian reefs of the GB are bathed by the MOW, characterized by warmer and saltier waters (S = 35.8–36.2 psu). On the contrary, the sponge aggregations seem to follow a distinct pattern, with those of Desmosponges related to the NACW at uppermost levels in the PoL, whereas those of Hexactellinids in the GoC related to intermediate waters MOW-AAIW. The main difference in the water masses of these two types of sponge aggregations is the T values (Figure 13).

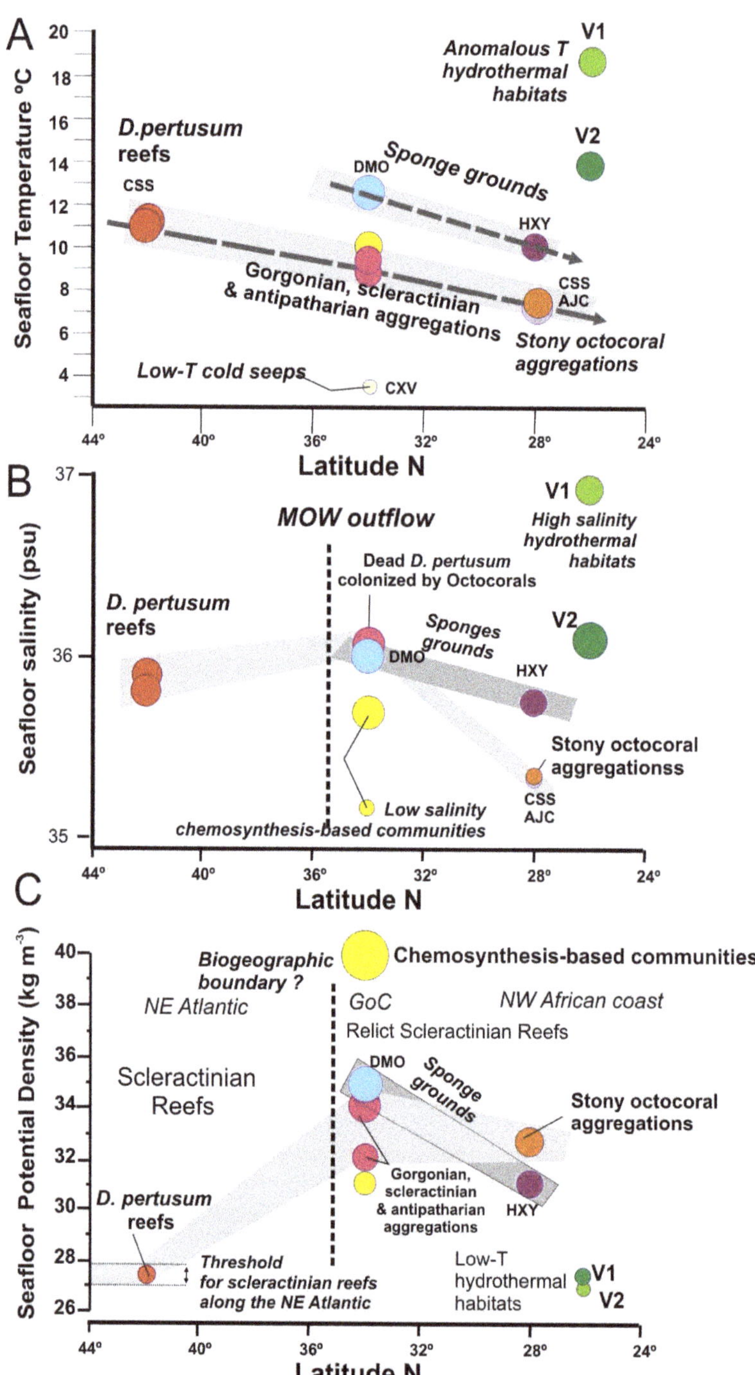

Figure 15. Seafloor water mass parameters vs. latitudes for each vulnerable deep-sea habitat detected in this study: (**A**) Temperature; (**B**) Salinity; and (**C**) Potential Density. Colour legend is the same as in Figure 13.

Chemosynthesis-based habitats of the GoC are also located under the influence of the MOW, except for the deepest mud volcano (e.g., Bomjardim MV), which is located under the influence of the NADW. The values for low-T hydrothermal habitats of the summit of the Tagoro volcano with high S and T are disturbed by the influence of hydrothermal fluids not reflecting its location at the thermocline. In contrast, benthic habitats of volcanic caves at the flanks of Tagoro are bathed by the NACW reflecting the major influence of water masses (Figure 13).

5.2.1. Water Mass Temperatures

Water temperature has been traditionally considered as one of the most important factors influencing the distribution of deep-sea organisms due to their different physiological tolerances [12,56]. Thermal thresholds for deep-sea coral habitats (including those conformed by scleractinians, stony octocorals, gorgonians, and black corals) are well constrained to temperature ranges of 7.3 and 11.5 °C within the depth range of 800–1400 m (Figure 13A). Within this threshold, all these deep-sea corals appear distributed into three levels of seafloor temperatures: (i) scleractinians conforming reefs (*Desmophyllum pertusum*, and *Madrepora oculata*) located at the maximum temperature tolerance of 11–11.5 °C; (ii) Gorgonians, scleractinian and antipatharians conforming aggregations at intermediate temperatures of 8.8–10 °C; and finally iii) Stony octocoral aggregations at the lowest temperatures of 7.3–7.7 °C (Figure 14A).

The temperature range of the distribution of the reef-forming scleractinians *D. pertusum* and *M. oculata* (4–12 °C) across the five case studies matches the temperature range of cold North Atlantic Intermediate waters and the temperature range where those reef-forming species have been detected in the past [56,57].

On the other hand, some aggregations of fragile octocorals (*Acanella, Radicipes*) appear at the lowest temperatures of 7.3–7.7 °C (Figure 14A) in the case studies but at a higher temperature range (1.5–6.1 °C) than that detected previously for the North Atlantic, so the present study provides new data for case studies located southwards of those considered [58]. Moreover, the low temperature values could point out that stony octocoral gardens living along the PoL are somehow influenced by the cold Antarctic Intermediate Waters (AAIW) that flows between the easternmost Canary Islands and the African margin (Figures 12 and 13), which was totally unexpected previously in this study.

Otherwise, desmosponges aggregations dominated by *Geodia* sp. and *Phakellia* sp., occurring within the crater of Mercator MV (Figure 5D) at shallow depths of 350–370 mbsl, tolerate high sea-floor temperatures up to 12.5 °C (Figure 14A) as also detected in the northern sector of the GoC, where the MOW influence is higher [54]. On the contrary, hexactinellid sponge aggregations with *Pheronema* and *Hyalonema* detected along the flanks of the Algacel MV at 820–840 mbsl (Figure 5E) live within the thresholds of the aggregations of gorgonians, scleractinians, and antipatharians, colonizing MDACs around methane seeps (Figure 14A). Indeed, these sponges were found at similar depths and environmental conditions in the northern part of the GoC [59,60]. On the other hand, deep-sea chemosynthesis-based habitats do not show seafloor temperature constraints, varying from 3.2 to 12.5 °C, as function of the seeps location. In the case of the large mussels *Bathymodiolus mauritanicus*, the seafloor temperature of 10.1 °C may fluctuate as function of intensification of the hydrocarbon seeps. No information on the effect of temperature on the distribution of methane seeps fauna has been found, but the fact that the same chemosymbiotic-species may occur in cold seeps with different oceanographic conditions in methane seeps of the northern and southern GoC may indicate a lower dependence on water mass properties for these chemosymbiotic organisms than for the above mentioned cnidarians and sponges [60,61]. In the case of the hydrothermal-related habitats (V1 = hornitos-like hydrothermal chimneys and V2 = lava cavities in Figure 14A), temperatures reflect thermal anomalies associated with venting hydrothermal fluids. Therefore, maximum measured temperature of expelled fluids in Tagoro volcano was 39 °C, while the ambient seawater range between 15 and 19 °C with local temperature anomalies of +2.7 °C associated with CO_2 emissions [50].

5.2.2. Salinity and Potential Density

Although salinity variations are very low in the deep sea, the distribution of deep-sea scleractinians and octocorals in some areas of the NE Atlantic has been related to potential density, which represents a parameter defined by salinity and temperature [62]. Indeed, reef-forming scleractinians of the selected sites are controlled by a narrow salinity range of 35.35–36.1 psu (Figure 14B). As seafloor temperatures, habitats conformed by deep-sea corals appear also distributed into three levels of salinity: (i) Scleractinian reefs (*D. pertusum* and *M. oculata*) occurred at salinity ranges between 35.8 and 36.1 psu; (ii) Gorgonian aggregations at salinity ranges between 35.6 and 36.1 psu; and (iii) Stony octocoral aggregations at the lowest salinity values between 35 and 35.2 psu. These data point out that there is an influence of the MOW waters bathing the scleractinian reefs, and of the mixing between the saltier MOW waters and the lower salinity AAIW waters for some gorgonian and stony octocoral aggregations (Figure 13). Indeed, the MOW flow seems to play an important role in the distribution and connectivity of deep-sea corals across the NE Atlantic, with a high diversity of deep-sea corals along the MOW pathway [63].

Deep-sea desmosponges aggregations at 1400 mbsl appear associated with salty water masses of 36 psu (Figure 13B) related to the export of bathyal benthos to the Atlantic through the MOW outflow as detected in other areas of the GoC [60]. On the contrary, shallow water desmosponges grounds associated with methane seeps of the Mercator MV (Figure 5C), located at 350 mbsl above the MOW influence, show the lowest levels of salinity of 35.75 psu (Figure 14B). As occurred for temperature values, large hexactinellid sponge aggregations occurred within the same salinity range as for deep-sea corals (Figure 14B). The role of the MOW exporting deep-sea sponge larvae from the Mediterranean to the Atlantic Ocean and some NE locations close to the GoC has been demonstrated recently, and this may partly explain the occurrence of some sponges in salty water masses related to the MOW [64].

The occurrence of scleractinian reefs along the NE Atlantic margins has been related to values of potential density, a parameter defined by salinity and temperature. Thus, *D. pertusum* and *M. oculata* reefs, were firstly constrained to a narrow potential density range (potential density; δ_θ = 27.35–27.65 kg m^{-3}) [62] which further, in other North Atlantic regions, proved to be larger than initially suggested as ranging δ_θ = 27.10–27.84 kg m^{-3} [63]. This range is similar for the *D. pertusum* and *M. oculata* reefs identified in the Galicia Bank, which falls within this suggested range with values of δ_θ = 27.353–27.431 kg m^{-3} (Figure 14C). In the case of the Galicia Bank, these conditions seem to be caused by the turbulent mixing between the MOW and the ENACW water masses promoted by the MOW current impinging the topography of the Galicia Bank [46]. The formation of turbulent eddies of MOW flow facilitates the downward transport of nutrients, influencing the present distribution of the scleractinian reefs and mounds mapped along its flanks and on the summit (Figure 3B).

On the contrary, the gorgonian and stony octocorals aggregations do not fit with this potential density threshold (Figure 14C). Thus, stony octocoral aggregations in the subtropical latitudes occur at potential density values δ_θ = 32.725 kg m^{-3}, whereas gorgonian-dominated aggregations occur at values of δ_θ = 32.050–34.900 kg m^{-3}, much higher than those for scleractinian reefs. Thus, according our data, the potential density range proposed for scleractinian reefs [62,63,65] would be only valid for the northernmost NE Atlantic margins, and more data on the presence of these species southwards is then needed for estimating the appropriate density range. One explanation is the potential influence of intrusion of AAIW flow along the NW Africa margin towards the GoC (Figure 12).

Desmosponges and hexactillenid aggregations also show different potential densities. Therefore, desmosponge aggregations show higher potential density ranges (δ_θ = 32.050–34.900 kg m^{-3}) than the studied hexactillenid sponge aggregations

(δ_θ = 31.058 kg m^{-3}) (Figure 14C) Hydrography plays an important role in shaping the distribution of sponge aggregations in the Atlantic, and several authors have noted the association of sponges with particular water masses due to their temperature and salinity characteristics or hydrodynamic conditions, such as tides or internal waves which enhance the food supply. Thus, the influence of the NACW (Figure 13) in the study area allows the presence of desmosponges assemblages with a wide Atlanto-Mediterranean distribution [60,64].

5.2.3. Dissolved Oxygen Concentrations

Dissolved oxygen (DO) values are close to saturation in most deep-sea areas, ranging from 2.5 mL·L^{-1} off SW Africa to 6 mL·L^{-1} in the NW Atlantic [65]. The values of DO for deep-sea corals in the study area are limited to values ranging from 4.3 to 5.5 mL·L^{-1} (Figure 14D). As occurred with temperature and salinities values, these habitats show three levels of DO for deep-sea coral habitats living at 800–1400 mbsl: (i) High oxygenation values of 5.44–5.55 mL·L^{-1} for scleractinian reefs; (ii) Intermediate oxygen values of 4.73–5.25 mL·L^{-1} for gorgonian aggregations; and (iii) low oxygen values of 4.32 mL·L^{-1} for stony octocorals aggregations. Otherwise, desmoponges sponge grounds show similar values of DO to aggregations of gorgonians and stony octocorals (Figure 14D). Deep-sea corals and sponges, like other organisms, are not very sensitive to small variations in DO unless it drops a certain threshold [12].

Otherwise, hypoxic or anoxia conditions are mainly related to submarine areas with methane seeps or hydrothermal vents. Seafloor anoxic conditions may be generated by intense activity of methane seeps, hydrothermal vents or submarine eruption plumes. Fluid emissions are not continuos but may be reactivated periodically on focused submarine vents or remain diffuse during long time [66]. Very low DO values of 0.64 mL·L^{-1} observed on deep-water mud volcanoes as Bonjardim in the GoC are closely to anoxia suggesting the intense activity of methane seeps (Figure 14D). If methane seeps are intense and/or undercurrents are weak in such deep waters, then low oxygen conditions prevail above seafloor and chemosynthesis-based habitats are only composed by colonies of Siboglinidae tubes [67]. However, in areas of methane seeps influenced by strong intermediate undercurrents (Figure 14D), co-occurrence of chemosynthesis-based and scleractinian-gorgonian aggregations has been detected as in this study and other MVs of the Gulf of Mexico [68]. This is the case for the extensive beds of deep-sea mussels Bathymodiulus mauritanicus beds living around bubbling methane seeps (Figure 5D) and MDACs patches colonyzed by gorgonians, antipatharians, and scleractinians in Algacel MV (Figure 5D). In such methane seeps, the siboglinids tubeworms play an important role in the connectivity between methane seeps and deep-sea corals by filtering sulphur reducing bacteria (SRB) allowing the predominance of anaerobic oxidation of methane (AOM) by archaeas building up carbonates patches and allows deep-sea coral larvae to grow isolated from the surrounding highly acidic methane muds [69].

5.3. Distribution of Vulnerable Deep-Sea Habitats from Subtropical to North Latitudes

The biodiversity of deep-sea habitats and their biogeographical affinities can vary over space and time due to regional and basin-scale changes in the oceanographic conditions [65]. However, active geological process affecting the ocean's seafloor such as methane seeps, hydrothermal vents or submarine eruptions can modulate the spatial distribution of deep-sea habitats as detected in this study. The water properties for deep-sea habitats show trends with latitude which are related to oceanographic conditions, but also anomalous values associated with methane seeps and hydrothermal vents (Figure 15). Therefore, deep-sea coral habitats show a notable decreasing trend in the suitable temperatures from north to south (Figure 15A): (i) 11–11.5 °C for scleractinian reefs habitats in the Galicia Bank; (ii) 8.8–10 °C for gorgonian and scleractinian aggregations associated with mud volcanoes of the GoC; and (ii) 7.4–7.7 °C for stony octocoral gardens of the PoL. Otherwise, the sponge grounds show a similar decreasing trend but at higher temperature ranges.

Thus, desmosponges habitats of the GoC are related to temperatures thresholds as high as 12.5 °C, whereas hexactillined sponge grounds of the Pol show lower temperatures of 7.3–7.33 °C (Figure 15A). This negative trend identified in the sensitive temperatures for these habitats may be caused by the cooling of the AAIW intermediate waters and it may affect the composition of these communities as identified on T-S diagram (Figure 13). The AAIW flows along the NW African coast and extends to the GoC, where it interacts with the MOW outflow, perhaps restricting the spreading of the later [70].

The values of salinity from 24° to 44° N show a different trend than the temperature ones (Figure 15B). In this case, the deep-sea habitats supporting maximum salinity values are located on the mud volcanoes of the GoC, that are strongly influenced by the MOW undercurrent (Figures 12 and 13). Therefore, gorgonian and scleractinian aggregations growing on MDACs around methane seeps show maximum salinity values up 36 psu. Both to the north and to the south, the salinity threshold for deep-sea corals diminishes. Thus, the scleractinian reefs of the Galicia Bank (42° N) show slightly lower salinity values of 35.9 psu and stony octocoral aggregations located on the PoL (28.30°–29° N) show the lowest salinity values of 35.35 psu (Figure 15B). Considering this, the low salinity and temperature values bathing the stony octocoral aggregations of the PoL are due to the influence of the AAIW flowing along the NW African (Figures 12 and 13).

The salinity threshold for the sponge aggregations follows the same descending trend from 34 to 28° N at regional scale (Figure 15B). Desmosponge grounds living close to methane seeps in the GoC are living under similar high salinity values to gorgonian and scleractinian aggregations (Figure 15B). However, hexactillenid sponge grounds living at shallower depths on the summits of the PoL mounds support higher salinity values than stony octocoral habitats (Figure 15B). This is interpreted as a result of the influence of periodic southwards intrusions of salty MOW waters into the PoL at water depths shallower than the AAIW undercurrents which may influence the development of those habitats.

5.4. Temporal Variability of AAIW Latitudinal Extension along the Northern Atlantic Ocean Might Have Caused the Massive Mortality of CWC Reefs?

At present live *D. pertusum* and *M. oculata* frameworks are very scarce in the GoC and generally substituted by other deep-sea corals (e.g., different species of gorgonians, antipatharians, etc.). Besides, a geographical boundary between scleractinian reefs and aggregations of other deep-sea corals is observed (Figure 15B,C). A distribution possibly related to the presence of the AAIW waters (Figure 13). Evidence for such a relation is provided by the occurrence of giant dead mounds up to 30 m in height conformed by dead scleractinians (*D. pertusum* and *M. oculata*), such as in the Pompeia Coral Province in the GoC (Figure 4), pointing to a massive mortality of scleractinian reefs off northwest Moroccan [69,71–73] and along the Mauritanian margins [74].

Based on radiocarbon data of scleractinians along the Western Mediterranean Sea during the last deglaciation times [75], we hypothesize that the beginning of intrusion of AAIW waters into the GoC during the last deglacial times may have been one of the causes of a massive mortality of scleractinians by shifting northwards the biogeographical boundary between some scleractinians and stony octocoral aggregations (Figure 15B). This point out that stony octocorals (*Corallium tricolor* and *C. niobe*), antipatharians, and gorgonians are more suitable to AAIW conditions than the scleractinians *D. pertusum* and *M. oculata*.

5.5. The role of Methane Seeps Driving Distribution of Chemosynthesis and Non-Chemosynthesis-Based Habitats

Seabed features formed by seafloor venting of hydrocarbon-enriched fluids are generically termed cold seeps, which are associated with high geological, geochemical and, biological activity [51,52]. Submarine MVs are one of the main seabed morphological expressions of cold seeps formed by violent eruptions, followed by progressive degassing of hydrocarbon-enriched fluids and sediments onto the seafloor [66].

5.5.1. Drivers Controlling Distribution of Habitats in Methane Seeps: Acidic Muds vs. Carbonates

Based on the habitat types identified, ROV-mounted CTD parameters and CH_4 water analyses along MVs of the GoC, two main drivers controlling their formation and distribution should be highlighted: (i) the rate of release of deep-seated methane-enriched fluids, and (ii) the formation of hard substrates such as MDACs (chimneys, slabs or pavements) by AOM processes [27].

The upward migration of methane is transformed by AOM into large amounts of sulphide on the surface [76]. This extremely acid seafloor is buffered by large populations of sulphur-oxidizing siboglinid tubeworms and sulphur-oxidizing bacterial mats, allowing formation of hotspots on the surface by consuming the AOM-sourced sulphur [69]. The occurrence of siboglinid tubeworms in the anoxic-oxic zone seems to be essential for the formation and the non-dissolution of carbonates at the seabed and, moreover for the progressive colonization by deep-sea corals and other suspension feeders on these hard carbonated substrates. Thus, some scleractinians, gorgonians, antipatharians, bamboo corals, and demosponges can colonize the upper parts of MDACs blocks, slabs and pavements, up to 1 m in diameter. In areas with extensive hydrocarbon seeps, massive MDACs are the dominant substrate for coral colonization and reef formation as detected in other MVs of the GoC [54,59].

5.5.2. CWC Mounds and Methane Seeps

The formation of giant carbonate mounds up to 30 m high built up by colonial scleractinians has been identified related to methane seeps in the GoC [69,71,72,77]. In the case of the Northern Pompeia Coral Ridge (westwards Algacel MV), we identified values of methane ranging from 41.93 to 43.24 nM and patches of shells of chemosymbiotic bivalves, mainly *L. asapheus* with some *Thyasira vulcolutre*, and scattered bacterial mats (sulfur-oxidizing-like *Beggiatoa*) indicating active methane seeps throughout the carbonate mounds. It has been proposed a direct relationship between carbonate reefs and fluid seepages by fertilizing waters sourced from methane seeps [78]. Recently, a new hypothesis on the formation of carbonate mounds conformed by scleractinians has been proposed in relation to methane seeps [69]. Thus, scleractinian larvae may colonize MDACs hotspots only if siboglinid tubeworms previously reduce the concentration of sulfide in the anoxic-oxic zone allowing skeletal calcification of scleractinians in MDACs surrounded by the highly acidic muds of methane seeps. Transition from active methane seeps, carbonate mounds and deep-sea coral colonization stages has also been proposed as evolutionary stages of MVs in the GoC [79].

Deep-sea coral colonizing MDACs hard substrates in areas of methane seeps have been reported throughout all margins of the world: The Gulf of México [80,81], Hikurangi Margin in New Zealand [82], Brazilian margin [83], the Darwin Mounds in the northern Rockall Trough [63], the Norwegian shelf [78], and the Porcupine Basin, west of Ireland [84]. Furthermore, the co-existence of chemosymbiotic vestimentiferan worms and bacterial mats with deep-sea corals in cold seeps has been reported in the Gulf of Mexico [81]. The co-occurrence of MDACs with annelids has also been reported from hydrocarbon seeps along the US northern and mid-Atlantic margin [85]. Both MDACs and chemosymbiotic deep-sea mussels in the U.S. Atlantic margin seeps shows average $\delta^{13}C$ signature of $-47‰$, a value consistent with microbially driven anaerobic oxidation of methane-rich fluids occurring at or near the sediment-water interface [85]. In these seep areas, macrofaunal densities dominated by annelid families Dorvilleidae, Capitellidae, and Tubificidae were four times greater than those present in deep-sea mussel beds habitats. These differences between chemosynthesis-based habitats have also been observed in the GoC [59,61]. In this way, it has been suggested that the difference between habitats in such extremophile environments is driven by quality and source of organic matter [86].

5.5.3. Type of Habitats and Methane Concentration

Based on this study, a relationship between the type of habitat and methane concentration in the water column may occur: (i) High concentrations of methane 97.60 nM with bubbling: extensive beds of living chemosymbiotic bivalves *B. mauritanicus, Lucinoma asapheus, Acharax gadirae,* or *Thyasira vulcolutre* fueled by bubbling gas through fissures and cracks; (ii) Intermediate methane concentrations ranging between 65 and 92 nM: AOM-formed MDACs accumulations colonized by aggregations of gorgonian and antipatharians (*Bathypathes, Leiopathes, Stichopathes*), bamboo corals (*Chelidonisis, Acanella*), stony octocorals (*Corallium*) and, sometimes even scleractinians (*D. pertusum*) as well as of demosponges; (iii) Low concentration ranging 20–30 nM allowing colonisation of typical non-chemosymbiotic bathyal soft bottom species such as sea-pens (*Kophobelemnon* sp., *Pennatula, Anthoptilum*), cerianthids, holothurians and decapods.

The influence of strong undercurrents bathing the flanks of the MVs favours the occurrence of non-chemosymbiotic fauna such as large sponge grounds (*Geodia* sp., *Phakellia* sp.) that colonize the scattered MDAC slabs, as well as sea-pens (*Funiculina quadrangularis*), cerianthids and annelids (*Hyalinoeciatubicola*) on the soft, muddy sediments.

5.6. Potential Ecological Restoration of Deep-Sea Habitats after Submarine Eruptions in the Macaronesia Region

Volcanism and associated hydrothermal systems are relevant processes for the evolution of the ocean basins, due to their impact on the geochemistry of the oceans and their potential to form significant deposits [87]. Low-T hydrothermal vents after violent submarine volcanic eruptions generate long-term CO_2 inputs to oceans due to the continuous degasification of the magmatic systems mainly placed on hot-spot volcanic islands like Hawaii or Canary Islands. This is due to the high contents in C bearing in the thick oceanic sediments below the submarine volcanoes that are expulsed by low-T hydrothermal vent systems [88].

In the Tagoro volcano, recolonization of pioneering fauna such as small oysters, serpulids and mobile species (e.g., shrimps, eels) was detected, representing first colonizers of the newly formed substrates in this volcanic environment. Newly formed habitats were also detected [89] together with the burial of extensive areas with aggregations of antipatharians and some deep-sea corals. Some authors indicated that the first colonizers at Tagoro volcano included a large proportion of common suspension feeders and predators of circalittoral and bathyal hard bottoms of the Macaronesian fauna, which could have exploited the uncolonized hard bottoms and the post eruptive fertilization of water masses [35,36]. Along the flanks of this volcano, caves show high values up 6.44 mL·L^{-1} of DO related to active CO_2 degasification two years after the volcanic eruption [50]. The trend representing the DO range for deep-sea complex habitats from 40° N to 26° N (Figure 16) shows the potential pathway from the present parameters to reach suitable DO conditions for the settlement and development of slow-growing organisms such as octocorals (gorgonians and soft corals) as well as antipatharians as detected in other areas of El Hierro Island [89].

We suggest that the comparison between habitats growing after submarine eruptions at different ages might be used to infer the rate of biological colonization and the natural ecological succession of habitats after a violent submarine eruption. Previously, a similar approach has been successfully done for understanding geological and biological evolutionary stages in methane seeps displaying different scenarios of rates of fluid venting [79] and hydrodynamics [21]. Therefore, within the Macaronesia volcanic archipelagos, several recent submarine eruptions have taken place in Capelinhos (west Faial Island, Azores) in 1958–59 or south of El Hierro Island (Canary Islands) in 2011–2012 [34]. The recent discovery of soft coral gardens dominated by *Alcyonacea* in the Azores Archipelago [90] colonizing the volcanic substrate created from the eruption of Capelinhos in 19581–959, opens new studies of succession and survivorship of habitat-forming species in the Macaronesia volcanic areas as they have already been developed in the Hawaiian Islands [91].

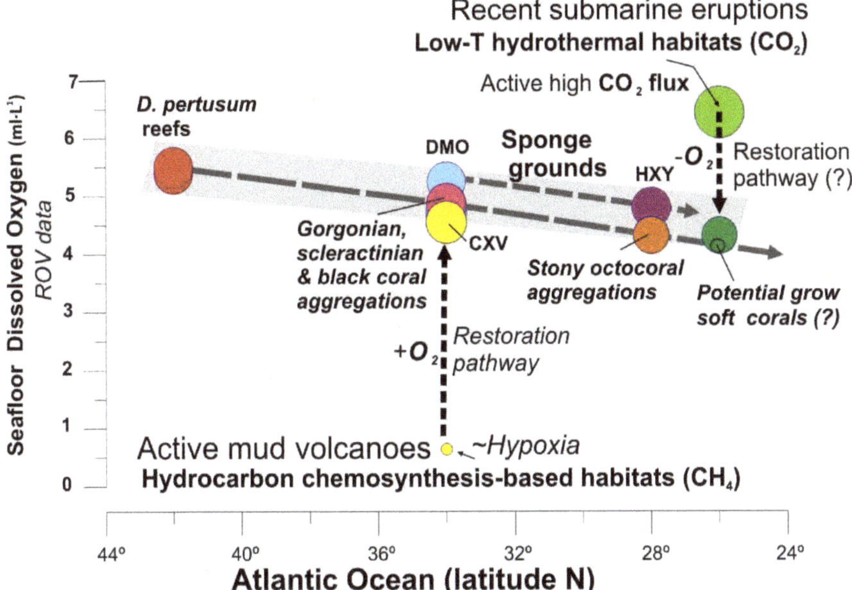

Figure 16. Seafloor dissolved oxygen values vs. latitude for each deep-sea habitat. Grey pattern shows the oxygen threshold for CWC reef and gardens, and deep-sea sponges along the northeast Atlantic Ocean. Dashed black lines show the potential pathways from conditions of cold seeps and recent submarine eruptions to be suitable for non-chemosynthesis-based habitats according the level of dissolved oxygen on the benthic layer. Data are fully listed in Supplementary Table S3.

This fact would support the hypothesis that the type of seabed substrate and the occurrence of fluid flow (CO_2, methane, sulfide, iron) together with some water mass properties (e.g., current speed, salinity, and temperature) might control the evolution of the type of habitats in such volcanic environments, i.e., (i) soft octocoral gardens developed after recent volcanic eruptions (~50 years ago) with latent hydrothermal CO_2 degassing and Fe fertilization but causing seabed acidification unsuitable for carbonates; (ii) Coral gardens and sponge aggregations with a wide variety of sessile species (e.g., octocorals, hexactinellid and lithistid sponges, black corals, bamboo corals and large gorgonians) supported by non-degassing volcanic rocks (up thousands of years old) as the submarine ridges of PoL allowing occurrence of carbonate-bearing benthic fauna.

5.7. Tools for Future Management of Vulnerable Marine Ecosystems

The integration of high-resolution bathymetry, measurements of in-situ water mass properties, and identification of deep-sea habitats ground truthing by ROV at regional scale represents an important tool for increasing the knowledge and protection of deep-water marine ecosystems, especially those areas with VMEs, affected by the ongoing climate changes and/or contemporary anthropogenic impacts. The future scenario for the impact in deep-sea species and habitats, such as scleractinian reefs, coral gardens or deep-sea sponge aggregations, with potential increases in CO_2 and CH_4 emissions or rising in the seawater temperature of the oceans at global scale, might be compared with the changes produced, at local scale, by natural emissions of methane in cold seeps or carbon dioxide in recent submarine eruptions.

Otherwise, these integrated studies are also of paramount importance for the efficient management of natural marine resources, such as the ferromanganese crusts bearing seamounts with high contents in Co, Ni, and other Rare Earth Elements (REE), which are presently targets for mineral exploration in the Area under contracts with the International Seabed Authority (ISA). These seamounts host VMEs, as detected in this and previous studies [92] as well as mineral resources [93] and, thus, spatial management plans are needed to address potential conflicts between deep seabed mining interests and the conservation of the deep-sea biodiversity presented in this study. The same applies for the hydrocarbon-bearing fluid habitats of the studied methane seeps, which are potential target areas for hydrocarbon exploitation in the subsurface and contain singular habitats with endemic species such as the chemosymbiotic bivalves of the GoC.

The five case studies presented here might be considered Ecologically or Biologically Significant Marine Areas (EBSAs) as considered by the United Nations Convention of Biological Diversity (according to different criteria of Annex I, decision IX/20). Unfortunately, significant ecosystem components of those areas (mainly benthic ones) may be threatened by human activities (e.g., fishing activities, hydrocarbon exploitation, or seabed mining), at present or in a near future. Therefore, these selected areas comprising a wide variety of deep-sea habitats in the North Atlantic should be protected and used as pilot areas for further mapping and monitoring of habitats and VMEs, contributing to the environmental management of the North Atlantic Ocean seafloor.

6. Conclusions

Five case studies located between 90 and 2500 mbsl depth, along the NE and Central Atlantic Ocean from 23° N to 42° N latitudes, are presented in this work. These were targeted with MBES mapping, ROV-mounted CTD data of the benthic layer (temperatures, salinity, potential density, dissolved oxygen, potential density, CO_2, CH_4 concentration) and identification and sampling of deep-sea habitats. Based on these datasets, we conclude that:

(1) The distribution of deep-sea habitats along the northeastern Atlantic Ocean is somehow influenced by the water mass properties of the benthic layer at basin scale. However, the water mass properties drivers are modulated by the effects, at regional and local scale, of methane seeps or low-temperature hydrothermal fields after submarine eruptions.

(2) Scleractinian reefs and aggregations of gorgonians, antipatharians, and scleractinians are well constrained to thresholds of seafloor temperature (7.31–11.5 °C), salinity (35.13–36.1 psu) and DO (4.325–5.5 mL·L^{-1}) within a depth range of 800–1400 m. These habitats appear distributed into three levels according to their values of temperature, salinity, potential density and DO: (i) *Desmophyllum pertusum* reefs with the higher values in temperature, salinity and DO values; (ii) Aggregations of gorgonians, scleractinians, and antipatharians with intermediate levels, and (iii) Aggregations of stony octocorals with the lower values in temperature, salinity and DO.

(3) The reef-forming *D. pertusum* identified in the Galicia Bank (NE Atlantic) matches the temperatures, salinity, and potential density values of other scleractinian reefs along the NE Atlantic margins as related to cold North Atlantic waters. However, our data show that other deep-sea corals living along the southern NE Atlantic between the NW African margin and Canary Islands are influenced by the mixing between the intermediate waters MOW and AAIW. This fact highlights the importance of the turbulent interaction between flows of intermediate water masses for deep-sea coral growth along the NE Atlantic Ocean.

(4) Based on these new data, we suggest that the limit between northward intrusions of the AAIW mass waters and the MOW outflow could represent a biogeographical boundary between habitats dominated by colonial scleractinians and with those dominated by gorgonians and stony octocorals. South of 35° N, deep-sea corals are represented by aggregations of stony octocorals (*Corallium tricolor* and *C. niobe*), antipatharians, and gorgonians instead of living scleractinian reefs and banks of the NE Atlantic. Along the NW Africa margin, giant coral ridges and mounds of mostly dead scleractinians (*D. pertusum* and *M. oculata*) has been reported off Morocco and the Mauritanian coast along the shelf break. This is an example of the dramatic loss of deep-sea scleractinian reefs along the northeast Atlantic Ocean since the last glacial.

(5) The co-occurrence of deep-sea coral and sponge habitats and chemosynthesis-based habitats was detected in methane seeps of the Gulf of Cádiz. Thus, extensive beds of living deep-sea mussels *Bathymodiolus mauritanicus* and other chemosymbiotic bivalves around focused bubbling CH_4 vents are placed close to aggregations of sponges, scleractinians, gorgonians, and antipatharians colonizing newly formed MDACs carbonates. Colonies of *Sibloginid* sp. tubeworms play an important role in the connectivity between seeps and some of these suspension feeders by generating pavement patches of MDACs and allowing them to be isolated from the acidic seafloor [69].

(6) Our data support that some deep-sea habitats can be very sensitive to local variations in dissolved oxygen (DO) concentrations within the benthic layer.

Supplementary Materials: The following are available online at https://www.mdpi.com/article/10.3390/oceans2020021/s1. Table S1: Cruise details, MBEs system and configurations, oceanographic data and sampling methods; Table S2: Seabed substrate, morphology, oceanographic variables, VME Indicator Taxa and seafloor micro-morphologies and habitat types; Table S3: ROV-mounted CTD mass water data.

Author Contributions: L.S. was the chief scientist of the SUBVENT-2 cruise and led the overall organization of the text. J.L.R. directed the collection of biological observations and the biological text of the manuscript. O.S.-G. contribute to the overall text of the manuscript and biological descriptions. P.M. lead the ROV "Luso" operations. B.R.-T. directed the microbiological analyses. D.P., R.L. and L.M.F.-S. processed bathymetry, provide multibeam maps and contributed to geomorphological description of the manuscript. T.M. and M.d.C.F.-P. contributed to the geology of the manuscript. F.J.G. and E.M. contributed to the petrological parts of the manuscript. E.L.-P. and E.S. contribute to physic-chemical data. J.T.V. was the chief scientist of the DRAGO511 cruise. All authors contributed to the writing of a draft of the manuscript. All authors have read and agreed to the published version of the manuscript.

Funding: Support for this work comes from the EXPLOSEA (CTM201675947-R), SUBVENT (CGL2012-39524-CO2), and EXARCAN (CTM2010-09496-E) projects funded by the Ministerio de Ciencia e Innovación of Spain. This study also benefits from the Atlantic Seabed Mapping International Working Group (ASMIWG) as part of the Atlantic Ocean Research Alliance Coordination and Support Action (AORA-CSA). This study is also a contribution to the EMODNET-Geology project (EASME/EMFF/2018/1.3.1.8-Lot 1/SI2.811048) and the European project H2020 GeoERA-MINDeSEA (Grant Agreement No. 731166, project GeoE.171.001).

Data Availability Statement: The data presented in this study are available on request from the corresponding author.

Acknowledgments: The authors thank captains, crew and ship parties of all scientific cruises aboard R/V *"Hespérides"*, *"Sarmiento de Gamboa"* and *"Miguel Oliver"*. Special thanks to the ROV "Luso" pilots team Antonio Calado, Andreia Afonso, Miguel Souto and Renato Bettencourt that have been the basis for this paper. Special thanks to the personnel of the Instituto Hidrográfico de la Marina (IHM) for their cooperation in the processing of multibeam bathymetry data.

Conflicts of Interest: The authors declare no conflict of interest.

References

1. Da Ros, Z.; Dell'Anno, A.; Morato, T.; Sweetman, A.K.; Carreiro-Silva, M.; Smith, C.J.; Papadopoulou, N.; Corinaldesi, C.; Bianchelli, S.; Gambi, C.; et al. The deep sea: The new frontier for ecological restoration. *Mar. Policy* **2019**, *108*, 103642. [CrossRef]
2. Mayer, L.; Jakobsson, M.; Allen, G.; Dorschel, B.; Falconer, R.; Ferrini, V.; Lamarche, G.; Snaith, H.; Weatherall, P. The Nippon Foundation—GEBCO seabed 2030 project: The quest to see the world's oceans completely mapped by 2030. *Geosciences* **2018**, *8*, 63. [CrossRef]
3. Harris, P.; Baker, E.K. *Seafloor Geomorphology as Benthic Habitat: GeoHab Atlas of Seafloor Geomorphic Features and Benthic Habitats;* Elsevier: Amsterdam, The Netherlands, 2012; p. 936.
4. International Hydrographic Organization; Intergovernmental Oceanographic Commission. *Standardization of Undersea Feature Names-Guidelines, Proposal Form, Terminology;* Edition 4.1.0 English/Spanish Version; IHO–IOC Publication B-6; Monaco International Hydrographic Bureau: Monaco, 2017; p. 23.

5.   Ercilla, G.; Casas, D.; Vázquez, J.T.; Iglesias, J.; Somoza, L.; Juan, C.; Medialdea, T.; León, R.; Estrada, F.; García-Gil, S.; et al. Imaging the recent sediment dynamics of the Galicia Bank region (Atlantic, NW Iberian Peninsula). *Mar. Geophys. Res.* **2011**, *32*, 99–126. [CrossRef]

6.   Serrano, A.; González-Irusta, J.M.; Punzón, A.; García-Alegre, A.; Lourido, A.; Ríos, P.; Blanco, M.; Gómez-Ballesteros, M.; Druet, M.; Cristobo, J.; et al. Deep-sea benthic habitats modeling and mapping in a NE Atlantic seamount (Galicia Bank). *Deep-Sea Res. Part I Oceanogr. Res. Pap.* **2017**, *126*, 115–127. [CrossRef]

7.   González, F.J.; Somoza, L.; Hein, J.R.; Medialdea, T.; León, R.; Urgorri, V.; Reyes, J.; Martín-Rubí, J.A. Phosphorites, Co-rich Mn nodules, and Fe-Mn crusts from Galicia Bank, NE Atlantic: Reflections of Cenozoic tectonics and paleoceanography. *Geochem. Geophys. Geosyst.* **2016**, *17*, 346–374. [CrossRef]

8.   Iorga, M.; Lozier, M. Signatures of the Mediterranean outflow from a North Atlantic climatology: 2. Diagnostic velocity fields. *J. Geophys. Res.* **1999**, *104*, 26011–26029. [CrossRef]

9.   Ambar, I.; Howe, M.R. Observations of the Mediterranean outflow—I mixing in the Mediterranean outflow. *Deep Sea Res.* **1979**, *26*, 535–554. [CrossRef]

10.   Machín, F.; Pelegrí, J.L. Northward penetration of Antarcticintermediate water off Northwest Africa. *J. Phys. Oceanogr.* **2009**, *39*, 512–535. [CrossRef]

11.   Prieto, E.; González-Pola, C.; Lavín, A.; Sanchez, R.F.; Ruiz-Villarreal, M. Seasonality of intermediate waters hydrography west of the Iberian Peninsula from an 8 yr semiannual time series of an oceanographic section. *Ocean Sci.* **2013**, *9*, 411–429. [CrossRef]

12.   Puerta, P.; Johnson, C.; Carreiro-Silva, M.; Henry, L.A.; Kenchington, E.; Morato, T.; Kazanidis, G.; Rueda, J.L.; Urra, J.; Ross, S.; et al. Influence of Water Masses on the Biodiversity and Biogeography of Deep-Sea Benthic Ecosystems in the North Atlantic. *Front. Mar. Sci.* **2020**, *7*, 239. [CrossRef]

13.   Gardner, J.M. Mud volcanoes on the Moroccan Margin. *EOS Trans. AGU* **1999**, *80*, 483.

14.   Ivanov, M.K.; Woodside, J.M.; Kenyon, N.H.; TTR Shipboard Party. Principal Scientific Results of the TTR. First Decade. In *Geological Processes on Deep-Water European Margins, Proceedings of the International Conference and Ninth Post-Cruise Meeting of the Training-Through-Research Programme, Moscow-Mozhenka, Russia, 28 January–2 February 2001*; Akhmanov, G., Suzyumov, A., Eds.; Intergovernmental Oceanographic Commission: Paris, France, 2001.

15.   Somoza, L.; Gardner, J.M.; Diaz-del-Rio, V.; Vazquez, T.; Pinheiro, L.; Hernández-Molina, F.J.; TASYO/ANASTASYA Shipboard Scientific Parties. Numerous methane gas related seafloor structures identified in the Gulf of Cádiz. *EOS Trans. AGU* **2002**, *83*, 541–547. [CrossRef]

16.   Somoza, L.; Díaz-del-Río, V.; León, R.; Ivanov, M.; Fernández-Puga, M.C.; Gardner, J.M.; Hernández-Molina, F.J.; Pinheiro, L.M.; Rodero, J.; Lobato, A.; et al. Seabed morphology and hydrocarbon seepage in the Gulf of Cádiz mud volcano area: Acoustic imagery, multibeam and ultrahigh resolution seismic data. *Mar. Geol.* **2003**, *195*, 153–176. [CrossRef]

17.   Pinheiro, L.M.; Ivanov, M.K.; Sautkin, A.; Akhmanov, G.; Magalhães, V.H.; Volkonskaya, A.; Monteiro, J.H.; Somoza, L.; Gardner, J.; Hamouni, N.; et al. Mud volcanism in the Gulf of Cádiz: Results from the TTR-10 cruise. *Mar. Geol.* **2003**, *195*, 131–151. [CrossRef]

18.   Van Rensbergen, P.; Depreiter, D.; Pannemans, B.; Moerkerke, G.; Van Rooij, D.; Marsset, B.; Akhmanov, G.; Blinova, V.; Ivanov, M.; Rachidi, M.; et al. The Arraiche mud volcano field at the Moroccan Atlantic slope, Gulf of Cádiz. *Mar. Geol.* **2005**, *219*, 1–17. [CrossRef]

19.   Medialdea, T.; Somoza, L.; Pinheiro, L.M.; Fernández-Puga, M.C.; Vázquez, J.T.; León, R.; Ivanov, M.K.; Magalhães, V.H.; Díaz-del-Río, V.; Vegas, R. Tectonics and mud volcano development in the Gulf of Cádiz. *Mar. Geol.* **2009**, *261*, 48–63. [CrossRef]

20.   León, R.; Somoza, L.; Medialdea, T.; Vázquez, J.T.; González, F.J.; López-González, N.; Casas, D.; Mata, M.P.; Fernández-Puga, M.C.; Giménez-Moreno, C.J.; et al. New discoveries of mud volcanoes on the Moroccan Atlantic continental margin (Gulf of Cádiz): Morpho-structural characterization. *Geo Mar. Lett.* **2012**, *32*, 473–488. [CrossRef]

21.   Palomino, D.; López-González, N.; Vázquez, J.T.; Fernández-Salas, L.M.; Rueda, J.L.; Sanchez-Leal, R.; Díaz-del-Río, V. Multidisciplinary study of mud volcanoes and diapirs and their relationship to seepages and bottom currents in the Gulf of Cádiz continental slope (northeastern sector). *Mar. Geol.* **2016**, *378*, 196–212. [CrossRef]

22.   Hensen, C.; Nuzzo, M.; Hornibrook, E.; Pinheiro, L.M.; Bock, B.; Magalhães, V.H.; Brückmann, W. Sources of mud volcano fluids in the Gulf of Cádiz-indications for hydrothermal imprint. *Geochim. Cosmochim. Acta* **2007**, *71*, 1232–1248. [CrossRef]

23.   Medialdea, T.; Vegas, R.; Somoza, L.; Vázquez, J.T.; Maldonado, A.; Díaz-del-Río, V.; Maestro, A.; Córdoba, D.; Fernández-Puga, M.C. Structure and evolution of the "Olistostrome" complex of the Gibraltar Arc in the Gulf of Cádiz (eastern Central Atlantic): Evidence from two long seismic cross-sections. *Mar. Geol.* **2004**, *209*, 173–198. [CrossRef]

24.   Toyos, M.H.; Medialdea, T.; León, R.; Somoza, L.; González, F.J.; Meléndez, N. Evidence of episodic long-lived eruptions in the Yuma, Ginsburg, Jesús Baraza and Tasyo mud volcanoes, Gulf of Cádiz. *Geo-Mar. Lett.* **2016**. [CrossRef]

25.   Díaz-del-Río, V.; Somoza, L.; Martínez-Frías, J.; Mata, M.P.; Delgado, A.; Hernandez-Molina, F.J.; Lunar, R.; Martín-Rubí, J.A.; Maestro, A.; Fernández-Puga, M.C.; et al. Vast fields of hydrocarbon-derived carbonate chimneys related to the accretionary wedge/olistostrome of the Gulf of Cádiz. *Mar. Geol.* **2003**, *195*, 177–200. [CrossRef]

26.   Magalhães, V.H.; Pinheiro, L.M.; Ivanov, M.K.; Kozlova, E.; Blinova, V.; Kolganova, J.; Vasconcelos, C.; McKenzie, J.A.; Bernasconi, S.M.; Kopf, A.J.; et al. Formation processes of methane-derived authigenic carbonates from the Gulf of Cádiz. *Sed. Geol.* **2012**, *243–244*, 155–168. [CrossRef]

27. Boetius, A.; Ravenschlag, K.; Schubert, C.J.; Rickert, D.; Widdel, F.; Gieseke, A.; Amann, R.; Jørgensen, B.B.; Witte, U.; Pfannkuche, O. A marine microbial consortium apparently mediating anaerobic oxidation of methane. *Nature* **2000**, *407*, 623–626. [CrossRef] [PubMed]
28. González, F.J.; Somoza, L.; Lunar, R.; Martínez-Frías, J.; Martín-Rubí, J.M.; Torres, T.; Ortiz, J.E.; Díaz-del-Río, V.; Pinheiro, L.M.; Magalhães, V.H. Hydrocarbon-derived ferromanganese nodules in carbonate-mud mounds from the Gulf of Cadiz: Mud-breccia sediments and clasts as nucleation sites. *Mar. Geol.* **2009**, *261*, 64–81. [CrossRef]
29. Ryan, W.F.B.; Carbotte, S.M.; Coplan, J.O.; O'Hara, S.; Melkonian, A.; Arko, R.; Weissel, R.A.; Ferrini, V.; Goodwillie, A.; Nitsche, F.; et al. Global multi-resolution topography synthesis. *Geochem. Geophys. Geosyst.* **2009**, *10*, 1525–2027. [CrossRef]
30. Acosta, J.; Uchupi, E.; Muñoz, A.; Herranz, P.; Palomo, C.; Ballesteros, M. ZEE Working Group Salt Diapirs, Salt Brine Seeps, Pockmarks and Surficial Sediment Creep and Slides in the Canary Channel off NW Africa. *Mar. Geophys. Res.* **2003**, *24*, 41–57. [CrossRef]
31. Vázquez, J.T.; Palomino, D.; Fernández-Puga, M.C.; Fernández-Salas, L.M.; Fraile-Nuez, E.; Medialdea, T.; Sánchez-Guillamón, O.; Somoza, L.; SUBVENT Team. Seafloor geomorphology of the Passage of Lanzarote (West Africa Margin): Influences of the oceanographic processes. In *2nd Deep-Water Circulation Congress*; Van Rooij, D., Rüggeberg, A., Eds.; VLIZ Special Publication: Ghent, Belgium, 2014; pp. 125–126. [CrossRef]
32. Hernández-Guerra, A.; Fraile-Nuez, E.; Borges, R.; López-Laatzen, F.; Vélez-Belchí, P.; Parrilla, G.; Müller, T.J. Transport variability in the Lanzarote passage (eastern boundary current of the North Atlantic subtropical Gyre). *Deep Sea Res. Part I* **2003**, *50*, 189–200. [CrossRef]
33. Machín, F.; Hernández-Guerra, A.; Pelegrí, J.L. Mass fluxes in the Canary Basin. *Prog. Oceanogr.* **2006**, *70*, 416–447. [CrossRef]
34. Somoza, L.; González, F.J.; Barker, S.J.; Madureira, P.; Medialdea, T.; de Ignacio, C.; Lourenço, N.; León, R.; Vázquez, J.T.; Palomino, D. Evolution of submarine eruptive activity during the 2011 El Hierro event as documented by hydroacoustic images and remotely operated vehicle observations. *Geochem. Geophys. Geosyst.* **2017**, *18*. [CrossRef]
35. Fraile-Nuez, E.; González-Dávila, M.; Santana-Casiano, J.M.; Arístegui, J.; Alonso-González, I.J.; Hernández-León, S.; Blanco, M.J.; Rodríguez-Santana, A.; Hernández-Guerra, A.; Gelado-Caballero, M.D.; et al. The submarine volcano eruption at the island of El Hierro: Physical-chemical perturbation and biological response. *Sci. Rep.* **2012**, *2*, 486. [CrossRef]
36. Santana-Casiano, J.M.; González-Dávila, M.; Fraile-Nuez, E.; de Armas, D.; González, A.G.; Domínguez-Yanes, A.; Escánez, J. The natural ocean acidification and fertilization event caused by the submarine eruption of El Hierro. *Sci. Rep.* **2013**, *3*, 1140. [CrossRef]
37. Palomino, D.; Vázquez, J.T.; Somoza, L.; León, R.; López-González, N.; Medialdea, T.; Fernández-Salas, L.M.; González, F.J.; Rengel, J.A. Geomorphological features in the southern Canary Island Volcanic Province: The importance of volcanic processes and massive slope instabilities associated with seamounts. *Geomorphology* **2016**, *255*, 125–139. [CrossRef]
38. Bogaard, P. The origin of the Canary Island Seamount Province—New ages of old seamounts. *Sci. Rep.* **2013**, *3*, 1–7.
39. Marino, E.; González, F.J.; Somoza, L.; Lunar, R.; Ortega, L.; Vázquez, J.T.; Reyes, J.; Bellido, E. Strategic and rare elements in Cretaceous-Cenozoic cobalt-rich ferromanganese crusts from seamounts in the Canary Island Seamount Province (northeastern tropical Atlantic). *Ore Geol. Rev.* **2017**, *87*, 41–61. [CrossRef]
40. Marino, E.; González, F.J.; Lunar, R.; Reyes, J.; Medialdea, T.; Castillo-Carrión, M.; Bellido, E.; Somoza, L. High-Resolution Analysis of Critical Minerals and Elements in Fe–Mn Crusts from the Canary Island Seamount Province (Atlantic Ocean). *Minerals* **2018**, *8*, 285. [CrossRef]
41. Marino, E.; González, F.J.; Kuhn, T.; Madureira, P.; Wegorzewski, A.V.; Mirao, J.; Medialdea, T.; Oeser, M.; Miguel, C.; Reyes, J.; et al. Hydrogenetic, diagenetic and hydrothermal processes forming ferromanganese crusts in the Canary Island Seamounts and their influence in the metal recovery rate with hydrometallurgical methods. *Minerals* **2019**, *9*, 439. [CrossRef]
42. Brandt, P.; Hormann, V.; Körtzinger, A.; Visbeck, M.; Krahmann, G.; Stramma, L.; Lumpkin, R.; Schmid, C. Changes in the ventilation of the oxygen minimum zone of the tropical North Atlantic. *J. Phys. Oceanogr.* **2010**, *40*, 1784–1801. [CrossRef]
43. Somoza, L.; Vázquez, J.T.; Campos, A.; Afonso, A.; Calado, A.; Fernández-Puga, M.C.; González, F.J.; Fernández-Salas, L.M.; Ferreira, M.; Sanchez-Guillamón, O.; et al. Informe Científico-Técnico Campaña SUBVENT-2, 2014, p. 43. Available online: http://info.igme.es/SidPDF/166000/941/166941_0000001.pdf (accessed on 1 December 2020).
44. Boyer, T.P.; Baranova, O.K.; Coleman, C.; García, H.E.; Grodsky, A.; Locarnini, R.A.; Mishonov, A.V.; Paver, C.R.; Reagan, J.R.; Seidov, D.; et al. World Ocean Database 2018. Available online: https://www.ncei.noaa.gov/sites/default/files/2020-04/wod_intro_0.pdf (accessed on 1 December 2020).
45. Schlitzer, R. Ocean Data View 2017. Available online: http://odv.awi.de (accessed on 1 December 2020).
46. Somoza, L.; Ercilla, G.; Urgorri, V.; León, R.; Medialdea, T.; Paredes, M.; González, F.J.; Nombela, M.A. Detection and mapping of cold-water coral mounds and living *Lophelia* reefs in the Galicia Bank, Atlantic NW Iberia margin. *Mar. Geol.* **2014**, *349*, 73–90. [CrossRef]
47. Urgorri, V.; Troncoso, J.S. A second record of *Laevipilinarolani* Warén&Bouchet, 1990 (Mollusca: Monoplacophora) from the Northwestof Spain. *J. Molluscan Stud.* **1994**, *60*, 157–163.
48. Vázquez, J.T.; Ercilla, G.; Somoza, L.; Palomino, D.; Alonso, B.; Casas, D.; Estrada, F.; Fernández-Puga, M.C.; Fernández-Salas, L.M.; León, R.; et al. Giant depressions on Atlantic continental margins: Relationship with diapirs. In *Book of Proceedings of the IX Symposium MIA2018*; Cunha, P.P., Dias, J., Veríssimo, H., Duarte, L.V., Dinis, P., Lopes, F.C., Bessa, A.F., Carmo, J.A., Eds.; Universidade de Coimbra: Coimbra, Portugal, 2018; pp. 251–252.

49. Rivera, J.; Lastras, G.; Canals, M.; Acosta, J.; Arrese, B.; Hermida, N.; Micallef, A.; Tello, O.; Amblas, D. Construction of an oceanic island: Insights from the El Hierro (Canary Islands) 2011–2012 submarine volcanic eruption. *Geology* **2013**, *41*, 355–358. [CrossRef]

50. González, F.J.; Rincón-Tomás, B.; Somoza, L.; Santofimia, E.; Medialdea, T.; Madureira, P.; López-Pamo, E.; Hein, J.R.; Marino, E.; de Ignacio, C.; et al. Low-temperature, shallow-water hydrothermal vent mineralization following the recent submarine eruption of Tagoro volcano (El Hierro, Canary Islands). *Mar. Geol.* **2020**, *430*, 106333. [CrossRef]

51. Dando, P.R.; Hovland, M. Environmental effects of submarine seeping natural gas. *Cont. Shelf Res.* **1992**, *12*, 1197–1207. [CrossRef]

52. Judd, A.; Hovland, M. *Seabed Fluid Flow. Impact on Geology, Biology, and the Marine Environment*; Cambridge University Press: Cambridge, UK, 2007.

53. Levin, L.A.; Baco, A.R.; Bowden, D.A.; Colaco, A.; Cordes, E.E.; Cunha, M.R.; Demopoulos, A.W.J.; Gobin, J.; Grupe, B.M.; Le, J.; et al. Hydrothermal vents and methane seeps: Rethinking the sphere of influence. *Front. Mar. Sci.* **2016**, *3*, 72. [CrossRef]

54. Rueda, J.L.; González-García, E.; Krutzky, C.; López-Rodríguez, F.J.; Bruque-Carmona, G.; López-González, N.; Palomino, D.; Sánchez-Leal, R.F.; Vázquez, J.T.; Fernández-Salas, L.M.; et al. From chemosynthesis-based communities to cold-water corals: Vulnerable deep-sea habitats of the Gulf of Cádiz. *Mar. Biodivers.* **2016**, *46*, 473–482. [CrossRef]

55. Lozano, P.; Rueda, J.L.; Gallardo-Núñez, M.; Urra, C.F.J.; Vila, Y.; López-González, N.; Palomino, D.; Sánchez-Guillamón, O.; Vázquez, J.T.; Fernández-Salas, L.M. Chapter 52—Habitat distribution and associated biota in different geomorphic features within a fluid venting area of the Gulf of Cádiz (Southwestern Iberian Peninsula, Northeast Atlantic Ocean). In *Seafloor Geomorphology as Benthic Habitat*, 2nd ed.; Harris, P., Baker, E., Eds.; Elsevier: Amsterdam, The Netherlands, 2020; pp. 847–861.

56. Yasuhara, M.; Danovaro, R. Temperature impacts on deep-sea biodiversity. *Biol. Rev.* **2014**, *91*, 275–287. [CrossRef]

57. Freiwald, A.; Roberts, J.M. *Cold-Water Corals and Ecosystems*; Springer: Berlin/Heidelberg, Germany, 2005.

58. Buhl-Mortensen, L.; Olafsdottir, S.H.; Buhl-Mortensen, P.; Burgos, J.M.; Ragnarsson, S.A. Distribution of nine cold-water coral species (Scleractinia and Gorgonacea) in the cold temperate North Atlantic: Effects of bathymetry and hydrography. *Hydrobiologia* **2015**, *759*, 39–61. [CrossRef]

59. Ramalho, L.V.; López-Fé, C.M.; Mateo-Rodríguez, A.; Rueda, J.L. Bryozoa from Deep-sea habitats of the northern Gulf of Cádiz (Northeastern Atlantic). *Zootaxa* **2020**, *4768*, 451–478. [CrossRef]

60. Sitjà, C.; Maldonado, M.; Farias, C.; Rueda, J.L. Export of bathyal benthos to the Atlantic through the Mediterranean outflow: Sponges from the mud volcanoes of the Gulf of Cadiz as a case study. *Deep Sea Res. Part I* **2020**, *163*, 103326. [CrossRef]

61. Rodrigues, C.F.; Hilário, A.; Cunha, M.R. Chemosymbiotic species from the Gulf of Cadiz (NE Atlantic): Distribution, life styles and nutritional patterns. *Biogeosciences* **2013**, *10*, 2569. [CrossRef]

62. Dullo, W.C.; Flögel, S.; Rüggerberg, A. Cold-water coral growth in relation to the hydrography of the Celtic and Nordic European continental margin. *Mar. Ecol. Prog. Series* **2008**, *371*, 165–176. [CrossRef]

63. Huvenne, V.A.; Masson, D.G.; Wheeler, A.J. Sediment dynamics of a sandy contourite: The sedimentary context of the Darwin cold-water coral mounds, Northern Rockall Trough. *Int. J. Earth Sci.* **2009**, *98*, 865–884. [CrossRef]

64. Xavier, J.; van Soest, R. Demosponge fauna of Ormonde and Gettysburg Seamounts (Gorringe Bank, north-east Atlantic): Diversity and zoogeographical affinities. *JMBA J. Mar. Biol. Assoc. UK* **2007**, *87*, 1643–1654. [CrossRef]

65. Watling, L.; Guinotte, J.; Clark, M.R.; Smith, C.R. A proposed biogeography of the deep ocean floor. *Prog. Oceanogr.* **2013**, *111*, 91–112. [CrossRef]

66. Ceramicola, S.; Dupré, S.; Somoza, L.; Woodside, J. Cold seep systems. In *Submarine Geomorphology*; Micallef, A., Krastel, S., Savini, A., Eds.; Springer International Publishing: Cham, Switzerland, 2018; pp. 367–388.

67. Rincón-Tomás, B.; González, F.J.; Somoza, L.; Sauter, K.; Madureira, P.; Medialdea, T.; Carlsson, J.; Reitner, J.; Hoppert, M. Siboglinidae Tubes as an Additional Niche for Microbial Communities in the Gulf of Cádiz—A Microscopical Appraisal. *Microorganisms* **2020**, *8*, 367. [CrossRef]

68. Demopoulos, A.W.; Bourque, J.R.; Frometa, J. Biodiversity and community composition of sediment macrofauna associated with deep-sea *Lophelia pertusa* habitats in the Gulf of Mexico. *Deep Sea Res. Part I* **2014**, *93*, 91–103. [CrossRef]

69. Rincón-Tomás, B.; Duda, J.P.; Somoza, L.; González, F.J.; Schneider, D.; Medialdea, T.; Reitner, J. Cold-water corals and hydrocarbon-rich seepage in Pompeia Province (Gulf of Cádiz)–Living on the edge. *Biogeosciences* **2019**, *16*, 1607–1627. [CrossRef]

70. Roque, D.; Parras-Berrocal, I.; Bruno Mejías, M.; Sanchez Leal, R.F.; Hernández-Molina, F.J. Seasonal variability of intermediate water masses in the Gulf of Cádiz: Implications of the Antarctic and subarctic seesaw. *Ocean. Sci.* **2019**, *15*, 1381–1397. [CrossRef]

71. Foubert, A.; Depreiter, D.; Beck, T.; Maignien, L.; Pannemans, B.; Frank, N.; Blamart, D.; Henriet, J.-P. Carbonate mounds in a mud volcano province off north-west Morocco: Key to processes and controls. *Mar. Geol.* **2008**, *248*, 74–96. [CrossRef]

72. Wienberg, C.; Hebbeln, D.; Fink, H.G.; Mienis, F.; Dorschel, B.; Vertino, A.; López Correa, M.; Freiwald, A. Scleractinian cold-water corals in the Gulf of Cádiz -first clues about their spatial and temporal distribution. *Deep-Sea Res.* **2009**, *56*, 1873–1893. [CrossRef]

73. Hebbeln, D.; Van Rooij, D.; Wienberg, C. Good neighbours shaped by vigorous currents: Cold-water coral mounds and contourites in the North Atlantic. *Mar. Geol.* **2016**, *378*, 171–185. [CrossRef]

74. Wienberg, C.; Titschack, J.; Freiwald, A.; Frank, N.; Lundälv, T.; Taviani, M.; Beuck, L.; Schröder-Ritzrau, A.; Krengel, T.; Hebbeln, D. The giant Mauritanian cold-water coral mound province: Oxygen control on coral mound formation. *Quat. Sci. Rev.* **2018**, *185*, 135–152. [CrossRef]

75. De la Fuente, M.; Skinner, L.; Ercilla, G.; d´Acremont, E.; Somoza, L.; González, F.J.; Lo Iacono, C.; Corbera, G.; Pena, L.D.; Sadekov, A.; et al. Inferring deglacial ventilation ages in Western Mediterranean waters using cold-water corals. In *EGU General Assembly Conference Abstracts*; EGU2020-20171; EGU: Viena, Austria, 2020. [CrossRef]

76. Hinrichs, K.U.; Boetius, A. The anaerobic oxidation of methane: New insights inmicrobial ecology and biogeochemistry. In *Ocean Margin Systems*; Wefer, G., Billett, D., Hebbeln, D., Jørgensen, B.B., Schlüter, M., van Weering, T.C.E., Eds.; Springer: Berlin/Heidelberg, Germany, 2002; pp. 457–477.

77. Van Rooij, D.; Blamart, D.; De Mol, L.; Mienis, F.; Pirlet, H.; Wehrmann, L.M.; Barbieri, R.; Maignien, L.; Templer, S.P.; de Haas, H.; et al. Cold-water coral mounds on the Pen Duick Escarpment, Gulf of Cádiz: The MiCROSYSTEMS project approach. *Mar. Geol.* **2011**, *282*, 102–117. [CrossRef]

78. Hovland, M. Do carbonate reefs form due to fluid seepage? *Terra Nova* **1990**, *2*, 8–18. [CrossRef]

79. León, R.; Somoza, L.; Medialdea, T.; González, F.J.; Díaz-del-Río, V.; Fernández-Puga, M.; Maestro, A.; Mata, M.P. Sea-floor features related to hydrocarbon seeps in deepwater carbonate-mud mounds of the Gulf of Cádiz: From mud flows to carbonate precipitates. *Geo-Mar. Lett.* **2007**, *27*, 237–247. [CrossRef]

80. Becker, E.L.; Cordes, E.E.; Macko, S.A.; Fisher, C.R. Importance of seep primary production to *Lophelia pertusa* and associated fauna in the Gulf of Mexico. *Deep Sea Res. Part I* **2009**, *56*, 786–800. [CrossRef]

81. Demopoulos, A. Biodiversity, biogeography, and connectivity of seeps and cold-water coral communities in the Gulf of Mexico. *Beyond Horiz.* **2011**, *11*, 37.

82. Liebetrau, V.; Eisenhauer, A.; Linke, P. Cold seep carbonates and associated cold-water corals at the Hikurangi Margin, New Zealand: New insights into fluid pathways, growth structures and geochronology. *Mar. Geol.* **2010**, *272*, 307–318. [CrossRef]

83. Gomes-Sumida, P.Y.; Yoshinaga, M.Y.; Madureira, L.A.S.-P.; Hovland, M. Seabed pockmarks associated with deep water corals off SE Brazilian continental slope, Santos Basin. *Mar. Geol.* **2004**, *207*, 159–167. [CrossRef]

84. Henriet, J.P.; Van Rooij, D.; Huvenne, V.; De Mol, B.; Guidard, S. Mounds and sediment drift in the Porcupine Basin, west of Ireland. In *European Continental Margin Sedimentary Processes: An Atlas of Side-Scan Sonar and Seismic Images*; Mienert, J., Weaver, P., Eds.; Springer: Berlin/Heidelberg, Germany, 2003; pp. 217–220.

85. Bourque, J.R.; Robertson, C.M.; Brooke, S.; Demopoulos, A.W. Macrofaunal communities associated with chemosynthetic habitats from the US Atlantic margin: A comparison among depth and habitat types. *Deep Sea Res. Part II* **2017**, *137*, 42–55. [CrossRef]

86. Prouty, N.G.; Sahy, D.; Ruppel, C.D.; Roark, E.B.; Condon, D.; Brooke, S.; Ross, S.W.; Demopoulos, A.W.J. Insights into methane dynamics from analysis of authigenic carbonates and chemosynthetic mussels at newly-discovered Atlantic Margin seeps. *Earth Planet. Sci. Lett.* **2016**, *449*, 332–344. [CrossRef]

87. Alt, J.C. Hydrothermal fluxes at mid-ocean ridges and on ridge flanks. *Geochemistry* **2003**, *335*, 853–864. [CrossRef]

88. Medialdea, T.; Somoza, L.; González, F.J.; Vázquez, J.T.; de Ignacio, C.; Sumino, H.; Sánchez-Guillamón, O.; Orihashi, Y.; León, R.; Palomino, D. Evidence of a modern deep water magmatic hydrothermal system in the Canary Basin (eastern central Atlantic Ocean). *Geochem. Geophys. Geosyst.* **2017**, *18*. [CrossRef]

89. Sotomayor-García, A.; Rueda, J.L.; Sánchez-Guillamón, O.; Urra, J.; Vázquez, J.T.; Palomino, D.; Fernández-Salas, L.M.; López-González, N.; González-Porto, M.; Santana-Casiano, J.M.; et al. First Macro-Colonizers and Survivors Around Tagoro Submarine Volcano, Canary Islands, Spain. *Geosciences* **2019**, *9*, 52. [CrossRef]

90. Somoza, L.; Medialdea, T.; González, F.J.; Calado, A.; Afonso, A.; Albuquerque, M.; Asensio-Ramos, M.; Bettencourt, R.; Blasco, I.; Candón, J.A.; et al. Multidisciplinary Scientific Cruise to the Northern Mid-Atlantic Ridge and Azores Archipelago. *Front. Mar. Sci.* **2020**. [CrossRef]

91. Putts, M.R.; Parrish, F.A.; Trusdell, F.A.; Kahng, S.E. Structure and development on Hawaiian Deep-water coral communities on Mauna Loa lava flows. *Mar. Ecol. Prog. Ser.* **2019**, *630*, 69–82. [CrossRef]

92. Ramiro-Sánchez, B.; González-Irusta, J.M.; Henry, L.A.; Cleland, J.; Yeo, I.; Xavier, J.R.; Carreiro-Silva, M.; Sampaio, Í.; Spearman, J.; Victorero, L.; et al. Characterization and mapping of a deep-sea sponge ground on the Tropic Seamount (Northeast Tropical Atlantic): Implications for Spatial Management in the High Seas. *Front. Mar. Sci.* **2019**, *6*, 278. [CrossRef]

93. Hein, J.R.; Mizell, K.; Koschinsky, A.; Conrad, T.A. Deep-ocean mineral deposits as a source of critical metals for high- and green-technology applications: Comparison with land-based resources. *Ore Geol. Rev.* **2013**, *51*, 1–14. [CrossRef]

Review

Offshore Geological Hazards: Charting the Course of Progress and Future Directions

Gemma Ercilla [1,*], David Casas [1], Belén Alonso [1], Daniele Casalbore [2], Jesús Galindo-Zaldívar [3], Soledad García-Gil [4], Eleonora Martorelli [5], Juan-Tomás Vázquez [6], María Azpiroz-Zabala [7], Damien DoCouto [8], Ferran Estrada [1], Mª Carmen Fernández-Puga [9], Lourdes González-Castillo [3], José Manuel González-Vida [10], Javier Idárraga-García [11], Carmen Juan [12], Jorge Macías [13], Asier Madarieta-Txurruka [3], José Nespereira [14], Desiree Palomino [6], Olga Sánchez-Guillamón [6], Víctor Tendero-Salmerón [3], Manuel Teixeira [15,16,17], Javier Valencia [18] and Mariano Yenes [14]

[1] Continental Margins Group, Consejo Superior de Investigaciones Científicas (CSIC), Instituto de Ciencias del Mar, Paseo Marítimo de la Barceloneta 37–49, 08003 Barcelona, Spain; davidcasas@icm.csic.es (D.C.); belen@icm.csic.es (B.A.); festrada@icm.csic.es (F.E.)

[2] Dipartimento di Scienze della Terra, Università di Roma "Sapienza", Piazzale Aldo Moro 5, 00185 Rome, Italy; daniele.casalbore@uniroma1.it

[3] Departamento de Geodinámica, Instituto Andaluz de Ciencias de la Tierra (IACT)–CSIC, Universidad de Granada, 18071 Granada, Spain; jgalindo@ugr.es (J.G.-Z.); lgcastillo@ugr.es (L.G.-C.); amadatxu@ugr.es (A.M.-T.); vtendero@ugr.es (V.T.-S.)

[4] BASAN, Centro de Investigación Mariña, Departamento de Geociencias Marinas, Universidade de Vigo, 36200 Vigo, Spain; sgil@uvigo.es

[5] Istituto di Geologia Ambientale e Geoingegneria, Sede Secondaria di Roma, Consiglio Nazionale delle Ricerche (CNR), Piazzale A. Moro, 5, 00185 Rome, Italy; eleonora.martorelli@cnr.it

[6] Instituto Español de Oceanografía—IEO, 29640 Malaga, Spain; juantomas.vazquez@ieo.es (J.-T.V.); desiree.palomino@ieo.es (D.P.); olga.sanchez@ieo.es (O.S.-G.)

[7] Applied Geology, Civil Engineering and Geosciences, Delft University of Technology, Stevinweg 1, 2628 Delft, The Netherlands; emeazeta@gmail.com

[8] UMR 7193, Institut des Sciences de la Terre Paris (ISTeP), CNRS, Sorbonne Université, F-75005 Paris, France; damien.do_couto@sorbonne-universite.fr

[9] Departamento de Ciencias de la Tierra, Facultad de Ciencias del Mar y Ambientales, INMAR (Instituto Universitario de Investigación Marina), Universidad de Cádiz, 11510 Cadiz, Spain; mcarmen.fernandez@uca.es

[10] Departamento de Matemática Aplicada, Universidad de Málaga, Escuela de Ingenierías Industriales, Campus de Teatinos, s/n, 29071 Malaga, Spain; jgv@uma.es

[11] Departamento de Física y Geociencias, Universidad del Norte, 081007 Barranquilla, Colombia; idarragaj@uninorte.edu.co

[12] Laboratoire d'Oceanologie et de Géosciences, UMR, Université de Lille, CNRS, Universite Littoral Côte d'Opale, F 59000 Lille, France; carmen.juanval@gmail.com

[13] Departamento de Matemática Aplicada, Facultad de Ciencias, Universidad de Málaga, Campus de Teatinos, s/n, 29080 Malaga, Spain; jmacias@uma.es

[14] Departamento de Geología, Universidad de Salamanca, 37008 Salamanca, Spain; jnj@usal.es (J.N.); myo@usal.es (M.Y.)

[15] IDL (Laboratório Associado)—Instituto Dom Luíz, Campo Grande, Faculdade de Ciências da Universidade de Lisboa, Edifício C1, 1749-016 Lisboa, Portugal; mane.teixeira@gmail.com

[16] Departamento de Geologia, Campo Grande, Faculdade de Ciências da Universidade de Lisboa, Edifício C6, 1749-016 Lisboa, Portugal

[17] Divisão de Geologia Marinha e Georecursos Marinhos, IPMA—Instituto Português do Mar e da Atmosfera, Rua C do Aeroporto, 1749-077 Lisboa, Portugal

[18] LYRA, Engineering Consulting, 01015 Gazteiz, Spain; javi.valencia.m@gmail.com

* Correspondence: gemma@icm.csic.es

Citation: Ercilla, G.; Casas, D.; Alonso, B.; Casalbore, D.; Galindo-Zaldívar, J.; García-Gil, S.; Martorelli, E.; Vázquez, J.-T.; Azpiroz-Zabala, M.; DoCouto, D.; et al. Offshore Geological Hazards: Charting the Course of Progress and Future Directions. *Oceans* **2021**, 2, 393–428. https://doi.org/10.3390/oceans2020023

Academic Editor: Pere Masqué

Received: 16 December 2020
Accepted: 24 May 2021
Published: 31 May 2021

Publisher's Note: MDPI stays neutral with regard to jurisdictional claims in published maps and institutional affiliations.

Abstract: Offshore geological hazards can occur in any marine domain or environment and represent a serious threat to society, the economy, and the environment. Seismicity, slope sedimentary instabilities, submarine volcanism, fluid flow processes, and bottom currents are considered here because they are the most common hazardous processes; tsunamis are also examined because they are a secondary hazard generated mostly by earthquakes, slope instabilities, or volcanic eruptions. The hazards can co-occur and interact, inducing a cascading sequence of events, especially in certain contexts, such as tectonic indentations, volcanic islands, and canyon heads close to the coast. We analyze the key characteristics and main shortcomings of offshore geological hazards to identify their present and future directions for marine geoscience investigations of their identification and characterization. This review establishes that future research will rely on studies including a high level of multidisciplinarity. This approach, which also involves scientific and technological challenges, will require effective integration and interplay between multiscale analysis, mapping, direct deep-sea observations and testing, modelling, and linking offshore observations with onshore observations.

Keywords: seismic faults; slope instabilities; submarine volcanism; fluid-flow processes; bottom currents; tsunamis; canyon heads; tectonic indentation; multidisciplinary approach

1. Introduction

Geological processes occurring within or at the surface of the Earth may lead to natural disasters causing loss of life, environmental damage, and major impacts on the economy and food security. Human behavior may also trigger natural disaster processes where no hazards existed before or increase the risks where they do exist. Knowledge of the geological elements likely to produce a disaster and their distribution, as well as the understanding of the mechanisms (conditioning factors and triggers), is critical for the prevention and mitigation of catastrophic events [1–4]. The time scales of those mechanisms are highly variable, from geologic to human scale, which contrasts greatly with the sudden impact of such events when they are triggered.

Like in any other Earth domain, the marine environment is associated with potentially hazardous geological processes that may represent serious threats to society, the economy, and the environment (http://www.esf.org/fileadmin/Public_documents/Publications/Natural_Hazards.pdf, accessed on 28 May 2020) (Figure 1). Such processes are present in all physiographic domains; even features and events that occur on the continental shelf and in deep-sea areas may have catastrophic effects on large areas in coastal environments [5]. Coastal areas are highly populated around the world and are the sites of most megacities [6]. A growing population, currently approximately 2.4 billion people, lives within 100 km of the coast (oceanconference.un.org, accessed on 28 May 2020). The expansion of urban coastal areas and coastal and offshore industries (e.g., communications, energy, and mineral extraction) has greatly increased the exposure and risk of large subaerial and submarine infrastructure. Contrary to the social perception, submarine areas host different and active features that present frequent geohazards.

Figure 1. Sketch showing the main offshore geological hazards. Inspired by [7].

Avoidance of hazardous areas is not always possible. Therefore, to reduce the risk of an area, conceptualized as the product of hazards and vulnerability, it is imperative to reduce vulnerability and obtain better knowledge of hazardous processes through new approaches and the development of new techniques and tools [8,9].

The field of offshore geohazards is wide because it covers different topic areas, such as identification and mapping of the hazards, risks, vulnerability, predictions, and warnings; covering all of these topics is beyond the scope of this paper. Specifically, this paper aims to provide the summary of the course and progress in the research of geoscientists regarding the main offshore hazardous geological processes, namely: seismicity, slope sedimentary instabilities, submarine volcanism, fluid flow processes, bottom currents, and tsunamis. Some geological settings, where interactions arise from cascading effects and thus create multihazard scenarios, are also shown. We analyze the key characteristics and main shortcomings of those geohazards, enabling the identification of present and future directions for marine geosciences focusing on offshore geohazards.

2. Definition and Classification of Offshore Geological Hazards

There are several definitions of geological hazards (e.g., [10–12]). As a synthesis, geological hazards can be considered all phenomena or conditions (on land and offshore), natural or induced by human activity, that can produce damage and that geological information can be used to predict, prevent, or correct.

The classification of offshore geological hazards may vary depending on their implications. From an engineering point of view, they are generally classified based on the problems they can generate during the exploration, installation, and operation of structures [13]. In contrast, marine geoscientists are more interested in understanding the features and processes that define a hazard; therefore, they classify them according to their causes.

Offshore geohazards arise from different geomorphological and geological features that produce scenarios in which diverse processes may act alone or in combination with others, triggering a chain of events. Morphological characteristics, such as relief (negative or positive) or overstepped slopes, may be indicators of processes that can generate hazards; however, depending on the activity planned, they may be considered a threat themselves.

The most widespread offshore geohazards are seismicity, slope instabilities, submarine volcanism, and processes related to fluid flow and bottom currents (e.g., [5,7,9,14]; https://niwa.co.nz/natural-hazards/marine-geological-hazards;https://marineboard.eu/marine-geohazards-blue-economy, accessed on 3 October 2020) (Figure 1). Tsunamis deserve special mention because they are a secondary hazard derived from or generated by another event, especially earthquakes, slope instabilities, or volcanic eruptions ([5,15,16], among others). Additionally, many of these processes are associated with intense submarine erosion that may be responsible for the general topography and microtopography of the seafloor (e.g., [17]). Seismicity has a direct impact on the ground due to vibrations that affect infrastructure and buildings. Moreover, seismicity can also produce indirect effects, such as liquefaction and slope instabilities [18]. Earthquakes are related to the presence of seismogenic faults.

Submarine slope instabilities, resulting in products collectively referred to as landslides, mass-transport deposits, or mass-transport complexes, are capable of damaging infrastructure resting on or fixed to the seafloor, such as vertical foundations, communication cables, and pipelines, due to the associated impact, dragging, excessive burial, or undermining effects [19,20]. Active submarine volcanoes are significant geological hazards because of their violent and explosive eruptions and related earthquakes, collapses of their summits (i.e., caldera) and fluid emissions; moreover, volcanic activity can trigger secondary hazards such as tsunamis and landslides [5].

Fluid flow processes, such as seepage of light hydrocarbons, migration of overpressurized muddy fluids forming volcanoes and diapirs (Figure 1), and gas hydrate formation and breakdown, constitute a main type of potential geohazard [7,21]. They are commonly a threat to navigation and offshore infrastructure, during both installation and operation, because they can cause damage or uncontrolled release of gas that in turn induces explosions or landslides.

The persistent action of bottom currents over the seafloor may create large areas affected by erosion and scouring and highly active seafloor conditions (Figure 1). Similarly, reworked or transported marine sediment is subsequently deposited, forming mounds in areas with high sedimentation rates. Bottom currents may also affect the sedimentological and geotechnical properties of seafloor and subsurface sediments, affecting their stability [22]. Thus, bottom currents may represent a hazardous process to subsurface and seafloor installations and infrastructure crossing the water column, because surface and intermediate currents can induce stress on them.

Human Activities in Submarine Environments

Human activities on the seafloor have increased sharply since the last half of the last century, accompanied by significant technological advances (Figure 2). Therefore, the understanding of offshore geological hazardous processes is fundamental for seafloor management. The main important offshore activities potentially exposed to offshore geohazards are as follows:

a. Submarine telecommunication cables are important offshore infrastructure and funnel 95% of all telephone and data communication. Approximately 378 subsea cables (total length of 1.2 million kilometers) (https://www.mapfreglobalrisks.com, accessed on 12 October 2020) rest on the seafloor, forming complex inter-continental, inter-peninsular, and island-continent networks https://www.submarinecablemap.com/, accessed on 3 February 2020). Some new deployments are designed to bury cables in the seafloor to protect them from trawlers, anchors, and turbidity currents [23].

b. Ports and industrial installations, airports, residential and recreation buildings, artificial islands, wind farms, and fish harming, among others, are human-made structures occupying subaerial and submarine surfaces, and they will increase due to human expansion. These structures may be affected by geological processes, but they may also be affected by potential human-induced hazards because of the interaction between seafloor structures and environmental processes.

c. Deep-sea mining has the potential to be an important submarine activity in the near future. This activity involves prospecting, exploitation, and extraction [24], and all three stages are subject to hazardous geological processes.

d. Fisheries and transport are critical economic activities around the world. Fishing grounds and commercial routes (navigation) may be locally affected by active geological processes occurring on the seafloor.

e. Hydrocarbon exploitation and transportation are performed by 53 countries on continental shelves and adjacent slopes, where the deployed infrastructure is placed on the seabed and interacts with geological processes during installation and operation [13].

f. Gas and oil pipelines, in contrast to the exploitation platforms whose activities focus on the local seabed, cross different physiographic regions on the continental margins and are therefore affected by different hazardous geological processes, which may deform and rupture them. In 2016, operators planned nearly 4000 miles of offshore pipelines through 2020 (https://www.offshore-mag.com/pipelines/article/16754997/, accessed on 28 October 2020).

g. Other common activities, such as sand recovery for the artificial nourishment of beaches, may represent hazards themselves because they may modify the sedimentary environment and natural processes.

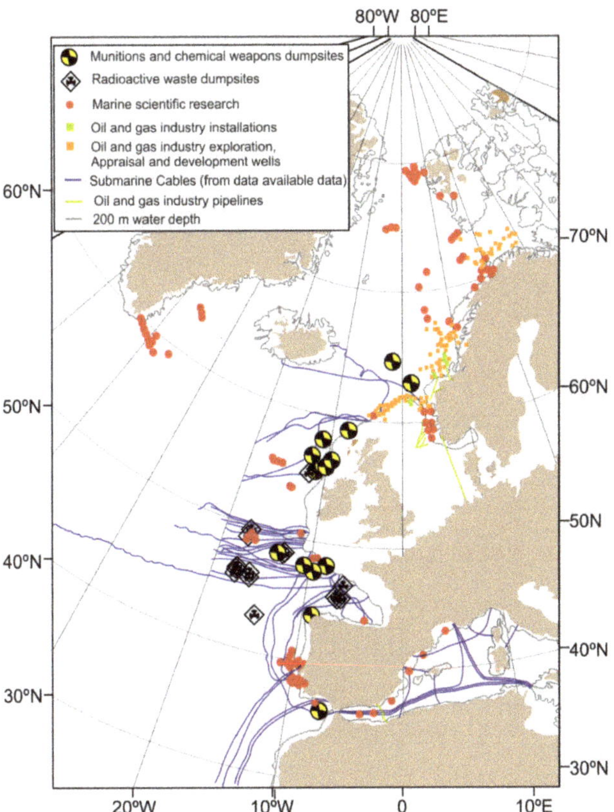

Figure 2. Some of the main human activities in deep-sea areas (>200 m water depth) in the NE Atlantic and western Mediterranean. The figure was used as idea and base to create a new one including more information from [25].

3. Some Prehistorical and Historical Cases of Offshore Geohazard Events

Offshore geological hazard events are recognized in all cultures and in all seas and oceans but are most common on highly active continental margins associated with tectonic plate boundaries (https://www.emdat.be/, accessed on 28 October 2020). Table 1 presents a small percentage of them, with only some of the widely known case events. All these disasters generated by prehistoric and historic marine geological events affected coastal populations, infrastructure, and the environment, highlighting the great vulnerability of these areas as well as the scarce knowledge regarding the triggering mechanisms of the events in particular or of the hazards in general. They have also revealed that active geological features do not recognize political frontiers and that hazard assessment must cross national boundaries. Furthermore, their occurrences have revealed that even small events may have a large impact on intensively exploited coasts (e.g., industry or tourism).

Table 1. Some prehistorical and historical cases of marine geohazard events.

Offshore Geohazard	Prehistorical and Historical Cases	Consequences
Earthquakes related to seismogenic faults	The 2011 Japan earthquake (North Pacific Ocean) of magnitude 9 (Mw) [26].	It caused an up to 30-m-high tsunami that flooded 110 km of coastline. Nearly 16,000 people were killed, and more than 400,000 buildings collapsed.
	The 2010 Chile earthquake (South Pacific Ocean) magnitude of 8.8 Mw [27].	It caused a tsunami with wave heights up to 30 m in the Chilean coastal region. It is the largest event along the South American Subduction Zone in half a century and produced 648 casualties.
	The 2004 Indian earthquake (Indian Ocean) of magnitude of 9.3 Mw [28].	It caused an up to 34 m-high-tsunami that produced an estimated 228 k casualties. This is one of the ten worst earthquakes in recorded history.
	The Al-Hoceima earthquake (SW Mediterranean) 1993–1994, 2004, and 2016 seismic crisis [29].	This event killed 464 people and caused 11.9 million Euros of economic losses in Spain.
	The 1908 Messina earthquake (NW Mediterranean) of magnitude 7.1 (Mw) with the epicenter in the Messina Strait graben [30].	It produced a local tsunami. It is the most destructive 20th and 21st century earthquake in Europe, with >80,000 deaths.
	The 1906 San Francisco earthquake (North Pacific Ocean) of magnitude 8.3 (Mw) with the epicenter located on the San Andreas Fault [31].	The economic impact was tremendous. The impact is assessed as US $524 million, and the earthquake left more than 3000 people dead and more than 28,000 buildings destroyed.
Slope instabilities	The 1979 Lomblen landslide (between the Indian Ocean and the Pacific Ocean) that generated a strong tsunami with heights of 7–9 m [32].	It caused 539 causalities and another 700-missing people.
	The 1979 Nice submarine landslide (NW Mediterranean) related to the construction of the new Nice harbor [33].	It generated a tsunami (wave heights to 3 m) and is probably one of the most important geological events to have occurred in France within the last 20 years. It caused casualties and considerable material damage [34].
	The 1929 Grand Banks slide (Northern Atlantic Ocean) was triggered by an earthquake (magnitude of 7.2 Ms) [35].	It generated a tsunami that killed 28 people and severed several submarine communication cables.
	The Storegga Slide (Norwegian Sea), approximately 8200 years ago [36,37] off the Norwegian coast.	It generated a tsunami that hit the west coast of Norway (run up 10–12 m), Scotland (4–6 m), Shetland (approx. 20–30 m), and the Faroes (0–10 m) [38].
Volcanism eruptions and slope instabilities on volcano flanks	The 1950 AD Santorini active volcanic eruptions (Aegean Sea) [39].	They produced debris flows on the flanks of Santorini Island that produced damage and causalities.
	On active Hawaiian volcanoes (Pacific Ocean), large, rapid flank movements often co-occur with large earthquakes. They were observed four times during the 19th and 20th centuries, each spaced approximately 50 years apart [40].	They affected the quality of life of local people living on the islands and impacted on the islands' economies.
	The 2011 Hierro submarine eruption [41].	It affected the quality of life of local people living on the island and impacted on the island's economy, which was based primarily on tourism.
	Active Azores volcanoes are affected by diffuse CO_2 emissions related to hydrothermal activity [42].	They may represent a public health risk, and occasionally family houses were evacuated when CO_2 concentrations in the air reached 8 mol%

Table 1. Cont.

Offshore Geohazard	Prehistorical and Historical Cases	Consequences
Fluid flow (gas, mud, and salt diapirs)	Events associated with active pockmarks (up to 15 m deep) on the seafloor of the off Patras and Aigion (northern Peleponnesos, Greece) [43].	These pockmarks were found to be venting gas prior to the earthquake (the M 5.4) on 14 July 1993.
	Catastrophic gas escape during the exploration drilling in the German Bight of the North Sea in 1963, the J. Storm II in 1972 [44] and in the North Sea in 1990.	Gas escape formed large, deep pockmarks over very short periods.
Erosion, scour, and seabed mobility by bottom currents	The Arklow Bank Wind Farm, the best wind resources in the Irish Sea was subjected to overall seabed movement [45].	Movement of the sandbank, channel migration, and overall erosion and accretion. Scouring was caused by the strong currents that flowed over the sandbank, often over 2 m/s.
	In the gravity-based foundations of the Frigg TP1 GBS, installed in fine sand soil at 104 m of water depth, in the North Sea [46,47].	2 m deep scour erosion at two corners.
	Several submarine pipeline failures in the Mississippi River delta and the Gulf of Mexico [48,49].	Seabed erosion by scouring around the pipe under the influence of currents caused the pipeline to be unsupported.

4. Offshore Geohazards and Their Main Key Questions

4.1. Tectonic Earthquakes: Seismogenic Faults

Internal geodynamic processes constitute the main engine that determines the present-day Earth's configuration. Most of the plate boundaries are located in offshore areas where seismically active faults constitute a principal marine threat [50,51]. In addition, some of them continue onshore, providing a good chance for direct observations (e.g., San Andreas Fault, [52]; the Alpine Fault, [53]). Progressive and continuous plate motion is accommodated by deformation along the plate boundaries where most of the active faults are located [54,55]. Some faults may undergo creep and are aseismic [56]. However, in sectors where two large fault blocks become coupled by asperities on the fault surface, elastic deformation, and stresses may increase, reaching the strength of the rocks [57]. The above factors drive sudden slip, producing an earthquake and, as a result, seafloor shaking, liquefaction, and permanent deformation [58].

Thrust faults related to subduction zones, including those associated with the Ring of Fire surrounding the Pacific Ocean and in the Indian Ocean, produce the most intense earthquakes [59–61] (Figure 3a). In these regions, although deep seismicity occurs, the most devastating earthquakes are located at shallow depths close to the coastlines. Examples include the Alaska (1964, Mw 9.2), Sumatra (2004, Mw 9.0–9.3.1), Chile (20101960, Mw 8.89.5), and Japan (2011, Mw 9) earthquakes and the related tsunamis [62] (https://earthquake.usgs.gov/, accessed on 11 September 2020). In addition, transcurrent faults can accumulate high stresses, eventually resulting in major strike-slip earthquakes, such as in Cape Mendocino in California (1992, Mw, 7.2) [63]. However, such pure strike-slip events generally do not produce vertical displacements of the seafloor in flat areas, although they may be significant when steep slopes are affected [64]. Regardless, seafloor deformation occurs in transpressional (e.g., Shackleton fracture zone, Figure 3b, [65,66]) and transtensional faults (e.g., Incrisis-Al Idrissi faults, Figure 3c, [29]). A particular setting occurs at the tips of such faults, where vertical displacement may trigger tsunamis [67]. Seismogenic normal faults are relatively scarce in marine environments and are most likely related to the isostatic response of the Earth's crust to ice loads close to coastal areas [68].

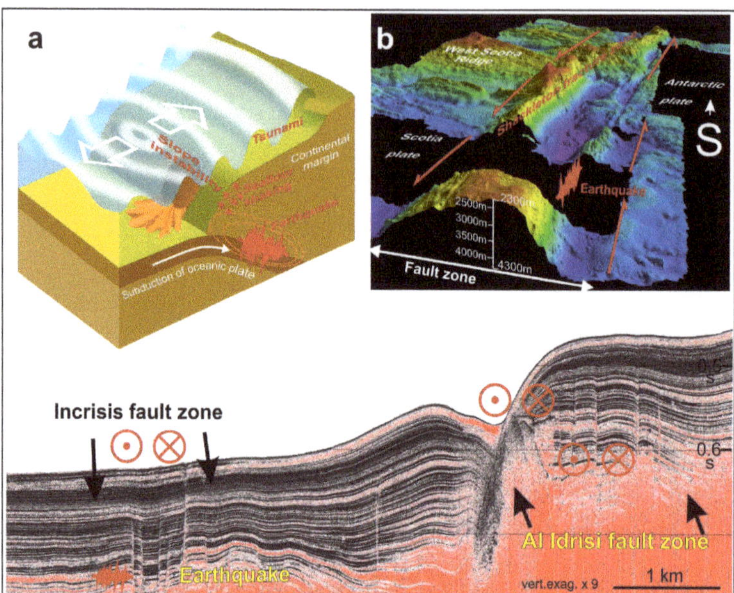

Figure 3. Examples of active seismogenic faults: (**a**) Subduction earthquakes in convergent plate margins have the highest magnitudes and produce seafloor shaking, submarine landslides, and tsunamis; (**b**) oblique-view perspective of the Shackleton fracture zone (Antarctica) that constitutes the transpressive sinistral active margin between the Scotia and Antarctic plates. Bathymetric features are determined by the permanent deformation of the seafloor; more details on this fracture zone can be found in [65,66]; (**c**) high-resolution seismic profile of the Incrisis and Al Idrisi sinistral fault zones in the central Alboran Sea, western Mediterranean. The tectonic activity of the Al Idrisi fault zone, now mostly covered by recent sediments, has been transferred to the new Incrisis fault zone, where the active faults affect the most recent sediments; more details on these fault zones in [29].

Detailed knowledge of fault features and their seismic characteristics is essential to preventing the effects of these geological hazards. The main target is to constrain the fault geometry, kinematics, dynamics, and seismic behavior to determine the related maximum magnitude of the seismic events and their recurrence interval. The numerical modelling of seafloor deformation and the propagation of seismic waves constitute one of the main tools to establish the direct impact of seismic waves on coastal populations and to determine the potential to produce submarine landslides and tsunamis. The determination of these key fault features and seismicity is developed either by coring, which allows the study of deposits related to the main events [69], or by geophysical methods. Seismological observations in areas surrounding active seismic zones are generally far from land, and the coverage of active seismic zones is limited, decreasing the accuracy of the locations of active seismogenic faults. Ocean bottom seismometers (OBSs) can record seismicity over long periods, but there are large delays in data recovery. Seismic reflection and acoustic techniques highlight the geometry of faults and improve the analysis of their activity. However, present-day standard techniques fail to produce a complete and detailed image of faults.

To improve the accurate analysis of the location and geometry of active faults, seafloor mapping of large parts of the oceans with new multibeam and sonar equipment is mandatory. The study of the continuity of faults at depth will require new seismic acquisition and data processing techniques, including 3D seismic methods for complex areas reaching depths of up to 12 km for crustal faults. The accuracy of seismological observations will increase with the installation of seafloor seismological observatories in wired networks to provide real-time data. In addition, a denser network of OBSs will be necessary to achieve good coverage in active regions [50]. The characterization of fault behavior over long periods is also highlighted as new insight. In this sense, the improvement of submarine geomorphological indexes will likewise contribute to better identification and analysis of the active faults. Shallow coring techniques and detailed high-resolution seismic parametric profiles will improve the analysis of the geometry and age of the related deposits. These observations will help to determine the fault slip and the recurrence periods of the main submarine palaeoearthquake records. Deep coring techniques may enhance the significance of fluids in fault activity. In addition, the study of marine faults with offshore to onshore continuity will be essential for collecting direct observations of fault zones (fault surfaces, fault striations, fault gouges, and related deposits) and for determining their kinematics. Geodetic networks (global positioning system (GPS) and high-precision levelling (HPL)) and satellite-based Earth observations will also be mandatory to measure their present-day activity and constrain isostatic rebound in glacial margins close to the coastlines.

The integration of these new data will improve the probabilistic seismic hazard models in offshore tectonically active areas and the evaluation of their importance as secondary triggering factors of other hazards, such as slope sedimentary instabilities and tsunamis.

4.2. Submarine Slope Instabilities

Sedimentary instabilities are common processes in all submarine environments, where the largest slope instabilities in the Earth occur [70] (Figures 1 and 4). They may be classified according to different approaches, such as mechanical behavior, particle support mechanisms, sediment concentrations, and longitudinal changes in their deposits, or according to the relationship between source areas, dimensions, and geometries of deposits [71–76]. Based on the mechanical properties and rheology of the processes, two main groups can be defined: (i) slides/slumps/spreads and (ii) gravity flows. These two groups, with important differences in their pre- and post-failure behavior, occur in all physiographic environments and are efficient transporters of sediment, organic carbon, nutrients, and pollutants [77–81]. They are scale-invariant processes that range greatly in size from the meter scale to many km across (Figure 4).

Slides/slumps are movements of sediment or rock along a surface of rupture that develops in a layer with low shear strength or a weak layer [82]. They are elastoplastic movements that include translational and rotational movements (Figure 4a–c). Spreads, the submarine characterization of which has increased during the last decade, are sediment or blocks of consolidated sediment moving over liquefied underlying material and not a basal shear plane [83]. Depending on the mechanical behavior or the energy available, all these mentioned movements of sediments may evolve into a sediment flow, but a flow may also develop directly if the sediment is completely remolded.

A wide range of flow types can occur because of the interplay of rheology, grain size composition, and concentration (Figure 4). Flows in general have viscoplastic behavior and can be divided into cohesive (e.g., mud flows and debris flows) (Figure 4d) and non-cohesive flows (grain flows), depending on the amount of fine-grained matrix [84]. One type of cohesionless flow involving large volumes of failing masses is debris/rock avalanches. Usually, such failures originate from deep rotational failures on high-gradient slopes and in volcanic environments [85]. Turbidity currents are a type of Newtonian flow in which fluid turbulence is key to supporting the sediment and keeping it in suspension. Turbidity currents can transport up to hundreds of cubic kilometers of sediment at high velocities (up to 19 m/s) over thousands of kilometers [33,86] (Figure 4c).

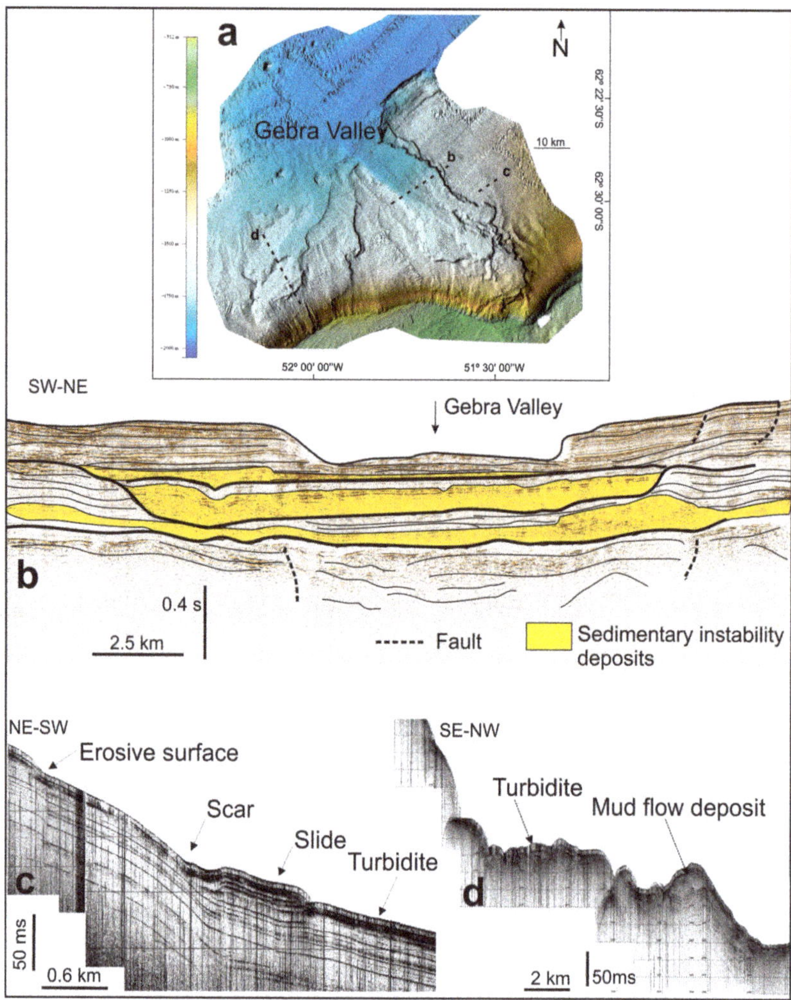

Figure 4. Examples of submarine slope instabilities: (**a**) multibeam bathymetry displayed in the Gebra Valley area, Bransfield Basin (Antarctic); more details regarding this valley can be found in [87]. The dashed lines show the cross-sections of seismic records in b, c, and d; (**b**) Airgun seismic record displaying the repeated large-scale slope failure events that were responsible for the cut-and-fill features forming the Gebra Valley; (**c,d**) parametric seismic records showing the different slope instabilities affecting the external margins of the Gebra Valley; modified from [88].

Field monitoring of turbidity currents has increased in recent years. However, these processes and failures are still primarily recorded in nature, and most of the knowledge acquired is through the interpretation of the resulting geomorphology and the study of the final deposits. This situation leaves unanswered key questions and uncertainties about all the mechanisms involved in sedimentary instabilities, which need further study. These questions will have to be addressed at different scales through repeated very high-resolution bathymetric surveys, high-resolution (and 3D) seismic surveys, new direct deep-sea monitoring and mobility sensors, in situ geotechnical tests, and experimental and numerical models. Better field observations and models will help to achieve an improved understanding of numerous aspects, such as the rates of seafloor changes, the role of preconditioning factors, the impact of triggers, how rapidly a slope failure can develop, and the volumes of sediment involved and reworked. Improved knowledge of the governing mechanisms, evolution, and transformation of these submarine sedimentary instabilities will be crucial to understanding the hazard they represent [89–94]. Their modelling will also help to assess their consequences.

When landslide hazard assessment is considered, concepts such as distribution, time, and magnitude must be considered [95–97]. Regional inventories and magnitude-frequency relationships, including events triggered over a long period of time or almost instantaneously, could provide critical information [98]. Nevertheless, accurate chronological constraints (ideally combining biostratigraphic and radiometric techniques) will be essential for hazard evaluation.

4.3. Submarine Volcanism

Volcanoes are vents in the Earth's crust through which molten rock, hot rock fragments, and gases can erupt. Magma can rise along conduits to the surface, forming lava that either continuously flows out or shoots upward. Furthermore, the lava can break into pieces that are thrown into the air or into the sea due to decompression of the gases it contains (https://www.britannica.com/science/volcano, accessed on 3 September 2020).

From a marine point of view, volcanic eruptions affecting the sea water column can be grouped into four basic types (Figure 5):

(i)   Subaerial eruptions close to the coastline affect the marine environment in different ways, as they can produce changes in the coastal configuration when lava flows pour into the sea forming a lava delta (e.g., [99]) (Figure 5b), collapse the volcanic edifice, or enter the sea of pyroclastic flows (e.g., [100]) (Figure 5a). Moreover, volcanic eruptions and dike intrusions can even cause slope sedimentary instabilities that enter the sea and trigger tsunami waves (e.g., [101] and references therein).

(ii)   Shallow-water eruptions (<200 m water depth, mwd) are commonly characterized by violent explosions, especially when they approach the water-air interface (Figure 5c), as observed for the first time at Surtsey in 1963 [102].

(iii)   Intermediate-water eruptions (approximately 300–600 mwd) are rarely observed, but they can be characterized by a peculiar eruptive style characterized by floating lava balloons or pumice emissions (Figure 5d). During these eruptions, lava globes can be expelled in a successive way that occurred in the recent submarine eruptions of Serreta (Terceira, Azores; [103] or Tagoro (Canary Islands) (Figure 5e).

(iv)   Deepwater eruptions (>600 mwd) are mostly effusive, and the associated lavas represent the most widespread surficial igneous rocks on Earth. Related studies have focused on basaltic lavas emplaced in mid-oceanic ridges, back-arc basins, intraplate seamounts, ocean volcanic islands, and plateaus. Three main types of submarine lavas can be distinguished according to their morphology and flow rates: pillow, lobate, and sheet [104,105]. For basaltic lavas, another important deposit is hyaloclastite occurring in both shallow and deep waters. The 2012 Havre eruption exhibited explosive activity in a deep-water sector (between 900 and 1100 mwd), producing a pumice raft approximately 400 km^2 in size and an abundance of fine ash on the seafloor over the course of one day [106].

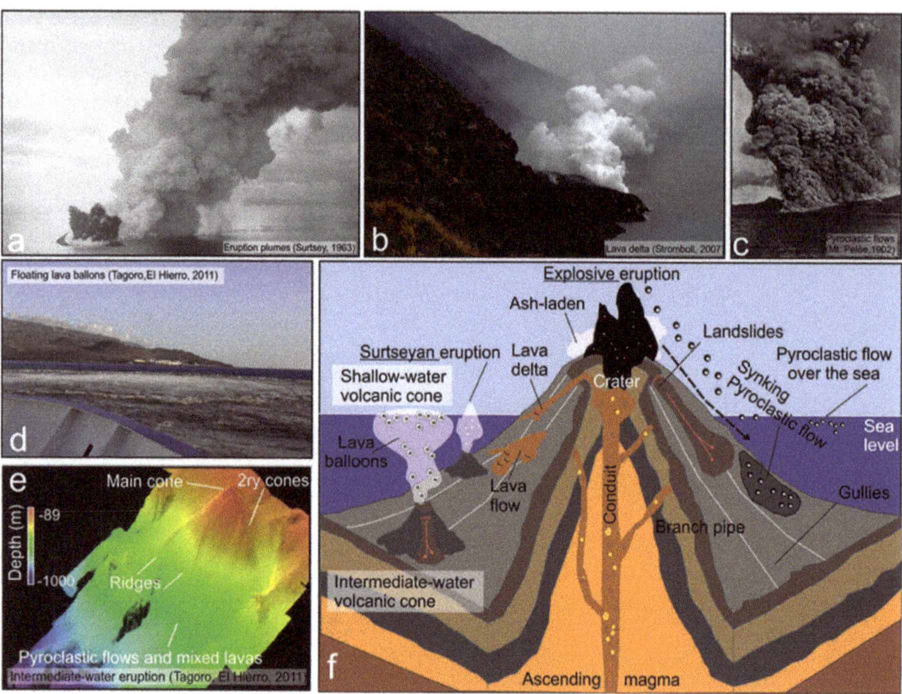

Figure 5. Main types of eruptions examples: (**a**) Historical image of surtseyan eruption plumes from Surtsey (Iceland) in 1963 (image from National Oceanic and Atmospheric Administration, from https://commons.wikimedia.org/wiki/File: Surtsey_eruption_2.jpg, accessed on 03 June 2021); (**b**) emplacement of the 2007 lava delta at Stromboli (image from Rolf Cosar, downloaded from https://commons.wikimedia.org/wiki/File:Stromboli-Lavadelta-2007-03-10.JPG, accessed on 03 June 2021); more details on the 2007 submarine eruption at Stromboli can be found in [107]; (**c**) pyroclastic flow occurred on 16 December 1902, during a volcanic eruption of Mt. Pelée in Martinique (image from A. Lacroix, https://commons.wikimedia.org/wiki/File:Photographie_du_nuage_noir_du_16_d%C3%A9cembre_1902_lors_ de_l%27%C3%A9ruption_de_la_Montagne_Pel%C3%A9e_en_Martinique.jpg, accessed on 3 June 2021); (**d**) floating volcaniclastic materials and gas emissions during the Tagoro volcanic eruption offshore El Hierro in 2011 (image courtesy of Eugenio Fraile, IEO); (**e**) 3D bathymetric map of the Tagoro volcano showing the main morphological characteristics; more details on this submarine eruption can be found in [108,109]; (**f**) 3D simplified sketch of the main types of eruptions affecting insular volcanoes.

Another important hazardous phenomenon associated with volcanic eruptions that can become a hazard is gas emissions. The most common volcanic gases are water vapor, carbon dioxide, carbon monoxide, sulfur dioxide, and hydrogen sulfide; to a lesser extent, methane, hydrogen, helium, nitrogen, hydrogen chloride, or hydrogen fluoride can also be emitted (https://www.britannica.com/science/volcano, accessed on 1 October 2020). These gas emissions can cause loss of life due to suffocation if they reach the surface and can kill fauna present in the environment surrounding the underwater eruption. Such eruptions can also induce strong acidification of seawater, resulting in the subsequent loss of habitats around the eruption. For example, in the eruption of the Tagoro volcano, the pH dropped to 5, and the partial pressure of dissolved carbon dioxide increased almost 1000 times [108,110].

Despite the large number and volume of submarine volcanic eruptions, our understanding of such processes and associated landforms is still limited, especially compared with their subaerial counterparts ([111–113] and reference therein). Many concepts are still based on the interpretation of ancient deposits and on theory (e.g., [114] and references therein). However, the growing availability of detailed digital elevation models (also used to depict seafloor changes associated with eruptions through repeated surveys) integrated with hydroacoustic monitoring and in situ observations of volcanic settings will exponentially increase their detection (considering that volcanic eruptions only occasionally reach the sea surface) and our knowledge of submarine eruptive processes (e.g., [115–120]).

A challenge-based study will provide knowledge to understand the processes that take place in the evolution of a submarine volcano at different depths. The main effort should focus on monitoring these processes using a variety of instrumentation (including on-land seismometers and marine stations with OBSs, hydrophones, pressure sensors, CTD (conductivity, temperature, and depth) instruments, and geochemical parameter sensors to control emissions) to allow study of the eruptive pulses and the content of emissions. Some of this instrumentation will be able to be connected by optical cables to laboratories onshore for online monitoring (e.g., [121]), and profiles can be made with a towed oceanographic rosette (tow-yo). Another challenge will be understanding the changes in seafloor morphology through time-lapse high-resolution bathymetry surveys, taking advantage of the use of autonomous underwater vehicles (AUVs) for deep volcanoes.

4.4. Fluid Flow Processes

Seepage is a global process that occurs in different geodynamic contexts in both active and passive continental margins. Generally, this process includes the leakage of hydrocarbons (particularly methane as both dissolved and free gas), water, and/or sediment [21,122] (Figure 6). The gas in shallow marine sediments [123,124] is mainly composed of methane, and its origin is attributed to either biogenic or thermogenic processes. The escape of fluid from the sediment may occur as micro-seeps or as sudden violent escapes (cold seeps), producing diverse types of morphologies on the sea floor (Figure 6a) or in the subsurface [21]. Some features have positive relief (e.g., mounds, methane-derived authigenic carbonate, gas hydrates, mud volcanoes), and others have negative topographies on the seafloor (e.g., pockmarks, collapses) (Figure 6). Gas can migrate through unlithified sediments along bedding planes, faults, and fractures (Figure 6b) driven by buoyancy forces and pressure gradients [125,126]. Glacial-isostatic and tectonic events may reactivate fractures and faults, producing temporal variability in spatially heterogeneous fluid flow [127].

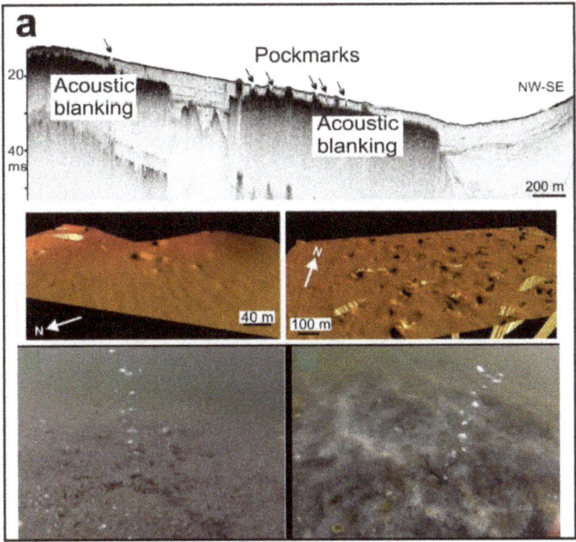

Figure 6. *Cont.*

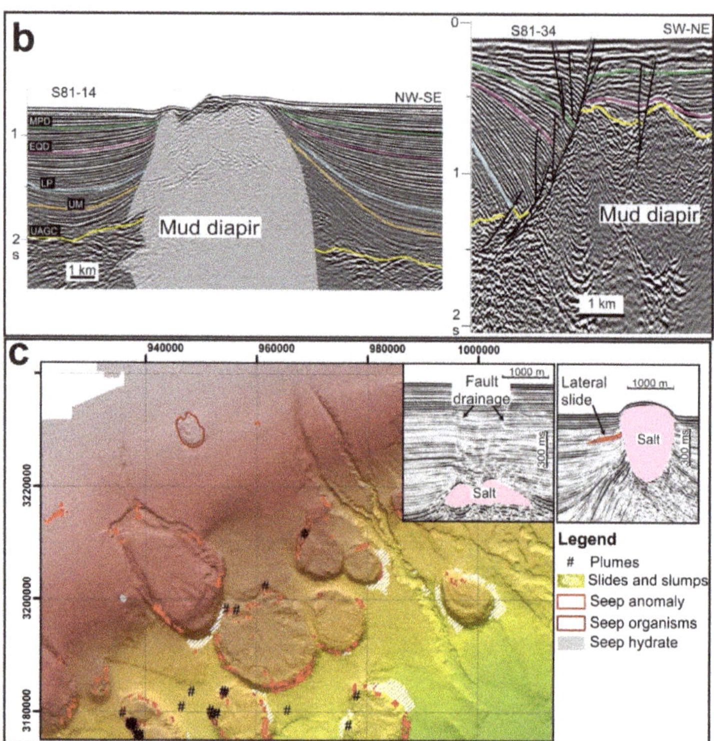

Figure 6. Examples of hazardous features related to fluid flow processes: (**a**) At the top: very high-resolution seismic profile (3.5 kHz) in the Ria de Vigo showing an acoustically blank zone related to the presence of shallow gas accumulations near the seabed (less than 1 meter below the seabed). It also shows the presence of pockmarks and the main paths of gas escape. In the middle: multibeam echosounder images from Ría de Vigo displaying depressions related to gas escapes (pockmarks) and mounds formed by the accumulation of debris (mud and bivalve shells) from mussel rafts. At the bottom: methane-gas bubbles escaping from the sandy and muddy seabed of the Ría de Vigo; (**b**) mud diapirs related to the compressive regime (left figure) and to listric faults (right figure); (**c**) high-resolution bathymetric map of the northeastern Gulf of Mexico showing salt diapirs (rounded positive structures) piercing the seafloor and their associated seep anomalies associated with oil and gas seepages, slides and slumps, and gas hydrates to a lesser extent (images from https://www.boem.gov/oil-gas-energy/mapping-and-data/map-gallery/boem-northern-gulf-mexico-deepwater-bathymetry-grid-3d; accessed on 15 September 2020; https://www.boem.gov/oil-gas-energy/mapping-and-data/map-gallery/seismic-water-bottom-anomalies-map-gallery, accessed on 16 September 2020; https://metadata.boem.gov/geospatial/WBA_Metadata.xml, accessed on 16 September 2020); more details on the northeastern Gulf of Mexico salt diapirs in [128,129]). Enclosed seismic records show: a detailed view of salt-related normal faults acting as drainage pathways for deep fluids and lateral slides associated with overburden salt diapirs in a 2D seismic line located in the Gulf of Lion, western Mediterranean Sea.

The mud fluidization and degassing processes associated with overpressure contribute to the formation of mud volcanoes. This fluidization is mainly due to the overpressure generated by tectonic stresses or by lithostatic pressure in regions with high sedimentation rates [130]. These types of structures are one of the most important methane sources in the hydrosphere and atmosphere [131–133]. Diapirs are gravitational/tectonic structures

(Figure 6b) produced mainly by salt, clays, or a mix of these lithologies that, currently, have little consideration in submarine hazard models. These intrusive bodies of relatively low density tend to migrate upward, deforming and piercing the overlying sedimentary sequences. They can appear with other fluid migration structures, such as mud volcanoes or pockmarks. In an extensive context, the development of diapirs commonly occurs close to deep listric faults (Figure 6b) that act as escape migration routes. The formation of mud diapirs is more frequent in compressive settings, as highly over-pressurized diapiric material (e.g., [134]) can move upward from subsurface depths up to 3–4 km to the seafloor. In various tectonic regimes, diapir activity may also trigger slope instabilities (Figure 6c) because the deformation and elevation of these structures favor seafloor oversteepening and seismicity [135].

Gas hydrate is an ice-like crystalline solid form of water and low molecular weight gas (e.g., methane, ethane, and carbon dioxide) [136]. Methane hydrates can form at any depth where the geothermal conditions are colder than the hydrate stability curve. This occurs in the upper hundred meters of marine sediment at water depths greater than 500 m. Nevertheless, certain conditions, such as the presence of saline pore waters or clays, can inhibit gas hydrate formation, while a high fluid flux can promote gas hydrate formation [137,138]. Seismic reflection techniques are used to determine the areal extent of gas hydrates in marine locations mainly by the identification of bottom-simulating reflectors (BSRs) (e.g., [139]). However, gas hydrates have been recovered from sites without BSRs. This is because the saturation of methane hydrate in the pore space must exceed approximately 40% for the seismic velocity to be altered significantly enough to generate a BSR [140]. Gas hydrates may be a significant hazard because they alter sea floor sediment stability and can lead to collapses and landslides that may trigger tsunamis [141–145], and their breakdown and release to the water column and atmosphere may have a strong influence on the environment and climate [146].

The presence of all these fluid flow features usually denotes subsurface hydraulic activity, over-pressurization, fluidization, and degassing processes, as well as sudden fluid (gas and/or liquid) release that may produce gas explosions, slope sedimentary instabilities, and an uplifting/subsiding seafloor [122,147–149] (Figure 6c). These processes can have major impacts on seabed infrastructures and on those requiring piles that are driven into the seafloor. Therefore, it will be necessary to extend systematic investigations to identify the locations of fluid dynamic processes in areas where their activity remains unknown currently. Thus, heat flow studies will need to be increased in order to detect and map new subseafloor marine fluid flows and understand their regimes. High-resolution 3D seismic surveys will also allow an accurate acoustic characterization and distribution assessment of the different fluid dynamic features and definition of their origins. They also have the potential to document and characterize in more detail the different types and timing of deformation patterns in areas close to diapirs and related mud volcanoes, with the goal of accurately determining the timing of fluid flow processes. Additionally, studies on microseismicity would allow the detection of fluid injection. Monitoring of fluid flows should also increase in active cold seeps; systematic sediment and gas sampling for biogeochemical analysis will aid the understanding of the general physical and geochemical characteristics of the escaping gas. Likewise, improved numerical models of gas hydrate formation, stability, quantification, and role in the shear strength of the host sediment will lead to progress in understanding the impact of gas hydrates on safety and seafloor stability.

4.5. Bottom Currents

The term "bottom currents" refers to all persistent currents flowing near the seabed that resuspend, transport, and/or control sediment deposition [150]. Although bottom currents are semipermanent features, they are characterized by high variability over a range of time scales (from daily to geological timescales; [151]). They may occur in shelf, slope (Figure 7) and deep basin settings. In shelf settings, wind and tidal forcings are common and produce different hydraulic regimes (e.g., tide- and current-dominated regimes). In deep-water settings, bottom currents are mostly related to thermohaline circulation. These currents are driven by density gradients (e.g., the North Atlantic Deep Water) and typically flow subparallel to the bathymetric contours with velocities of 1–20 cm/s [152]. In particular settings (e.g., narrow gateways), bottom currents strongly intensify, reaching velocities of 50–300 cm/s (e.g., the Mediterranean outflow water that spills over the Gibraltar sill, e.g., [153,154]). In deep-water settings, submarine canyons can be swept by focused tidal currents with up and down flows and velocities of 25–50 cm/s [142]. Moreover, an increasing number of studies (e.g., [142,155–158]) have highlighted that several intermittent oceanographic processes can affect the seabed: eddies, gyres, helical flows, benthic storms, cascading dense shelf water, internal waves, and currents related to extreme, cyclonic, and tsunami waves.

As most of these processes can produce complex flow conditions, an increase in the velocity of bottom currents and additional shear stress may produce considerable sediment resuspension and seabed erosion [155] (Figure 7f–g). All these observations highlight that a variety of oceanographic processes are able to exert a significant impact on shaping the seafloor when bottom currents are active for a prolonged period of time (Figure 7a). At small spatial scales, they generate various erosional and depositional bedforms, ranging in size from centimeters to kilometers (see, e.g., [151,159]) (Figure 7c,f–h) and whose identification is particularly relevant for geohazard assessments of seabed infrastructure. At a larger scale, bottom currents with persistent activity on a geological time scale (e.g., thermohaline-induced currents) may form regionally extensive contourite depositional systems [158,160–163], including a variety of depositional elements (contourite drifts) (Figure 7a) and erosional elements (contourite channels, furrows, moats, and erosive terraces).

Evaluating the action of bottom currents is crucial for hazard assessment because intense seabed erosion may locally favor slope instability [164] (Figures 7a–e,i). Moreover, migrating bedforms (e.g., sand waves) and erosion (Figure 7f–h) can have major impacts on seabed infrastructure [7]. This can be extremely relevant in narrow straits swept by powerful tidal currents and by internal waves (e.g., the Messina Strait) that can create a dangerous setting for submarine cables and pipelines [165]. Other crucial areas are canyons swept by strong tidal bottom currents, topographic highs (e.g., seamounts and ridges) where bottom currents interact with topography [163,166–168], areas affected by tidal forcing and associated internal waves [169], areas of local upwelling [170,171], seasonal fluctuations in the main circulation pattern [172], or areas of sinking dense water [173], which may trigger slope sedimentary instabilities (e.g., [174,175]). Finally, it is noteworthy that contourite deposits can be prone to becoming unstable (e.g., [176]), as several predisposing factors (e.g., mounded morphology on steep slopes and the low shear strength related to high sedimentation rates) may favor slope failures [22] (Figure 7i).

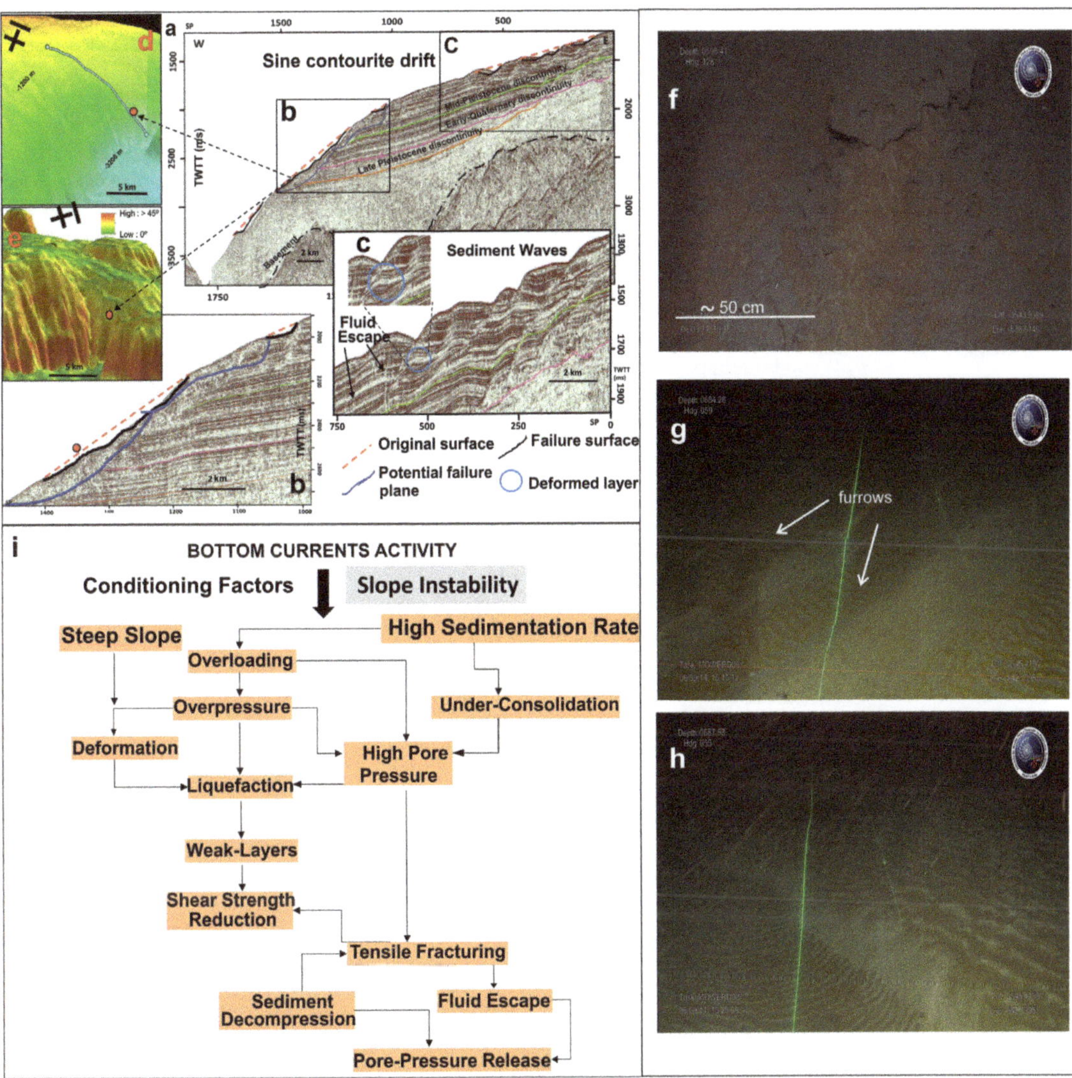

Figure 7. Examples of hazardous features related to bottom current action of the Mediterranean outflow water around Iberia: (**a**) Sines contourite drift (Portugal margin, NE Atlantic) is affected by slide scars on the seafloor and failure surfaces in the subsurface. Additionally, upslope-migrating sediment waves occur in the near-surface sediments; (**b**) details of slide scars and potential failure surfaces; (**c**) details of sediment waves; (**d**) details of the bathymetric map with the track of the seismic profile shown in (**a**). The red dot marks the headscarp of a landslide scar; (**e**) detail of the slope gradient map; (**f–h**) ROV images showing bedforms in the continental slope of the Gulf of Cadiz (NE Atlantic): (**f**) Indurated muddy outcrop (grey in color) indicates intense current flow erosion; the mudstone is covered partially by sand with ripples. The two laser lines are separated 50 cm; (**g**) erosive furrows excavated on the muddy seafloor, covered partially by sandy sediment with starved rectilinear to sinuous asymmetric ripples. This starvation allows for the exposure of the mudstone surface over which the ripples are moving; (**h**) sinuous sand wave with superimposed linguoid to sinuous asymmetrical ripples on the stoss side and rectilinear to sinuous ripples in the trough area. The two laser lines are separated 50 cm; (**i**) diagram showing the intrinsic and hazardous properties of contourite deposits that may contribute to slope failures. Figure 7a–e,i was adapted with permission from [158]. Copyright 2019, Elsevier.

Recognition of the potential hazard of deep-water bottom currents is increasing because new and large seafloor areas of contourites have intensively eroded during the last 15 years due to mobile seafloors and slope sedimentary instabilities, which have been mapped (e.g., [150,155,158,160,162,177]). The assessment of the role of bottom current activity as a hazardous process is challenging for geoscientists due in part to the need to establish a dialogue with physicist oceanographers (e.g., [162]); as it is necessary to define flow conditions, induced bed shear stress and effects on morpho-sedimentary processes affecting the seabed, as well as their evolution over time. Thus, a multidisciplinary approach, including oceanographic, morphologic, sedimentologic, stratigraphic, and geotechnical studies, should be used. This will require multidisciplinary surveys and the integration of complex datasets, including oceanographic data (CTDs, acoustic Doppler current profilers (ADCPs), and transmissometers to measure the water properties and velocity not only at the near-bottom but also throughout the water column), multibeam bathymetry data, sub-bottom profile data, seafloor samples, sediment cores, and remotely operated vehicle (ROV) videos. Data integration enabled characterization of the different oceanographic processes, their interactions and timing, and their influences on the near-bottom flows acting on the seafloor. AUV surveys will also be required for high-resolution geophysical surveys in deep-water environments. Repeated bathymetric surveys and seafloor observatory systems will be required to define seabed evolution over time. Finally, hydrosedimentary modelling will be very helpful in assessing seabed changes and bed shear stress over a defined time period (e.g., the lifetime of the infrastructure).

4.6. Tsunamis

The main mechanisms for tsunami generation are earthquakes caused by seismogenic fault movement (96% of events), slope sedimentary instabilities, and volcanic eruptions (Table S1). In addition, with the release of large volumes of gas from seafloor sediment, atmospheric disturbances (meteotsunamis), or even cosmic impacts can also produce tsunamis [178] (Figure 8). Finally, anthropogenically induced submarine slope sedimentary instabilities have triggered local tsunamis [179]. To perform tsunami hazard assessment, three main components of the phenomena must be addressed: the generating mechanism, wave propagation in the open sea, and coastal inundation.

Seismotectonic tsunamis are triggered by the coseismic vertical displacement of the seafloor impacted by an earthquake and the transmission of this movement to the water column [180,181] (see Video S1 in the Supplementary Materials). Normal fault movement causes the water masses to sink toward the formed depression, generating an initial large sine wave at the surface of water mass (Figure 8a). In contrast, reverse faults move the seafloor upward, and the water column is pushed upward (Figure 8b), forming an initial large crest wave on the sea surface. Tsunami wave generation by seismotectonics is controlled by the rupture velocity (mostly slow velocities), fault type, slip and average vertical displacement, width, length, and segmentation of the rupture zone of the fault in the seafloor [181].

Submarine slope instabilities (such as slumps, slides, debris/rock avalanches, debris flows) generally involve large volumes of sediments and rocks, and the associated movement within the water body generates a dipole-like water wave that can eventually generate major tsunami waves [15,182–184]. The initial shape of the tsunami wave is defined by a depression–uplift pair in the water surface (Figure 8c). The water depression is due to the sudden sediment vacuum that occurs at the slide scar (Figure 8c, number 1), which becomes occupied by sea water, and the uplift (Figure 8c, number 2) is due to the pressure force exerted upwards by the fast-moving slide material. Several aspects of submarine slope instabilities as tsunami sources are currently being discussed, namely, the rheology (because it influences the deformation of the sliding material during their runout) [84,85], acceleration, and volume. These aspects are considered to be key factors controlling the geometry (depression and uplift) of the generated tsunami wave [185].

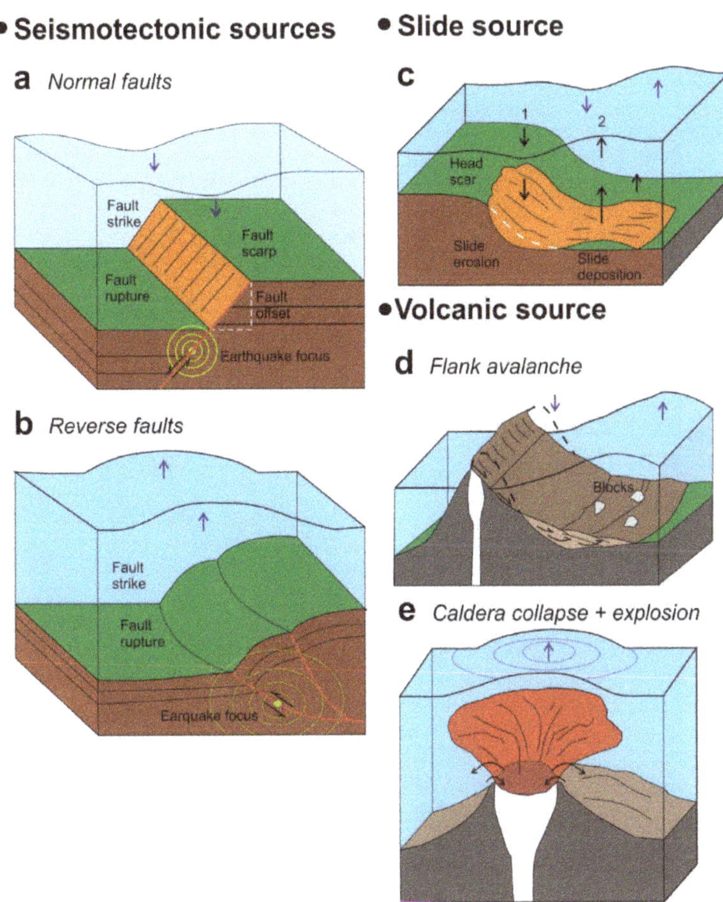

Figure 8. Sketch of the main tsunamigenic sources of geological origin: (**a**) Normal fault activity; (**b**) reverse fault activity; (**c**) submarine landslide. Numbers 1 and 2 refer to the time-sequenced development of the initial tsunami shape in the water surface; (**d**) collapse flanks of volcanoes; (**e**) volcanic explosion related to eruptions and caldera collapses.

Volcanic eruptions can also induce subaerial and submarine slides, slumps, debris/rock avalanches, or debris flows on the flanks of volcanoes that, in turn, can produce tsunami waves (Figure 8d) and they can also generate caldera explosions that can cause the complete collapse of the edifice [186–188]. These explosions produce waves (Figure 8e) that are generated first by the explosion itself and then by the sinking of the volcanically mobilized material.

Tsunami waves move in all directions from the source area, affecting the entire water column [189]. Their initial propagation, especially their direction, is conditioned by the geometric and deformation characteristics of the source structure. Tsunami waves are initially characterized by their large wavelengths (tens or hundreds of kilometers), small heights in the open sea (on the scale of centimeters), and large velocities. The wave velocity is greater in deeper waters (700 km/h at depths > 4000 m), and when depth decreases, the velocity also decreases to 30–50 km/h at the coast. Simultaneously, the wavelength decreases and the wave increases in height to balance the kinetic energy with potential energy (e.g., [190]).

Tsunamis can have significant impacts on coastal communities, depending on regional and local bathymetry and coastal geomorphology variability [191,192]. The occurrence of reefs, human infrastructure, the geometry of the coastline and beaches, and the presence of bays, estuaries or deltas at river mouths can influence the size, appearance, and impact of tsunamis when they arrive at the coast. Typically, a tsunami reaches the coast as a series of successive crests and valleys, sometimes separated by several or tens of minutes, and can reach the coast as a rapid flood, more rarely as a wall of water, or, sometimes, as an initial withdrawal of the sea (e.g., [178]). Thus, the destruction caused by a tsunami on the coast can be very different at relatively short distances. The long wavelength of tsunamis gives them more momentum such that they can flood areas hundreds of meters and even kilometers from the coast. The maximum height above sea level that a tsunami reaches on the coast is known as the runup and mostly ranges from 1 m to 30 m, with extreme heights > 500 m, as in the case of the tsunami that occurred in Lituya Bay [184], when the coast is very close to the source area or where the coastal geomorphology amplifies the tsunami effects (resonant effects) (e.g., [181]).

The challenges in tsunami hazard research should focus on several aspects. First, an accurate definition of tsunamigenic sources is needed because it will help reduce the uncertainty in the triggering mechanism and will be important for studying the events themselves (e.g., the frequency of reoccurrence, potential areas to be impacted). Other aspects should include establishing their recurrence intervals by identifying past events in sediment cores (palaeoearthquakes, palaeoslides, and palaeotsunamis), and applying in situ measurements plus long-term monitoring. For seismotectonic sources, faults have to be described in terms of their tectonic style, dynamics, present-day activity, fault zone geometry, fault offsets of sedimentary units, and fault surface. For slope sedimentary instability-generated tsunamis, knowledge of the seafloor geometry, slope failure processes, and their early post-failure evolution is fundamental to determining their triggering potential. To analyze these concerns, high- and ultrahigh-resolution bathymetric data and 3D seismic reflection profiles, in situ seismicity measurements and observations, long-term monitoring, and longer sediment cores will be fundamental.

Additional challenges that will also be important for studying tsunami events and their impacts include the development of increasingly realistic mathematical models of the tsunami generation process, propagation through the water masses, and the impact on the coastal zones and the establishment of tsunami early warning systems (TEWSs). This work will include increasing the model resolution, developing more efficient and faster than real-time (FTRT) codes, and using future exa-scale computational architectures. Probabilistic tsunami hazard analyses (PTHAs) will have to be conducted in different areas of the world at global, regional, and even local scales with the aim of understanding tsunami hazards and developing tsunami risk reduction activities. PTHA increases the knowledge of the potential tsunamigenic threats at different scales by estimating the probability of exceeding specific levels of tsunami metrics, such as the maximum inundation height or runup within a certain period of time, around determined locations. Furthermore, probabilistic tsunami forecasting (PTF) attempts to address the uncertainty in tsunami forecasts by formulating a probability density function (PDF). The use of PTF in the context of rapid hazard assessment and in TEWSs is also a major challenge.

5. Scenarios with Multiple Geological Hazards

Following the above arguments and characteristics, the understanding of the different geohazard factors also needs to recognize the distribution of the main hazardous features, how they can interact, and their potential to generate cascading events. The most common of this type of event comprises an earthquake that triggers a landslide, both of which can produce a tsunami. Additionally, bottom currents can scour an overstepped seafloor, thereby reducing the shear strength of the unaffected sediments upslope and leading to their failure, forming a landslide that may produce a tsunami. Furthermore, the breakdown of sub-bottom gas hydrates can increase the pore pressure of the sediment bearing the released gas, which may lead to tsunamigenic slope sedimentary instability. Despite the highly scattered distribution of these factors along continental margins, they commonly coincide in certain specific environments or geological contexts, which should be monitored by the scientific community. Diverse settings, such as fjords, active river prodeltas, canyon-fan systems, subduction areas, or even high-latitude open slopes, may be critical. Among them, three settings are highlighted for their multiple hazardous features.

5.1. Tectonic Indentation Areas

Tectonic indentation areas in marine settings are significant because the tectonic structures developed in a framework of continental collision. The consequences of this tectonic activity are the presence of source areas that can produce multiple geohazards, such as earthquakes, sedimentary instabilities and, to a lesser extent, tsunamis. The continental indentation structures in marine areas occur in convergent continental margins related to plate corners (Taiwan, [193]) and accretionary wedges (Manila Trench, [194]) and in areas of early continental collision, such as the westernmost Mediterranean (Alboran Sea, [195]; Aguilas Arc in the Gulf of Vera, [196]). Particularly, the central Alboran Sea and the Aguilas Arc/Gulf of Vera (Figure 9) are key areas for understanding the link between indentation and geological hazards in a land–marine transition context; this is because although both exhibit similar hazardous features (seismic faults and slope instability deposits), their degree of development is different. Continental indentation influences the tectonics of the adjacent oceanic areas, as occurs during the northward indentation of the Arabian plate in Eurasia, which determines the westward motion of the Anatolian Block related to the development of the Aegean Sea [197]. The indenter blocks are generally bounded by lateral seismogenic strike-slip faults that permit displacement during the process of collision [198].

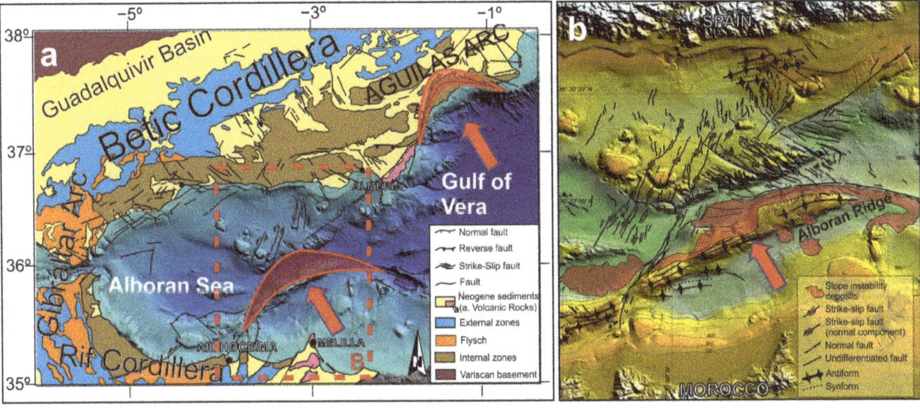

Figure 9. Example of multiple geological hazards in a tectonic indentation area in the western Mediterranean: (**a**) Geologic map showing the main tectonic features of the westernmost Mediterranean and the tectonic indentation zones in the central Alboran and Aguilas Arc in the Gulf of Vera; (**b**) detailed tectonic features of the Central Alboran Basin (for more details see [195]) and areas of generalized slope instability deposits (red areas). Legend: the red arrow indicates the direction of tectonic indentation and the red shading indicates the location of tectonic indentation.

Indentation structures simultaneously develop fault sets, folds, and block tilting, which can generate submarine slope sedimentary instabilities. These structures require integrated analysis by 3D analogue modelling, which has improved over time from early models [198] to recent models [199]. Future research will need to determine the stage of development of tectonic indentations. Additionally, the future development of new generations of numerical modelling is required. Another key question to address in the study of marine indentation zones is the tsunamigenic potential of strike-slip-related faults. In general, this type of fault is not considered tsunamigenic because it does not significantly displace the seafloor. However, new data in the central Alboran Sea contradict this theory and indicate the need to investigate other strike-slip faults in similar geological frameworks [195].

5.2. Canyon Heads Close to Coast

Submarine canyons, especially their shallower parts, are commonly very active geomorphological features that should be highlighted because of their association with multiple hazards (e.g., [200–202]). In general, canyons are located on the edge between a continental shelf and the continental slope, but some excavate the shelf to the point that their heads are only a few hundred meters from coastal towns, for example, the Garrucha (Figure 10a,e) and Gioia canyons in the Mediterranean [97–198,203,204] and the Congo and Capbreton canyons in the eastern Atlantic [19,205]. In these scenarios, changes can occur due to interactions among coastal processes (deposition and erosion) (e.g., [205]), river discharge (e.g., [206]), oceanographic processes (e.g., [207]), and seismicity related to tectonic processes (e.g., [208]). These activities can also produce favorable conditions for sedimentary instabilities (Figure 10a–d), which may produce tsunamis. Likewise, canyon heads close to the coast strongly influence tsunami propagation and runup (e.g., [209]).

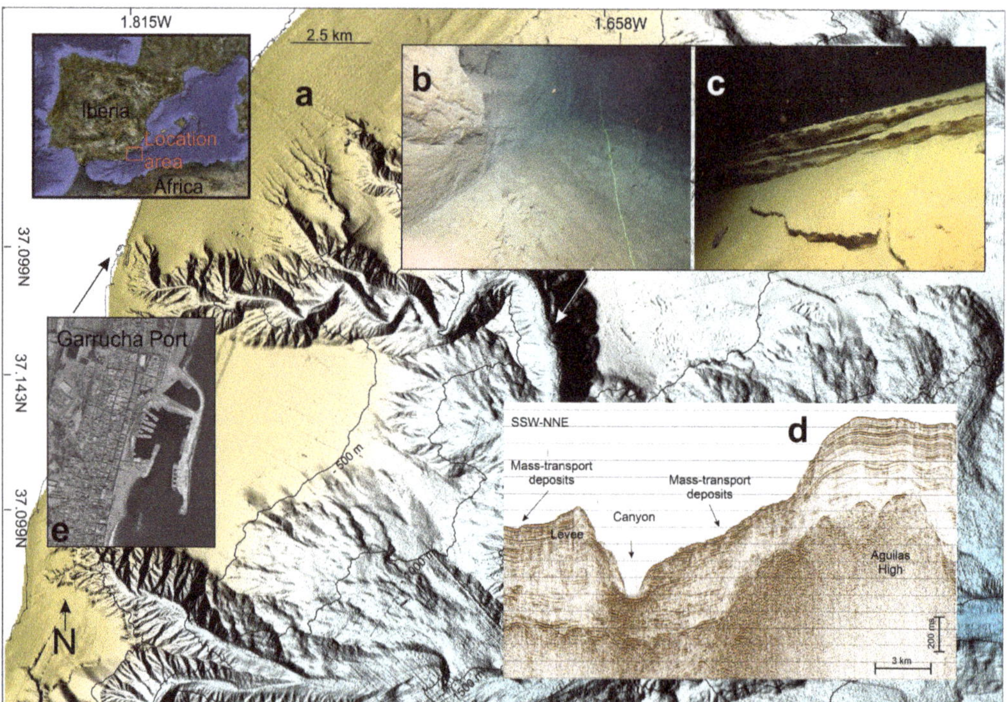

Figure 10. Example of multiple geological hazards in the Garrucha Canyon, SW Mediterranean: (**a**) Bathymetric map displaying a canyon head affected by intense gullying of its two main tributaries; (**b**,**c**) ROV (remotely operated vehicle) images displaying slope failures affecting the canyon walls; (**d**) seismic profile illustrating the occurrence of mass-transport processes that contribute to the erosion of the canyon walls; (**e**) photograph of the Garrucha port, which is located at the canyon head, where the sedimentary instability processes that contribute to canyon-head retrogradation affect its pier.

Therefore, key scientific issues to be addressed include obtaining a better understanding of each submarine and coastal geological process and the oceanographic and climatic processes that govern the retrogradation, incision, and enlargement of canyon heads (e.g., [210]). To achieve this goal, detailed 2D, 3D, and 4D geomorphological visualization will be needed. This work should be carried out not only at the head of the canyons but also in the adjacent areas of the continental shelf, open continental slope, and infralittoral zone (Figure 10a). This visualization will allow us to typify and map with high precision the different morphological elements (both erosive and depositional) of the integrated canyon-head-margin system [101]. Within the canyon heads, the morphometric, chronostratigraphic, sedimentological, and geotechnical characterizations of submarine slope instabilities will be crucial; the integrated results will allow us to estimate the recurrence of events and to assess and model the potential canyon-head stability [204]. New insights will be fundamental to establishing the spatiotemporal relationship between slope sedimentary instabilities, tectonics, and oceanography and to defining areas that may be prone to failure (e.g., [211]).

5.3. Volcanic Islands

Volcanic islands and their submarine portions merge multiple geohazards, mainly associated with volcanic eruptions, flank collapses, slope instabilities (Figures 5a–e and 11), and associated tsunamis, although strong volcano-tectonic subsidence [212], retrogressive erosion at canyon heads [213,214], and earthquake swarms (e.g., [215]) deserve special attention for hazard assessment. Flank collapses and slope instabilities ([216] and reference therein) (Figures 5g and 11) represent a common hazardous process during the evolution of many insular volcanoes, which are often able to mobilize volumes up to thousands of cubic kilometers (e.g., [186,217–219]). For instance, the 100–400×10^6 m^3 flank collapse affecting Anak Krakatau in 2018 generated a tsunami with a runup of up to 13 m along the Sunda Strait. Despite the high tsunamigenic potential associated with these large-scale events, their hazard is relatively low because they have recurrence times on the order of thousands of years. In contrast, small- and medium-sized slope instabilities affecting active volcanic flanks are more hazardous because they have markedly shorter recurrence times and are able to generate local but devastating tsunamis [220,221]. One of the best examples is recognizable at Stromboli Island, where five tsunamigenic landslides over just the last century have been reconstructed [222]. In the case of highly explosive eruptions, the entrance of pyroclastic currents into the sea can also generate tsunami waves or travel (their upper and dilute parts) over the sea for distances of tens of kilometers before impacting surrounding coastal communities, as described during the 1883 Krakatoa eruption [223].

Considering the multiple often closely related hazards affecting volcanic islands, the key scientific recommendation for an effective hazard assessment in such areas should include the use of an integrated and multidisciplinary approach encompassing both the submarine and subaerial flanks of the island. High-resolution mapping will be fundamental to understanding the variability in volcanic edifices and associated landforms (Figure 11) and to performing systematic parametrization to provide insights into the complex interplay between the volcanic, tectonic, erosive-depositional, and eustatic processes (for shallow-water areas) that control the genesis of volcanic islands ([113] and references therein). This mapping (Figure 11) will also be the basis for planning more detailed surveys with seismic methods, ROV dives, and seafloor sampling and for successive bathymetric comparisons aimed at understanding what occurs during eruptive crises. In this regard, the availability of multiple time-lapse bathymetric surveys has been proven to be a very effective tool for monitoring seafloor changes associated with volcanic and/or failure events occurring in both shallow water (e.g., [107,115,224]) and deep water (e.g., [225,226]). In particular, the integration of repeated bathymetric surveys with acoustic monitoring and/or ROV dives will increase our ability to detect and understand eruption dynamics in submarine environments [120,227,228].

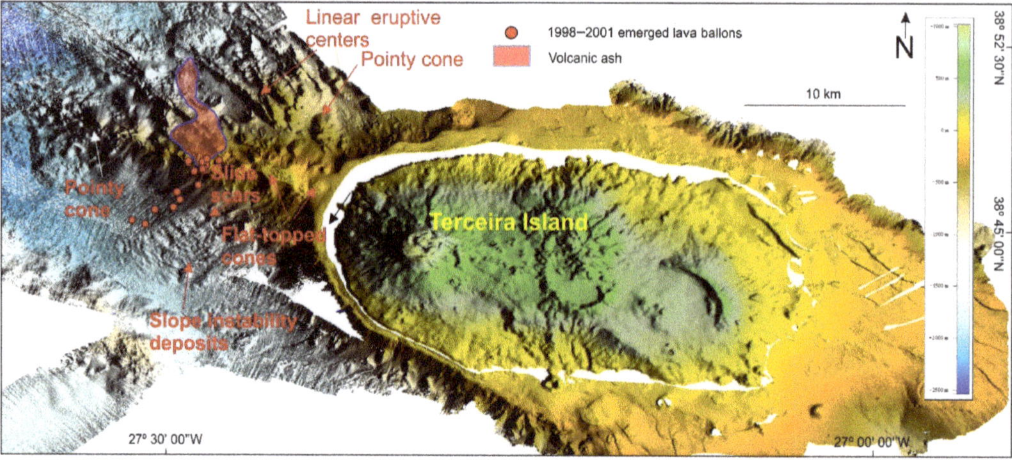

Figure 11. Main characteristics of Terceira Island (Azores; North Atlantic Ocean) and the bathymetry surrounding the island. The map also shows the area where lava balloons clusters and volcanic ash were observed at the sea surface during the 1998–2001 Serreta eruptions; more details on these eruptions can found in [118,229].

6. Conclusions: A Distinctive Multidisciplinary Approach to Study Offshore Geological Hazards

Offshore geological hazards include convulsive and persistent geological processes and are mainly represented by seismicity, slope sedimentary instabilities, submarine volcanism, fluid flows, and bottom currents; tsunamis are also mentioned because they are commonly a secondary hazard generated mostly by earthquakes, slope instabilities, or volcanic eruptions. They can occur in any domain or environment in the oceans and seas and represent a real and serious threat to society, the economy, and the environment.

Despite the progress in data acquisition and the establishment of evolutionary models for the different hazardous features, each dedicated section has identified knowledge gaps and how these gaps can be addressed. We also note that hazardous processes can interact and potentially generate cascading events.

This review establishes that the challenges for improving outcomes in offshore geohazard research can be addressed with multidisciplinary approach studies. This approach requires cross-disciplinary research to bring together multiscale analysis, mapping, direct deep-sea observations and testing, and modelling in scenarios with individual, but mainly multiple geohazards. This approach will lead to multicriteria decisions for understanding hazardous processes and their causative factors.

A qualitative step in the multiscale analysis involves the acquisition of long-term geological records, such as seismic profiles with different degrees of resolution and penetration and geophysical data (e.g., magnetometer and gravimeter data) (Figure 12), all acquired simultaneously in surveys using emerging technology and applying advanced tools for processing and geophysical modelling (Figure 12). The long-term records also provide the opportunity to study seismic profiles and sediment cores (the longer the better) (Figure 12) to improve our understanding of the magnitude and frequency of hazardous processes. Advancing techniques in sediment core analysis and age dating will contribute to reducing the uncertainty between stratigraphic correlations and increase their temporal resolution. This will facilitate the attainment of more accurate information about the sediment age and the recurrence interval of hazardous events. The success of these observations from these conventional but continuously advancing techniques will be closely tied to seafloor mapping, direct deep-sea observations and testing, and modelling.

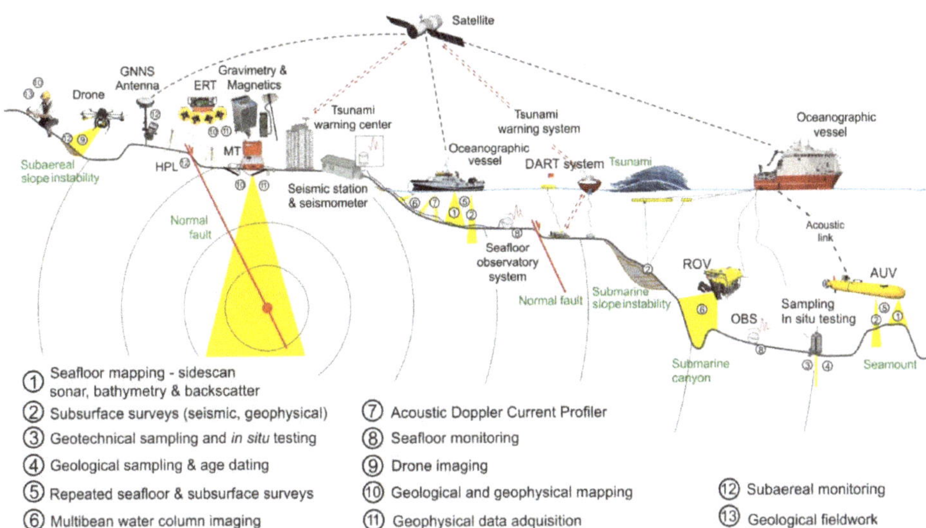

1. Seafloor mapping - sidescan sonar, bathymetry & backscatter
2. Subsurface surveys (seismic, geophysical)
3. Geotechnical sampling and *in situ* testing
4. Geological sampling & age dating
5. Repeated seafloor & subsurface surveys
6. Multibean water column imaging
7. Acoustic Doppler Current Profiler
8. Seafloor monitoring
9. Drone imaging
10. Geological and geophysical mapping
11. Geophysical data adquisition
12. Subaereal monitoring
13. Geological fieldwork

Figure 12. A distinctive methodological approach to study offshore geological hazards. The main future directions for studying offshore geohazards will be the implementation of multiscale and multidisciplinary approaches joining conventional and emerging tools for monitoring, mapping, direct observations, in situ testing, and modelling. Scenarios can be affected by multiple hazardous features, some in a land-marine transition context, and the integration of offshore and onshore observations is essential. The figure was used as idea and base to create a new one including more information from [79].

Mapping is not a new approach but is still needed to fill the existing gaps along many continental margins and basinal plains and to provide higher-resolution seafloor maps from shallow to deep-sea areas, thus providing more details on the occurrence of hazardous features. This task is mandatory for taking the next step in geological hazard assessment. The capacity to perform repeated high-resolution multibeam bathymetric surveys is also a very effective tool for monitoring seafloor changes, and it has to be implemented for a better understanding of hazardous processes. In this sense, the use of AUVs with multibeam sonars and sub-bottom profilers in repeated surveys and of ROVs for direct observations are essential to map the active hazardous seafloor features with maximum resolution, even in deep sea environments (Figure 12); the temporal resolution of their mapping will be important to report their dynamic evolution. Seismic record acquisition also plays an important role in mapping prior hazardous processes. In this sense, a greater use of 3D seismic data is expected to offer new and unprecedented sub-bottom geomorphologic information, enabling an accurate delimitation and characterization of active structures, especially in complex geological settings. However, the success of future efforts to map hazardous seafloor features will require confident geomorphological models and mapping standards for the correct understanding and recognition of features. There is still a long way to go before the scientific community reaches an agreement on standards for marine geohazard mapping, but this is a requirement for future multiple hazard maps and catalogues.

Direct observations and testing are also technical challenges because both are linked to the development of new sensors, techniques, protocols, and infrastructures, such as seafloor observatories (Figure 12). The development of new seafloor observatory systems with capabilities and facilities for remote and real-time recording over long periods of time will lead to qualitative advances. Direct observations of active structures and measurements of smaller-scale, but highly recurrent, events will enhance our understanding of larger processes and also provide important data for small-scale models. Moreover, direct observations and measurements from the water column via the optimization of mooring systems, CTDs, ADCPs, and transmissometers will be essential to understand the physics of the environment (e.g., bottom currents, turbidity, etc.). Additionally, direct seafloor monitoring by seafloor network systems, including OBSs with longer standing periods than are available in the present and DART (Deep-ocean Assessment and Reporting of Tsunamis) buoy systems, or the use of submarine cables to detect earthquakes, will provide better seismic data with which to define active faults, surface ruptures, volcanic activity, and tsunamis (Figure 12). Direct seafloor observations and measurements, together with inland seismic stations, will allow us to define seismogenic and aseismic faults and estimate realistic peak ground acceleration (PGA) values, which depend on the epicentral distance and earthquake magnitude (Figure 12). Seismic loading is critical to defining the factor of safety (F) and the susceptibility to slope failure of different seafloor areas. Integrated seafloor and inland monitoring will be a key element to increase the reliability and timeliness of the information used by early warning systems (Figure 12). However, future generations of AUVs and ROVs, which have become more widely accessible to the scientific community and easier to manage, will also provide new opportunities for in situ sampling of sediments, monitoring the rate of seafloor mobility, fluid seepage characterization, and measuring in situ geotechnical parameters. The in situ geotechnical properties (Figure 12) involve the goal of obtaining contact measurements (seafloor and sub-bottom) using advanced static (e.g., cone penetrometer, pressuremeter, heat flow), dynamic (e.g., XBP) and combined systems. The measurement of the dynamic effect of seismic events or other cyclic sources, such as storm waves and internal waves on the sediment, especially on the pore pressure, is another rarely performed technical approach that will need to be enhanced (e.g., dynamic simple shear tests). Special in situ tests can reveal the real effect of high sedimentation rates or fluid flow dynamics (gas emissions) on the sediment and its geomechanical characteristics. In addition, in situ geotechnical property measurements will be closely tied to the assessment of potential seafloor stability by the application of probabilistic methods, for which GIS is an adequate and very powerful tool.

The success of future efforts for an effective seismic hazard assessment will be reliant on new and better geological models taking advantage of developments in artificial intelligence. Consequently, the future of the field of geohazards is coupled to enhanced computational capabilities. This is also because this field will face a massive volume of datasets (i.e., the so-called big data problem). Datasets will be generated from new hull-mounted and towed methodological instruments as well as autonomous and permanent observational systems recording multiple hazard datasets with higher temporal and spatial resolution. This means that in the future, a common but key challenge will be the ability to efficiently manage and analyze (also in real time) massive data; therefore, the need to train geoscientists and build the capacity to operate advanced computational systems are needed. Massive data linked to the advances in artificial intelligence will open new lines of research for the use of advanced deep learning and machine learning algorithms, for example, for automatic detection and classification of different variables (e.g., slope gradients, roughness, backscatter signal amplitudes, grain size, density) involved in the identification of geohazard features. The development of complex neural networks trained to detect variables of interest (e.g., in seismograms, bathymetries, cores) will offer an important advancement both qualitatively (elements could be detected automatically that could go unnoticed by the most trained analyst) and quantitatively (the analysis of multiple datasets as well as their spatiotemporal relationships will increase exponentially). Such advances will lead to a much greater understanding of hazardous processes and will have significant effect on the probabilistic methods for assessing geological hazards with more robust models from which early warning systems will benefit.

Furthermore, there is also the need to enhance multidisciplinary studies in the geological scenarios with multiple hazardous processes that can interact and generate cascading events. These scenarios can be affected by hazardous features that are connected from sea to land. Therefore, the integration of on-land information, e.g., Global Navigation Satellite System (GNSS), high-precision levelling, seismicity monitoring, multiple geophysical datasets (including magnetometer, gravimeter, magnetotelluric (MT), electrical resistivity tomography (ERT), and HPL data, and field, drone, and satellite Earth observations), with submarine results is critical to realizing the correct assessment of hazards (Figure 12). In this sense, scientific and technical coordination of the research community working on subaerial and submarine hazardous structures with different datasets is a major task that is still in its infancy, and reinforcement is required in the future.

Supplementary Materials: The following are available online at https://www.mdpi.com/article/10.3390/oceans2020023/s1, Table S1: Okada's parameters [3] describing the two fault segments involving of the Horseshoe and Marques de Pombal faults as source for the tsunami modelling. Video S1: Video of tsunmai simulation in the Gulf of Cadiz.

Author Contributions: Conceptualization, G.E. and D.C. (David Casas); Writing—original draft: all authors based on their geological hazard expertise; Introduction by G.E. and D.C.; Human activities and historical cases and by G.E. and B.A.; Definition and classification of geohazards by G.E. and D.C. (David Casas); Tectonic earthquakes by J.G.-Z., F.E., L.G.-C., A.M.-T., and V.T.-S.; Landslides by D.C. (David Casas), M.A.-Z., and J.I.-G.; Volcanism by D.C. (Daniele Casalbore), O.S.-G., and J.-T.V.; Tsunamis by J.-T.V., J.M., J.M.G.-V., and D.P.; Fluid-flow by S.G.-G., D.D., M.C.F.-P., and G.E.; Bottom currents by E.M., M.T., and C.J.; Scenarios involving multiple geological hazards by G.E., D.C. (David Casas)., D.C. (Daniele Casalbore), F.E., and J.G.-Z.; Conclusions by G.E., D.C. (David Casas), J.N., J.V., and M.Y.; Integration—review-reediting: G.E. and D.C. (David Casas). All authors have read and agreed to the published version of the manuscript.

Funding: This study was supported by the following scientific projects: FAUCES (MINECO/FEDER, CMT2015-65461-C2-R), DAMAGE (MINECO/FEDER, CGL2016-80687-R); B-RNM-301-UGR18, P18-RT-3275 and RNM148 from Junta de Andalucía; GRACE (Eurofleets: EFP_SEA02-024); VULCANO (CTM2012-36317); RIGEL (IEO); MEGAFLOW (MEC/FEDER, RTI2018-096064-B-C21); and ChEESE (UE (H2020-INFRAEDI-2018-1, PROJECT ID: 823844). This study was also carried out in the framework of the IGCP 640–S4LIDE. In addition, the research was also conducted in collaboration with the Andalusian PAIDI Research Group RNM 328 (Coastal and Marine Geology and Geophysics).

Acknowledgments: The ICM-CSIC authors acknowledge Severo Ochoa funding from the Spanish government through the "Severo Ochoa Centre of Excellence" accreditation (CEX2019-000928-S). We also thank IHS-Markit for the Kingdom Suite educational license. This article is dedicated to the memory of Albert Figueras, who devoted the last years of his life as director of the UTM-CSIC.

Conflicts of Interest: The authors declare no conflict of interest.

References

1. Bell, F.G. *Geological Hazards: Their Assessment, Avoidance and Mitigation*; CRC Press: Boca Raton, FL, USA, 2003.
2. Nadim, F. Challenges to geo-scientists in risk assessment for submarine slides. *Nor. J. Geol.* **2006**, *86*, 351–362.
3. Hough, G.; Green, J.; Fish, P.; Mills, A.; Moore, R. A geomorphological mapping approach for the assessment of seabed geohazards and risk. *Mar. Geophys. Res.* **2011**, *32*, 151–162. [CrossRef]
4. Ercilla, G.; Casas, D. *Submarine Mass Movements: Sedimentary Characterization and Controlling Factors*; Intech: London, UK, 2012.
5. Chiocci, F.L.; Cattaneo, A.; Urgeles, R. Seafloor mapping for geohazard assessment: State of the art. *Mar. Geophys. Res.* **2011**, *32*, 1–11. [CrossRef]
6. Neumann, B.; Vafeidis, A.T.; Zimmermann, J.; Nicholls, R.J. Future coastal population growth and exposure to sea-level rise and coastal flooding—A global assessment. *PLoS ONE* **2015**, *10*, e0118571. [CrossRef] [PubMed]
7. Camargo, J.; Silva, M.; Ferreira, J.A.; Araújo, T. Marine geohazards: A bibliometric-based review. *Geosciences* **2019**, *9*, 100. [CrossRef]
8. Campbell, K.J. Deepwater geohazards: How significant are they? *Lead. Edge* **1999**, *18*, 514–519. [CrossRef]
9. Yonggang, J.I.A.; Chaoqi, Z.H.U.; Liping, L.I.U.; Dong, W. Marine geohazards: Review and future perspective. *Acta Geol. Sin.* **2016**, *90*, 1455–1470. [CrossRef]
10. Bouma, A.H. *Offshore Geologic Hazards: A Short Course Presented at Rice University, May 2–3, 1981 for the Offshore Technology Conference*; Bouma, A., Sangrey, D., Coleman, J., Prior, D., Trippet, A., Dunlap, W., Hooper, J., Eds.; American Association of Petroleum Geologists: Tulsa, OK, USA, 1981; Volume 18, pp. 48–101.
11. Bates, R.L.; Jackson, J.A. *Glossary of Geology*; American Geological Institute: Alexandria, VA, USA, 1987.
12. Ayala, F.J. *Introducción a Los Riesgos Geológicos. En IGME (1988) Riesgos Geológicos*; Serie Geología Ambiental; Instituto Geológico y Minero de España (IGME): Madrid, Spain, 1998.

13. Calarco, M.; Zolezzi, F.; Johnson, W.J. Offshore Geohazards-Industry Implications and Geoscientist Role. In Proceedings of the Near Surface Geoscience 2014—20th European Meeting of Environmental and Engineering Geophysics, Athens, Greece, 14–18 September 2014; European Association of Geoscientists & Engineers: Houten, The Netherlands, 2014; pp. 1–5.
14. Camerlenghi, A. Addressing Submarine Geohazards Through Scientific Drilling. In Proceedings of the EGU General Assembly 2009, Vienna, Austria, 19–24 April 2009; Center for Astrophysics: Vienna, Ausutria, 2009; p. 3151.
15. Harbitz, C.B.; Løvholt, F.; Bungum, H. Submarine landslide tsunamis: How extreme and how likely? *Nat. Hazards* **2014**, *72*, 1341–1374. [CrossRef]
16. Sassa, S.; Takagawa, T. Liquefied gravity flow-induced tsunami: First evidence and comparison from the 2018 Indonesia Sulawesi earthquake and tsunami disasters. *Landslides* **2018**, *16*, 195–200. [CrossRef]
17. Espuela, J.A.; Ercilla, G.; Farran, M.L. Submarine erosion in North Spanish Atlantic sea. *Hydro Int.* **2005**, *9*, 54–57.
18. Keller, E.A.; Pinter, N. *Active Tectonics*; Prentice Hall: Upper Saddle River, NJ, USA, 1996.
19. Cooper, C.K.; Wood, J.; Andrieux, O. Turbidity Current Measurements in the Congo Canyon. In Proceedings of the Offshore Technology Conference, Houston, TX, USA, 6–9 May 2013; pp. 1–12.
20. Carter, L.; Gavey, R.; Talling, P.; Liu, J. Insights into Submarine geohazards from breaks in subsea telecommunication cables. *Oceanography* **2014**, *27*, 58–67. [CrossRef]
21. Hovland, M.; Judd, A.G. *Seafloor Pockmarks and Seepages, Impact on Geology, Biology and the Marine Environment*; Graham & Trotman: London, UK, 1988.
22. Laberg, J.S.; Camerlenghi, A. Chapter 25 The Significance of Contourites for Submarine Slope Stability. In *Developments in Sedimentology*; Rebesco, M., Camerlenghi, A., Eds.; Elsevier: New York, NY, USA, 2008; Volume 60, pp. 537–556.
23. Carter, L. *Submarine Cables and the Oceans: Connecting the World (No. 31)*; UNEP/Earthprint: Nairobi, Kenya, 2010.
24. Halfar, J.; Fujita, R.M. Precautionary management of deep-sea mining. *Mar. Policy* **2002**, *26*, 103–106. [CrossRef]
25. Benn, A.R.; Weaver, P.P.; Billet, D.S.M.; van den Hove, S.; Murdock, A.P.; Doneghan, G.B.; Le Bas, T. Human activities on the deep seafloor in the North East Atlantic: An assessment of spatial extent. *PLoS ONE* **2010**, *5*, e12730. [CrossRef] [PubMed]
26. Dunbar, P.; McCullough, H.; Mungov, G.; Varner, J.; Stroker, K. 2011 Tohoku earthquake and tsunami data available from the National oceanic and atmospheric administration/National geophysical data center. *Geomat. Nat. Hazards Risk* **2011**, *2*, 305–323. [CrossRef]
27. Fritz, H.M.; Petroff, C.M.; Catalán, P.A.; Cienfuegos, R.; Winckler, P.; Kalligeris, N.; Weiss, R.; Barrientos, S.E.; Meneses, G.; Valderas-Bermejo, C.; et al. Field survey of the 27 February 2010 Chile tsunami. *Pure Appl. Geophys.* **2011**, *168*, 1989–2010. [CrossRef]
28. Hawkes, A.D.; Bird, M.; Cowie, S.; Grundy-Warr, C.; Horton, B.P.; Hwai, A.T.S.; Law, L.; Macgregor, C.; Nott, J.; Ong, J.E.; et al. Sediments deposited by the 2004 Indian ocean tsunami along the Malaysia-Thailand Peninsula. *Mar. Geol.* **2007**, *242*, 169–190. [CrossRef]
29. Galindo-Zaldivar, J.; Ercilla, G.; Estrada, F.; Catalán, M.; d'Acremont, E.; Azzouz, O.; Casas, D.; Chourak, M.; Vazquez, J.T.; Chalouan, A.; et al. Imaging the growth of recent faults: The case of 2016–2017 seismic sequence sea bottom deformation in the Alboran sea (Western Mediterranean). *Tectonics* **2019**, *37*, 2513–2530. [CrossRef]
30. Meschis, M.; Roberts, G.P.; Mildon, Z.K.; Robertson, J.; Michetti, A.M.; Walker, J.F. Slip on a mapped normal fault for the 28th December 1908 Messina earthquake (Mw 7.1) in Italy. *Sci. Rep.* **2019**, *9*, 6481. [CrossRef]
31. Stover, C.W.; Coffman, J.L. *Seismicity of the United States, 1568–1989 (Revised)*; U.S. Geological Survey Professional Paper; U.S. Government Printing Office: Washington, DC, USA, 1993.
32. Brune, S.; Ladage, S.; Babeyko, A.Y.; Müller, C.; Kopp, H.; Sobolev, S.V. Submarine landslides at the eastern Sunda margin: Observations and tsunami impact assessment. *Nat. Hazards* **2009**, *54*, 547–562. [CrossRef]
33. Gennesseaux, M.; Mauffret, A.; Pautot, G. Les glissements sous-marins de la pente continentale niçoise et la rupture de câbles en mer ligure (Mediterrenée Occidentale). *Comptes Rendus l'Académie Sciences Paris* **1980**, *290*, 959–962.
34. Dan, G.; Sultan, N.; Savoye, B. The 1979 nice harbour catastrophe revisited: Trigger mechanism inferred from geotechnical measurements and numerical modelling. *Mar. Geol.* **2007**, *245*, 40–64. [CrossRef]
35. Piper, D.J.W.; Shor, A.N.; Clarke, J.E.H. The 1929 "grand banks" earthquake, slump, and turbidity current. In *Sedimentologic Consequences of Convulsive Geologic Events*; Clifton, H.E., Ed.; Geological Society of America: Boulder, CO, USA, 1988; Volume 229, pp. 77–92.
36. Bugge, T. Submarine slides on the Norwegian continental margin with special emphasis on the storegga slide. *IKU Rep.* **1983**, *110*, 1–152.
37. Bryn, P.; Berg, K.; Stoker, M.S.; Haflidason, H.; Solheim, A. Contourites and their relevance for mass wasting along the Mid-Norwegian margin. *Mar. Pet. Geol.* **2005**, *22*, 85–96. [CrossRef]
38. Bondevik, S.; Løvholt, F.; Harbitz, C.; Mangerud, J.; Dawson, A.; Svendsen, J.I. The Storegga slide tsunami—Comparing field observations with numerical simulations. *Mar. Pet. Geol.* **2003**, *22*, 195–208. [CrossRef]
39. Forte, G.; De Falco, M.; Santangelo, N.; Santo, A. Slope stability in a multi-hazard eruption scenario (Santorini, Greece). *Geosciences* **2019**, *9*, 412. [CrossRef]
40. Wyss, M. A proposed source model for the great Kau, Hawaii, earthquake of 1868. *Bull. Seismol. Soc. Am.* **1988**, *78*, 1450–1462.
41. Carracedo, J.C.; Torrado, F.P.; González, A.R.; Soler, V.; Turiel, J.L.F.; Troll, V.R.; Wiesmaier, S. The 2011 submarine volcanic eruption in El Hierro (Canary islands). *Geol. Today* **2012**, *28*, 53–58. [CrossRef]

42. Hildenbrand, A.; Marques, F.O.; Catalão, J. Large-scale mass wasting on small volcanic islands revealed by the study of Flores island (Azores). *Sci. Rep.* **2018**, *8*, 13898. [CrossRef] [PubMed]

43. Hovland, M. Seabed pockmarks on the Helike detal front. In *Ancient Helike and Aigialeia, Proceedings of the Second International Conference, Aigion, Greece, 1–3 December 1995*; Katsonopoulou, D., Schildardi, D., Soter, S., Eds.; Helike Society: Athens, Greece, 1995; pp. 461–471.

44. Worzel, J.L.; Watkins, J.S. Location of a Lost Drilling Platform. In Proceedings of the Offshore Technology Conference, Houston, TX, USA, 5–7 May 1974; pp. 771–772.

45. Brook, H.; Cook, R.; Harris, J. Competitive concrete gravity base foundation for offshore wind farm. In *Coast, Marine Structured and Breakwater, Adapting to Change*; Allsop, W., Ed.; Thomas Telford Books: London, UK, 2010; pp. 62–73.

46. Whitehouse, R.J.S.; Sutherland, J.S.; O'Brien, D. Seabed Scour Assessment for Offshore Windfarm. In Proceedings of the 3rd International Conference on Scour and Erosion, Gouda, The Netherlands, 3–5 July 2007; pp. 1–320.

47. Dahlberg, R. Observations of scour around offshore structures. *Can. Geotech. J.* **1983**, *20*, 617–628. [CrossRef]

48. Demars, K.R.; Vanover, E.A. Measurement of wave-induced pressures and stresses in a sandbed. *Mar. Geotechnol.* **1985**, *6*, 29–59. [CrossRef]

49. Herbich, J.B. Hydromechanics of submarine pipelines: Design problems. *Can. J. Civ. Eng.* **2011**, *12*, 863–874. [CrossRef]

50. Lubick, N. Earthquakes from the ocean: Danger zones. *Nature* **2011**, *476*, 391–392. [CrossRef]

51. Bird, P.; Kagan, Y.Y.; Jackson, D.D. Plate tectonics and earthquake potential of spreading ridges and oceanic transform faults. In *Plate Boundary Zones*; Stein, S., Freymueller, J.T., Eds.; AGU: Washington, DC, USA, 2002; pp. 203–218.

52. Beeson, J.W.; Johnson, S.Y.; Goldfinger, C. The transtensional offshore portion of the northern San Andreas fault: Fault zone geometry, late pleistocene to holocene sediment deposition, shallow deformation patterns, and asymmetric basin growth. *Geosphere* **2017**, *13*, 1173–1206. [CrossRef]

53. Barnes, P.M.; Sutherland, R.; Delteil, J. Strike-slip structure and sedimentary basins of the southern alpine fault, Fiordland, New Zealand. *Geol. Soc. Am. Bull.* **2005**, *117*, 411–435. [CrossRef]

54. Frohlich, C.; Apperson, K.D. Earthquake focal mechanisms, moment tensors, and the consistency of seismic activity near plate boundaries. *Tectonics* **1992**, *11*, 279–296. [CrossRef]

55. Baraza, J.; Ercilla, G.; Nelson, C.H. Potential geologic hazards on the eastern Gulf of Cadiz slope (SW Spain). *Mar. Geol.* **1999**, *155*, 191–215. [CrossRef]

56. Gahalaut, V.K.; Kundu, B.; Laishram, S.S.; Catherine, J.; Kumar, A.; Singh, M.D.; Tiwari, R.P.; Chadha, R.K.; Samanta, S.K.; Ambikapathy, A.; et al. Aseismic plate boundary in the Indo-Burmese wedge, Northwest Sunda Arc. *Geology* **2013**, *41*, 235–238. [CrossRef]

57. Avouac, J.-P. From geodetic imaging of seismic and aseismic fault slip to dynamic modeling of the seismic cycle. *Annu. Rev. Earth Planet. Sci.* **2015**, *43*, 233–271. [CrossRef]

58. Wang, K.; Hu, Y.; He, J. Deformation cycles of subduction earthquakes in a viscoelastic earth. *Nature* **2012**, *484*, 327–332. [CrossRef]

59. Cloos, M. Thrust-type subduction-zone earthquakes and seamount asperities: A physical model for seismic rupture. *Geology* **1992**, *20*, 601–604. [CrossRef]

60. Polet, J.; Kanamori, H. Shallow subduction zone earthquakes and their tsunamigenic potential. *Geophys. J. Int.* **2000**, *142*, 684–702. [CrossRef]

61. Kaneda, Y.; Kawaguchi, K.; Araki, E.; Matsumoto, H.; Nakamura, T.; Kamiya, S.; Ariyoshi, K.; Hori, T.; Baba, T.; Takahashi, N. Development and Application of an Advanced Ocean Floor Network System for Megathrust Earthquakes and Tsunamis. In *Seafloor Observatories: A New Vision of the Earth from the Abyss*; Favali, P., Beranzoli, L., De Santis, A., Eds.; Springer: Berlin/Heidelberg, Germany, 2015; pp. 643–662.

62. Tajima, F.; Mori, J.; Kennett, B.L.N. A review of the 2011 Tohoku-Oki earthquake (Mw 9.0): Large-scale rupture across heterogeneous plate coupling. *Tectonophysics* **2013**, *586*, 15–34. [CrossRef]

63. Choy, G.L.; McGarr, A. Strike-slip earthquakes in the oceanic lithosphere: Observations of exceptionally high apparent stress. *Geophys. J. Int.* **2002**, *150*, 506–523. [CrossRef]

64. Tanioka, Y.; Satake, K. Tsunami generation by horizontal displacement of ocean bottom. *Geophys. Res. Lett.* **1996**, *23*, 861–864. [CrossRef]

65. Galindo-Zaldívar, J.; Jabaloy, A.; Maldonado, A.; Martínez-Martínez, J.M.; de Galdeano, C.S.; Somoza, L.; Surinach, E. Deep crustal structure of the area of intersection between the Shackleton fracture zone and the West Scotia Ridge (Drake Passage, Antarctica). *Tectonophysics* **2000**, *320*, 123–139. [CrossRef]

66. Maldonado, A.; Balanyá, J.C.; Barnolas, A.; Galindo-Zaldívar, J.; Hernández, J.; Jabaloy, A.; Livermore, R.; Martínez, J.M.; Rodríguez-Fernández, J.; de Galdeano, C.S.; et al. Tectonics of an extinct ridge-transform intersection, Drake Passage (Antarctica). *Mar. Geophys. Res.* **2000**, *21*, 43–68. [CrossRef]

67. Estrada, F.; González-Vida, J.M.; Peláez, J.A.; Galindo-Zaldívar, J.; Ercilla, G.; Vázquez, J.T. Tsunamigenic Risk Associated to Vertical Offset in Transcurrent Fault Termination: The Case of the Averroes Fault (Alboran Sea). In Proceedings of the Fault2SHA—4th Workshop, Barcelona, Spain, 3–5 June 2019; ICM-CSIC: Barcelona, Spain, 2019; pp. 14–15.

68. Pisarska-Jamroży, M.; Belzyt, S.; Börner, A.; Hoffmann, G.; Hüneke, H.; Kenzler, M.; Obst, K.; Rother, H.; van Loon, A.J. Evidence from seismites for glacio-isostatically induced crustal faulting in front of an advancing land-ice mass (Rügen island, SW Baltic sea). *Tectonophysics* **2018**, *745*, 338–348. [CrossRef]

69. Goldfinger, C.; Nelson, C.H.; Morey, A.E.; Johnson, J.E.; Patton, J.R.; Karabanov, E.; Gutiérrez-Pastor, J.; Eriksson, A.T.; Gràcia, E.; Dunhill, G.; et al. *Turbidite Event History—Methods and Implications for Holocene Paleoseismicity of the Cascadia Sub-duction Zone*; U.S. Geological Survey: Reston, VA, USA, 2012.

70. Masson, D.G.; Kenyon, N.H.; Weaver, P.P.E. Slides, debris Flows, and Turbidity Currents. In *Oceanography: An Illustrated Guide*; Summerhayes, C.P., Thorpe, S.A., Eds.; Manson Publishing: London, UK, 1996; pp. 136–151.

71. Nardin, T.R.; Hein, F.J.; Gorsline, D.S.; Edwards, B.D. A Review of Mass Movement Processes, Sediment and Acoustic Character-istics, and Contrasts in Slope and Base-Of-Slope Systems Versus Canyon-Fan-Basin Floor Systems. In *Geology of Continental Slopes*; Doyle, L.J., Pilkey, O.H., Eds.; SEPM Society for Sedimentary Geology: Tulsa, OK, USA, 1979; Volume 27, pp. 61–73.

72. Mulder, T.; Cochonat, P. Classification of offshore mass movements. *J. Sediment. Res.* **1996**, *66*, 43–57. [CrossRef]

73. Shanmugam, G. 50 years of the turbidite paradigm (1950s–1990s): Deep-water processes and facies models—A critical perspective. *Mar. Pet. Geol.* **2000**, *17*, 285–342. [CrossRef]

74. Tripsanas, E.K.; Piper, D.J.; Jenner, K.A.; Bryant, W.R. Submarine mass-transport facies: New perspectives on flow processes from cores on the eastern North American margin. *Sedimentology* **2008**, *55*, 97–136. [CrossRef]

75. Frey-Martínez, J.; Cartwright, J.; James, D. Frontally confined versus frontally emergent submarine landslides: A 3D seismic characterisation. *Mar. Pet. Geol.* **2006**, *23*, 585–604. [CrossRef]

76. Moscardelli, L.; Wood, L. New classification system for mass transport complexes in offshore Trinidad. *Basin Res.* **2008**, *20*, 73–98. [CrossRef]

77. Canals, M.; Puig, P.; de Madron, X.D.; Heussner, S.; Palanques, A.; Fabres, J. Flushing submarine canyons. *Nature* **2006**, *444*, 354–357. [CrossRef] [PubMed]

78. Talling, P.J.; Paull, C.K.; Piper, D.J.W. How are subaqueous sediment density flows triggered, what is their internal structure and how does it evolve? Direct observations from monitoring of active flows. *Earth Sci. Rev.* **2013**, *125*, 244–287. [CrossRef]

79. Clare, M.A.; Vardy, M.E.; Cartigny, M.J.; Talling, P.J.; Himsworth, M.D.; Dix, K.; Harris, J.M.; Whitehouse, R.J.S.; Belal, M. Direct monitoring of active geohazards: Emerging geophysical tools for deep-water assessments. *Near Surf. Geophys.* **2017**, *15*, 427–444. [CrossRef]

80. Gwiazda, R.; Paull, C.K.; Ussler, W.; Alexander, C.R. Evidence of modern fine-grained sediment accumulation in the Monterey fan from measurements of the pesticide DDT and its metabolites. *Mar. Geol.* **2015**, *363*, 125–133. [CrossRef]

81. Azpiroz-Zabala, M.; Cartigny, M.J.B.; Talling, P.J.; Parsons, D.R.; Sumner, E.J.; Clare, M.A.; Simmons, S.M.; Cooper, C.; Pope, E.L. Newly recognized turbidity current structure can explain prolonged flushing of submarine canyons. *Sci. Adv.* **2017**, *3*, e1700200. [CrossRef]

82. Locat, J.; Leroueil, S.; Locat, A.; Lee, H. Weak Layers: Their Definition and Classification from a Geotechnical Perspective. In *Submarine Mass Movements and their Consequences*; Krastel, S., Behrmann, J.-H., Völker, D., Stipp, M., Berndt, C., Urgeles, R., Chaytor, J., Huhn, K., Strasser, M., Harbitz, C.B., Eds.; Springer: Cham, Switzerland, 2014; pp. 3–12.

83. Hutchinson, J. General Report: Morphological and Geotechnical Parameters of Landslides in Relation to Geology and Hydrogeol-ogy. In Proceedings of the 5th International Symposium on Landslides, Lausanne, Switzerland, 10–15 July 1988; Bonnard, C., Ed.; August Aimé Balkema: Rotterdam, The Netherlands, 1988; Volume 5, pp. 3–35.

84. Mulder, T.; Alexander, J. The physical character of subaqueous sedimentary density flows and their deposits. *Sedimentology* **2001**, *48*, 269–299. [CrossRef]

85. Masson, D.G.; Watts, A.B.; Gee, M.J.R.; Urgeles, R.; Mitchell, N.C.; Le Bas, T.P.; Canals, M. Slope failures on the flanks of the western Canary islands. *Earth Sci. Rev.* **2002**, *57*, 1–35. [CrossRef]

86. Normark, W.R.; Piper, D.J.W. Initiation Processes and Flow Evolution of Turbidity Currents: Implications for the Depositional Record. In *From Shoreline to Abyss: Contributions in Marine Geology in Honor of Francis Parker Shepard*; Osborne, R.H., Ed.; SEPM Society for Sedimentary Geology: Tulsa, OK, USA, 1991; Volume 46, pp. 207–230.

87. Casas, D.; Ercilla, G.; García, M.; Yenes, M.; Estrada, F. Post-rift sedimentary evolution of the Gebra Debris Valley. A submarine slope failure system in the Central Bransfield basin (Antarctica). *Mar. Geol.* **2013**, *340*, 16–29. [CrossRef]

88. Casas, D.; García, M.; Bohoyo, F.; Maldonado, A.; Ercilla, G. The Gebra–Magia complex: Mass-transport processes reworking trough-mouth fans in the Central Bransfield basin (Antarctica). *Geol. Soc. Lond. Spec. Publ.* **2018**, *461*, 61–75. [CrossRef]

89. De Blasio, F.V. Dynamics of Mass Flows and of Sediment Transport in Amazonis Planitia, Mars. In Proceedings of the European Planetary Science Congress 2013, London, UK, 8–13 September 2013; EPSC: London, UK, 2013; Volume 8, pp. 1–2.

90. Urlaub, M.; Talling, P.; Zervos, A. A Numerical Investigation of Sediment Destructuring as a Potential Globally Widespread Trigger for Large Submarine Landslides on Low Gradients. In *Submarine Mass Movements and their Consequences*; Krastel, S., Behrmann, J.-H., Völker, D., Stipp, M., Berndt, C., Urgeles, R., Chaytor, J., Huhn, K., Strasser, M., Harbitz, C.B., Eds.; Springer: Cham, Switzerland, 2014; pp. 177–188.

91. Stevenson, C.J.; Jackson, C.A.L.; Hodgson, D.M.; Hubbard, S.M.; Eggenhuisen, J.T. Deep-water sediment bypass. *J. Sediment. Res.* **2015**, *85*, 1058–1081. [CrossRef]

92. Hizzett, J.L.; Clarke, J.E.H.; Sumner, E.J.; Cartigny, M.J.B.; Talling, P.J.; Clare, M.A. Which triggers produce the most erosive, frequent, and longest runout turbidity currents on deltas? *Geophys. Res. Lett.* **2017**, *45*, 855–863. [CrossRef]

93. Paull, C.K.; Talling, P.J.; Maier, K.L.; Parsons, D.; Xu, J.; Caress, D.W.; Gwiazda, R.; Lundsten, E.M.; Anderson, K.; Barry, J.P.; et al. Powerful turbidity currents driven by dense basal layers. *Nat. Commun.* **2018**, *9*, 4114. [CrossRef] [PubMed]

94. Miramontes, E.; Eggenhuisen, J.T.; Jacinto, R.S.; Poneti, G.; Pohl, F.; Normandeau, A.; Campbell, D.C.; Hernández-Molina, F.J. Channel-levee evolution in combined contour current-turbidity current flows from flume tank experiments. *Geology* **2020**, *48*, 353–357. [CrossRef]

95. Guzzetti, F.; Malamud, B.D.; Turcotte, D.L.; Reichenbach, P. Power-law correlations of landslide areas in central Italy. *Earth Planet. Sci. Lett.* **2002**, *195*, 169–183. [CrossRef]

96. Thomas, S.; Hooper, J.; Clare, M. Constraining Geohazards to the Past: Impact Assessment of Submarine Mass Movements on Seabed Developments. In *Submarine Mass Movements and Their Consequences*; Mosher, D.C., Shipp, R.C., Moscardelli, L., Chaytor, J.D., Baxter, C.D.P., Lee, H.J., Urgeles, R., Eds.; Springer: Dordrecht, The Netherlands, 2010; pp. 387–398.

97. Casas, D.; Chiocci, F.; Casalbore, D.; Ercilla, G.; de Urbina, J.O. Magnitude-frequency distribution of submarine landslides in the Gioia basin (Southern Tyrrhenian sea). *Geo-Mar. Lett.* **2016**, *36*, 405–414. [CrossRef]

98. Stark, C.P.; Hovius, N. The characterization of landslide size distributions. *Geophys. Res. Lett.* **2001**, *28*, 1091–1094. [CrossRef]

99. Di Traglia, F.; Nolesini, T.; Solari, L.; Ciampalini, A.; Frodella, W.; Steri, D.; Allotta, B.; Rindi, A.; Marini, L.; Monni, N.; et al. Lava delta deformation as a proxy for submarine slope instability. *Earth Planet. Sci. Lett.* **2018**, *488*, 46–58. [CrossRef]

100. Trofimovs, J.; Amy, L.; Boudon, G.; Deplus, C.; Doyle, E.; Fournier, N.; Hart, M.B.; Komorowski, J.C.; Le Friant, A.; Lock, E.J.; et al. Submarine pyroclastic deposits formed at the Soufrière Hills volcano, Montserrat (1995–2003): What happens when pyroclastic flows enter the ocean? *Geology* **2006**, *34*, 549–552. [CrossRef]

101. Casalbore, D.; Clementucci, R.; Bosman, A.; Chiocci, F.L.; Martorelli, E.; Ridente, D. Widespread mass-wasting processes off NE Sicily (Italy): Insights from morpho-bathymetric analysis. *Geol. Soc. Lond. Spec. Publ.* **2020**, *500*, 393–403. [CrossRef]

102. Kokelaar, P. Magma-water interactions in subaqueous and emergent basaltic. *Bull. Volcanol.* **1986**, *48*, 275–289. [CrossRef]

103. Kueppers, U.; Nichols, A.R.L.; Zanon, V.; Potuzak, M.; Pacheco, J.M.R. Lava balloons—Peculiar products of basaltic submarine eruptions. *Bull. Volcanol.* **2012**, *74*, 1379–1393. [CrossRef]

104. Perfit, M.R.; Chadwick, W.W. Magmatism at mid-ocean ridges: Constraints from volcanological and geochemical investigations. *Geophys. Monogr. Am. Geophys. Union* **1998**, *106*, 59–116. [CrossRef]

105. McClinton, J.T.; White, S.M.; Colman, A.; Rubin, K.H.; Sinton, J.M. The role of crystallinity and viscosity in the formation of submarine lava flow morphology. *Bull. Volcanol.* **2014**, *76*, 854. [CrossRef]

106. Jutzeler, M.; Marsh, R.; Carey, R.J.; White, J.D.L.; Talling, P.J.; Karlstrom, L. On the fate of pumice rafts formed during the 2012 havre submarine eruption. *Nat. Commun.* **2014**, *5*, 3660. [CrossRef]

107. Bosman, A.; Casalbore, D.; Romagnoli, C.; Chiocci, F.L. Formation of an 'a'ā lava delta: Insights from time-lapse multibeam bathymetry and direct observations during the stromboli 2007 eruption. *Bull. Volcanol.* **2014**, *76*, 838. [CrossRef]

108. Fraile-Nuez, E.; González-Dávila, M.; Santana-Casiano, J.M.; Arístegui, J.; Alonso-González, I.J.; Hernández-León, S.; Blanco, M.J.; Rodríguez-Santana, A.; Hernández-Guerra, A.; Gelado-Caballero, M.D.; et al. The submarine volcano eruption at the island of El Hierro: Physical-chemical perturbation and biological response. *Sci. Rep.* **2012**, *2*, 486. [CrossRef] [PubMed]

109. Vázquez, J.T.; Palomino, D.; Sánchez-Guillamón, O.; Fernández-Salas, L.M.; Fraile-Nuez, E.; Gómez-Ballesteros, M.; López-González, N.; Tello, O.; Lozano, P.; Santana-Cassiano, J.M.; et al. Preliminary geomorphological analysis of the Tagoro Volcano underwater eruption (submarine slope of El Hierro Island). In *Libro de Resúmenes V Simposio Internacional de Ciencias del Mar*; Valle, C., Aguilar, J., Arechavala, P., Asensio, L., Blasco, J., Cabrera, R., Cobelo, A., Corbí, H., Cravo, A., De-La-Ossa, J.A., et al., Eds.; Universidad de Alicante: Alicante, Spain, 2016; pp. 335–336.

110. Santana-Casiano, J.M.; González-Dávila, M.; Fraile-Nuez, E.; de Armas, D.; González, A.G.; Domínguez-Yanes, J.F.; Escánez, J. The natural ocean acidification and fertilization event caused by the submarine eruption of El Hierro. *Sci. Rep.* **2013**, *3*, 1140. [CrossRef]

111. White, J.D.L.; McPhie, J.; Soule, S.A. Submarine Lavas and Hyaloclastite. In *The Encyclopedia of Volcanoes*, 2nd ed.; Sigurdsson, H., Ed.; Academic Press: Amsterdam, The Netherlands, 2015; Chapter 19; pp. 363–375.

112. White, J.D.L.; Schipper, C.I.; Kano, K. Submarine Explosive Eruptions. In *The Encyclopedia of Volcanoes*, 2nd ed.; Sigurdsson, H., Ed.; Academic Press: Amsterdam, The Netherlands, 2015; Chapter 31; pp. 553–569.

113. Casalbore, D. Volcanic Islands and Seamounts. In *Submarine Geomorphology*; Micallef, A., Krastel, S., Savini, A., Eds.; Springer: Cham, Switzerland, 2018; pp. 333–347.

114. Cas, R.A.F.; Giordano, G. Submarine volcanism: A review of the constraints, processes and products, and relevance to the cabo de gata volcanic succession. *Ital. J. Geosci.* **2014**, *133*, 362–377. [CrossRef]

115. Rivera, J.; Lastras, G.; Canals, M.; Acosta, J.; Arrese, B.; Hermida, N.; Micallef, A.; Tello, O.; Amblas, D. Construction of an oceanic island: Insights from the El Hierro (Canary islands) 2011–2012 submarine volcanic eruption. *Geology* **2013**, *41*, 355–358. [CrossRef]

116. Embley, R.W.; Merle, S.G.; Baker, E.T.; Rubin, K.H.; Lupton, J.E.; Resing, J.A.; Dziak, R.P.; Lilley, M.D.; Chadwick, W.W.; Shank, T.; et al. Eruptive modes and hiatus of volcanism at West Mata seamount, NE Lau basin: 1996–2012. *Geochem. Geophys. Geosystems* **2014**, *15*, 4093–4115. [CrossRef]

117. Somoza, L.; González, F.J.; Barker, S.J.; Madureira, P.; Medialdea, T.; de Ignacio, C.; Lourenço, N.; León, R.; Vázquez, J.T.; Palomino, D. Evolution of submarine eruptive activity during the 2011–2012 El Hierro event as documented by hydroacoustic images and remotely operated vehicle observations. *Geochem. Geophys. Geosyst.* **2017**, *18*, 3109–3137. [CrossRef]

118. Casas, D.; Pimentel, A.; Pacheco, J.; Martorelli, E.; Sposato, A.; Ercilla, G.; Alonso, B.; Chiocci, F. Serreta 1998–2001 submarine volcanic eruption, offshore Terceira (Azores): Characterization of the vent and inferences about the eruptive dynamics. *J. Volcanol. Geotherm. Res.* **2018**, *356*, 127–140. [CrossRef]

119. Tepp, G.; Chadwick, W.W.; Haney, M.M.; Lyons, J.J.; Dziak, R.P.; Merle, S.G.; Butterfield, D.A.; Young, C.W. Hydroacoustic, seismic, and bathymetric observations of the 2014 submarine eruption at Ahyi seamount, Mariana Arc. *Geochem. Geophys. Geosystems* **2019**, *20*, 3608–3627. [CrossRef]

120. Le Saout, M.; Bohnenstiehl, D.R.; Paduan, J.B.; Clague, D.A. Quantification of eruption dynamics on the north rift at Axial seamount, Juan de Fuca Ridge. *Geochem. Geophys. Geosystems* **2020**, *21*, e2020GC009136. [CrossRef]

121. Kelley, D.S.; Delaney, J.R.; Juniper, S.K. Establishing a new era of submarine volcanic observatories: Cabling Axial seamount and the endeavour segment of the Juan de Fuca Ridge. *Mar. Geol.* **2014**, *352*, 426–450. [CrossRef]

122. Judd, A.; Hovland, M. *Seabed Fluid Flow: The Impact on Geology, Biology and the Marine Environment*; Cambridge University Press: Cambridge, UK, 2007.

123. Martínez-Carreño, N.; García-Gil, S. The holocene gas system of the Ría de Vigo (NW Spain): Factors controlling the location of gas accumulations, seeps and pockmarks. *Mar. Geol.* **2013**, *344*, 82–100. [CrossRef]

124. García-Gil, S.; Cartelle, V.; de Blas, E.; de Carlos, A.R.; Díez, R.; Durán, R.; Ferrín, A.; García-Moreiras, I.; García-García, A.; Iglesias, J.; et al. Gas somero en el margen continental Ibérico. *Bol. Geol. Min.* **2015**, *126*, 575–608.

125. Cartwright, J. The impact of 3D seismic data on the understanding of compaction, fluid flow and diagenesis in sedimentary basins. *J. Geol. Soc.* **2007**, *164*, 881–893. [CrossRef]

126. Sun, Q.; Wu, S.; Cartwright, J.; Dong, D. Shallow gas and focused fluid flow systems in the Pearl river mouth basin, Northern South China sea. *Mar. Geol.* **2012**, *315–318*, 1–14. [CrossRef]

127. Andreassen, K.; Hubbard, A.; Winsborrow, M.; Patton, H.; Vadakkepuliyambatta, S.; Plaza-Faverola, A.; Gudlaugsson, E.; Serov, P.; Deryabin, A.; Mattingsdal, R.; et al. Massive blow-out craters formed by hydrate-controlled methane expulsion from the Arctic seafloor. *Science* **2017**, *356*, 948–953. [CrossRef]

128. BOEM. *Gulf of Mexico Deepwater Bathymetry with Hillshade*; Bureau of Ocean Energy Management: Washington, DC, USA, 2016.

129. BOEM. *BOEM Seismic Water Bottom Anomalies—Gulf of Mexico—Gulf of Mexico NAD27*; Bureau of Ocean Energy Management: Washington, DC, USA, 2016.

130. Milkov, A.V. Global Distribution of Mud Volcanoes and Their Significance in Petroleum Exploration as a Source of Methane in the Atmosphere and Hydrosphere and as a Geohazard. In *Mud Volcanoes, Geodynamics and Seismicity*; Martinelli, G., Panahi, B., Eds.; Springer: Dordrecht, The Netherlands, 2005; pp. 29–34.

131. Etiope, G.; Klusman, R.W. Geologic emissions of methane to the atmosphere. *Chemosphere* **2002**, *49*, 777–789. [CrossRef]

132. Ciais, P.; Chris, S.; Govindasamy, B.; Bopp, L.; Brovkin, V.; Canadell, J.; Chhabra, A.; Defries, R.; Galloway, J.; Heimann, M. Carbon and Other Biogeochemical Cycles. In *Climate Change 2013: The Physical Science Basis*; Stocker, T.F., Ed.; Cambridge University Press: Cambridge, UK, 2013; pp. 465–570.

133. Etiope, G. *Natural Gas Seepage: The Earth's Hydrocarbon Degassing*; Springer: Cham, Switzerland, 2015.

134. Fernández-Puga, M.C.; Vázquez, J.T.; Somoza, L.; del Rio, V.D.; Medialdea, T.; Mata, M.P.; León, R. Gas-related morphologies and diapirism in the Gulf of Cádiz. *Geo-Mar. Lett.* **2007**, *27*, 213–221. [CrossRef]

135. Eiby, G.A. Earthquakes and tsunamis in a region of diapiric folding. *Tectonophysics* **1982**, *85*, T1–T8. [CrossRef]

136. Kvenvolden, K.A. A primer on the geological occurrence of gas hydrate. *Geol. Soc. Lond. Spec. Publ.* **1998**, *137*, 9–30. [CrossRef]

137. Soloviev, V.; Ginsburg, G.D. Formation of submarine gas hydrates. *Bull. Geol. Soc. Den.* **1994**, *41*, 86–94.

138. Hesse, R. Pore water anomalies of submarine gas-hydrate zones as tool to assess hydrate abundance and distribution in the subsurface what have we learned in the past decade? *Earth Sci. Rev.* **2003**, *61*, 149–179. [CrossRef]

139. Cunningham, R.; Lindholm, R. Seismic Evidence for Widespread Gas Hydrate Formation, Offshore West Africa. In *Petroleum Systems of South Atlantic Margins*; AAPG: Tulsa, OK, USA, 2000; pp. 93–105. [CrossRef]

140. IUSGS. Gas Hydrates-Primer. Available online: https://www.usgs.gov/centers/whcmsc/science/gas-hydrates-primer (accessed on 12 November 2020).

141. Nixon, M.F.; Grozic, J.L. Submarine slope failure due to gas hydrate dissociation: A preliminary quantification. *Can. Geotech. J.* **2007**, *44*, 314–325. [CrossRef]

142. Shanmugam, G. *New Perspectives on Deep-Water Sandstones: Origin, Recognition, Initiation, and Reservoir Quality*; Elsevier: Amsterdam, The Netherlands, 2012.

143. Mienert, J.; Vanneste, M.; Bünz, S.; Andreassen, K.; Haflidason, H.; Sejrup, H.P. Ocean warming and gas hydrate stability on the Mid-Norwegian margin at the storegga slide. *Mar. Pet. Geol.* **2005**, *22*, 233–244. [CrossRef]

144. Meinert, J. *Methane Hydrate and Submarine Slides*; Elsevier: Amsterdam, The Netherlands, 2009.

145. Elger, J.; Berndt, C.; Rüpke, L.; Krastel, S.; Gross, F.; Geissler, W.H. Submarine slope failures due to pipe structure formation. *Nat. Commun.* **2018**, *9*, 715. [CrossRef] [PubMed]

146. Judd, A.G.; Hovland, M.; Dimitrov, L.I.; Gil, S.G.; Jukes, V. The geological methane budget at continental margins and its influence on climate change. *Geofluids* **2002**, *2*, 109–126. [CrossRef]

147. Garcia-Gil, S.; Vilas, F.; Garcia-Garcia, A. Shallow gas features in incised-valley fills (Ría de Vigo, NW Spain): A case study. *Cont. Shelf Res.* **2002**, *22*, 2303–2315. [CrossRef]

148. Garcia-Gil, S. A natural laboratory for shallow gas: The Ras Baixas (NW Spain). *Geo-Mar. Lett.* **2003**, *23*, 215–229. [CrossRef]

149. Sharke, A.; Ruppel, C.; Kodis, M.; Brothers, D.; Lobecker, E. Widespread methane leakage from the sea floor on the Northern US Atlantic margin. *Nat. Geosci.* **2014**, *7*, 657–661. [CrossRef]

150. Rebesco, M.; Camerlenghi, A.; Van Loon, A.J. Contourite research: A field in full development. *Dev. Sedimentol.* **2008**, *60*, 1–10. [CrossRef]
151. Stow, D.A.V.; Hernández-Molina, F.J.; Llave, E.; Sayago-Gil, M.; del Río, V.D.; Branson, A. Bedform-velocity matrix: The estimation of bottom current velocity from bedform observations. *Geology* **2009**, *37*, 327–330. [CrossRef]
152. Hollister, C.D.; Heezen, B.C. Geological Effects of Ocean Bottom Currents: Western North Atlantic. In *Studies in Physical Oceanography*; Gordon, A.L., Ed.; Gordon and Breach: New York, NY, USA, 1972; Volume 2, pp. 37–66.
153. Llave, E.; Schönfeld, J.; Hernández-Molina, F.J.; Mulder, T.; Somoza, L.; del Río, V.D.; Sánchez-Almazo, I. High-resolution stratigraphy of the Mediterranean outflow contourite system in the Gulf of Cadiz during the late pleistocene: The impact of Heinrich events. *Mar. Geol.* **2006**, *227*, 241–262. [CrossRef]
154. Kaboth, S.; de Boer, B.; Bahr, A.; Zeeden, C.; Lourens, L.J. Mediterranean outflow water dynamics during the past ~570 kyr: Regional and global implications. *Paleoceanography* **2017**, *32*, 634–647. [CrossRef]
155. Rebesco, M.; Hernández-Molina, F.J.; Van Rooij, D.; Wåhlin, A. Contourites and associated sediments controlled by deep-water circulation processes: State-of-the-art and future considerations. *Mar. Geol.* **2014**, *352*, 111–154. [CrossRef]
156. Hernández-Molina, F.J.; Wåhlin, A.; Bruno, M.; Ercilla, G.; Llave, E.; Serra, N.; Rosón, G.; Puig, P.; Rebesco, M.; Van Rooij, D.; et al. Oceanographic processes and morphosedimentary products along the Iberian margins: A new multidisciplinary approach. *Mar. Geol.* **2016**, *378*, 127–156. [CrossRef]
157. Ercilla, G.; Casas, D.; Hernández-Molina, F.J.; Roque, C. Generation of Bedforms by the Mediterranean Outflow Current at the Exit of the Strait of Gibraltar. In *Atlas of Bedforms in the Western Mediterranean*; Guillén, J., Acosta-Yepes, J., Chiocci, F.L., Palanques, A., Eds.; Springer: Cham, Switzerland, 2017; pp. 273–280.
158. Teixeira, M.; Terrinha, P.; Roque, C.; Rosa, M.; Ercilla, G.; Casas, D. Interaction of alongslope and downslope processes in the Alentejo Margin (SW Iberia)—Implications on slope stability. *Mar. Geol.* **2019**, *410*, 88–108. [CrossRef]
159. Kuijpers, A.; Nielsen, T. Near-bottom current speed maxima in North Atlantic contourite environments inferred from current-induced bedforms and other seabed evidence. *Mar. Geol.* **2016**, *378*, 230–236. [CrossRef]
160. Hernández-Molina, F.J.; Llave, E.; Stow, D.A.V.; García, M.; Somoza, L.; Vázquez, J.T.; Lobo, F.J.; Maestro, A.; del Río, V.D.; León, R.; et al. The contourite depositional system of the Gulf of Cádiz: A sedimentary model related to the bottom current activity of the Mediterranean outflow water and its interaction with the continental margin. *Deep Sea Res. Part II Top. Stud. Oceanogr.* **2006**, *53*, 1420–1463. [CrossRef]
161. Martorelli, E.; Bosman, A.; Casalbore, D.; Falcini, F. Interaction of down-slope and along-slope processes off Capo Vaticano (southern Tyrrhenian sea, Italy), with particular reference to contourite-related landslides. *Mar. Geol.* **2016**, *378*, 43–55. [CrossRef]
162. Ercilla, G.; Juan, C.; Hernández-Molina, F.J.; Bruno, M.; Estrada, F.; Alonso, B.; Casas, D.; Farran, M.l.; Llave, E.; García, M.; et al. Significance of bottom currents in deep-sea morphodynamics: An example from the Alboran sea. *Mar. Geol.* **2016**, *378*, 157–170. [CrossRef]
163. Juan, C.; Ercilla, G.; Estrada, F.; Alonso, B.; Casas, D.; Vázquez, J.T.; d'Acremont, E.; Medialdea, T.; Hernández-Molina, F.J.; Gorini, C.; et al. Multiple factors controlling the deep marine sedimentation of the Alboran sea (SW Mediterranean) after the Zanclean Atlantic mega-flood. *Mar. Geol.* **2020**, *423*, 106138. [CrossRef]
164. Casas, D.; Yenes, M.; Urgeles, R. Submarine mass movements around the iberian Peninsula. The building of continental margins through hazardous processes. *Bol. Geol. Min.* **2015**, *126*, 257–278.
165. Droghei, R.; Falcini, F.; Casalbore, D.; Martorelli, E.; Mosetti, R.; Sannino, G.; Santoleri, R.; Chiocci, F.L. The role of internal solitary waves on deep-water sedimentary processes: The case of up-slope migrating sediment waves off the Messina strait. *Sci. Rep.* **2016**, *6*, 36376. [CrossRef]
166. Palomino, D.; Vázquez, J.T.; Ercilla, G.; Alonso, B.; López-González, N.; Díaz-del-Río, V. Interaction between seabed morphology and water masses around the seamounts on the Motril Marginal Plateau (Alboran sea, Western Mediterranean). *Geo-Mar. Lett.* **2011**, *31*, 465–479. [CrossRef]
167. Falcini, F.; Martorelli, E.; Chiocci, F.L.; Salusti, E. A general theory for the effect of local topographic unevenness on contourite deposition around marine capes: An inverse problem applied to the Italian continental margin (Cape Suvero). *Mar. Geol.* **2016**, *378*, 74–80. [CrossRef]
168. Miramontes, E.; Garreau, P.; Caillaud, M.; Jouet, G.; Pellen, R.; Hernández-Molina, F.J.; Clare, M.A.; Cattaneo, A. Contourite distribution and bottom currents in the NW Mediterranean sea: Coupling seafloor geomorphology and hydrodynamic modelling. *Geomorphology* **2019**, *333*, 43–60. [CrossRef]
169. Puig, P.; Palanques, A.; Guillén, J.; El Khatab, M. Role of internal waves in the generation of nepheloid layers on the Northwestern Alboran slope: Implications for continental margin shaping. *J. Geophys. Res.* **2004**, *109*, C09011. [CrossRef]
170. Séranne, M.; Abeigne, C.R.N. Oligocene to holocene sediment drifts and bottom currents on the slope of Gabon continental margin (West Africa): Consequences for sedimentation and southeast Atlantic upwelling. *Sediment. Geol.* **1999**, *128*, 179–199. [CrossRef]
171. Sarhan, T. Upwelling mechanisms in the Northwestern Alboran sea. *J. Mar. Syst.* **2000**, *23*, 317–331. [CrossRef]
172. Macias, D.; Garcia-Gorriz, E.; Stips, A. The seasonal cycle of the Atlantic jet dynamics in the Alboran sea: Direct atmospheric forcing versus Mediterranean thermohaline circulation. *Ocean Dyn.* **2016**, *66*, 137–151. [CrossRef]
173. Foglini, F.; Campiani, E.; Trincardi, F. The reshaping of the South West Adriatic margin by cascading of dense shelf waters. *Mar. Geol.* **2016**, *375*, 64–81. [CrossRef]

174. Ercilla, G.; Casas, D.; Estrada, F.; Vázquez, J.T.; Iglesias, J.; García, M.; Gómez, M.; Acosta, J.; Gallart, J.; Maestro-González, A. Morphosedimentary features and recent depositional architectural model of the Cantabrian continental margin. *Mar. Geol.* **2008**, *247*, 61–83. [CrossRef]

175. Juan, C.; Ercilla, G.; Hernández-Molina, F.J.; Estrada, F.; Alonso, B.; Casas, D.; García, M.; Farran, M.L.; Llave, E.; Palomino, D.; et al. Seismic evidence of current-controlled sedimentation in the Alboran sea during the pliocene and quaternary: Palaeoceanographic implications. *Mar. Geol.* **2016**, *378*, 292–311. [CrossRef]

176. Laberg, J.S.; Baeten, N.J.; Vanneste, M.; Forsberg, C.F.; Forwick, M.; Haflidason, H. Sediment Failure Affecting Muddy Contourites on the Continental Slope Offshore Northern Norway: Lessons Learned and Some Outstanding Issues. In *Submarine Mass Movements and their Consequences*; Lamarche, G., Mountjoy, J., Eds.; Springer: Cham, Switzerland, 2016; pp. 281–289.

177. Ercilla, G.; Galindo-Zaldivar, J.; Valencia, J.; Tendero-Salmeron, V.; Estrada, F.; Casas, D.; Alonso, B.; d'Acremont, E.; Comas, M.; Tomas Vazquez, J.; et al. Understanding the Geomorphology of the Gulf of Vera (Western Mediterranean): Clues from Offshore and Onland Structures. In *EGU General Assembly Conference Abstracts*; EGU: Munich, Germany, 2019; Volume 21, p. 13411.

178. Röbke, B.R.; Vött, A. The tsunami phenomenon. *Prog. Oceanogr.* **2017**, *159*, 296–322. [CrossRef]

179. Assier-Rzadkieaicz, S.; Heinrich, P.; Sabatier, P.C.; Savoye, B.; Bourillet, J.F. Numerical modelling of a landslide-generated tsunami: The 1979 nice event. *Pure Appl. Geophys.* **2000**, *157*, 1707–1727. [CrossRef]

180. Geist, E.L. Local Tsunamis and Earthquake Source Parameters. In *Advances in Geophysics*; Dmowska, R., Saltzman, B., Eds.; Elsevier: Amsterdam, The Netherlands, 1997; Volume 39, pp. 117–209.

181. Bryant, E.A. *Tsunami: The Underrated Hazard*; Springer: Berlin/Heidelberg, Germany, 2014.

182. Tappin, D.R. Mass Transport Events and Their Tsunami Hazard. In *Submarine Mass Movements and their Consequences*; Mosher, D.C., Shipp, R.C., Moscardelli, L., Chaytor, J.D., Baxter, C.D.P., Lee, H.J., Urgeles, R., Eds.; Springer: Dordrecht, The Netherlands, 2010; pp. 667–684.

183. Tappin, D.R. Tsunamis from submarine landslides. *Geol. Today* **2017**, *33*, 190–200. [CrossRef]

184. González-Vida, J.M.; Macías, J.; Castro, M.J.; Sánchez-Linares, C.; de la Asunción, M.; Ortega-Acosta, S.; Arcas, D. The Lituya bay landslide-generated mega-tsunami—Numerical simulation and sensitivity analysis. *Nat. Hazards Earth Syst. Sci.* **2019**, *19*, 369–388. [CrossRef]

185. Løvholt, F.; Pedersen, G.; Harbitz, C.B.; Glimsdal, S.; Kim, J. On the characteristics of landslide tsunamis. *Philos. Trans. R. Soc. A Math. Phys. Eng. Sci.* **2015**, *373*, 20140376. [CrossRef]

186. Moore, J.G.; Normark, W.R.; Holcomb, R.T. Giant Hawaiian landslides. *Annu. Rev. Earth Planet. Sci.* **1994**, *22*, 119–144. [CrossRef]

187. Masson, D.G. Catastrophic collapse of the volcanic island of Hierro 15 ka ago and the history of landslides in the Canary islands. *Geology* **1996**, *24*, 231–234. [CrossRef]

188. Keating, B.H.; McGuire, W.J. Island edifice failures and associated tsunami hazards. *Pure Appl. Geophys.* **2000**, *157*, 899–955. [CrossRef]

189. Ayala-Carcedo, F.J.; Olcina Cantos, J. *Riesgos Naturales*; Editorial Ariel: Barcelona, Spain, 2002.

190. Levin, B.W.; Nosov, M.A. *Physics of Tsunamis*; Springer: Berlin/Heidelberg, Germany, 2016.

191. Macías, J.; Vázquez, J.T.; Fernández-Salas, L.M.; González-Vida, J.M.; Bárcenas, P.; Castro, M.J.; Díaz-del-Río, V.; Alonso, B. The Al-Borani submarine landslide and associated tsunami. A modelling approach. *Mar. Geol.* **2015**, *361*, 79–95. [CrossRef]

192. Pattiaratchi, C. Influence of ocean topography on tsunami propagation in Western Australia. *J. Mar. Sci. Eng.* **2020**, *8*, 629. [CrossRef]

193. Lu, C.Y.; Malavieille, J. Oblique convergence, indentation and rotation tectonics in the Taiwan mountain belt: Insights from experimental modelling. *Earth Planet. Sci. Lett.* **1994**, *121*, 477–494. [CrossRef]

194. Jiabiao, L.; Xianglong, J.; Aiguo, R.; Shimin, W.; Ziyin, W.; Jianhua, L. Indentation tectonics in the accretionary wedge of middle Manila Trench. *Chin. Sci. Bull.* **2004**, *49*, 1279–1288. [CrossRef]

195. Estrada, F.; Galindo-Zaldívar, J.; Vázquez, J.T.; Ercilla, G.; D'Acremont, E.; Alonso, B.; Gorini, C. Tectonic indentation in the Central Alboran sea (Westernmost Mediterranean). *Terra Nova* **2018**, *30*, 24–33. [CrossRef]

196. Coppier, G.; Griveaud, P.; de Larouziere, F.D.; Montenat, C.; d'Estevou, P.O. Example of neogene tectonic indentation in the eastern betic cordilleras: The Arc of Aguilas (Southeastern Spain). *Geodin. Acta* **1989**, *3*, 37–51. [CrossRef]

197. Cavalié, O.; Jónsson, S. Block-like plate movements in Eastern Anatolia observed by InSAR. *Geophys. Res. Lett.* **2014**, *41*, 26–31. [CrossRef]

198. Davy, P.; Cobbold, P. Indentation tectonics in nature and experiment. 1. Experiments scaled for gravity. *Bull. Geol. Inst. Univ. Upps.* **1988**, *14*, 129–141.

199. Lu, C.Y.; Chen, C.T.; Malavieille, J. Oblique Convergence, Indentation and Bivergent Tectonics in Taiwan: Insights from Field Observations and Analog Models. In *Geophysical Research Abstracts*; EGU: Munich, Germany, 2019; Volume 21.

200. Chiocci, F.L.; Ridente, D. Regional-scale seafloor mapping and geohazard assessment. The experience from the Italian project MaGIC (Marine Geohazards along the Italian Coasts). *Mar. Geophys. Res.* **2011**, *32*, 13–23. [CrossRef]

201. Iacono, C.L.; Sulli, A.; Agate, M.; Lo Presti, V.; Pepe, F.; Catalano, R. Submarine canyon morphologies in the Gulf of Palermo (Southern Tyrrhenian sea) and possible implications for geo-hazard. *Mar. Geophys. Res.* **2011**, *32*, 127–138. [CrossRef]

202. Huang, Z.; Nichol, S.L.; Harris, P.T.; Caley, M.J. Classification of submarine canyons of the Australian continental margin. *Mar. Geol.* **2014**, *357*, 362–383. [CrossRef]

203. Ercilla, G.; Juan, C.; Periáñez, R.; Alonso, B.; Abril, J.M.; Estrada, F.; Casas, D.; Vázquez, J.T.; d'Acremont, E.; Gorini, C.; et al. Influence of alongslope processes on modern turbidite systems and canyons in the Alboran Sea (southwestern Mediterranean). *Deep Sea Res. I Oceanogr. Res. Pap.* **2019**, *144*, 1–16. [CrossRef]

204. Casas, D.; Ercilla, G.; Alonso, B.; Yenes, M.; Nespereira, J.; Estrada, F.; Ceramicola, S. Submarine Mass Movements Affecting the Almanzora-Alías-Garrucha Canyon System (SW Mediterranean)(SKT). In Proceedings of the 34th International Association of Sedimentologists (IAS) Meeting of Sedimentology, Sedimentology to Face Societal Challenges on Risk, Resources and Record of the Past, Rome, Italy, 10–13 September 2019.

205. Mazières, A.; Gillet, H.; Castelle, B.; Mulder, T.; Guyot, C.; Garlan, T.; Mallet, C. High-resolution morphobathymetric analysis and evolution of Capbreton submarine canyon head (Southeast bay of Biscay—French Atlantic Coast) over the last decade using descriptive and numerical modeling. *Mar. Geol.* **2014**, *351*, 1–12. [CrossRef]

206. Puig, P.; Palanques, A.; Martín, J. Contemporary sediment-transport processes in submarine canyons. *Annu. Rev. Mar. Sci.* **2014**, *6*, 53–77. [CrossRef] [PubMed]

207. Gómez-Ballesteros, M.; Druet, M.; Muñoz, A.; Arrese, B.; Rivera, J.; Sánchez, F.; Cristobo, J.; Parra, S.; García-Alegre, A.; González-Pola, C.; et al. Geomorphology of the avilés canyon system, Cantabrian sea (Bay of Biscay). *Deep Sea Res. II Top. Stud. Oceanogr.* **2014**, *106*, 99–117. [CrossRef]

208. Mountjoy, J.J.; Barnes, P.M.; Pettinga, J.R. Morphostructure and evolution of submarine canyons across an active margin: Cook Strait sector of the Hikurangi margin, New Zealand. *Mar. Geol.* **2009**, *260*, 45–68. [CrossRef]

209. Aranguiz, R.; Shibayama, T. Effect of submarine canyons on tsunami propagation: A case study of the biobio canyon, Chile. *Coast. Eng. J.* **2013**, *55*, 1350016-1–1350016-23. [CrossRef]

210. Casalbore, D.; Chiocci, F.L.; Mugnozza, G.S.; Tommasi, P.; Sposato, A. Flash-flood hyperpycnal flows generating shallow-water landslides at Fiumara mouths in Western Messina Strait (Italy). *Mar. Geophys. Res.* **2011**, *32*, 257–271. [CrossRef]

211. Micallef, A.; Mountjoy, J.J.; Barnes, P.M.; Canals, M.; Lastras, G. Geomorphic response of submarine canyons to tectonic activity: Insights from the cook strait canyon system, New Zealand. *Geosphere* **2014**, *10*, 905–929. [CrossRef]

212. Anzidei, M.; Bosman, A.; Carluccio, R.; Casalbore, D.; Caracciolo, F.D.; Esposito, A.; Nicolosi, I.; Pietrantonio, G.; Vecchio, A.; Carmisciano, C.; et al. Flooding scenarios due to land subsidence and sea-level rise: A case study for Lipari island (Italy). *Terra Nova* **2017**, *29*, 44–51. [CrossRef]

213. Chiocci, F.L.; Casalbore, D. Unexpected fast rate of morphological evolution of geologically-active continental margins during quaternary: Examples from selected areas in the Italian seas. *Mar. Pet. Geol.* **2017**, *82*, 154–162. [CrossRef]

214. Casalbore, D.; Romagnoli, C.; Bosman, A.; Anzidei, M.; Chiocci, F.L. Coastal hazard due to submarine canyons in active insular volcanoes: Examples from Lipari island (Southern Tyrrhenian sea). *J. Coast. Conserv.* **2018**, *22*, 989–999. [CrossRef]

215. Carracedo, J.C.; Troll, V.R.; Zaczek, K.; Rodríguez-González, A.; Soler, V.; Deegan, F.M. The 2011–2012 submarine eruption off El Hierro, Canary islands: New lessons in oceanic island growth and volcanic crisis management. *Earth Sci. Rev.* **2015**, *150*, 168–200. [CrossRef]

216. Casalbore, D.; Clare, M.A.; Pope, E.L.; Quartau, R.; Bosman, A.; Chiocci, F.L.; Romagnoli, C.; Santos, R. Bedforms on the submarine flanks of insular volcanoes: New insights gained from high resolution seafloor surveys. *Sedimentology* **2020**. [CrossRef]

217. Urgeles, R.; Canals, M.; Baraza, J.; Alonso, B.; Masson, D. The most recent megalandslides of the Canary islands: El Golfo debris avalanche and Canary debris flow, West El Hierro island. *J. Geophys. Res. Solid Earth* **1997**, *102*, 20305–20323. [CrossRef]

218. Romagnoli, C.; Casalbore, D.; Chiocci, F.L.; Bosman, A. Offshore evidence of large-scale lateral collapses on the eastern flank of Stromboli, Italy, due to structurally-controlled, bilateral flank instability. *Mar. Geol.* **2009**, *262*, 1–13. [CrossRef]

219. Coombs, M.L.; White, S.M.; Scholl, D.W. Massive edifice failure at Aleutian Arc volcanoes. *Earth Planet. Sci. Lett.* **2007**, *256*, 403–418. [CrossRef]

220. Casalbore, D.; Bosman, A.; Chiocci, F.L. Study of Recent Small-Scale Landslides in Geologically Active Marine Areas Through Repeated Multibeam Surveys: Examples from the Southern Italy. In *Submarine Mass Movements and Their Consequences*; Yamada, Y., Kawamura, K., Ikehara, K., Ogawa, Y., Urgeles, R., Mosher, D., Chaytor, J., Strasser, M., Eds.; Springer: Dordrecht, The Netherlands, 2012; pp. 573–582.

221. Casalbore, D.; Passeri, F.; Tommasi, P.; Verrucci, L.; Bosman, A.; Romagnoli, C.; Chiocci, F.L. Small-scale slope instability on the submarine flanks of insular volcanoes: The case-study of the Sciara del Fuoco slope (Stromboli). *Int. J. Earth Sci.* **2020**, *109*, 2643–2658. [CrossRef]

222. Maramai, A.; Graziani, L.; Tinti, S. Tsunamis in the Aeolian islands (Southern Italy): A review. *Mar. Geol.* **2005**, *215*, 11–21. [CrossRef]

223. Carey, S.; Sigurdsson, H.; Mandeville, C.; Bronto, S. Volcanic Hazards from Pyroclastic Flow Discharge into the Sea: Examples from the 1883 Eruption of Krakatau, Indonesia. In *Volcanic Hazards and Disasters in Human Antiquity*; McCoy, F.W., Heiken, G., Eds.; Geological Society of America: Boulder, CO, USA, 2000; Volume 345, pp. 1–14.

224. Chiocci, F.L.; Romagnoli, C.; Tommasi, P.; Bosman, A. The stromboli 2002 tsunamigenic submarine slide: Characteristics and possible failure mechanisms. *J. Geophys. Res.* **2008**, *113*, B10102. [CrossRef]

225. Caress, D.W.; Clague, D.A.; Paduan, J.B.; Martin, J.F.; Dreyer, B.M.; Chadwick, W.W.; Denny, A.; Kelley, D.S. Repeat bathymetric surveys at 1-metre resolution of lava flows erupted at Axial seamount in April 2011. *Nat. Geosci.* **2012**, *5*, 483–488. [CrossRef]

226. Chadwick, W.W.; Dziak, R.P.; Haxel, J.H.; Embley, R.W.; Matsumoto, H. Submarine landslide triggered by volcanic eruption recorded by in situ hydrophone. *Geology* **2012**, *40*, 51–54. [CrossRef]

227. Carey, R.; Soule, S.A.; Manga, M.; White, J.; McPhie, J.; Wysoczanski, R.; Jutzeler, M.; Tani, K.; Yoerger, D.; Fornari, D.; et al. The largest deep-ocean silicic volcanic eruption of the past century. *Sci. Adv.* **2018**, *4*, e1701121. [CrossRef] [PubMed]
228. Chadwick, W.W.; Rubin, K.H.; Merle, S.G.; Bobbitt, A.M.; Kwasnitschka, T.; Embley, R.W. Recent eruptions between 2012 and 2018 discovered at West Mata submarine volcano (NE Lau Basin, SW Pacific) and characterized by new ship, AUV, and ROV data. *Front. Mar. Sci.* **2018**, *6*, 495. [CrossRef]
229. Gaspar, J.L.; Queiroz, G.; Pacheco, J.M.; Ferreira, T.; Wallenstein, N.; Almeida, M.H.; Coutinho, R. Basaltic Lava Balloons Produced During the 1998–2001 Serreta Submarine Ridge Eruption (Azores). In *Explosive Subaqueous Volcanism*; White, J.D.L., Smellie, J.L., Clague, D.A., Eds.; American Geophysical Union: Washington, DC, USA, 2003; pp. 205–212.

Article

A Conflict between the Legacy of Eutrophication and Cultural Oligotrophication in Hiroshima Bay

Tamiji Yamamoto [1,*,†], **Kaori Orimoto** [1], **Satoshi Asaoka** [1], **Hironori Yamamoto** [2] and **Shin-ichi Onodera** [3]

[1] Graduate School of Biosphere Science, Hiroshima University, 1-4-4 Kagamiyama, Higashi-Hiroshima 739-8528, Japan; volumins02orimo-keko@n.vodafone.ne.jp (K.O.); stasaoka@hiroshima-u.ac.jp (S.A.)

[2] Fukken Co., Ltd., 2-10-11 Hikarimachi, Higashi-Hiroshima 732-0052, Japan; h-yamamoto@fukken.co.jp

[3] Graduate School of Integrated Arts and Sciences, Hiroshima University, 1-7-1 Kagamiyama, Higashi-Hiroshima 739-8521, Japan; sonodera@hiroshima-u.ac.jp

*   Correspondence: tamyama@hiroshima-u.ac.jp; Tel.: +81-82-434-6717

† Current Address: Center for Restoration of Basin Ecosystem and Environment (CERBEE), 4-3-13 Takaya-Takamigaoka, Higashi-Hiroshima 739-2115, Japan.

Abstract: Although the water quality in Hiroshima Bay has improved due to government measures, nutrient reduction has sharply decreased fisheries production. The law was revised in 2015, where the nutrient effluents from the sewage treatment plants were relaxed, yet no increase in fishery production was observed. Herein, we investigate the distribution of C, N, S, and P within Hiroshima Bay. Material loads from land and oyster farming activity influenced the C and S distributions in the bay sediments, respectively. Natural denitrification caused N reduction in areas by the river mouths and the landlocked areas whose sediments are reductive. The P content was high in the areas under aerobic conditions, suggesting metal oxide-bound P contributes to P accumulation. However, it was low in the areas with reducing conditions, indicating P is released from the sediments when reacting with H_2S. In such reductive sediments, liberated H_2S also consumes dissolved oxygen causing hypoxia in the bottom layer. It was estimated that $0.28\ km^3$ of muddy sediment and 1.8×10^5 ton of P accumulated in Hiroshima Bay. There remains conflict between the 'Legacy of Eutrophication' in the sediment and 'Cultural Oligotrophication' in the surface water due to 40 years of reduction policies.

Keywords: carbon; eutrophic; Hiroshima Bay; phosphorus; nitrogen; sulfur; sediment

Citation: Yamamoto, T.; Orimoto, K.; Asaoka, S.; Yamamoto, H.; Onodera, S.-i. A Conflict between the Legacy of Eutrophication and Cultural Oligotrophication in Hiroshima Bay. *Oceans* **2021**, *2*, 546–565. https://doi.org/10.3390/oceans2030031

Academic Editor: Michael W. Lomas

Received: 30 December 2020
Accepted: 10 August 2021
Published: 16 August 2021

Publisher's Note: MDPI stays neutral with regard to jurisdictional claims in published maps and institutional affiliations.

1. Introduction

Estuaries function as filters/traps or removal devices for nitrogen (N) and phosphorus (P) to the offshore [1]. At the river mouth, flocculation of colloidal iron oxides occurs upon the mixing of riverine freshwater and seawater, and orthophosphate is trapped as Fe(III) (oxyhydr)oxides-P and sedimented on the bottom [2]. In addition to this chemical process, particulate matter tends to be sedimented at the innermost area of estuary due to estuarine circulation, which is induced by river discharge [3,4]. Phytoplankton grows by taking up nutrients supplied from land through rivers and other point sources and the sea. Phytoplankton will sink to the bottom as dead cells or feces of filter feeders along with other particulate matter. In the succession of mineralization on and in the sediments, organic-bound P can be liberated, and orthophosphate is taken up by phytoplankton and bacteria again in the water column. This is one main framework of the filtering or trapping function of P in estuarine systems.

However, the P cycle in the sediments and overlying water is much more complex. Iron phosphate is formed in rich Fe and phosphate conditions. Fe(III) (oxyhydr)oxides form at the oxic/anoxic boundary between sediment and overlying water [5]. During oxygen consumption via decomposition of organic matter supplied from the upper water column, Fe and P are released to the bottom of the water column. In anoxic sediments under non-sulfide conditions, such as lakes, ferrous phosphate may significantly contribute to P retention in the sediments [6]. However, the production of H_2S in reducing sediments in estuaries counteracts the functioning of Fe(III) (oxyhydr)oxides in trapping P by forming iron sulphides (FeS_x) instead [7]. Simultaneously, Fe(III) (oxyhydr)oxides-P may undergo reductive dissolution in the presence of S^{2-} and release orthophosphate [8]. Subsequently, the increased liberation of orthophosphate, Fe^{2+} and Mn^{2+} over long periods forms authigenic phosphorites known as apatite [9–11] and vivianite [12–15]. Thus, in general, P is permanently removed from the water column and buried in the sediment, and not all sedimented P returns to the water column [16–18]. The importance of each diagenetic process is different in different areas and sites, depending on the depositional environment, sedimentation rate, and prevailing redox conditions [5].

N is removed in estuaries. The main pathway of N removal is denitrification, accounting for ~50% of N loaded through rivers [19–22], while anaerobic ammonium oxidation (anammox) and burial in sediments are negligible [23]. Nitrification can be the major nitrate source for denitrification, supplying >80% of the nitrate demand in the St. Lawrence Estuary [24,25]. When the bottom is always reductive with no nitrate supply, denitrification depends on riverine nitrate supply or oxic seawater intrusion, as observed in Lakes Shinji and Nakaumi, Japan [26]. Under anoxic conditions and during bottom water hypoxia, dissimilatory nitrate reduction to ammonium (DNRA) competes with denitrification, and it may recycle the N in the system and contribute to an increase in primary production [27–32]. This varies across different areas and sites depending on the level of eutrophication; the ratio of nitrate to organic matter is a primary factor which overcomes denitrification or DNRA in the system [33]. Furthermore, the presence of reduced sulfur in the sediments also influences denitrification, DNRA [34], and coupled nitrification-denitrification [35]. Furthermore, a larger fraction of nitrate can sometimes be retained in the system in the presence of sulfides [27,35]. Overall, estuaries can work naturally as N removal devices and phosphorus traps.

The Japanese government established the 'Law Concerning Special Measures for Conservation of the Environment of the Seto Inland Sea' in 1978, following its predecessor, the 'Law Concerning Tentative Measures for Conservation of the Environment of the Seto Inland Sea' in 1973 [36]. As part of the measures taken for the implementation of this law, reductions in P and N contents in discharge water from all workplaces/factories, which discharge >50 ton day^{-1} of water, began in 1980 and 2005, respectively. Thus, the long-term reduction of nutrients (40 years for P and 15 years for N) has successfully contributed to improving the water quality, as evidenced by the apparent increase in water transparency [37]. However, such simple reductions of P and N do not render the desired effect to the Seto Inland Sea ecosystem because the response of the ecosystem often shows non-linearities characterized by hysteresis [38–41]. Fishery production in the Seto Inland Sea has sharply decreased with decreasing primary production since the late 1980s without tracing the same trajectory as it had traveled during the eutrophication period [41].

In Hiroshima Bay, which is in the western part of the Seto Inland Sea, oyster culture has been conducted intensively for more than 60 years, accounting for 60% of the total production in Japan [42]. However, the oyster production in the bay is in a crisis due to the lack of feed phytoplankton, and recently, it has been observed that the oyster larvae cannot survive due to the lack of small-sized phytoplankton (<5 μm in diameter) suitable for their feed [43]. In contrast, oysters, through their feeding activity, stimulate biogeochemical processes in the sediments by increasing the sedimentation of organic matter from the water column [44,45]. Therefore, the sediment quality is muddy and contains vast amounts of refractory organic matter which have been historically deposited from the oyster culture via both oyster feces and dead oyster meat, and organisms attached on the shells sink to the bottom. In the sediment of such reduced conditions, sulfide is produced along with the anaerobic decomposition of refractory organic matter. Therefore, acid volatile sulfide (AVS), water content (WC), and ignition loss (IL) of the sediment are higher in the Hiroshima Bay compared to the neighboring area with less aquaculture activity [46]. Oxygen depletion in the bottom layer (hypoxia) is still observed even after 40 years of the P reduction measure. This can be attributed to the oxygen consumption by reductive substances such as sulfide produced in the sediment, not by freshly produced organic matter decomposition. This can be referred to as a 'Legacy of Eutrophication' or 'Legacy of Hypoxia', which has been referenced in the previous studies [47–50]. That is, H_2S and other reductive matter produced in the anaerobic sediment, in which an abundance of refractory organic matter accumulated in the bottom sediments during the former eutrophication period, consume the bottom dissolved oxygen (DO) and liberate phosphate into the bottom water even several decades after the peak eutrophication period.

The fisheries production has been decreasing since the 1980s when the measure was implemented [41]. All kinds of bioproduction, not only seaweed culture, bivalve culture, and capture fisheries, have apparently decreased [41,51,52]; the fish catch and shellfish production in the Seto Inland Sea in 2014 was one-third and one-sixth of their peaks in 1980s and 1970s, respectively. Therefore, the central government revised the above-mentioned law in 2015 [53]. The major revision was the relaxation of nutrient removal from sewage treatment water. However, no signs of improvement in any aquaculture production and capture fisheries were observed until 2021. The challenge in the Seto Inland Sea including Hiroshima Bay is a 'conflict' between the 'Legacy of Eutrophication' of the sediments and 'Cultural Oligotrophication' of the surface water by the excess nutrient reduction, taken as the measure to improve eutrophication. The sediment quality of the Seto Inland Sea, including those in Hiroshima Bay, have been surveyed thrice thus far by the government [54–56] with the standard methods [57]; however, there are few analyses and discussions in relation to fishery production.

In this study, we investigate the sediment quality in Hiroshima Bay, describe the differences in the distribution patterns of sediment carbon (C), N, P, and sulfur (S) contents, and discuss the possible causes of the distribution patterns of these elements in the sediments from physical, chemical, and biological process perspectives. A discussion on the conflict between the eutrophic sediment state and the oligotrophic surface water condition, which has been induced by the long-term nutrient reduction measures is also presented.

2. Materials and Methods

2.1. Characteristics of Hiroshima Bay

Hiroshima Bay is in the western part of the Seto Inland Sea, Japan. A strict nutrient reduction measure of P and N for the Seto Inland Sea has been mandated by the government since 1980 and 1995, respectively. Both the total phosphorus (TP) and total nitrogen (TN) discharge loads in 2014 were approximately 1/2 of their peak loads of in 1976 and 1994, respectively [52]. Oyster culture is intensively conducted in the bay, producing ~60% of the total production in Japan [42]. Although the nutrients transported from the land well cherished the oyster production through feed phytoplankton growth in the bay, the nutrient reduction measure strongly damaged the oyster production, resulting a decrease in oyster production of fresh meat to 1.6–2.0×10^4 ton yr^{-1} in recent years from the peak production of 3×10^4 ton yr^{-1} in 1986 [42]. While cultured oysters may suppress excess phytoplankton blooms by their feeding activity, the biodeposits deteriorate the sediment quality [44]. In addition to feces, pseudofeces, dead oysters, and other creatures grown on the raft culture system sink onto the bottom when farmers are handling the oysters, deteriorating the sediment quality. Thus, the sediment quality of the bay is poor compared to that of the neighboring area in terms of oxidation-reduction potential (ORP), IL, and AVS [46].

Hiroshima Bay has an area of 1043 km^2, average depth of 25.8 m, volume of 26.9×10^9 m^3, and catchment area of 3743 km^2 [36]. It is geographically divided into two areas: the northern Hiroshima Bay (nHB), north of the narrow strait (Nasabi Strait) between Itsukushima Island and Etajima Island, and the southern Hiroshima Bay (sHB), south of the channel (Figure 1). The nHB area is 141.2 km^2, excluding Etajima Bay and Kure Bay, and the remaining is that of sHB [58]. In nHB, pronounced stratification is established due to large freshwater input from the Ohta River system; the average total discharge including the other small rivers is 7.5×10^6 m^3 day^{-1} and the contribution of the Ohta River is ~90% [58]. The freshwater residence time in nHB is estimated to be 27 d in the study [58]. Thus, the stratification is a 'halocline,' ultimately reducing vertical mixing [58], thereby contributing to the expansion of hypoxia in the bottom layer [59]. The freshwater from the Ohta River system entrains approximately seven times volume saline bottom water on average from the sHB [58], which may transport both suspended and resolved matter from the sHB. Meanwhile, stratification of the water column is not as strong in sHB, except the western part where there is freshwater input from the Nishiki River and the Oze River (Figure 1). Kure Bay and Etajima Bay are stagnant; therefore, hypoxia with DO concentration <2 mg L^{-1} in the bottom water is observed in September in these bays in addition to the innermost area of nHB [59].

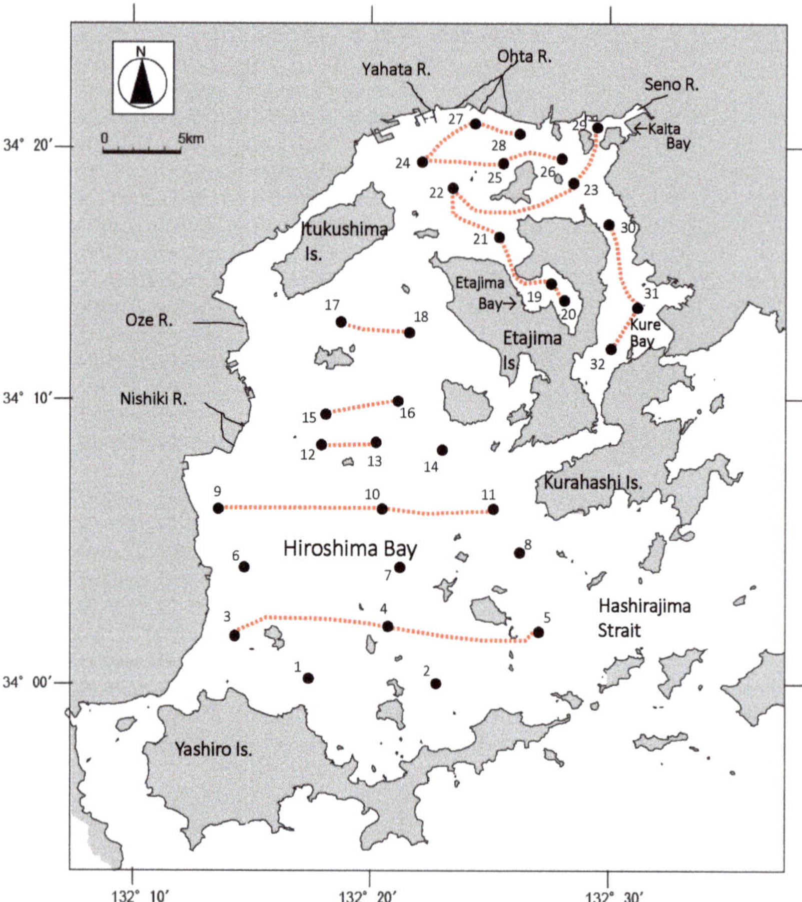

Figure 1. Map showing the monitoring lines of sediment thickness (dotted line) and core sampling stations (filled circle). 5–8 October 2010.

2.2. Field Observations

Field observations were conducted during a research cruise at 14 stations in nHB and 18 stations in sHB during 5–8 October 2010 (Figure 1). The thickness of muddy sediments in Hiroshima Bay was monitored throughout the cruise at a ship speed of 4–5 knots, using a sounding device (SH-20, Senbon Electric Co., Ltd., Numazu, Japan) emitting two acoustic frequency beams (15 and 200 kHz) [60]. As the device is graduated in 10 cm intervals, we measured 1 cm between the tick by the eyes. The sediment thickness data was used to estimate the accumulated P amounts in the sediments in Section 2.4.

At every station, five sediment core samples were collected using a KK-type core sampler with an acrylic tube inner diameter of 4.8 cm. Three of the sediment cores were selected for analyses based on their appearance and lack of artificial disturbance. First, the overlying water temperature was measured with a bar thermometer. The temperature range was small, between 22.7 (Stations 24 and 32) and 25.8 °C (Stations 10 and 19) because the water column had already mixed vertically due to surface cooling. After measuring the overlying water temperature, the core samples were sliced at 5-cm intervals from the surface to 20 cm depth, and at 10-cm intervals below this until the depth that the sediment was collected with the device was reached. The slicing of the sediment core samples was rapidly performed on the ship deck after the sediments were collected at each station. Although the surface of the sediment may have partially oxidized during the slicing process, we aimed to not oxidize the sediments by storing each sediment sample in a gas-tight plastic container 7.8 cm in diameter and 4.6 cm tall and placed them in a dark, cool (0–5 °C) container filled with nitrogen gas until various analyses are conducted. Sediment core sample collection failed at Station 5 because the bottom was composed of sandy sediments.

2.3. Chemical Analyses

Immediately after sediment core sampling, the ORP was measured for the samples collected from 0–5 cm using an ORP meter (PS-112C, DKK-TOA Corp., Tokyo, Japan). Temperature was recorded using the ORP meter. The Eh value was calculated from the ORP and temperature values, where Eh (mV) = ORP (mV) + 206 − 0.7 (t − 25). The AVS concentration was determined with a detection tube (Hedrotech-S, Gastech Corp.) for the sediment samples collected from 0–5 and 5–10 cm [61]. By adding acid to the sediment sample, H_2S gas was detected as the color of the reagent in the test tube changed (brown); $H_2S + Pb(CH_3COO)_2$ (white) → PbS (brown) + $2CH_3COOH$. The detection limit of the detection tube was 0.0002 mg, and the Coefficient of variation (CV) was 5%. Here, AVS is an operationally defined property that is the H_2S evolved from the acid leaching of sediment. A detail review on what compounds are measured by the acid leaching method of AVS is described in the literature [62]. Dividing sediment into aqueous phase and solid phase, the major dissolved species in AVS are HS^-, H_2S, and the aqueous FeS under the condition of reduced sediments with pH < 7. In the solid phase of sediment, FeS_2 (pyrite) may not contribute to AVS. Notably, part of the FeS in the aqueous phase consists of nanoparticles of mackinawite due to sample handling usually through a 0.2 μm pore size filter. The hydrogen sulfide (H_2S) concentration was determined with a detection tube (Kitagawa-type, Komyo Rikagaku, 200SA, 200SB, Kawasaki, Japan; detection limit 0.1 mg L^{-1}, accuracy ± 15%) [63] for the interstitial water of sediment core samples collected from 0–5 and 5–10 cm after centrifugation at 3000 rpm for 15 min. The head space of the centrifugation tube was filled with nitrogen gas, and both AVS and H_2S analyses were conducted within 3 min to minimize oxidation of the sample. This quick measurement procedure does not give significant results [64]. The detection tube was calibrated beforehand by H_2S standard solution by dissolving an aliquot of $Na_2S_9H_2O$ (Nacalai Tesque, Kyoto, Japan) in 3% NaCl solution to correct a salinity error. The H_2S oxidation rate is dependent on the initial concentration of H_2S and salinity but not on water temperature and pH [64].

All sediment samples were stored in a cool (0–5 °C) dark box and brought back to the land laboratory. After thoroughly mixing each sample with a spatula, the mud wet density (g cm^{-3}) was determined using a weighing balance. The sample dry weight was determined after drying the wet mud samples for 12 h at 110 °C in an oven. The WC (%) of each mud sample was determined as the difference between the wet and dry weights. The dried mud samples were combusted at 600 °C in a muffle furnace for 4 h and weighed again. IL was determined as the difference in weights before and after combustion [65].

After adding 2N HCl solution to a portion of the dried mud sample to remove inorganic carbon, samples were dried, and total organic carbon (TOC), TN, and total sulfur (TS) were determined using a CHN analyzer (CHNS/0 2400II, Perkin Elmer, Waltham, MA, USA). The TP was determined using an inductively coupled plasma-atomic emission spectroscopy (ICP-AES, Optima 7300DV, Perkin Elmer; The Ministry of the Environment of Japan [57]) with a detection limit of 0.1%. The measurements were conducted in duplicate in the first run, and then the second run was taken place in case the two values in the first run are apart. We adopted the average value of the two closest values.

2.4. Estimation of the Sediment P Content

First, we created a depth contour map showing the thickness of muddy sediments. All the depths of the surface muddy sediment measured with the acoustic device were characterized by WC (average: 61.4% +/− standard deviation: 3.6%) and IL (average: 9.90% +/− standard deviation: 0.91%). Here, the data were selected for 18 stations where the demarcation between the soft, muddy surface sediment and the hard sediment below it was clear. The demarcation level was used to estimate the thickness of the surface muddy sediment in the other stations.

Then, the mud volume (m^3) was calculated by multiplying the area (m^2) of every 5-cm-thick sample. A horizontal distribution map of each element concentration was produced at 5-cm depth intervals, and areas (m^2) with the same concentration were summed. Then, total amount of P in the Hiroshima Bay sediments was estimated using the area (m^2), average concentration ($mg\ g^{-1}$), mud wet density ($g\ cm^{-3}$), and thickness of each layer.

3. Results

3.1. Sediment C, N, S, and P Contents

Vertical profiles and horizontal distributions of TOC, TN, TS, and TP are illustrated in Figures 2–6. The vertical profiles of each element showed high values in the surface layer and decreased gradually with depth (Figure 2). Only the TS content was high in the depth of 10–20 cm at the Ohta River mouth (Station 27) and Kure Bay (Stations 31 and 32). The average values, standard variations, and coefficient of variations are summarized in Table 1. In the average value in the table, only TS has a peak value at 10–15 cm depth. Small variation in the vertical profile of TP values was clear (Figure 2), which is evidenced by the low CV (6–11%) as summarized in Table 1.

Table 1. Average (Avg) and standard deviation (Stdev) of TOC, TN, TS, and TP in the Hiroshima Bay sediments observed in 5–8 October 2010. Coefficient of variation (CV, %) is also shown.

Depth (cm)	TOC (mg C g dry^{-1})			TN (mg N g dry^{-1})			TS (mg S g dry^{-1})			TP (mg P g dry^{-1})		
	Avg	Stdev	CV (%)	Avg	Stdev	CV (%)	Avg	Stdev	CV (%)	Avg	Stdev	CV (%)
2.5	15	3	17	1.6	0.5	32	3.9	1.4	35	0.57	0.04	6
7.5	14	3	21	1.4	0.4	31	3.9	1.3	32	0.52	0.04	9
12.5	14	3	24	1.2	0.4	33	4.1	1.7	41	0.50	0.06	11
17.5	13	3	25	1.2	0.4	35	3.7	1.6	43	0.46	0.03	6
25	12	4	30	1.2	0.4	36	3.6	1.3	37	0.46	0.04	8

TOC was high in nHB, particularly at the Ohta River mouth (Stations 27) with the highest value (>20 mg g dry^{-1}) followed by Etajima Bay (Station 20) and Kure Bay (Station 32) with values ~15–20.0 mg g dry^{-1} (Figures 2 and 3). In sHB, TOC was slightly high in the western part, where the Nishiki River water enters, with values of 10–15 mg g dry^{-1} (Figure 3) which were high at the surface (Figure 2).

The highest TN values were observed in the central area, both in nHB (2.9 mg g dry^{-1} at Station 22) and sHB (3.0 mg g dry^{-1} at Stations 10) as shown in Figure 4. However, it showed lower values (1–2 mg g dry^{-1}) below the surface (Figure 2). In contrast, the values at the Ohta River mouth (Station 27) and the Nishiki River mouth (Station 9) were

<1.5 mg g dry^{-1}, as shown in Figures 2 and 4, which were different from the distribution of TOC, as shown in Figure 3.

The highest TS value in the surface 0–5 cm layer, 7.3 mg g dry^{-1}, was observed at the mouth of Etajima Bay (Station 21) as shown in Figure 5; however, the concentration was low in the deeper layer at this station (not shown in Figure 2). The highest values (9.3–9.4 mg g dry^{-1}) were detected at 10–20 cm depth at Station 20 in Etajima Bay (Figure 2). Except those high values in and around Etajima Bay and the western part of sHB (5.9 mg g dry^{-1} at Station 9) in the surface layer, the TS concentration, including in the deeper layer, did not differ much between nHB (2.0–7.1 mg g dry^{-1}) and sHB (1.1–5.9 mg g dry^{-1}).

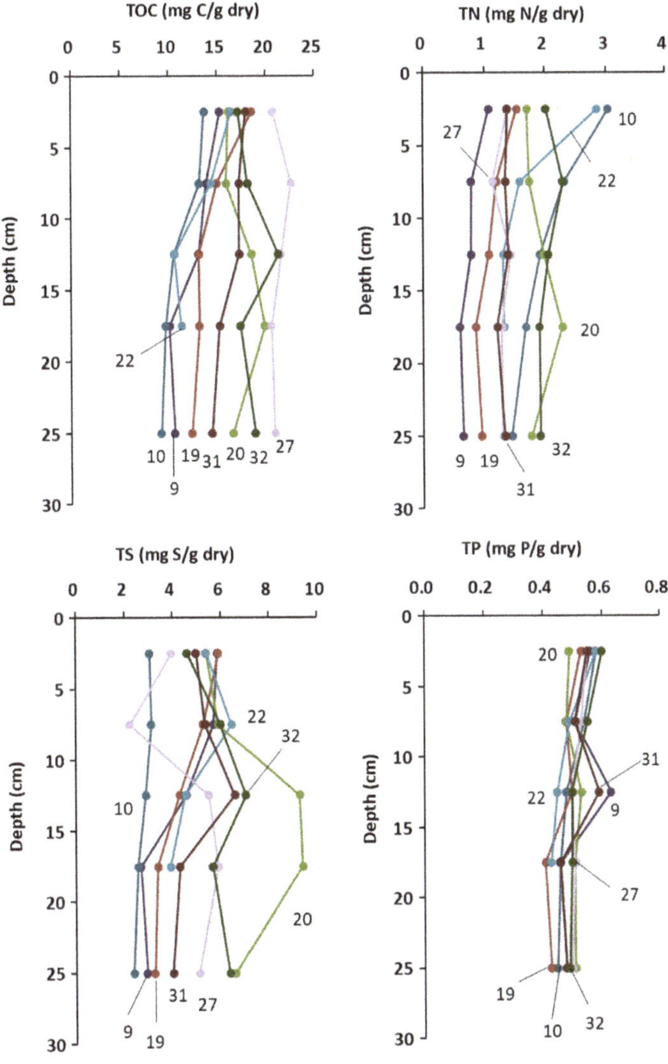

Figure 2. Vertical distributions of total organic carbon (TOC), total nitrogen (TN), total sulfur (TS), and total phosphorus (TP) at selected stations. 5–8 October 2010. The number in the figure indicates the station number of the selected stations; 27 (the Ohta River mouth), 9 (the Nishiki River mouth), 22 (the center of nHB), 10 (the center of sHB), 31, 32 (Kure Bay), and 19, 20 (Etajima Bay), respectively.

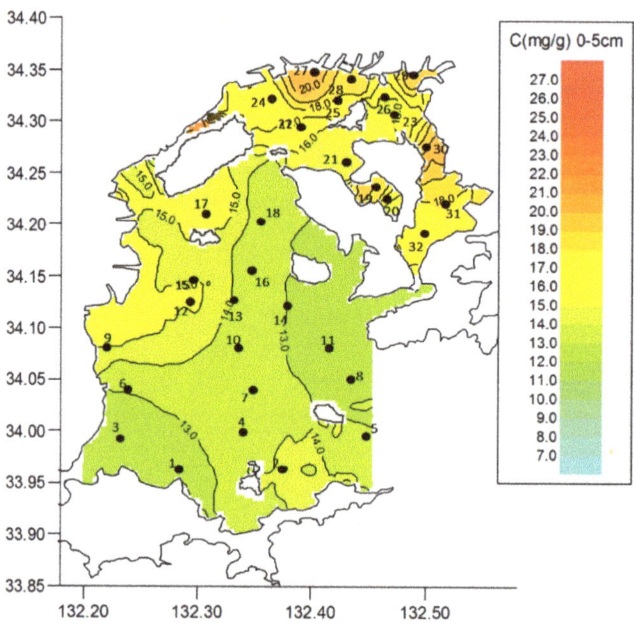

Figure 3. Spatial distributions of the total organic carbon content of muddy surface sediments (0–5 cm) in Hiroshima Bay in 5–8 October 2010.

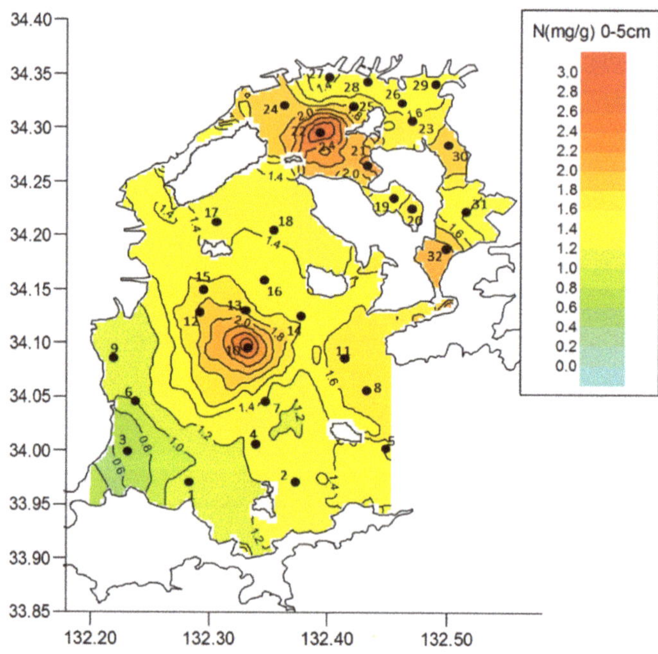

Figure 4. Spatial distributions of the nitrogen content of muddy surface sediments (0–5 cm) in Hiroshima Bay in 5–8 October 2010.

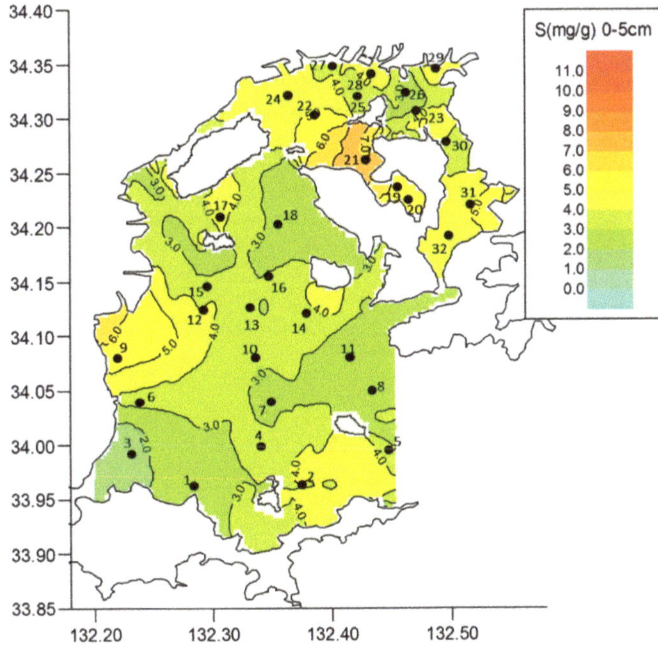

Figure 5. Spatial distributions of the sulfur content of muddy surface sediments (0–5 cm) in Hiroshima Bay in 5–8 October 2010.

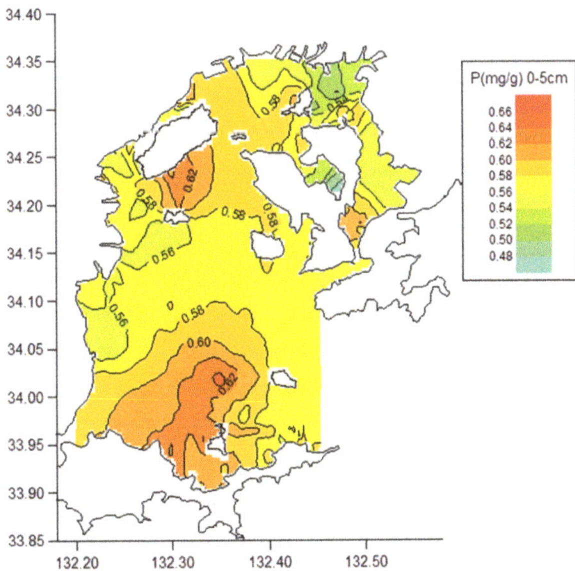

Figure 6. Spatial distributions of the phosphorus content of muddy surface sediments (0–5 cm) in Hiroshima Bay in 5–8 October 2010.

The spatial distribution of TP was somewhat homogeneous and was different from the other elements, with values of 0.37–0.67 mg g dry^{-1} in both nHB and sHB (Figures 2 and 6).

The values at the northeastern part of nHB along with the Ohta River mouth (Station 27) and the Nishiki River mouth (Station 9) were slightly low, as shown in Figure 6.

As shown in Figure 7, while the thickness of muddy sediments in nHB was 0.2–0.4 m at the mouths of the Ohta River and the Yahata River, it was 0.3–0.5 m thick in Kure Bay, where the water is stagnant due to the landlocked configuration (Figure 1). In sHB, the muddy sediment thickness was also high (0.4–0.5 m) in the southernmost area north of Yashiro Island and locally in western area. The volume of muddy sediments was estimated to be approximately 2.82×10^8 m^3 for the entire HB. The amount of P accumulated in the Hiroshima Bay sediment, estimated using thickness data of the muddy sediment and the P content, was 1.8×10^5 ton. Accumulation processes will be discussed later.

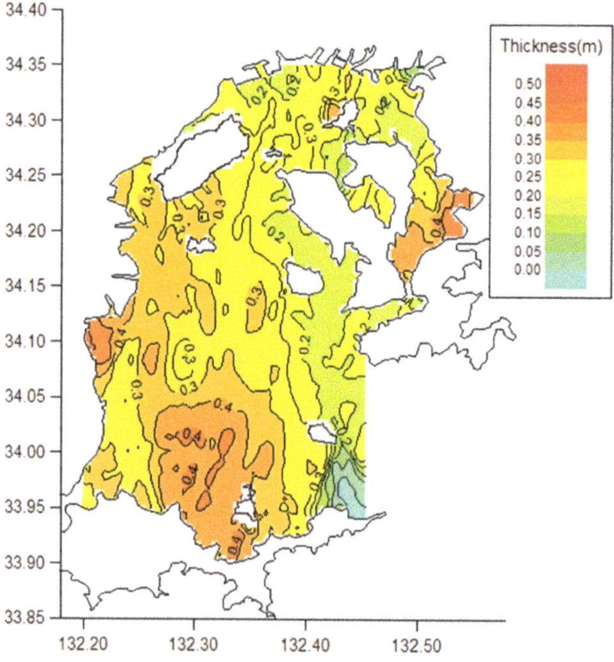

Figure 7. Spatial distributions of the thickness of muddy sediments in Hiroshima Bay estimated for the observed data in 5–8 October 2010.

3.2. Subsidiary Parameters Relating to C, N, S and P Cycles

Observed parameters (Eh, WC, IL, AVS, and H$_2$S) for the soft, muddy surface sediment whose depths are different at every station, which relate to C, N, S and P cycles, are summarized in Table 2. The Eh in nHB was from −150 to −27 mV. The lowest area was Etajima Bay, with values ranging from −150 to −120 mV, indicating that sediment in Etajima Bay was under severe reducing conditions. In sHB, the sediment conditions were also generally reducing (−153 to −25 mV), except for some sandy areas (Stations 1 and 3: +23 and +45 mV, respectively). The WC ranged 69.8–82.5% in nHB and 46.6–81.8% in sHB. The IL values were high in nHB (10.9–12.7%), and particularly high in Etajima Bay, Kaita Bay, and Kure Bay, compared to those in sHB (4.8–11.3%). The AVS was higher in nHB, with values of 0.17–1.14 mg g^{-1}, than in sHB (0.04–0.35 mg g^{-1}). The highest AVS value was observed at the mouth of the Ohta River (Station 27). H$_2$S in the interstitial water of the sediments was detected at most stations. The highest value (23 mg S L^{-1}) was observed both in Kure Bay (Station 32) and the center of sHB (Station 13), and the second highest value (19 mg S L^{-1}) was in Etajima Bay (Station 20).

Table 2. Summary of observed parameters (Eh, water content, ignition loss, acid volatile sulfide, and hydrogen sulfide) in the surface (0–5 cm) sediments of Hiroshima Bay. 5–8 October 2010. H_2S is the concentration in the interstitial water of sediment. N and S mean northern and southern Hiroshima Bay, respectively.

Parameters	Area	Value Range
Eh (mV)	N	−150−−27
	S	−153−+45
WC (%)	N	69.8–82.5
	S	46.6–81.8
IL (%)	N	10.9–12.7
	S	4.8–11.3
AVS (mg g dw^{-1})	N	0.17–1.14
	S	0.04–0.35
H_2S (mg L^{-1})	N	0–19
	S	0–23

4. Discussion

The C content was generally high at the mouths of Ohta River, Yahata River, and Seno River (Kaita Bay). This may be due to large inputs of organic matter from the cities of Hiroshima and Kure, which have large populations (Hiroshima: 1.2 million, Kure: 0.23 million). Even in sHB, the C content in the western part was high, which can be explained by the material load of the Nishiki River flowing through Iwakuni City (population: 0.14 million). One of the major sources of particulate organic matter is the direct load from the rivers. According to Kittiwanich et al. [66], the 10-years (1991–2000) average riverine particulate organic nitrogen (PON) and particulate phosphorus (PP) loadings are 14.5 mg N m^{-2} day^{-1} and 2.73 mg P m^{-2} day^{-1}, respectively for nHB, and 0.6 mg N m^{-2} day^{-1} and 0.27 mg P m^{-2} day^{-1} for sHB, respectively. Notably, other large PON and PP loads are those conveyed by the estuarine circulation, which is driven by the riverine discharge; those for nHB are 21.0 mg N m^{-2} day^{-1} and 2.59 mg P m^{-2} day^{-1}, respectively, and for sHB they are 1.0 mg N m^{-2} day^{-1} and -0.26 mg P m^{-2} day^{-1}, respectively. The minus PP value in sHB indicates PP was transported from sHB to outside the bay. These values indicate the material load by the estuarine circulation is comparative to the riverine direct N and P loads or more. Although the carbon loads from the rivers and the estuarine circulation were not estimated in their study, it is inferred that the same tendency in POC load being high in nHB and low in sHB.

The second largest source of particulate organic matter in Hiroshima Bay is biodeposits from oyster culture, which are intensively conducted particularly in nHB. According to the material budget calculation by Kittiwanich et al. [66], the amount of PON and PP produced as feces and pseudofeces by cultured oysters are 18.1 mg N m^{-2} day^{-1} and 2.09 mg P m^{-2} day^{-1} for nHB, respectively, and 1.3 mg N m^{-2} day^{-1} and 0.15 mg P m^{-2} day^{-1} for sHB, respectively. The particulate matter load by the biological process, particularly PON, is larger than those of direct riverine load both in nHB and sHB. The biodeposits from the cultured oysters contribute approximately 2/3 of the total sedimentation both in PON and PP in nHB. The remaining total biodeposits produced in the water column mainly comes from the feces of zooplankton. In nHB, including Etajima Bay, oyster farming is intensive; approximately 12,000 oyster rafts (of them, 1700 rafts in Etajima Bay) exist in nHB, equivalent to 5760 million individual oysters. According to Yamamoto et al. [44], 27.6 kg DW deposits raft^{-1} day^{-1} are estimated to be produced. Biodeposits of ~330 ton DW raft^{-1} day^{-1} sink down onto the seafloor of HB. This estimate does not include oyster meats, regardless if they were dead or alive, and other attached creatures, which may have fallen during handling by farmers, which may reflect almost the same amount of the feces and pseudofeces deposits.

Conversely, the distribution of N differed substantially from that of C, showing low values at the mouth of the Ohta River (Figure 4). Consequently, this area had high sediment C/N molar ratios (>11). In sHB, the C/N molar ratio was also high (12–23) in the western part near the Nishiki River than in the central part (5–11). The low N values in these river mouth areas can be explained by denitrification processes. The occurrence of denitrification is also evidenced by the low N/P molar ratio. The N/P value 5.6 recorded at the Ohta River mouth was much smaller than the Redfield ratio (16). It was reported that severe hypoxia (DO concentration <2 mg L^{-1}) of the bottom layer is observed in the north-east part of nHB usually in September and recovers in October every year [59]. The hypoxia is limited in the innermost north end of nHB (Stations 27 and 28) where riverine input is received, and the area landlocked by islands with narrow connections to the neighboring areas (Stations 19, 20, 23, 26, 29, 30, 31, and 32). It is reported that the annual scale of hypoxia in the main nHB, except in the landlocked area, is dependent on the DO supply accompanied by the advective lateral flow driven by the estuarine circulation, although the seasonal recovery of hypoxia occurs due to ventilation by vertical mixing of the water column as the season progresses [59].

Long-term monitoring data on the bottom DO concentration as well as the other water quality data are available from the homepage of Hiroshima Fisheries and Marine Technology Center [67]. As the number of stations monitored by the prefectural institution is limited, here we use the data collected at two stations (Stations 17 and 21 in the original data set) whose locations are close to our sampling stations (Stations 22 and 31) for discussion. Therefore, in Figure 8, we used our station numbers for these sites. The data indicate that the bottom DO concentration at Station 31 in Kure Bay is much lower than that at Station 22 (Figure 8). This is due to the water stagnancy in Kure Bay, while DO is supplied by the estuarine circulation at Station 22 as mentioned above. There is no fixed long-term trend in the bottom DO concentrations even though the nutrient reduction measure has been implemented for 40 years. The most serious eutrophication period was during the 1970s–1980s and before in terms of the number of harmful algal blooms; the record high was 1974 (14 times yr^{-1}) and the second record high was 1986 (10 times yr^{-1}) in Hiroshima Bay [67]. However, no harmful algal blooms are observed in recent couple of years. This can be the 'Legacy of Eutrophication', which has been mentioned in previous studies [47–50]. That is, H_2S and other reductive matter produced in the anaerobic sediment, in which an abundance of organic matter accumulated in the bottom sediments during eutrophication period, continue to consume the bottom DO and release phosphate into the bottom water, even several decades after the peak eutrophication period.

In Figure 8, we selected the lowest 10 values for DO and the highest 10 values for each nutrient species at Station 31 (Kure Bay) during 1988 to 2010 since the sampling interval is not regular before 1987. The bottom NH_4-N concentration was high during June–July, except in August 1987; meanwhile, the bottom NO_3-N was high during August–December with several exceptions (January 1989 and June 2004). This may imply that organic matter decomposition progresses in early summer (June–July) and is followed by nitrification (August–December). During the nitrification process followed by H2S generation, the DO consumption proceeds and forms hypoxia in September. Taking the difference between the NO_3-N peak value to the following lowest value which occurred 1–2 months later, the denitrification rate is ~6.6 mg N m^{-2} day^{-1} for Station 31 on average over the three decades. Correspondingly, it is estimated to be 4.6 mg N m^{-2} day^{-1} for Station 22. This is only a first-order estimation, and the values may be overestimated because the decrease in NO_3-N concentration is also caused by the other processes such as physical dilution and ammonification, etc. However, it is in the range of the reported values (0–23.2 mg N m^{-2} day^{-1}) determined by the acetylene inhibition method for the sediments collected at Station 22 in 1994–1995 [68], and close to the value estimated for the entire Hiroshima Bay using the data from April 1991 to March 1992 (7.9 mg N m^{-2} day^{-1}) [69] and slightly lower than the value estimated for the nHB using the data set of 1991–2000 (14 mg N m^{-2} day^{-1}) [70]. The latter two were those estimated by a material budget calculation using models. It is not so much different from the values estimated for the sediments of the eutrophic Stockholm archipelago, Baltic Sea (1.26–24.1 mg N m^{-2} day^{-1}, in the original study, 90–1723 μmol N m^{-2} day^{-1}) [50].

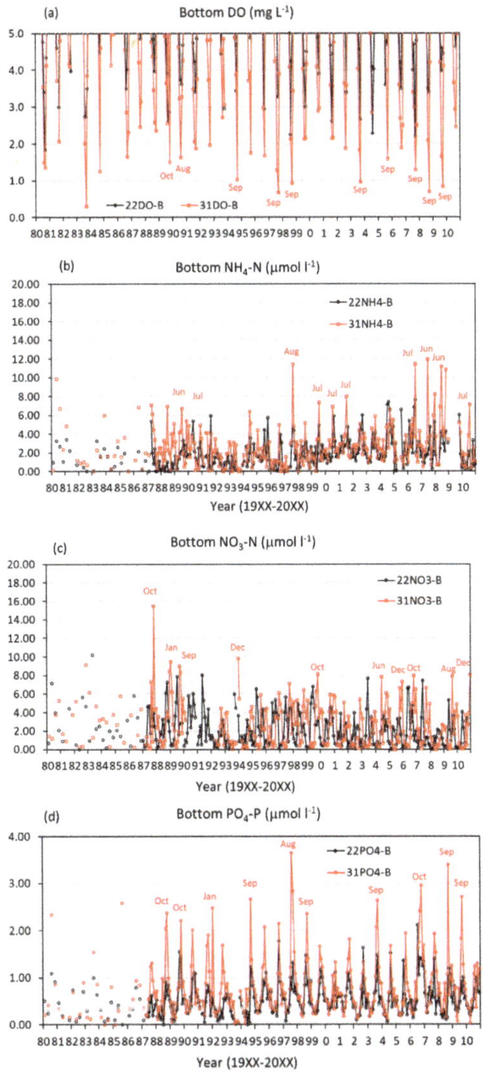

Figure 8. Temporal variations of bottom water (**a**) dissolved oxygen (DO), (**b**) PO$_4$-P, (**c**) NH$_4$-N and (**d**) NO$_3$-N concentrations. The samples were collected at 1 m above the bottom. Data are cited from Hiroshima Fisheries and Marine Technology Center [67]. Month names were given to the top 10 annual peak values for each nutrient species (lowest 10 for DO) at Station 31 (Kure Bay) during 1988 to 2010 since the data set was not complete before 1987.

The S and P distribution patterns differed substantially from those of C and N. For the S and P cycles, we should consider whether the sediment conditions were aerobic or anaerobic, similar to N. Sulfate ions are plentiful in seawater and are the source of sulfate reduction under anaerobic conditions in addition to organic matter. In other words, H$_2$S is the byproduct of sulfate reduction. As described above, the H$_2$S concentration and AVS, in which H$_2$S is contained, were high in Etajima Bay, where S originates not only from sulfate ions, but also from oyster feces. This may be the cause of the high S contents in the sediments of Etajima Bay (Figure 5). In Japan, the standard permissible values of AVS and H$_2$S for aquatic life are 2 mg g dw^{-1} and 0 mg L^{-1}, respectively [65]. Compared to the standards, all stations except Station 27 cleared the AVS standard, and Stations 1, 3, 8, 11, and 14 in sHB cleared the H_2S standard. In comparison to the values between the samples from 0–5 cm depth and from 5–10 cm depth, the AVS was higher at 0–5 cm depth (0.34 ± 0.21 mg g dw^{-1}) than 5–10 cm depth (0.21 ± 0.15 mg g dw^{-1}). This may be attributed to the formation of FeS and other compounds, which can be detected as AVS [62]. Meanwhile, the H_2S was higher in 5–10 cm (6.1 ± 5.4 mg L^{-1}) than 0–5 cm (4.6 ± 5.6 mg L^{-1}). This may indicate H_2S dissolved in the pore water is diffused to the overlying water and causes the consumption of DO. The volume-based percentage content of AVS in TS and that of H_2S in AVS were $7.3 \pm 5.0\%$ (0.8–28.6%) and $1.1 \pm 1.3\%$ (0.0–6.4%), respectively. Although the fraction of H_2S in the sediment TS is somewhat small, it should be continuously produced as long as the vast organic matter remains in the sediments and sulfate is supplied from seawater.

The observed P content was comparatively low at the stations located at the mouth of rivers and stagnant areas (Stations 19, 20, 26, 27, 29, and 31 in Figure 6), contrary to the high C content at these stations (Figure 3). In these areas, material load is high either from rivers or oyster culture. The concentration of the elements in the sediment is determined by the balance between the sedimentation flux and the release of dissolved matter from the sediment due to decomposition. Therefore, although the organic matter sedimentation is high, the relatively low sediment P values implies a high benthic flux owing to anaerobic conditions. The sediment Eh was low (-150 to -120 mV) in these areas, as described before, indicating reducing conditions. H_2S is contained in the sediments of all stations in nHB as described in Section 3.2. As shown in Figure 8, peak phosphate concentration in the bottom layer of Station 31 is usually observed in September, which coincides with the occurrence of hypoxia. This phenomenon is the same at Station 22. The phosphate under anaerobic conditions originates from metal oxide-bound P, such as $FeOOH{\equiv}PO_4$ in the reaction with H_2S, in the sediment. The production of H_2S in the sediments will continue if both sulfate and the organic matter are present in the sediments, even if it is refractory. This is the 'Legacy of Eutrophication' as previously mentioned [47–50].

An estimated 1.8×10^5 ton of P accumulated in the Hiroshima Bay sediment, equivalent to half the amount consumed in a year in Japan (3.5×10^5 ton) [71]. Even if P is liberated yearly from metal oxide-bound-P, the amount liberated from the anoxic nHB sediments is estimated to be 26 ton P yr^{-1} from the average concentration in the bottom as 2 μmol L^{-1}, assuming the bottom water depth containing the concentration is 3 m. If we assume the P accumulated in the sediment changed the form to those liberated yearly, it will take 6900 years until the end of release of all the accumulated P. Although this estimation is too simple and rough, it is certain that the annual hypoxia will last long even if a harsher nutrient reduction measure is taken.

Additional measures are required to remediate such deteriorated anoxic sediments. Two engineering methods have been implemented to remediate deteriorated anoxic sediments. One simply removes and dampens the deteriorated anoxic sediments in the other site. Because of the salt content, we cannot use the sediments in agriculture as fertilizer or anywhere else on land. Therefore, the most suitable place to dampen was the shallow coastal area along the coastline in terms of transportation cost. This kind of reclamation was effective during the rapid economic growth period because there was demand for land for industries and houses. However, such reclamation was restricted by the tentative law in 1973 because it damages coastal ecosystems. Another way is to remediate the sediment quality by adsorbing or oxidizing the reductive substances using materials that have such functions. Traditionally, natural sand was used for capping deteriorated coastal sediments. However, the collection of natural sand was restricted by the law because it may damage the ecosystems of original sites where the sand is collected regardless of whether it is land, river, or sea. Furthermore, natural sand has no function to reduce the reductive substances. Thus, we have developed several functional materials from industrial byproducts, which are certificated by the Ministry of the Environment of Japan [72], and hot-air dried oyster shell (HACOS), steel-making slag, and granulated coal ash are those in which the present authors' group has been deeply involved [73–78]. Their use in the future will be beneficial due to cost-effectiveness and the aspect of creating a recycle-oriented society.

Third, the recovery of the accumulated P as a resource can be an alternative. Currently, P is recovered from sewage at treatment plants and used in fertilizers and other products [79]. Abelson [80] indicated that phosphorus is a resource facing shortages on the Earth; therefore, techniques should be developed in the future to recover P that is accumulated in coastal sediments.

The Ministry of the Environment of Japan observed the sediment quality in the Seto Inland Sea, Japan, including Hiroshima Bay [54–56]. There were 46 stations in the first observation in 1982, and 23 in the second and third observations, which did not differ drastically from our number of sampling stations (32) in October 2010, showing an even distribution of the stations. The results are summarized along with our data from October 2010 in Table 3. The sampling depth in the observation by the Ministry of the Environment was the same as those of the first layer in our observations (0–5 cm), with data in deeper layer >5 cm lacking. Decreasing trends over time in TOC and TN were found in a report by Komai [81]; however, the discussion is limited. In contrast, only TP values in the present study were slightly higher than the average TP value from 2003. A possible cause is the trapping/adsorption of phosphate by iron oxides and manganese oxides that are formed under aerobic conditions. While the sediment P retention capacity is small when the bottom condition is anaerobic because of the reductive dissolution of phosphate detached from iron/manganese oxides [82,83], it increases via trapping by iron/manganese oxides as oxygen is supplied [84]. A record flooding during July 10–16 of 2010 [85], 3 months before our observations, might have increased the sediment P content.

Table 3. Comparison of TOC, TN, and TP values between those reported by Ministry of the Environment, Japan [54–56] and the present study (2010). The sediments were collected from the same depth in all samples (0–5 cm) in Hiroshima Bay. A similar number of sampling stations were included in the observations (46 in 1982, 23 in 1993 and 2003, and 32 in 2010), and were evenly distributed throughout the bay.

Year		TOC (mg g^{-1})	TN (mg g^{-1})	TP (mg g^{-1})	References
1982	Max	30	3.5	0.75	
	Min	9.6	1.0	0.34	
	Avg	20	2.5	0.59	[54]
	SD	5.4	0.66	0.11	
	CV (%)	27	27	18	
1993	Max	26	2.8	0.62	
	Min	2.8	0.36	0.26	
	Avg	18	2.1	0.53	[55]
	SD	5.7	0.62	0.090	
	CV (%)	32	30	17	
2003	Max	29	3.1	0.68	
	Min	4.4	0.60	0.20	
	Avg	17	2.2	0.51	[56]
	SD	6.8	0.78	0.12	
	CV (%)	39	36	23	
Oct 2020	Max	21	3.0	0.66	
	Min	9.4	0.62	0.49	
	Avg	15	1.6	0.57	This study
	SD	2.6	0.50	0.040	
	CV (%)	17	32	6	

To remediate the eutrophic conditions, the P and N reduction measure has been in place for 40 years for the Seto Inland Sea including Hiroshima Bay. Consequently, the water transparency has increased drastically [37]; however, the production of commercially important fish, bivalves, and cultured seaweed has almost collapsed in the Seto Inland

Sea [41,51,52]. We previously reported this could be because of 'cultural oligotrophication' [41,86]. Thus, we conclude that the Seto Inland Sea, including Hiroshima Bay, has been experiencing conflict between the oligotrophication of the surface water and the legacy of eutrophication of the bottom sediments. The recent decrease in the oyster production of Hiroshima Bay is a serious issue for farmers. The line length farmers use to hang oysters is 10 m. Therefore, the oysters are not affected by the hypoxia, which usually forms in the bottom layer. However, the anoxic sediment conditions are unsuitable for benthic animals, including bivalves, as their rearing environment. They should be an important feed for both benthic and pelagic fishes; for example, red sea bream, which is a highly regarded fish in Japan, prefers to feed on shrimp dwelling on the bottom. Thus, the hypoxia and sediment which contains deadly poisonous H_2S can cause the collapse of the total ecosystem in addition to the oligotrophication in the surface water.

To alleviate the oligotrophic conditions of the surface water and increase the bioproduction, the Japanese government revised the law in 2015. The measure was to increase the sewage effluent load to a level at which natural and farming seaweeds, bivalves, and major fish species can sufficiently grow [53]. In Hyogo Prefecture, in the eastern part of the Seto Inland Sea, they relaxed the sewage treatment effluent during the Nori growing season in the winter by suppressing the nitrification and denitrification processes at sewage treatment plants [87]. However, observations have proven the effect was spatially limited because of the diffusion feature of seawater [87]. Computer simulations, which combine physical and biological processes, also showed limited effects [88]. Another cause of the limitation of the effect was a permissible limit in the maximum concentrations of nitrogen and phosphorus of the effluents they can release. The nutrient concentration was still regulated by the policy in 2015. Then, the central government revised the law again to expand the target sites to any workplace which emits phosphorus and nitrogen. The revised bill has recently passed the House of Representatives on 3 June 2021 [89].

Since the oyster rafts are extensively located in nHB, the supply of nutrients by increasing effluents from workplaces may not impact the offshore rafts. Therefore, we are attempting to develop a fertilizer to enhance the growth of farmed oysters [90], which can also be used for other bivalves, Nori, and other farming organisms. The most important issue that we must address is the concentration and time we should increase the effluents from workplaces, how many fertilizers, and when it should be applied. As the environmental conditions and type of fisheries conducted are different in each area, each prefectural government must devise a practical and effective plan for individual sea areas. For example, Hiroshima Prefecture is assigned to plan for Hiroshima Bay. When increasing nutrient load to maintain the oyster growth, the prefectural government should consider the oyster growing season is October to March. However, nutrient increases must not induce harmful algal blooms. Simultaneously, we must consider the remediation of anoxic sediments. Researchers are expected to propose some acceptable scientific perspectives using a sophisticated simulation model, which includes integrated physical-biogeochemical processes in both the water column and sediments. Our group has previously established a model that can reproduce the bottom hypoxia and remediation effects of byproducts by applying them to the sediments [91]. The other models we have been developed can be used to measure the oligotrophication of the surface water and oyster production [43,66]. We must discuss the simulation outputs in detail as a future prospect with different stakeholders.

5. Conclusions

The C, N, S, and P showed different spatial distributions within Hiroshima Bay. C was influenced more by material loads from land, whereas N reduced by denitrification in river mouths and landlocked areas with reduced sediments. S was affected by oyster farming activities, which are intensively conducted in, for example, nHB and Etajima Bay, and sulfate reduction that may occur under reducing conditions. P was relatively low in reduced sediments of landlocked areas (northeast of nHB, Kure Bay and Etajima Bay), where P is released under reduced conditions, and distributed evenly in the vertical profile compared to the other elements. P was somewhat high compared to C and N referring to the Redfield ratio, suggesting that it can accumulate in sediments. Compared to the governmental monitoring data on sediment quality, the P value of the present study is higher than that of 2003, whereas decreasing trends in C and N were found. Approximately 1.8×10^5 tons of P is accumulated in 0.28 km^3 of muddy sediment in Hiroshima Bay, with a maximum thickness of 0.5 m.

The Seto Inland Sea, including Hiroshima Bay, is facing conflict between the 'Legacy of Eutrophication' in the bottom sediments and the 'Cultural Oligotrophication' of the surface water. Particularly, the lack of feed phytoplankton for farming oyster is a serious challenge for farmers in the Hiroshima Bay. In contrast, the hypoxia observed every September is unlikely to recover forever. A major cause is the biodeposits supplied from farming oysters. The oxygen consumption in the bottom layer is attributed to the reductive substances such as H_2S. Recovery of the sediment quality is important for animals which are feed for both benthic and pelagic fish. Therefore, we should apply materials that can adsorb the reductive substances to recover the total ecosystem, in addition to adding nutrients to the surface water. A science-based plan with several options with computer-aided perspectives, including cost performance, is warranted for the local government to facilitate proper decision-making.

Author Contributions: Conceptualization, T.Y. and S.-i.O.; methodology, T.Y. and H.Y.; software, H.Y.; validation, S.A.; formal analysis, K.O.; investigation, T.Y., K.O., S.A. and H.Y.; resources, T.Y.; data curation, S.A.; writing—original draft preparation, T.Y.; writing—review and editing, S.A. and H.Y.; visualization, K.O. and H.Y.; supervision, T.Y.; project administration, T.Y. and S.-i.O.; funding acquisition, S.-i.O. All authors have read and agreed to the published version of the manuscript.

Funding: This research was funded by JSPS Grant-in-Aid for Scientific Research (KAKENHI) (A), Grant Number JP21241011.

Data Availability Statement: The data presented in this study are available on request from the corresponding author. The data are not publicly available because they are collected in an university student's graduation project. The corresponding author who was the supervisor of the student keeps all the data in the forms of CD and printed matter.

Acknowledgments: We thank the captain, Kazumitsu Nakaguchi and the crew of the R/T vessel Toyoshio Maru, Hiroshima University for their help during the research cruise.

Conflicts of Interest: The authors declare no conflict of interest.

References

1. Asmala, E.; Carstensen, J.; Conley, D.; Slomp, C.P.; Stadmark, J.; Voss, M. Efficiency of the coastal filter: Nitrogen and phosphorous removal in the Baltic Sea. *Limnol. Oceanogr.* **2017**, *62*, S222–S238. [CrossRef]
2. Prastka, K.E.; Malcolm, S.J. Particulate phosphorus in the Humber estuary. *Neth. J. Aquat. Ecol.* **1994**, *28*, 397–403. [CrossRef]
3. Bowden, K.F.; Din, S.H.S.E. Circulation and mixing processes in the Liverpool Bay area of the Irish Sea. *Geophys. J. Int.* **1966**, *11*, 279–292. [CrossRef]
4. Bowden, K.F.; Gilligan, R.M. Characteristic features of estuarine circulation as represented in the Mersey Estuary. *Limnol. Oceanogr.* **1971**, *16*, 490–502. [CrossRef]
5. Ruttenberg, K.C.; Berner, R.A. Authigenic apatite formation and burial in sediments from non-upwelling, continental margin environments. *Geochim. Cosmochim. Acta* **1993**, *57*, 991–1007. [CrossRef]
6. Gachter, R.; Muller, B. Why the phosphorus retention of lakes does not necessarily depend on the oxygen supply to their sediment surface. *Limnol. Oceanogr.* **2003**, *48*, 929–933. [CrossRef]
7. Roden, E.E.; Edmonds, J.W. Phosphate mobilization in iron-rich anaerobic sediments: Microbial Fe(III) oxide reduction versus iron-sulfide formation. *Arch. Hydrobiol.* **1997**, *139*, 347–378. [CrossRef]
8. Sugawara, K.; Koyama, T.; Kamata, E. Recovery of precipitated phosphate from lake muds related to sulfate reduction. *J. Earth Sci. Nagoya Univ.* **1957**, *5*, 60–67.
9. Froelich, P.N.; Arthur, M.A.; Burnett, W.C.; Deakin, M.; Hensley, V.; Jahnke, R.; Kaul, L.; Kim, K.H.; Roe, K.; Soutar, A.; et al. Early diagenesis of organic matter in Peru continental margin sediments: Phosphorite precipitation. *Mar. Geol.* **1988**, *80*, 309–343. [CrossRef]
10. Slomp, C.P.; Epping, E.H.G.; Helder, W.; Raaphorst, W.V. A key role for iron-bound phosphorus in authigenic apatite formation in North Atlantic continental platform sediments. *J. Mar. Res.* **1996**, *54*, 1179–1205. [CrossRef]

11. Shulz, H.N.; Schulz, H.D. Large sulfur bacteria and the formation of phosphorite. *Science* **2005**, *307*, 416–418. [CrossRef] [PubMed]
12. Suess, E. Mineral phases formed in anoxic sediments by microbial decomposition of organic matter. *Geochim. Cosmochim. Acta* **1979**, *43*, 339–352. [CrossRef]
13. Egger, M.; Jilbert, T.; Behrends, T.; Rivard, C.; Slomp, C.P. Vivianite is a major sink for phosphorus in methanogenic coastal surface sediments. *Geochim. Cosmochim. Acta* **2015**, *169*, 217–235. [CrossRef]
14. Rothe, M.; Kleeberg, A.; Hupfer, M. The occurrence, identification and environmental relevance of vivianite in waterlogged soils and aquatic sediments. *Earth Sci. Rev.* **2016**, *158*, 51–64. [CrossRef]
15. Lenstra, W.K.; Egger, M.; van Helmond, N.A.G.M.; Kritzberg, E.; Conley, D.J.; Slomp, C.P. Large variations in iron input to an oligotrophic Baltic Sea estuary: Impact on sedimentary phosphorus burial. *Biogeosciences* **2018**, *15*, 6979–6996. [CrossRef]
16. Froelich, P.N.; Bender, M.L.; Luedtke, N.A. The marine phosphorus cycle. *Am. J. Sci.* **1982**, *4*, 105. [CrossRef]
17. Delaney, M.L. Phosphorus accumulation in marine sediments and the oceanic phosphorus cycle. *Global Biogeochem. Cycles* **1998**, *12*, 563–572. [CrossRef]
18. Bouwman, A.F.; Bierkens, M.F.P.; Griffioen, J.; Hefting, M.M.; Middelburg, J.J.; Middelkoop, H.; Slomp, C.P. Nutrient dynamics, transfer and retention along the aquatic continuum from land to ocean: Towards integration of ecological and biogeochemical models. *Biogeosciences* **2013**, *10*, 1–22. [CrossRef]
19. Knowles, R. Denitrification. *Microbiol. Rev.* **1982**, *46*, 43–70. [CrossRef]
20. Seitzinger, S.P. Denitrification in freshwater and coastal marine ecosystems: Ecological and geochemical significance. *Limnol. Oceanogr.* **1988**, *33*, 702–724. [CrossRef]
21. Seitzinger, S. Denitrification in aquatic sediments. In *Denitrification in Soil and Sediment*; Revsbech, N.P., Sørensen, J., Eds.; Plenum Press: New York, NY, USA, 1990; pp. 301–322.
22. Smith, C.J.; DeLaune, R.D.; Patrick, W.H., Jr. Fate of riverine nitrate entering an estuary: I. Denitrification and nitrogen burial. *Estuaries* **1985**, *8*, 15–21. [CrossRef]
23. Dalsgaard, T.; Thamdrup, B.; Canfield, D.E. Anaerobic ammonium oxidation (anammox) in the marine environment. *Res. Microbiol.* **2005**, *156*, 457–464. [CrossRef] [PubMed]
24. Poulin, P.; Pelletier, E.; Saint-Louis, R. Seasonal variability of denitrification efficiency in northern salt marshes: An example from the St. Lawrence Estuary. *Mar. Environ. Res.* **2007**, *63*, 490–505. [CrossRef]
25. Meyer, R.L.; Allen, D.E.; Schmidt, S. Nitrification and denitrification as sources of sediment nitrous oxide production: A microsensor approach. *Mar. Chem.* **2008**, *110*, 68–76. [CrossRef]
26. Senga, Y.; Okumura, M.; Seike, Y. Seasonal and spatial variation in the denitrifying activity in estuarine and lagoonal sediments. *J. Oceanogr.* **2010**, *66*, 155–160. [CrossRef]
27. Christensen, P.B.; Rysgaard, S.; Sloth, N.P.; Dalsgaard, T.; Schwærter, S. Sediment mineralization, nutrient fluxes, denitrification and dissimilatory nitrate reduction to ammonium in an estuarine fjord with sea cage trout farms. *Aquat. Microb. Ecol.* **2000**, *21*, 73–84. [CrossRef]
28. An, S.M.; Gardner, W.S. Dissimilatory nitrate reduction to ammonium (DNRA) as a nitrogen link, versus denitrification as a sink in a shallow estuary (Laguna Madre/Baffin Bay, Texas). *Mar. Ecol. Prog. Ser.* **2002**, *237*, 41–50. [CrossRef]
29. Nizzoli, D.; Carraro, E.; Nigro, V.; Viaroli, P. Effect of organic enrichment and thermal regime on denitrification and dissimilatory nitrate reduction to ammonium (DNRA) in hypolimnetic sediments of two lowland lakes. *Water Res.* **2010**, *44*, 2715–2724. [CrossRef]
30. Jäntti, H.; Hietanen, S. The effects of hypoxia on sediment nitrogen cycling in the Baltic Sea. *Ambio* **2012**, *41*, 161–169. [CrossRef]
31. Thamdrup, B. New Pathways and processes in the global nitrogen cycle. *Annu. Rev. Ecol. Evol. Syst.* **2012**, *43*, 407–428. [CrossRef]
32. Giblin, A.E.; Tobias, C.R.; Song, B.; Weston, N.; Banta, G.T.; Rivera-Monroy, V.H. The importance of dissimilatory nitrate reduction to ammonium (DNRA) in the nitrogen cycle of coastal ecosystems. *Oceanography* **2013**, *26*, 124–131. [CrossRef]
33. Burgin, A.J.; Hamilton, S.K. Have we overemphasized the role of denitrification in aquatic ecosystems? A review of nitrate removal pathways. *Front. Ecol. Environ.* **2007**, *5*, 89–96. [CrossRef]
34. Tobias, C.R.; Anderson, I.C.; Canuel, E.A.; Macko, S.A. Nitrogen cycling through a fringing marsh-aquifer ecotone. *Mar. Ecol. Prog. Ser.* **2001**, *210*, 25–39. [CrossRef]
35. Christensen, P.B.; Glud, R.N.; Dalsgaard, T.; Gillespie, P. Impacts of longline mussel farming on oxygen and nitrogen dynamics and biological communities of coastal sediments. *Aquaculture* **2003**, *218*, 567–588. [CrossRef]
36. Ministry of the Environment, Japan. Available online: http://www.env.go.jp/water/heisa.html (accessed on 28 December 2020).
37. Ministry of Land, Infrastructure, Transport and Tourism, Japan. Available online: http://www.pa.cgr.mlit.go.jp/chiki/suishitu/ (accessed on 28 December 2020).
38. Scheffer, M. Alternative stable states in eutrophic, shallow freshwater systems: A minimal model. *Hydrobiol. Bull.* **1989**, *23*, 73–83. [CrossRef]
39. Scheffer, M.; Carpenter, S.; Foley, J.A.; Folke, C.; Walker, B. Catastrophic shifts in ecosystems. *Nature* **2001**, *413*, 591–596. [CrossRef] [PubMed]
40. Duarte, C.M.; Conley, D.J.; Carstensen, J.; Sanchez-Camacho, M. Return to Neverland: Shifting baselines affect eutrophication restoration targets. *Estuar. Coasts* **2009**, *32*, 29–36. [CrossRef]
41. Yamamoto, T. The Set Inlad Sea—Eutrophic or oligotrophic? *Mar. Poll. Bull.* **2003**, *47*, 37–42. [CrossRef]

42. Hiroshima City Agriculture, Forestry and Fisheries Promotion Center. Available online: http://www.haff.city.hiroshima.jp/suisansc/kaki_rekisi.html (accessed on 28 December 2020).
43. Wahyudin Yamamoto, T. Modeling bottom-up and top-down controls on the low recruitment success of oyster larvae in Hiroshima Bay, Japan. *Aquaculture* **2020**, *529*, 735564. [CrossRef]
44. Yamamoto, T.; Maeda, H.; Matsuda, O.; Hashimoto, T. Effects of culture density on the growth and fecal production of the oyster Crassostrea gigas. *Nippon Suisan Gakkaishi* **2009**, *75*, 230–236, (In Japanese with English Abstract). [CrossRef]
45. Ray, N.E.; Fulweiler, R.W. Seasonal patterns of benthic-pelagic coupling in oyster habitats. *Mar. Ecol. Prog. Ser.* **2020**, *652*, 95–109. [CrossRef]
46. Yamamoto, T.; Hashimoto, T.; Matsuda, O.; Go, A.; Nakaguchi, K.; Haraguchi, K. Comparison of sediment quality between Hiroshima Bay and Suo Nada-Special reference to the seasonal variations of sediment quality parameters and their relationships. *Nippon Suisan Gakkaishi* **2008**, *74*, 1037–1042, (In Japanese with English Abstract). [CrossRef]
47. Conley, D.J.; Humborg, C.; Rahm, L.; Savchuk, O.P.; Wulff, F. Hypoxia in the Baltic Sea and basin-scale changes in phosphorus biogeochemistry. *Environ. Sci. Technol.* **2002**, *36*, 5315–5320. [CrossRef]
48. Turner, R.E.; Rabalais, N.N.; Justic, D. Gulf of Mexico hypoxia: Alternate states and a legacy. *Environ. Sci. Technol.* **2008**, *42*, 2323–2327. [CrossRef]
49. Hermans, M.; Lenstra, W.K.; van Helmond, N.A.G.M.; Behrends, T.; Egger, M.; Séguret, M.J.; Gustafsson, E.; Gustafsson, B.G.; Slomp, C.P. Impact of natural re-oxygenation on the sediment dynamics of manganese, iron and phosphorus in a euxinic Baltic Sea basin. *Geochim. Cosmochim. Acta* **2019**, *246*, 174–196. [CrossRef]
50. Van Helmond, N.A.; Robertson, E.K.; Conley, D.J.; Hermans, M.; Humborg, C.; Kubeneck, L.J.; Lenstra, W.K.; Slomp, C.P. Removal of phosphorus and nitrogen in sediments of the eutrophic Stockholm archipelago, Baltic Sea. *Biogeosciences* **2020**, *17*, 2745–2766. [CrossRef]
51. Abo, K.; Yamamoto, T. Oligotrophication and its measures in the Seto Inland Sea, Japan. *Bull. Jap. Fish. Res. Edu. Agen.* **2019**, *49*, 21–26.
52. E-Stat, Statistics of Japan. Available online: https://www.e-stat.go.jp/ (accessed on 28 December 2020).
53. Ministry of the Environment, Japan. The Partial Revision of the 'Law Concerning Special Measures for Conservation of the Environment of the Seto Inland Sea'. Available online: https://www.env.go.jp/water/heisa/setonaikai_law_rev.html (accessed on 8 June 2021).
54. Ministry of the Environment, Japan. *Research on the Basic Information of Environmental Conditions of the Seto Inland Sea, Japan (I)*; CD-ROM; Ministry of the Environment: Tokyo, Japan, 1988. (In Japanese)
55. Ministry of the Environment, Japan. *Research on the Basic Information of Environmental Conditions of the Seto Inland Sea, Japan (II)*; CD-ROM; Ministry of the Environment: Tokyo, Japan, 1997. (In Japanese)
56. Ministry of the Environment, Japan. *Research on the Basic Information of Environmental Conditions of the Seto Inland Sea, Japan (III)*; CD-ROM; Ministry of the Environment: Tokyo, Japan, 2006. (In Japanese)
57. Ministry of the Environment, Japan. The Methods of Sediment Analyses. 1998. Available online: http://www.env.go.jp/hourei/05/000178.html (accessed on 28 December 2020).
58. Yamamoto, T.; Yoshikawa, S.; Hashimoto, T.; Takasugi, Y.; Matsuda, O. Estuarine circulation processes in the northern Hiroshima Bay, Japan. *Bull. Coast. Oceanogr.* **2000**, *37*, 111–118, (In Japanese with English abstract).
59. Yamamoto, H.; Yamamoto, T.; Takada, T.; Mito, Y.; Takahashi, T. Dynamic analysis of oxygen-deficient water mass formed in the northern part of Hiroshima Bay using a pelagic-benthic coupled ecosystem model. *J. Jap. Soc. Water Envir.* **2011**, *34*, 19–28, (In Japanese with English Abstract). [CrossRef]
60. Nishimura, K.; Sayanagi, K.; Nagao, T.; Murakami, F.; Joshima, M.; Takatou, M. Digitization of an analog sub-bottom profiler. *J. Sch. Mar. Sci. Tech. Tokai Univ.* **2008**, *6*, 129–139, (In Japanese with English Abstract).
61. Arakawa, K. Teishitsu Chousa Hou (Methods of Sediment Quality Analyses). In *Suishitsu Odaku Chosa Shishin*; Japanese Fishery Resource Conservation Association, Ed.; Japanese Fishery Resource Conservation Association: Tokyo, Japan, 1980; pp. 237–272. (In Japanese)
62. Rickard, D.; Morse, J.W. Acid volatile sulfide (AVS). *Mar. Chem.* **2005**, *97*, 141–197. [CrossRef]
63. Asaoka, S.; Yamamoto, T.; Hayakawa, S. Removal of hydrogen sulfide using granulated coal ash. *J. Japan Soc. Wat. Environ.* **2009**, *32*, 363–368, (In Japanese with English Abstract). [CrossRef]
64. Asaoka, S.; Yamamoto, T.; Takahashi, Y.; Yamamoto, H.; Kim, K.H.; Orimoto, K. Development of an on-site simplified determination method for hydrogen sulfide in marine sediment pore water using a shipboard ion electrode with consideration of hydrogen sulfide oxidation rate. In *Interdisciplinary Studies on Environmental Chemistry—Environmental Pollution and Ecotoxicology*; Kawaguchi, M., Misaki, K., Sato, H., Yokokawa, T., Itai, T., Nguyen, T.M., Ono, J., Tanabe, S., Eds.; TERRAPUB: Tokyo, Japan, 2012; pp. 345–352.
65. Japan Fisheries Resource Conservation Association. *Standard for Fisheries Water*; Japanese Fishery Resource Conservation Association: Tokyo, Japan, 1995; 68p. (In Japanese)
66. Kittiwanich, J.; Yamamoto, T.; Kawaguchi, O.; Hashimoto, T. Analyses of phosphorus and nitrogen cyclings in the estuarine ecosystem of Hiroshima Bay by a pelagic and benthic coupled model. *Est. Coast. Shelf Sci.* **2007**, *75*, 189–204. [CrossRef]
67. Hiroshima Fisheries and Marine Technology Center. Available online: https://www.pref.hiroshima.lg.jp/soshiki/32/suigi-top.html (accessed on 31 May 2021).

68. Kim, D.H.; Matsuda, O.; Yamamoto, T. Nitrification, denitrification and nitrate reduction rates in the sediment of Hiroshima Bay, Japan. *J. Oceanogr.* **1997**, *53*, 317–324.
69. Yamamoto, T.; Takeshita, K.; Hiraga, N.; Hashimoto, T. An estimation of net ecosystem metabolism and net denitrification of the Seto Inland Sea, Japan. *Ecol. Model.* **2008**, *215*, 55–68. [CrossRef]
70. Yamamoto, T.; Kubo, A.; Hashimoto, T.; Nishii, Y. Long-term changes in net ecosystem metabolism and net denitrification in the Ohta River estuary of northern Hiroshima Bay—An analysis based on the phosphorus and nitrogen budgets. In *Progress in Aquatic Ecosystem Research*; Burk, A.R., Ed.; Nova Science Publisher, Inc.: New York, NY, USA, 2005; pp. 99–120.
71. Kuroda, A.; Takiguchi, N.; Kato, J.; Ohtake, H. Development of technolgies to save phosphorus resources in response to phosphate crisis. *J. Environ. Biotech.* **2005**, *4*, 87–94. (In Japanese)
72. Ministry of the Environment, Japan. Environmental Technology Verification (ETV) Program. Available online: https://www.env.go.jp/policy/etv/ (accessed on 4 June 2021).
73. Asaoka, S.; Yamamoto, T.; Kondo, S.; Hayakawa, S. Removal of hydrogen sulfide using crushed oyster shell from pore water to remediate organically enriched coastal marine sediments. *Biores. Technol.* **2009**, *100*, 4127–4132. [CrossRef]
74. Yamamoto, T.; Kondo, S.; Kim, K.H.; Asaoka, S.; Yamamoto, H.; Tokuoka, M.; Hibino, T. Remediation of muddy tidal flat sediments using hot air-dried crushed oyster shells. *Mar. Poll. Bull.* **2012**, *64*, 2428–2434. [CrossRef] [PubMed]
75. Asaoka, S.; Yamamoto, T. Blast furnace slag can effectively remediate coastal marine sediments affected by organic enrichment. *Mar. Poll. Bull.* **2010**, *60*, 573–578. [CrossRef] [PubMed]
76. Kim, K.H.; Asaoka, S.; Yamamoto, T.; Hayakawa, S.; Takeda, K.; Katayama, M.; Onoue, T. Mechanisms of hydrogen sulfide removal with steel making slag. *Environ. Sci. Technol.* **2012**, *46*, 10169–10174. [CrossRef]
77. Asaoka, S.; Yamamoto, T.; Yoshioka, I.; Tanaka, H. Remediation of coastal marine sediments using granulated coal ash. *J. Hazard. Mat.* **2009**, *172*, 92–98. [CrossRef]
78. Asaoka, S.; Hayakawa, S.; Kim, K.H.; Takeda, K.; Katayama, M.; Yamamoto, T. Combined adsorption and oxidation mechanisms of hydrogen sulfide on granulated coal ash. *J. Coll. Interface Sci.* **2012**, *377*, 284–290. [CrossRef] [PubMed]
79. Harada, Y. Outlook of Japan's policy in recycling phosphorus from sewage. *J. Japan Soc. Wat. Environ.* **2011**, *34*, 2–6. (In Japanese)
80. Abelson, P.H. A potential phosphate crisis. *Science* **1999**, *283*, 2015. [CrossRef]
81. Komai, Y. Changes of sediment quality and benthos in the Seto Inland Sea. In *Sediment Quality of the Seto Inland Sea, Japan*; Yanagi, T., Ed.; Kouseisha-Kouseikaku: Tokyo, Japan, 2008; pp. 43–60. (In Japanese)
82. Ingall, E.D.; Bustin, R.M.; Vancappellen, P. Influence of water column anoxia on the burial and preservation of carbon and phosphorus in marine shales. *Geochim. Cosmochim. Acta* **1993**, *57*, 303–316. [CrossRef]
83. Rozan, T.F.; Taillefert, M.; Trouwborst, R.E.; Glazer, B.T.; Ma, S.F.; Herszage, J.; Valdes, L.M.; Price, K.S.; Luther, G.W. Iron-sulfur-phosphorus cycling in the sediments of a shallow coastal bay: Implications for sediment nutrient release and benthic macroalgal blooms. *Limnol. Oceanogr.* **2002**, *47*, 1346–1354. [CrossRef]
84. Heidenreich, M.; Kleeberg, A. Phosphorus-binding in iron-rich sediments of a shallow Reservoir: Spatial characterization based on sonar data. *Hydrobiologia* **2003**, *506*, 147–153. [CrossRef]
85. Hiroshima Prefecture: Records of Huge Desarsters in Hiroshima Area. Available online: https://www.sabo.pref.hiroshima.lg.jp/portal/sonota/saigai/002dosya.htm (accessed on 20 June 2021).
86. Yamamoto, T.; Hanazato, T. *Oligotrophication in Seas and Lakes—No Fish. in Clear Water*; Yamamoto, T., Hanazato, T., Eds.; Chijin Shokan: Tokyo, Japan, 2015; p. 195. (In Japanese)
87. Harada, K.; Abo, K.; Kawasaki, S.; Takesako, F.; Miyahara, K. Influence of DIN derived from ports and treated sewage effluent on nori (*Pyropia*) farms of the northeastern part of Harima Nada. *Bull. Jpn. Soc. Fish. Oceanogr.* **2018**, *82*, 26–35, (In Japanese with English Abstract).
88. Abo, K.; Tarutani, K.; Harada, K.; Miyahara, K.; Nakayama, A.; Yagi, H. Effects of nutrient discharge on Nori aquaculture area in Kako River estuary. *J. JSCE Ser. B2 Coast. Eng.* **2012**, *68*, I1116–I1120, (In Japanese with English Abstract).
89. House of Representatives, Japan. The Diet, No. 204, Report on Discussions of The Partial Revision of the 'Law Concerning Special Measures for Conservation of the Environment of the Seto Inland Sea'. Available online: https://www.shugiin.go.jp/internet/itdb_gian.nsf/html/gian/keika/1DD1E7A.htm (accessed on 9 June 2021).
90. Yamamoto, T.; Ishida, S.; Nakahara, S.; Hiraoka, K.; Omichi, Y.; Mutsuda, H. Fertilizer application to enhance the growth of raft-cultured oysters. In Proceedings of the JSFS 85th Anniversary-Commemorative International Symposium "Fisheries Science for Future Generations", Tokyo, Japan, 22–24 September 2017. No. 05001, 2p.
91. Yamamoto, H.; Yamamoto, T.; Mito, Y.; Asaoka, S. Numerical evaluation of the use of granulated coal ash to reduce an oxygen-deficient water mass. *Mar. Poll. Bull.* **2016**, *107*, 188–205. [CrossRef] [PubMed]

Concept Paper

Facing the Forecaster's Dilemma: Reflexivity in Ocean System Forecasting

Nicholas R. Record [1,*] and Andrew J. Pershing [2]

[1] Bigelow Laboratory for Ocean Sciences, East Boothbay, ME 04544, USA
[2] Climate Central, Inc., Princeton, NJ 08542, USA; apershing@climatecentral.org
* Correspondence: nrecord@bigelow.org

Abstract: Unlike atmospheric weather forecasting, ocean forecasting is often reflexive; for many applications, the forecast and its dissemination can change the outcome, and is in this way, a part of the system. Reflexivity has implications for several ocean forecasting applications, such as fisheries management, endangered species management, toxic and invasive species management, and community science. The field of ocean system forecasting is experiencing rapid growth, and there is an opportunity to add the reflexivity dynamic to the conventional approach taken from weather forecasting. Social science has grappled with reflexivity for decades and can offer a valuable perspective. Ocean forecasting is often iterative, thus it can also offer opportunities to advance the general understanding of reflexive prediction. In this paper, we present a basic theoretical skeleton for considering iterative reflexivity in an ocean forecasting context. It is possible to explore the reflexive dynamics because the prediction is iterative. The central problem amounts to a tension between providing a reliably accurate forecast and affecting a desired outcome via the forecast. These two objectives are not always compatible. We map a review of the literature onto relevant ecological scales that contextualize the role of reflexivity across a range of applications, from biogeochemical (e.g., hypoxia and harmful algal blooms) to endangered species management. Formulating reflexivity mathematically provides one explicit mechanism for integrating natural and social sciences. In the context of the Anthropocene ocean, reflexivity helps us understand whether forecasts are meant to mitigate and control environmental changes, or to adapt and respond within a changing system. By thinking about reflexivity as part of the foundation of ocean system forecasting, we hope to avoid some of the unintended consequences that can derail forecasting programs.

Keywords: ocean forecasting; reflexivity; endangered species; fisheries; harmful algal blooms; coupled natural-human systems; Anthropocene ocean

Citation: Record, N.R.; Pershing, A.J. Facing the Forecaster's Dilemma: Reflexivity in Ocean System Forecasting. *Oceans* **2021**, *2*, 738–751. https://doi.org/10.3390/oceans2040042

Academic Editor: Diego Macías

Received: 6 July 2021
Accepted: 4 November 2021
Published: 12 November 2021

Publisher's Note: MDPI stays neutral with regard to jurisdictional claims in published maps and institutional affiliations.

1. Introduction

The convention of studying natural systems—atmosphere, biosphere, etc.—as separate from human systems, is beginning to change [1]. Earth System Science and Anthropocene Studies increasingly couple natural and social science. As these disciplines merge, and especially in the context of applications, viewing scientists as part of a system of study is a challenge.

In forecasting applications within the natural sciences, this challenge has generally meant that a forecast is assumed to have no direct bearing on the outcome of the event in question. For example, forecasting rain does not affect whether or not it actually rains. Forecasting the time of a lunar eclipse does not alter the time of the eclipse.

In social sciences, on the other hand, it is more common for scientists to view predictions as a dynamic part of the subject system. For example, the prediction of a stock market collapse can itself cause a market collapse if investors panic in response to the prediction. This could happen even if no collapse would have occurred in the absence of the prediction [2]. This phenomenon is known as "reflexivity" and is a common component of human systems (Figure 1). Reflexive prediction has been examined in many fields, such as economics, political science, and even studies of faith healing [3], and has a rich history of study in the social sciences [4,5]. Reflexivity itself is sometimes seen as the property that distinguishes the natural sciences from the social sciences [6,7].

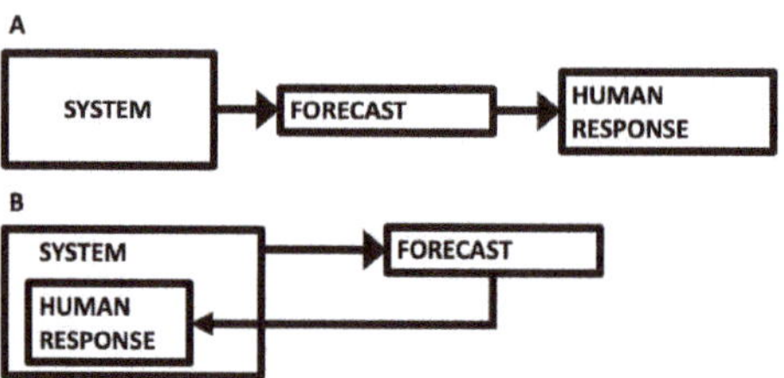

Figure 1. (**A**) The conventional forecasting scheme, where a system informs a forecast, which informs some human response. (**B**) A reflexive forecasting scheme where the human response is part of the system dynamics.

Natural systems forecasting has deep roots in weather forecasting, which is generally non-reflexive. However, many natural systems do have reflexive dynamics. For example, the dissemination of epidemic forecasts can alter human responses, changing the dynamics of the epidemic itself. A dire epidemic forecast could prompt a severe lockdown, thereby stifling the epidemic. Yet without the prediction, the lockdown might have come too late, and the dire outcome might have come to pass. There is evidence that the COVID-19 pandemic has reflexive dynamics and that taking these dynamics into account alters forecasts and outcomes [8].

Ocean system forecasting differs from weather forecasting in that many societally important forecasts deal with reflexive systems. Fisheries management often depends on a prediction of the stock size in future years. In turn, yearly fisheries forecasts can alter both fishing and management behavior, changing the mortality dynamics of the fish stocks. Similarly, endangered species management often relies on forecasts from population viability analysis. Management actions based on these forecasts are aimed at changing the predicted population trajectories. Even predictions of the global ocean climate system depend strongly on the human response to climate predictions themselves, where one of the explicit goals of making projections is to inform policy choices that will change the human forcing of the climate system.

Despite the prevalence of reflexivity in natural systems, it is typically not proactively taken into account in natural systems forecasting. The effects of reflexivity often lead to unexpected or unintended consequences, thus many fishery and endangered species management efforts require periodic yearly or multiannual iterative reevaluation of the objectives and targets, responding to missed targets reactively. For example, the Common Fisheries Policy of the European Commission has the requirement that any multiannual plan will "provide for its revision after an initial ex-post evaluation, in particular to take account of changes in scientific advice" [9]. Yet with increasingly more real-time forecasts available at people's fingertips, it is more urgent to understand reflexivity on a foundational level and to develop approaches to incorporate this understanding into forecasts proactively.

The feedback dynamic of reflexivity makes the phenomenon somewhat paradoxical, frustrating many attempts at prediction. A classic example is the 1948 U.S. election between Thomas Dewey and Harry Truman. Forecasts of a Dewey victory were so confident that the Chicago Tribune published the now infamous "Dewey Wins" headline on the day after the election. While it is possible that the mistaken forecast could have been due to a statistical or methodological error, the evidence suggests that the perception of a likely Dewey victory altered voter behavior [10]. That is, the forecast itself, and its dissemination, influenced (i.e., reversed) the outcome of the election. If that is true, then forecasts of a

Truman victory would have led to the opposite outcome—a Dewey win. The election was close—would it have been impossible to correctly forecast this event precisely because of the reversing effect of the forecast? This case illustrates the paradox of trying to make an accurate prediction in a reflexive system.

The paradoxical nature of reflexivity has been explored in literature, ranging from Oedipus to Ebenezer Scrooge to Asimov's Foundation, and has the same self-referential characteristic employed in philosophical cornerstones, such as Russell's Paradox and Gödel's Incompleteness Theorem [7]. The apparent paradox has led some to conclude that prediction in reflexive systems is not possible. The "Law of Forecast Feedback" [11] argues that a reliable prediction is not possible in a reflexive system. This pessimism is understandable, particularly when it comes to forecasting single binary or low-frequency events, such as elections or market collapses. Although natural sciences have often omitted reflexivity, they may offer an opportunity to address this paradox. Many natural systems forecasting programs involve high-frequency iterative forecasting, where forecasts are made and evaluated on a short time scale. The iterative nature of these applications provides an opportunity to examine how reflexivity works, and whether there are patterns that emerge or strategies that can be employed to make prediction successful despite reflexivity.

This paper examines the consequences of an iterative forecasting system having a reflexive component. It builds from a first-principles framework for prediction in ecology, adding a reflexive term to the dynamics. In particular, we incorporate two main elements of reflexive prediction: first, that the outcome would have been different without dissemination of the forecast, and second, that the forecast was believed and acted on [6]. We do not explicitly treat the mode of forecast dissemination. In practice, the mode of forecast dissemination is a key part of its influence on human behavior. For the purpose of illustrating some foundational properties of reflexivity in forecasting, we do not expand on the modes of forecast dissemination and the wide range of potential responses, but we recognize it as another important component to forecast implementation. By mapping previous ocean forecasting efforts into a biparametric time–time space, we explore how the iterative nature of many ocean forecasting endeavors can inform our understanding of reflexivity in forecasting, and we chart possible ways forward.

2. Theory

A generalized formulation of an ecosystem forecast can be written in terms of component parts as [12]:

$$Y_{t+1} = f(Y_t, X_t | \bar{\theta} + \alpha) + \epsilon_t \tag{1}$$

where Y_t is the state variable we are trying to forecast for time $t + 1$, X_t are environmental covariates, $\bar{\theta}$ and α represent model parameter mean and error, and ϵ_t represents process error. To analyze the effects of reflexive prediction in an iterative forecast, we will examine a simple example of this general formulation,

$$Y_{t+1} = \beta_1 Y_t + \beta_0 + \epsilon_t \tag{2}$$

This is the basic case for discontinuous (discrete) forecasting, where the state variable Y at time $t + 1$ is a linear function of the previous time step—essentially a linear autocorrelative model. The simplified formulation allows us to explore basic properties of iterative reflexive prediction. The general idea can be extended to a more complex model. To account for reflexivity, we separate the actual system trajectory, Y_t, from the disseminated forecast, Z_t. As pointed out earlier, we don't explore different modes of dissemination here, but we note that different modes of dissemination can lead to different forms of reflexivity.

Case 1: No reflexivity. In the simple model system for Y, the best forecast equation would be

$$Z_{t+1} = \beta_1 Y_t + \beta_0 \tag{3}$$

where the disseminated forecasted Z_{t+1} is identical to Y_{t+1} without the error term. A forecaster could use this equation to make reliable and unbiased predictions at each time step, with uncertainty described by ϵ plus whatever uncertainty exists around the parameter measurements. This case represents the conventional, non-reflexive point of view. The forecast has high accuracy, basically limited only by the magnitude of the error terms (Figure 2A).

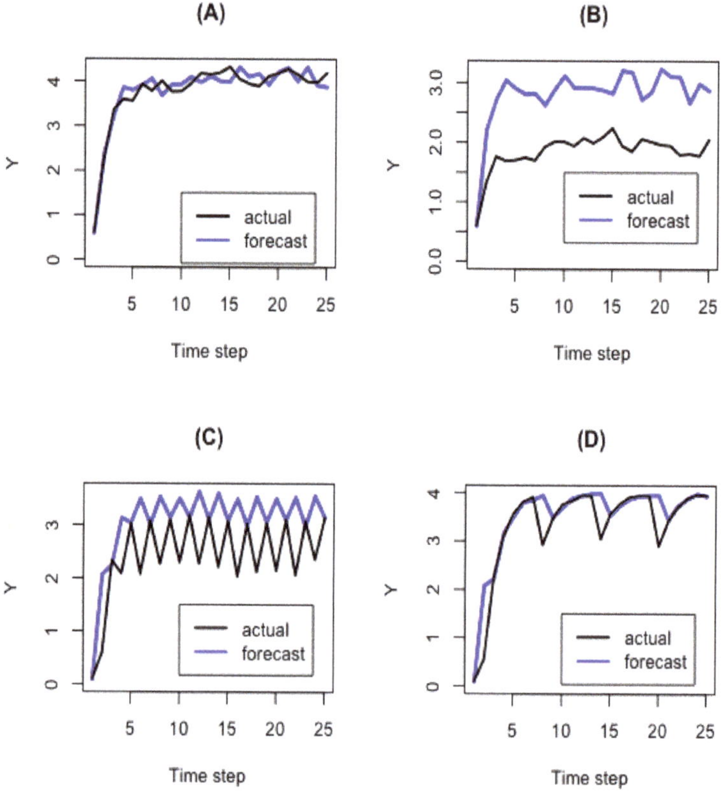

Figure 2. (**A**) Simulation with no reflexivity. (**B**) Simulation with reflexivity. (**C**) Simulation with reflexivity including a response to forecast accuracy. Error has been set to zero to make the cyclicity apparent. (**D**) Simulation with reflexivity including a response to forecast accuracy that includes memory of the accuracy over the past five time steps.

Case 2: Self-defeating reflexivity. In a reflexive prediction system, the outcome depends on the prediction. One way to express this is to add a reflexivity term to the general forecast equation:

$$Y_{t+1} = f(Y_t, X_t | \bar{\theta} + \alpha) + \epsilon_t + g(Z_{t+1}) \tag{4}$$

where g is some function of the disseminated forecast. This function is analogous to the "internal decision model" [13]. Here, the outcome of the event at time $t+1$ depends on what the forecast was for that time (i.e., $t+1$). There are two types of reflexive prediction: self-fulfilling and self-defeating (also referred to as "bandwagon" and "underdog" [4]). In a self-fulfilling reflexive system, forecasting a particular outcome makes that outcome more likely (e.g., the market collapse example). In a self-defeating reflexive system, forecasting a particular outcome makes that outcome less likely (e.g., the Truman election). Here we take the self-defeating reflexive prediction as the illustrative case.

For self-defeating forecasts, Y could be an index of some effect, such as the magnitude of an epidemic or the mortality rate of an endangered species—something that stakeholders would generally want to minimize. Dissemination of the forecast causes an inverse response, decreasing the magnitude of Y. In the linear model example, we add a response term to the forecast equation:

$$Y_{t+1} = \beta_1 Y_t + \beta_0 + \epsilon_t - \rho_t(Z_{t+1}) \tag{5}$$

where ρ_t is an increasing function of the disseminated forecast Z_{t+1}. The higher the forecast value Z_{t+1} is, the stronger the counteracting response will be, leading to a smaller value of Y_{t+1}. The equation for Z represents whatever the forecaster's strategy is for predicting the system. It could be the linear model that describes the non-reflexive Y_t, or something different.

While the scientific process behind a forecast aspires to objectivity, it exists in a broader system with subjective goals. These goals may be expressed in the kinds of processes that are forecasted (i.e., through funding choices) or in how forecasts are disseminated (e.g., the difference between forecasts of a hurricane path and communication strategies designed to get people out of harm's way). For the sake of simplicity, we will not separate the objective goals of scientific accuracy from the societal goals embedded in the application and dissemination of the forecast. Thus, when we refer to the forecaster's goals, we are essentially talking about the goals of the entire forecasting program.

Suppose the forecasting program were put in place with the ultimate aim of minimizing the value of Y (i.e., stop the epidemic or eliminate the mortality rate of the endangered species). Under this reflexive scenario, the naive strategy would be to always provide the direct forecast. For this example, we formulate a response term: $\rho_t = \rho_0 \tanh(Z_{t+1})$. In other words, a high forecast value for the next time step motivates a response that counters the expectation. Using the tanh functional form caps the magnitude of the response. The forecast would have low accuracy, but the desired outcome would be achieved. On the other hand, a high-accuracy forecast would not prompt the response that minimizes the negative effect Y.

The consequences are twofold. First, by responding to the forecast as a warning, the actual value of Y is driven down. This could be considered a case where the desired effect is achieved (Figure 2B). However, a second consequence is that the forecast is now never accurate. It always overshoots the actual by an interval equal to ρ (on average). For this scenario to actually work, the forecast users would have to never catch on to the fact that the forecast is always more dire than reality. To put this into real terms, it would be akin to forecasting a fishery collapse every year, and although none ever occurs, the fishery repeatedly reduces catches as though a collapse were always imminent. This contradiction between forecast accuracy and forecast utility (from the perspective of the desired societal outcome) is the central point to the Law of Forecast Feedback.

Case 3: Iterative self-defeating reflexivity. Realistically, people would lose trust in a consistently dire forecast due to its consistent lack of accuracy. This is where the iterative dynamic comes into play. To account for this, we introduce a scaling factor τ to the response term that represents the reliability of the forecast:

$$Y_{t+1} = \beta_1 Y_t + \beta_0 + \epsilon_t - \rho_t \tau_t \tag{6}$$

Here τ is an inverse function of the error in the forecast (i.e., $\frac{|Z_t - Y_t|}{Y_t}$), where $\tau = 1$ when accuracy is perfect and goes to zero for very low-accuracy forecasts. In other words, as the forecast becomes inaccurate, it also loses its influence. Now we have the simplest fully iteratively reflexive forecast model.

We can formulate this factor τ as $\tau_t = e^{\frac{-\tau_0 |Z_t - Y_t|}{Y_t}}$. For a perfect forecast, $\tau_t = 1$, yielding a full response to the forecast. For very high inaccuracy, τ_t decays to zero, zeroing out the response term. The parameter τ_0 shapes how quickly (as a function of forecast inaccuracy) the response term goes to zero. A high τ_0 would mean that only a small amount of inaccuracy is needed for people to stop believing in and responding to the forecast. The result is an oscillating pattern, where a reliable forecast is acted on, driving Y down, thus making the next forecast inaccurate, diminishing the response, and driving Y back up (Figure 2C). This is akin to the boom–bust reflexive dynamics seen in market systems [7].

Case 4: Iterative + learning self-defeating reflexivity. As a final note, there's no reason to assume that the response only depends on the previous time step. Depending on circumstances, it is possible that collective memory would evaluate the forecast reliability over multiple previous time steps. This can be added to the model using a number of time steps m, over which τ is computed and averaged. The result is a variably reliable forecast, with periodic lapses in accuracy (Figure 2D). From here, it is not difficult to imagine a wide range of periodic and quasi-periodic patterns that can occur depending on the form of τ_t and other properties of these equations. All of the richness of dynamical systems modeling could appear in the formulation of reflexivity.

3. The Forecaster's Dilemma

The question for the forecaster now becomes: how to deal with these opposing forces? On the one hand, a theoretically reliable forecast can alter behavior, making the forecast unreliable. On the other hand, consistently unreliable forecasts are likely to be ignored. The problem for the forecaster can be framed as the tension between two goals:

Goal 1: The accuracy directive. Conventionally, forecasters have tried to make predictions that accurately describe a future event. This also corresponds with goals of science to improve our understanding of the natural world. When the event comes to pass, a comparison between the forecast and the event serves as the assessment. This amounts to minimizing $\sum_t \frac{|Z_t - Y_t|}{Y_t}$.

Goal 2: The influence directive. The purpose of a forecast is usually to elicit some action. This generally corresponds with some practical societal goal. The Y variable represents a negative effect that the forecast is aspiring to diminish over time, so this amounts to minimizing $\sum_t Y_t$ (This could also be framed as maximizing a positive effect, such as species recovery).

A forecaster in a reflexive system should consider whether it is possible to meet these two goals simultaneously, and if so, what is the best forecasting strategy i.e., the choice of function for Z that accomplishes both directives? The example provided here is convergent in a recursive sense. That is, one can iteratively plug Y_{t+1} back into the equation as Z_{t+1}, and the forecast for the next time step will converge on a value that is both accurate and minimizes the negative effect, basically toeing a line between the two cases. However, most real-world examples will probably be more complicated, with more dynamic and complex $g(Z)$ functions.

4. Solving the Forecaster's Dilemma

Reflexivity is not just of academic interest. The coronavirus pandemic brought home the point that reflexivity in forecasts can have very real consequences. As people come to use and expect increasingly more real-time forecasting, the issue of reflexivity represents an emerging scientific challenge. With the field of ocean system forecasting growing rapidly and experiencing a "Cambrian Explosion" in applications [14], forecasters should consider the role of reflexivity when adopting a forecasting program. We suggest three components to guide consideration of reflexivity. In some cases, dealing with reflexivity will be a nonissue. When reflexivity is deeply embedded within a system, however, new scientific and pragmatic approaches will need to be developed.

4.1. Step 1: Identify How Reflexivity Could Occur in the System

Ocean systems are organized across a range of temporal, spatial, and biological scales. These scales are related such that to first order, organisms that are small tend to have lower trophic level and shorter generation times and vice versa due to the strong size structuring in ocean ecosystems [15]. Ocean system forecasts tend to align along similar temporal, spatial, and biological scales because of this natural hierarchal structuring (Figure 3).

Most ocean system forecasting programs reviewed here fall along the one-to-one line in Figure 3. This scale matching is probably not an accidental coincidence, as decision making and predictability often align with the dominant process scales in a system. At one end of the spectrum, long-term planning for long-lived protected or exploited species is informed by their comparatively longer reproductive schedules, which aligns with a longer forecasting range [16,17]. Similarly, plankton with rapid doubling times fall at the other end of the spectrum, with shorter forecast ranges [18]. However, there are some examples that deviate from this one-to-one line. In the upper left region are harmful planktonic taxa like *Vibrio* and *Alexandrium*, which are short-generation taxa for which longer forecasts have been made [19,20]. This can be motivated either by forecasting and monitoring limitations, or by a particular forecast use. In the lower right, there are examples of short-range forecasts for protected or endangered taxa, such as whales and sea turtles. These few short-range forecasts for higher trophic levels focus on forecasting location rather than abundance, making the generation length less relevant. This also represents a response to a particular forecast use.

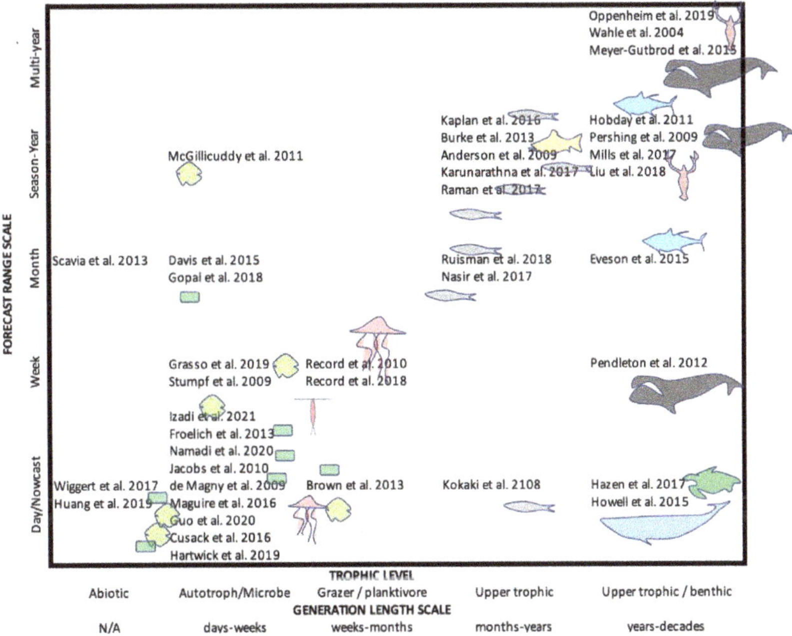

Figure 3. Forecast ranges of ocean system forecasting research programs and their associated biological time scales sources from literature examples, extended from the list provided in [14]. Positions are estimated based on descriptions in the texts [16–18,20–53].

This schematic needs to be shifted slightly to understand the role of reflexivity. Most of these forecasting studies do not consider the time scale of human response. Considering the coupled natural human system angle offers a slightly different lens, where we consider the dominant response time scales of the entire system, including the ecosystem and the human system together (Figure 4). If the response time is much longer than the forecast range (lower right of Figure 4), for example, reflexivity will be minimal to non-existent. In this scenario, many iterative forecasts would be made before any response takes place, so the human response would not affect the forecast. There are also cases where the human response has no bearing on the forecast. For example, a jellyfish forecast [46] might guide recreational activities, but probably would not influence the jellyfish populations themselves. Similarly, a forecast of the abundance of a harmful algal species might lead to a fishery closure, but the bloom would persist unaffected. In any of these scenarios, if forecasting is guiding monitoring efforts, there is the possibility of feedback even in non-reflexive systems that can lead to confirmation bias.

If the coupled system response time is similar to or shorter than the forecast range, then there is the potential for reflexivity. These cases would fall near the one-to-one line or upper-left triangle of Figure 4. Examples include short-range forecasts of harmful or toxic algal species, or mid- to long-range forecasts for protected or exploited species. If there is the potential for reflexivity, the next step is to try to measure and describe it.

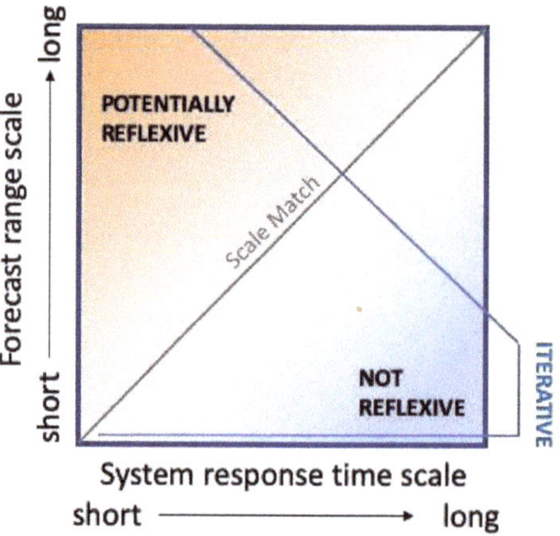

Figure 4. Schematic of forecast and system process scales. Here, the system includes the coupled natural human processes.

4.2. Step 2: Determine Whether Reflexivity Is Self-Defeating, Self-Fulfilling, or a Combination of Both

Once the potential for reflexivity has been characterized, the next step represents the empirical or data analysis side of the question. Detecting or measuring the human response can be challenging, especially if monitoring and modeling efforts have focused on the non-human parts of the system.

Consider, as an example, the management of endangered marine mammals. The Marine Mammal Protection act in the U.S. was amended in 1994 such that marine mammals are managed according to potential biological removal (PBR), which frames population viability as a forecast: i.e., "a certain probability of an event occurring in a given amount of time" [54,55]. The calculation of PBR includes a scaling factor F_R, which is set between 0 and 1. If a forecast looks particularly dire, setting F_R to zero sets human-caused mortality to zero (in theory) regardless of the value of other factors. F_R offers flexibility—and reflexivity—to the human response via the PBR calculation [56], and these numbers inform management decisions designed to reduce (or not) human-caused mortality of marine mammals. This places these forecasts toward the upper left of Figure 4, with long-range forecasts but short system response times.

If there is a way to quantify human response to the forecast, such as PBR, one can then examine whether the dynamics of this response is coupled to the dynamics of the natural system, such as in a dynamical systems framework. Two-way causality can be explored using tools such as Granger causality or convergent cross mapping [57], though there are cases where these tests are flawed [58]. In principle, it may be possible to detect reflexivity in ecological time series without measurement of the human response using Takens' theorem [59].

The species with the longest running PBR time series is the North Atlantic right whale, for which there are 25 years of data [60]. For the first decade of the 2000s, PBR was reduced to zero by setting F_R to zero in an attempt to restore this highly endangered species. The population climbed until 2010, at which point PBR was raised. Shortly thereafter, mortality rates began to rise (Figure 5A), and the species plummeted into crisis mode again [61]. When comparing PBR to human-caused mortality for this species, the strongest lagged correlation is with PBR leading mortality by four years (Figure 5B). Granger analysis supports this direction of lagged causality ($p < 0.05$, but see [58]), consistent with the notion that changing the forecast for the species has the following effect of changing the actual population trajectory of that species.

In some sense, this direction of causality shouldn't be surprising: the goal of the management approach is to have a following effect on the population. But the right whale example is a case of self-defeating reflexivity, and self-defeating reflexivity can cut two ways. A dire forecast can motivate recovery efforts, thereby improving the forecasted outlook, as intended—but a more favorable forecast can have the opposite effect. In the right whale time series, increases in PBR were followed a few years later by increases in human-caused mortality, reversing the recovery trend that the right whale population had been showing. While there are likely multiple factors at play in the sudden reversal of the right whale population trajectory, including climate and oceanographic changes [61,62], the pattern is consistent with the dynamics we would expect in a self-defeating reflexive forecasting system, leading to an unintended consequence. Ideally, we would like reflexive feedback to take effect when the population is struggling, but not when it's doing well—in other words, a concave-down curve in Figure 5B rather than concave up. Accounting for this kind of reflexive dynamic more deliberately probably requires a more mechanistic understanding of the reflexive term $g(Z)$.

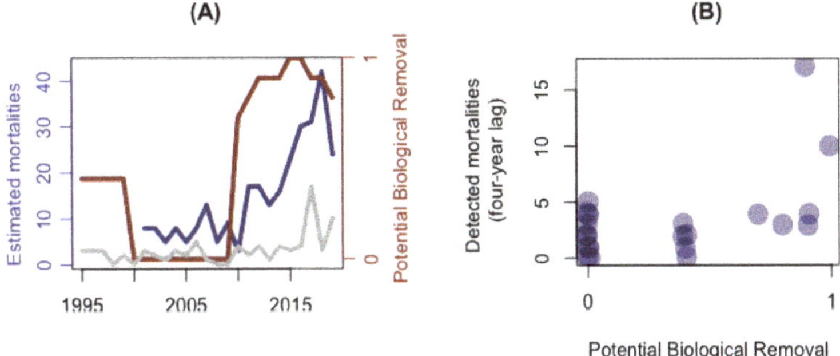

Figure 5. (**A**) Time series of PBR for the North Atlantic right whale (red) and two estimates of mortality: documented human-caused mortalities (grey) and the annual population change from the Pace model, subtracting out new calves (blue). (**B**) Lagged relationship between potential biological removal (PBR) and mortality. Data aggregated from NOAA and the North Atlantic Right Whale Consortium reports [60].

4.3. Step 3: Incorporate Human Response into a Forecast Model

If there is significant reflexivity in a forecasting system, with important consequences, the next step is to try to formulate that response mathematically and incorporate it into a model. This step represents an open area of scientific research and theory. A key question to answer here is: Can the accuracy and influence directives both be met?

Ocean forecasting up to now has largely followed the tradition of weather forecasting, combining mechanistic or processed based formulations, such as the advection-diffusion-reaction equation, with statistical formulations. More recently, machine learning algorithms have been replacing the statistical components, and to some extent the mechanistic components as well, to derive predictive rules from data. This framework has so far been mostly an elaboration of the $f(Z)$ term. The reflexive term $g(Z)$ represents a largely unexplored research opportunity where similar approaches could be used. In fisheries forecasting, stochastic models have been used to couple these components [63].

Coupled natural-human systems are complex. There probably is not an analog for the Navier-Stokes equations for the human part of the system. However, examining mathematical formulations in a theoretical context can help answer whether the accuracy and influence directives are at odds with each other. If they are, then it could be a sign that the forecast will do more harm than good—or at least not have the intended effect.

There is also the opportunity to think beyond dynamical systems and statistics. For example, because information is exchanged in at least two directions in reflexive systems, some have viewed forecasting in the context of game theory [64]. This view becomes particularly interesting when multiple agents, including forecasters and users, are exchanging wide ranging information about the system in question. On the ocean, forecast users can be capturing, processing, and exchanging real-time observations and knowledge, such as where a commercial fish species is found—knowledge that is not available to the forecaster but plays a role in human response [65]. Similarly, multiple forecasters might be using different knowledge and approaches, and exchanging some part of that information with each other. In this way, forecasting programs can be nested within networks of social-natural systems with complex information flow.

The challenge of reflexive systems forecasting highlights the need to be making more operational forecasts. The iterative property of these forecasting systems is key, providing the data and experience needed to build up an understanding of reflexive dynamics. The tendency is often to focus forecasting efforts on high-stakes problems, such as endangered species or health hazards, but lower stakes problems (e.g., nuisance species, ecotourism) could provide a safer arena for building up the datasets needed to analyze and understand reflexive dynamics in ocean systems in new ways.

5. Conclusions: Reflexivity in the Changing Ocean

Reflexivity highlights the significant human dimension and associated challenges in emerging forecasting programs. Traditionally, whether forecasting the weather or some ecosystem process, the natural system is viewed in an objective sense, separate from the human observer. In the context of ocean systems, reflexivity is an emerging challenge that has bearing both to how we understand and interact with the ocean, and how we understand and make use of algorithms.

Regarding human interactions with the ocean, the "Anthropocene Ocean" is described as a socio-material space [66] where physical and biological systems are interlinked with social and scientific systems. In this context, there are two common frameworks useful for understanding ocean system forecasting. One framework is that of planetary security— mitigating the risks of environmental damage due to human activities. In this context, forecasting would serve as an aid to monitoring and controlling environmental processes. Reflexivity is implicit in this construct, as the human response is the mechanism for influencing the environment. Here both the accuracy and influence directives are important for the forecast to be effective.

With the urgency around issues like climate change, there are practical limitations to the "measure and control" approach to dealing with the Anthropocene. An alternative emerging perspective is the idea of correlational and relational epistemologies, where management structures would sense and adapt to events in real-time [66,67], without a causal understanding or attempt to mitigate the dynamics. This perspective also relies on algorithmic and digital technologies, but in this context, forecasting serves as information connectivity and not as a means of system control. Reflexivity is not necessarily implicit in this perspective. When reflexivity is present, a forecast could potentially still be useful without both of the two directives met, if it is nested within a larger network of correlational human responses and feedbacks [67]. This view has drawbacks as well, and potential for unintended consequences. As each new forecasting system comes online, we should ask whether it is intended to understand (in a causal sense) and control aspects of the ocean system, or whether it is intended as part of a network of adaptation tools. This distinction will help narrow the scope of reflexivity in the forecasted system.

The question of whether the goal is to predict and control the environment, or to adapt and respond to a changing environment, is at the core of many discussions about big data, algorithms, and artificial intelligence. As forecasting algorithms become more widespread and embedded in our social relationship with Earth systems, ocean science can take lessons from the growing field of algorithmic accountability. Across applications ranging from resume sorting to prison sentencing, algorithms are replacing human decision making. The proliferation of algorithms in this way has led to many unintended consequences [68]. Ocean system forecasting shares this risk of unintended consequences—something that has already occurred in a few ocean forecasting programs [69,70]. For reflexive forecasts, when the accuracy and influence directives are at odds with each other, there is high potential for unintended consequences. The field of algorithmic accountability is developing method-ologies for addressing this, such as action plans for redress when unintended outcomes occur, which can be applied to ocean forecasting to help prevent unintended consequences or address them when they occur [71–73].

Despite the potentially confounding nature of reflexivity, the subject represents a rich area of scientific inquiry. The reflexive term in the forecasting equation—i.e., the $g(Z)$—captures an emerging challenge in natural systems forecasting. Many forecasting evaluation analyses choose not to treat the reflexive feedback dynamic [74], and others have simply ignored it [75]. Some leave the human response to the realm of policy, communications, or to forecast users, while others view this part of the equation as a focus for quantitative study and analysis in its own right [2]. The example developed here separates $f(Y)$ and $g(Z)$ into additive terms, but it is possible that they interact in more complex and nonlinear ways yet to be discovered. There is also another layer of complexity added to $g(Z)$ dynamics when considering the mode of forecast dissemination. People respond differently depending on how a forecast is communicated, and if the system is reflexive, then communication choices can feed back on the natural system dynamics $f(Z)$. By representing iterative system forecasting as a combination of two components, $f(Y)$ and $g(Z)$, we see a promising quantitative starting point for integrating natural sciences with social and behavioral sciences, as well as a pathway for using forecasts as a tool for navigating the complex interactions between humans and the ocean.

Author Contributions: Conceptualization, N.R.R. and A.J.P.; methodology, N.R.R.; software, N.R.R.; validation, N.R.R.; formal analysis, N.R.R.; investigation, N.R.R. and A.J.P.; resources, N.R.R. and A.J.P.; data curation, N.R.R.; writing—original draft preparation, N.R.R. and A.J.P.; writing—review and editing, N.R.R. and A.J.P.; visualization, N.R.R. and A.J.P.; supervision, N.R.R. and A.J.P.; project administration, N.R.R. and A.J.P.; funding acquisition, N.R.R. and A.J.P. All authors have read and agreed to the published version of the manuscript.

Funding: Support for this research came from institutional funds from the Tandy Center for Ocean Forecasting at Bigelow Laboratory for Ocean Sciences and the Otto Mønsteds Fond via Denmark Technical University (NRR) and from NSF grant OCE-1851866 (AJP).

Institutional Review Board Statement: Not applicable.

Informed Consent Statement: Not applicable.

Data Availability Statement: Code and data used in this study are stored in the following open repository: https://github.com/SeascapeScience/forecasters-dilemma (accessed on 1 November 2021).

Conflicts of Interest: The authors declare no conflict of interest.

References

1. Alberti, M.; Asbjornsen, H.; Baker, L.A.; Brozović, N.; Drinkwater, L.E.; Drzyzga, S.A.; Jantz, C.A.; Fragoso, J.; Holland, D.S.; Kohler, T.A.; et al. Research on Coupled Human and Natural Systems (Chans): Approach, Challenges, and Strategies. *Bull. Ecol. Soc. Am.* **2011**. [CrossRef]
2. Kopec, M. A More Fulfilling (and Frustrating) Take on Reflexive Predictions. *Philos. Sci.* **2011**, *78*, 1249–1259. [CrossRef]

3.	Henshel, R.L. The Boundary of the Self-Fulfilling Prophecy and the Dilemma of Social Prediction. *Br. J. Sociol.* **1982**, *33*, 511–528. [CrossRef]
4.	Simon, H. Bandwagon and Underdog Effects of Election Predictions,(1954,[1957]). *Models Man* **1954**, *18*, 245–253.
5.	Grunberg, E.; Modigliani, F. The Predictability of Social Events. *J. Political Econ.* **1954**, *62*, 465–478. [CrossRef]
6.	Stack, G.J. Reflexivity, Prediction and Paradox. *Dialogos* **1978**, *13*, 91–101.
7.	Soros, G. Fallibility, Reflexivity, and the Human Uncertainty Principle. *J. Econ. Methodol.* **2013**, *20*, 309–329. [CrossRef]
8.	Record, N.R.; Pershing, A.J. A Note on the Effects of Epidemic Forecasts on Epidemic Dynamics. *PeerJ* **2020**, *8*, e9649. [CrossRef]
9.	European Union. EU Regulation (Eu) No 1380/2013 of the European Parliament and of the Council of 11 December 2013 on the Common Fisheries Policy, Amending Council Regulations (Ec) No 1954/2003 and (Ec) No 1224/2009 and Repealing Council Regulations (Ec) No 2371/2002 and (Ec) No 639/2004 and Council Decision 2004/585/Ec. *J. Eur. Union Bruss.* **2013**, *354*, 22–61.
10.	McDonald, D.G.; Glynn, C.J.; Kim, S.-H.; Ostman, R.E. The Spiral of Silence in the 1948 Presidential Election. *Commun. Res.* **2001**, *28*, 139–155. [CrossRef]
11.	Smith, G.C. The Law of Forecast Feedback. *Am. Stat.* **1964**, *18*, 11–14.
12.	Dietze, M.C. Prediction in Ecology: A First-Principles Framework. *Ecol. Appl.* **2017**, *27*, 2048–2060. [CrossRef] [PubMed]
13.	Beinhocker, E.D. Reflexivity, Complexity, and the Nature of Social Science. *J. Econ. Methodol.* **2013**, *20*, 330–342. [CrossRef]
14.	Payne, M.R.; Hobday, A.J.; MacKenzie, B.R.; Tommasi, D.; Dempsey, D.P.; Fässler, S.M.; Haynie, A.C.; Ji, R.; Liu, G.; Lynch, P.D.; et al. Lessons from the First Generation of Marine Ecological Forecast Products. *Front. Mar. Sci.* **2017**, *4*, 289. [CrossRef]
15.	Andersen, K.H.; Berge, T.; Gonçalves, R.J.; Hartvig, M.; Heuschele, J.; Hylander, S.; Jacobsen, N.S.; Lindemann, C.; Martens, E.A.; Neuheimer, A.B.; et al. Characteristic Sizes of Life in the Oceans, from Bacteria to Whales. *Annu. Rev. Mar. Sci.* **2016**, *8*, 217–241. [CrossRef]
16.	Meyer-Gutbrod, E.L.; Greene, C.H.; Sullivan, P.J.; Pershing, A.J. Climate-Associated Changes in Prey Availability Drive Reproductive Dynamics of the North Atlantic Right Whale Population. *Mar. Ecol. Prog. Ser.* **2015**, *535*, 243–258. [CrossRef]
17.	Oppenheim, N.G.; Wahle, R.A.; Brady, D.C.; Goode, A.G.; Pershing, A.J. The Cresting Wave: Larval Settlement and Ocean Temperatures Predict Change in the American Lobster Harvest. *Ecol. Appl.* **2019**, *29*, e02006. [CrossRef]
18.	Froelich, B.; Bowen, J.; Gonzalez, R.; Snedeker, A.; Noble, R. Mechanistic and Statistical Models of Total Vibrio Abundance in the Neuse River Estuary. *Water Res.* **2013**, *47*, 5783–5793. [CrossRef]
19.	McGillicuddy, D.J. Models of harmful algal blooms: Conceptual, empirical, and numerical approaches. *J. Mar. Syst. J. Eur. Assoc. Mar. Sci. Tech.* **2010**, *83*, 105. [CrossRef]
20.	Gopal, K.; Shitan, M. Development of a Web Portal to Forecast the Monthly Mean Chlorophyll Concentration of the Waters off Peninsular Malaysia's West Coast. *Malays. J. Math. Sci.* **2018**, *12*, 99–119.
21.	Wahle, R.A.; Incze, L.S.; Fogarty, M.J. First Projections of American Lobster Fishery Recruitment Using a Settlement Index and Variable Growth. *Bull. Mar. Sci.* **2004**, *74*, 101–114.
22.	Anderson, J.J.; Beer, W.N. Oceanic, Riverine, and Genetic Influences on Spring Chinook Salmon Migration Timing. *Ecol. Appl.* **2009**, *19*, 1989–2003. [CrossRef]
23.	De Magny, G.C.; Long, W.; Brown, C.W.; Hood, R.R.; Huq, A.; Murtugudde, R.; Colwell, R.R. Predicting the Distribution of Vibrio Spp. in the Chesapeake Bay: A Vibrio Cholerae Case Study. *EcoHealth* **2009**, *6*, 378–389. [CrossRef]
24.	Pershing, A.J.; Record, N.R.; Monger, B.C.; Mayo, C.A.; Brown, M.W.; Cole, T.V.; Kenney, R.D.; Pendleton, D.E.; Woodard, L.A. Model-Based Estimates of Right Whale Habitat Use in the Gulf of Maine. *Mar. Ecol. Prog. Ser.* **2009**, *378*, 245–257. [CrossRef]
25.	Stumpf, R.P.; Tomlinson, M.C.; Calkins, J.A.; Kirkpatrick, B.; Fisher, K.; Nierenberg, K.; Currier, R.; Wynne, T.T. Skill Assessment for an Operational Algal Bloom Forecast System. *J. Mar. Syst.* **2009**, *76*, 151–161. [CrossRef] [PubMed]
26.	Jacobs, J.M.; Rhodes, M.M.R.; Brown, C.W.; Hood, R.R.; Leight, A.; Long, W.; Wood, R. *Predicting the Distribution of Vibrio Vulnificus in Chesapeake Bay*; NOAA Technical Memorandum NOS NCCOS 112; NOAA National Centers for Coastal Ocean Science, Center for Coastal Environmental Health and Biomolecular Research, Cooperative Oxford Laboratory: Oxford, MD, USA, 2010.
27.	Record, N.; Pershing, A.; Runge, J.; Mayo, C.; Monger, B.; Chen, C. Improving Ecological Forecasts of Copepod Community Dynamics Using Genetic Algorithms. *J. Mar. Syst.* **2010**, *82*, 96–110. [CrossRef]
28.	Hobday, A.J.; Hartog, J.R.; Spillman, C.M.; Alves, O. Seasonal Forecasting of Tuna Habitat for Dynamic Spatial Management. *Can. J. Fish. Aquat. Sci.* **2011**, *68*, 898–911. [CrossRef]
29.	McGillicuddy, D.J.; Townsend, D.; He, R.; Keafer, B.; Kleindinst, J.; Li, Y.; Manning, J.; Mountain, D.; Thomas, M.; Anderson, D.M. Suppression of the 2010 Alexandrium Fundyense Bloom by Changes in Physical, Biological, and Chemical Properties of the Gulf of Maine. *Limnol. Oceanogr.* **2011**, *56*, 2411–2426. [CrossRef]
30.	Pendleton, D.E.; Sullivan, P.J.; Brown, M.W.; Cole, T.V.; Good, C.P.; Mayo, C.A.; Monger, B.C.; Phillips, S.; Record, N.R.; Pershing, A.J. Weekly Predictions of North Atlantic Right Whale Eubalaena Glacialis Habitat Reveal Influence of Prey Abundance and Seasonality of Habitat Preferences. *Endanger. Species Res.* **2012**, *18*, 147–161. [CrossRef]
31.	Brown, C.W.; Hood, R.R.; Long, W.; Jacobs, J.; Ramers, D.; Wazniak, C.; Wiggert, J.; Wood, R.; Xu, J. Ecological Forecasting in Chesapeake Bay: Using a Mechanistic–empirical Modeling Approach. *J. Mar. Syst.* **2013**, *125*, 113–125. [CrossRef]
32.	Burke, B.J.; Peterson, W.T.; Beckman, B.R.; Morgan, C.; Daly, E.A.; Litz, M. Multivariate Models of Adult Pacific Salmon Returns. *PLoS ONE* **2013**, *8*, e54134. [CrossRef]
33.	Scavia, D.; Evans, M.A.; Obenour, D.R. A Scenario and Forecast Model for Gulf of Mexico Hypoxic Area and Volume. *Environ. Sci. Technol.* **2013**, *47*, 10423–10428. [CrossRef]

34. Eveson, J.P.; Hobday, A.J.; Hartog, J.R.; Spillman, C.M.; Rough, K.M. Seasonal Forecasting of Tuna Habitat in the Great Australian Bight. *Fish. Res.* **2015**, *170*, 39–49. [CrossRef]
35. Howell, E.A.; Hoover, A.; Benson, S.R.; Bailey, H.; Polovina, J.J.; Seminoff, J.A.; Dutton, P.H. Enhancing the Turtlewatch Product for Leatherback Sea Turtles, a Dynamic Habitat Model for Ecosystem-Based Management. *Fish. Oceanogr.* **2015**, *24*, 57–68. [CrossRef]
36. Dabrowski, T.; Lyons, K.; Nolan, G.; Berry, A.; Cusack, C.; Silke, J. Harmful Algal Bloom Forecast System for Sw Ireland. Part I: Description and Validation of an Operational Forecasting Model. *Harmful Algae* **2016**, *53*, 64–76. [CrossRef] [PubMed]
37. Kaplan, I.C.; Williams, G.D.; Bond, N.A.; Hermann, A.J.; Siedlecki, S.A. Cloudy with a Chance of Sardines: Forecasting Sardine Distributions Using Regional Climate Models. *Fish. Oceanogr.* **2016**, *25*, 15–27. [CrossRef]
38. Maguire, J.; Cusack, C.; Ruiz-Villarreal, M.; Silke, J.; McElligott, D.; Davidson, K. Applied Simulations and Integrated Modelling for the Understanding of Toxic and Harmful Algal Blooms (Asimuth): Integrated Hab Forecast Systems for Europe's Atlantic Arc. *Harmful Algae* **2016**, *53*, 160–166. [CrossRef]
39. Mills, K.E.; Pershing, A.J.; Hernández, C.M. Forecasting the Seasonal Timing of Maine's Lobster Fishery. *Front. Mar. Sci.* **2017**, *4*, 337. [CrossRef]
40. Hazen, E.L.; Palacios, D.M.; Forney, K.A.; Howell, E.A.; Becker, E.; Hoover, A.L.; Irvine, L.; DeAngelis, M.; Bograd, S.J.; Mate, B.R.; et al. WhaleWatch: A Dynamic Management Tool for Predicting Blue Whale Density in the California Current. *J. Appl. Ecol.* **2017**, *54*, 1415–1428. [CrossRef]
41. Nasir, N.; Samsudin, R.; Shabri, A. Forecasting of Monthly Marine Fish Landings Using Artificial Neural Network. *Int. J. Adv. Soft Compu. Appl.* **2017**, *9*, 75–89.
42. Raman, R.K.; Sathianandan, T.; Sharma, A.; Mohanty, B. Modelling and Forecasting Marine Fish Production in Odisha Using Seasonal Arima Model. *Natl. Acad. Sci. Lett.* **2017**, *40*, 393–397. [CrossRef]
43. Wiggert, J.D.; Hood, R.R.; Brown, C.W. Modeling hypoxia and its ecological consequences in chesapeake bay. In *Modeling Coastal Hypoxia*; Springer: New York, NY, USA, 2017; pp. 119–147.
44. Kokaki, Y.; Tawara, N.; Kobayashi, T.; Hashimoto, K.; Ogawa, T. Sequential Fish Catch Forecasting Using Bayesian State Space Models. In Proceedings of the 2018 24th International Conference on Pattern Recognition (ICPR), Beijing, China, 20–24 August 2018; pp. 776–781.
45. Liu, G.; Eakin, C.M.; Chen, M.; Kumar, A.; De La Cour, J.L.; Heron, S.F.; Geiger, E.F.; Skirving, W.J.; Tirak, K.V.; Strong, A.E. Predicting Heat Stress to Inform Reef Management: NOAA Coral Reef Watch's 4-Month Coral Bleaching Outlook. *Front. Mar. Sci.* **2018**, *5*, 57. [CrossRef]
46. Record, N.R.; Tupper, B.; Pershing, A.J. The Jelly Report: Forecasting Jellyfish Using Email and Social Media. *Anthr. Coasts* **2018**, *1*, 34–43.
47. Rusiman, M.S.; Sufahani, S.; Robi, N.A.I.M.; Abdullah, A.W.; Azmi, N.A. Predictive Modelling of Marine Fish Landings in Malaysia. *Adv. Appl. Stat.* **2018**, *53*, 123–135. [CrossRef]
48. Davis, B.J.; Jacobs, J.M.; Zaitchik, B.; DePaola, A.; Curriero, F.C. Vibrio Parahaemolyticus in the Chesapeake Bay: Operational in Situ Prediction and Forecast Models Can Benefit from Inclusion of Lagged Water Quality Measurements. *Appl. Environ. Microbiol.* **2019**, *85*, e01007-19. [CrossRef] [PubMed]
49. Grasso, I.; Archer, S.D.; Burnell, C.; Tupper, B.; Rauschenberg, C.; Kanwit, K.; Record, N.R. The Hunt for Red Tides: Deep Learning Algorithm Forecasts Shellfish Toxicity at Site Scales in Coastal Maine. *Ecosphere* **2019**, *10*, e02960. [CrossRef]
50. Hartwick, M.A.; Urquhart, E.A.; Whistler, C.A.; Cooper, V.S.; Naumova, E.N.; Jones, S.H. Forecasting Seasonal Vibrio Para-haemolyticus Concentrations in New England Shellfish. *Int. J. Environ. Res. Public Health* **2019**, *16*, 4341. [CrossRef] [PubMed]
51. Huang, P.; Trayler, K.; Wang, B.; Saeed, A.; Oldham, C.E.; Busch, B.; Hipsey, M.R. An Integrated Modelling System for Water Quality Forecasting in an Urban Eutrophic Estuary: The Swan-Canning Estuary Virtual Observatory. *J. Mar. Syst.* **2019**, *199*, 103218. [CrossRef]
52. Guo, J.; Dong, Y.; Lee, J.H. A Real Time Data Driven Algal Bloom Risk Forecast System for Mariculture Management. *Mar. Pollut. Bull.* **2020**, *161*, 111731. [CrossRef]
53. Namadi, P.; Deng, Z. Modeling and Forecasting Vibrio Parahaemolyticus Concentrations in Oysters. *Water Res.* **2021**, *189*, 116638. [CrossRef]
54. Taylor, B.L.; Wade, P.R.; De Master, D.P.; Barlow, J. Incorporating Uncertainty into Management Models for Marine Mammals. *Conserv. Biol.* **2000**, *14*, 1243–1252. [CrossRef]
55. Lonergan, M. Potential Biological Removal and Other Currently Used Management Rules for Marine Mammal Populations: A Comparison. *Mar. Policy* **2011**, *35*, 584–589. [CrossRef]
56. Barlow, J.; Swartz, S.L.; Eagle, T.C.; Wade, P.R. US Marine Mammal Stock Assessments: Guidelines for Preparation, Background, and a Summary of the 1995 Assessments. *NOAA Tech. Memo. NMFS-OPR* **1995**, *6*, 73.
57. Barraquand, F.; Picoche, C.; Detto, M.; Hartig, F. Inferring Species Interactions Using Granger Causality and Convergent Cross Mapping. *Theor. Ecol.* **2021**, *14*, 87–105. [CrossRef]
58. San Liang, X. Information Flow and Causality as Rigorous Notions Ab Initio. *Phys. Rev. E* **2016**, *94*, 052201. [CrossRef] [PubMed]
59. Ye, H.; Beamish, R.J.; Glaser, S.M.; Grant, S.C.; Hsieh, C.-h.; Richards, L.J.; Schnute, J.T.; Sugihara, G. Equation-Free Mechanistic Ecosystem Forecasting Using Empirical Dynamic Modeling. *Proc. Natl. Acad. Sci. USA* **2015**, *112*, E1569–E1576. [CrossRef] [PubMed]

60. Pettis, H.; Pace, R.; Hamilton, P. North Atlantic Right Whale Consortium 2020 Annual Report Card. In *Report to the North Atlantic Right Whale Consortium*; NOAA: Washington, DC, USA, 2021.
61. Davies, K.T.; Brillant, S.W. Mass Human-Caused Mortality Spurs Federal Action to Protect Endangered North Atlantic Right Whales in Canada. *Mar. Policy* **2019**, *104*, 157–162. [CrossRef]
62. Record, N.R.; Runge, J.A.; Pendleton, D.E.; Balch, W.M.; Davies, K.T.; Pershing, A.J.; Johnson, C.L.; Stamieszkin, K.; Ji, R.; Feng, Z.; et al. Rapid Climate-Driven Circulation Changes Threaten Conservation of Endangered North Atlantic Right Whales. *Oceanography* **2019**, *32*, 162–169. [CrossRef]
63. Rätz, H.-J.; Charef, A.; Abella, A.; Colloca, F.; Ligas, A.; Mannini, A.; Lloret, J. A Medium-Term, Stochastic Forecast Model to Support Sustainable, Mixed Fisheries Management in the Mediterranean Sea. *J. Fish Biol.* **2013**, *83*, 921–938. [CrossRef]
64. Bowden, R.J. Feedback Forecasting Games: An Overview. *J. Forecast.* **1989**, *8*, 117–127. [CrossRef]
65. Verweij, M.; Van Densen, W.; Mol, A. The Tower of Babel: Different Perceptions and Controversies on Change and Status of North Sea Fish Stocks in Multi-Stakeholder Settings. *Mar. Policy* **2010**, *34*, 522–533. [CrossRef]
66. Rothe, D. Jellyfish Encounters: Science, Technology and Security in the Anthropocene Ocean. *Crit. Stud. Secur.* **2020**, *8*, 145–159. [CrossRef]
67. Chandler, D.; Pugh, J. Islands and the Rise of Correlational Epistemology in the Anthropocene: Rethinking the Trope of the "Canary in the Coalmine". *Isl. Stud. J.* **2020**. [CrossRef]
68. Chen, L.; Ma, R.; Hannák, A.; Wilson, C. Investigating the Impact of Gender on Rank in Resume Search Engines. In Proceedings of the Proceedings of the 2018 CHI Conference on Human Factors in Computing Systems, Montreal, QC, Canada, 21–26 April 2018; pp. 1–14.
69. Pershing, A.J.; Mills, K.E.; Dayton, A.M.; Franklin, B.S.; Kennedy, B.T. Evidence for Adaptation from the 2016 Marine Heatwave in the Northwest Atlantic Ocean. *Oceanography* **2018**, *31*, 152–161. [CrossRef]
70. Hobday, A.J.; Hartog, J.R.; Manderson, J.P.; Mills, K.E.; Oliver, M.J.; Pershing, A.J.; Siedlecki, S. Ethical Considerations and Unanticipated Consequences Associated with Ecological Forecasting for Marine Resources. *ICES J. Mar. Sci.* **2019**, *76*, 1244–1256. [CrossRef]
71. Crawford, K.; Schultz, J. Big Data and Due Process: Toward a Framework to Redress Predictive Privacy Harms?(2014). *Boston Coll. Law Rev.* **2014**, *55*, 93.
72. Shah, H. Algorithmic Accountability. *Philos. Trans. R. Soc. A Math. Phys. Eng. Sci.* **2018**, *376*, 20170362. [CrossRef]
73. Grasso, I.; Russell, D.; Matthews, A.; Matthews, J.; Record, N.R. Applying Algorithmic Accountability Frameworks with Domain-Specific Codes of Ethics: A Case Study in Ecosystem Forecasting for Shellfish Toxicity in the Gulf of Maine. In Proceedings of the 2020 ACM-IMS on Foundations of Data Science Conference, New York, NY, USA, 19–20 October 2020; pp. 83–91.
74. Cramer, E.Y.; Ray, E.L.; Lopez, V.K.; Bracher, J.; Brennen, A.; Rivadeneira, A.J.C.; Gerding, A.; Gneiting, T.; House, K.H.; Huang, Y.; et al. Evaluation of Individual and Ensemble Probabilistic Forecasts of COVID-19 Mortality in the US. *medRxiv* **2021**. [CrossRef]
75. Ioannidis, J.P.; Cripps, S.; Tanner, M.A. Forecasting for COVID-19 Has Failed. *Int. J. Forecast.* **2020**. In Press. [CrossRef]

Review

Glow on Sharks: State of the Art on Bioluminescence Research

Laurent Duchatelet [1,*], Julien M. Claes [1], Jérôme Delroisse [2], Patrick Flammang [2] and Jérôme Mallefet [1,*]

1   Marine Biology Laboratory, Earth and Life Institute, University of Louvain—UCLouvain,
    1348 Louvain-la-Neuve, Belgium; julien.claes@gmail.com
2   Biology of Marine Organisms and Biomimetics Unit, Research Institute for Biosciences,
    University of Mons—UMONS, 7000 Mons, Belgium; Jerome.delroisse@umons.ac.be (J.D.);
    Patrick.Flammang@umons.ac.be (P.F.)
*   Correspondence: Laurent.Duchatelet@uclouvain.be (L.D.); Jerome.mallefet@uclouvain.be (J.M.)

Abstract: This review presents a synthesis of shark bioluminescence knowledge. Up to date, bioluminescent sharks are found only in Squaliformes, and specifically in Etmopteridae, Dalatiidae and Somniosidae families. The state-of-the-art knowledge about the evolution, ecological functions, histological structure, the associated squamation and physiological control of the photogenic organs of these elusive deep-sea sharks is presented. Special focus is given to their unique and singular hormonal luminescence control mechanism. In this context, the implication of the photophore-associated extraocular photoreception—which complements the visual adaptations of bioluminescent sharks to perceive residual downwelling light and luminescence in dim light environment—in the hormonally based luminescence control is depicted in detail. Similarities and differences between shark families are highlighted and support the hypothesis of an evolutionary unique ancestral appearance of luminescence in elasmobranchs. Finally, potential areas for future research on shark luminescence are presented.

Keywords: shark; luminescence; Etmopteridae; Dalatiidae; Somniosidae; photophore; hormonal control; counter-illumination

Citation: Duchatelet, L.; Claes, J.M.; Delroisse, J.; Flammang, P.; Mallefet, J. Glow on Sharks: State of the Art on Bioluminescence Research. *Oceans* **2021**, 2, 822–842. https://doi.org/10.3390/oceans2040047

Academic Editors: Antonio Bode and Eric Clua

Received: 1 September 2021
Accepted: 3 December 2021
Published: 17 December 2021

1. Introduction

Bioluminescence is the ability of living organisms to produce visible light [1]. Mention of this phenomenon dates back to antiquity with the description of "cold light" by Aristotle in his book "De Anima". Many centuries later, Charles Darwin, on board the Beagle, witnessed and described light in water as "milky sea" in his logbook. The first studies demonstrating mechanisms underlying bioluminescence appeared in 1667 with Robert Boyle, who depicted the oxygen requirement for luminescence production.

Bioluminescence is the result of a spontaneous exergonic chemical reaction involving the oxidation of a luciferin catalyzed by a luciferase [2], which produces a transitory excited state that finally relaxes by emitting a photon with oxyluciferin as final product [3–5]. Luminous systems involve either luciferase and luciferin as separate components or a complex molecule called "photoprotein" comprising a preoxidized luciferin and a luciferasic activity [5]. Luciferins are often common to different taxa while luciferases are thought to be typically species-specific [6] but exceptions to this rule have been recently highlighted [7–9]. Similar luciferases, first described in species from same phyla (within a clade), are now found between phylogenetically distant species supporting the existence of multiphyletic distribution of these luciferases [7–9].

Rare in terms of species number, luminescence nevertheless arose in over 700 genera from 13 phyla covering all kingdoms, except plants and archaea, e.g., [6,10,11]. This phylogenetic diversity results from numerous independent evolutions of light-producing capability—in fact, it is estimated that bioluminescence arose independently more than 90 times during evolution [1,3,9,10,12,13], which suggests that luminescence is of paramount ecological importance, but also that the acquisition of light emission capability is a relatively easy and quick process [1].

Eighty percent of luminous taxa inhabit marine environments, from coastal shallow waters to abyssal depths, and are mainly found in bacteria, protists, ctenophores, cnidarians, annelids, mollusks, chaetognaths, crustaceans, echinoderms, tunicates, and fishes [1,6,14,15]. Within the water column, the mesopelagic zone (i.e., 200–1000 m depth) has been estimated to host a high proportion of luminous organisms, e.g., 70% of mesopelagic bony fishes appear to be luminous [16]. Comparatively, terrestrial and freshwater luminous animals are only represented by earthworms, snails and limpets, fireflies, beetles, and some insect larvae, e.g., [17–23].

Although observations of shark luminescence have been reported for almost two centuries [24], the first research projects dedicated to shark luminescence were only initiated in 2005 and focused on the study of the ecological functions and physiological control of the photophores of a single species, the velvet belly lanternshark, *Etmopterus spinax* [25]. Since then, shark luminescence research has flourished, with detailed phylogenetical, ecological, histological, and physiological studies now available for numerous species, e.g., [26,27]. By synthesizing the findings of those studies, this article aims to provide not only a holistic view of current shark luminescence knowledge but also perspectives for future research.

2. Luminous Shark Diversity

Among cartilaginous fishes, only sharks have evolved the ability to emit light. Indeed, no report of bioluminescence exists for ratfishes (Chimaeriformes) apart from the mention of a luminous fluid on the rabbit fish, *Chimaera monstrosa*, in 1810 interpreted as coming from luminous bacteria due to the degradation of the harvested organism [28]. In addition, the anecdotal observation of putative photogenic organs in the deep-sea dark blind ray, *Benthobatis moresbyi* [29], has not been confirmed by more recent and detailed morphological studies [30,31]. Bioluminescence in sharks appears currently restricted to Squaliformes, where only three families (Dalatiidae, Somniosidae, and Etmopteridae) contain luminescent representatives (Figure 1). Indeed, although bioluminescence has once been suggested for the specific supralabial white band of the megamouth shark, *Megachasma pelagios*, recent work invalidated this assumption [32]. Although the luminescence reported for the species that belong to Etmopteridae and Dalatidae was initially thought to have been acquired independently [33,34], the recent discovery of luminescence from a somniosid, the velvet dogfish, *Zameus squamulosus* [35,36], now strongly suggests that the acquisition of luminescence capability in sharks represents a single evolutionary event, which occurred during the deep-sea radiation of Squaliformes at the end of the Cretaceous [35]. Although fossil studies estimate the Etmopteridae radiation around 90 million years ago [37], molecular data presents a separation of Etmopteridae from other Squaliformes during the Upper Cretaceous (i.e., 65–90 million years ago) [35,38]. The Dalatiidae family radiated later, during the Paleocene after the Cretaceous/Paleocene mass extinction, 65–105 million years ago, when they substituted extinct marine reptiles and fishes in the epipelagic fauna before reaching the deep sea [34,35,39].

Interestingly, the photogenic structures appear ubiquitous in Etmopteridae (four genera: *Trigonognathus*, *Aculeola*, *Centroscyllium* and *Etmopterus*; 52 species) and Dalatiidae (seven genera: *Dalatias*, *Isistius*, *Mollisquama*, *Euprotomicroides*, *Squaliolus*, *Euprotomicrus* and *Heteroscymnoides*; 10 species). Nevertheless, *Z. squamulosus* is the only somniosid shark known to possess photophores (Figure 1; Table S1). In parallel, luminescence has been observed in live in 15 species only, however, covering most clades, i.e., the blurred smooth lanternshark, *Etmopterus bigelowi* [40]; the southern lanternshark, *Etmopterus granulosus* [41]; the blackbelly lanternshark, *Etmopterus lucifer* [41]; the slendertail lanternshark, *Etmopterus molleri* [42,43]; the smooth lanternshark, *Etmopterus pusillus* [44]; *E. spinax* [45,46]; the splendid lanternshark, *Etmopterus splendidus* [47]; the green lanternshark, *Etmopterus virens* [40]; the smalleye pygmy shark, *Squaliolus aliae* [43,48]; the kitefin shark, *Dalatias licha* [41]; the taillight shark, *Euprotomicroides zantedeschia* [49]; the pygmy shark, *Euprotomicrus bispina-*

tus [50]; the cookiecutter shark, *Isistius brasiliensis* [51,52]; *Z. squamulosus* [36]; the viper dogfish, *Trigonognathus kabeyai* (Mallefet, unpublished data); Figures 1 and 2; Table S1. In addition, expected luminous species are encountered in Etmopteridae and Dalatiidae, based on the presence of photophore structures and/or flank marks in the holotype description [33,35,36,38,40–49,51–90]; Table S1. This led to a total number of 62 luminous sharks, i.e., ~11% of currently described species (550 in total [91]), while in comparison, luminescence is thought to have appeared in only ~5% of bony fishes [12].

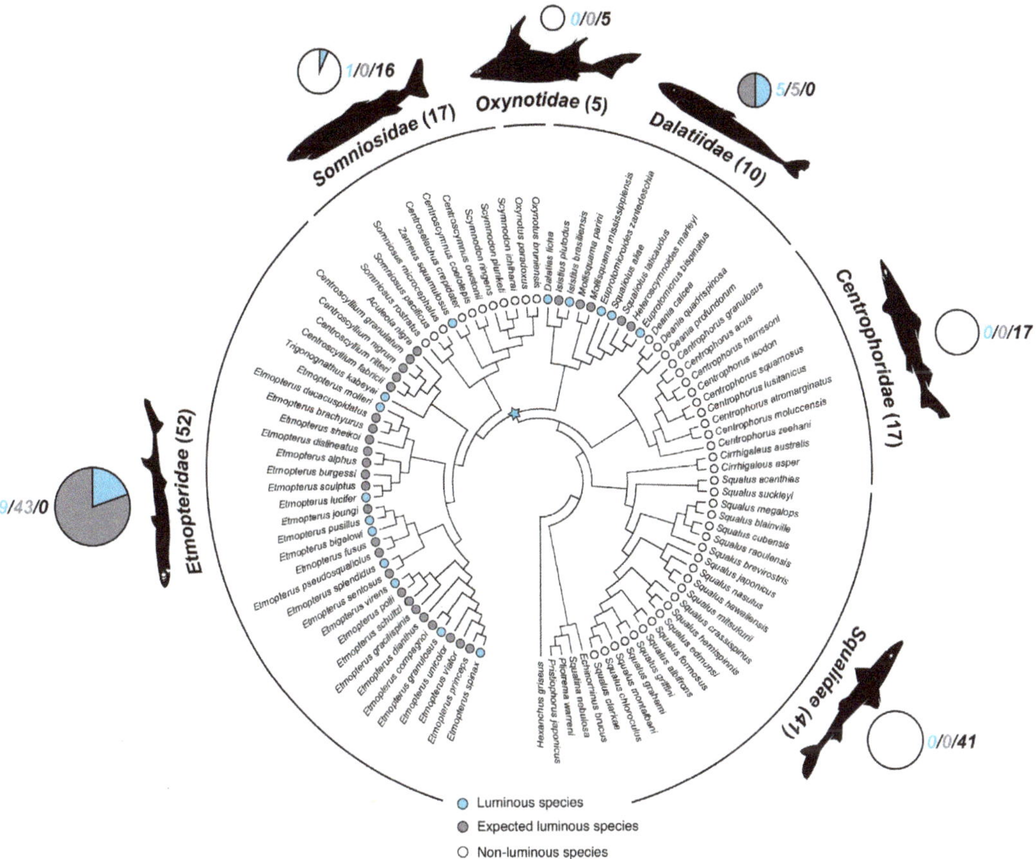

Figure 1. Shark luminescence distribution within Squaliformes families based on published phylogenies [35,38,53,92–96]. Circles inside the tree represent the luminous (blue), expected luminous (gray), and non-luminous (white) status of each represented species. Statuses are based on in vivo pictures, physiological studies (luminous), the presence of photophores or flank marks (expected luminous), and none of these criteria (non-luminous) (see Table S1). Circles outside the tree, scaled to the total number of species (number in brackets) within a given family [54,91], indicate the proportion of luminous (blue), expected luminous (gray), as well as non-luminous (white) species. For each family, total number of luminous/expected luminous/non-luminous species are given next to the outside circle. Blue star indicates the expected origin of luminescence in sharks.

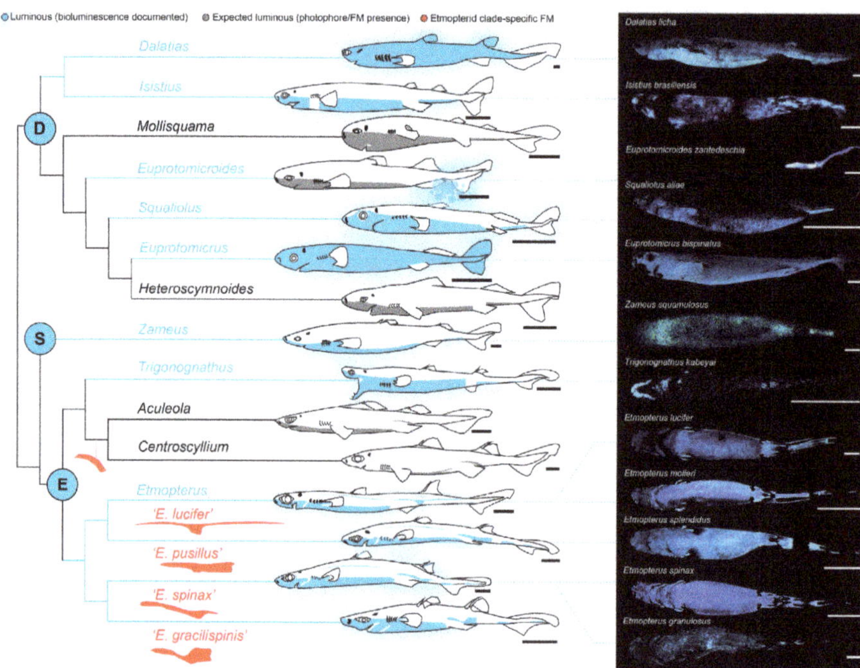

Figure 2. Shark luminescent patterns. Squaliform cladogram with taxonomic grouping where photophores/flank markings (gray shade) or live luminescence (blue shade) have been observed [35,36,40,41,49–74,97,98] and spontaneous luminescence of representative shark species lateral view for *D. licha*, *S. aliae*, *E. bispinatus*, excreted bioluminescent fluid for *E. zantedeschia* (blue bubbles on the cladogram), and ventral view for the others]. D, Dalatiidae; E, Etmopteridae; FM, flank markings; S, Somniosidae. Indicative scale bars, 5 cm. Shark drawings are modified from [34,74]. Photographs courtesy by D. Perrine (*E. bispinatus*), T. Raczynski (*E. zantedeschia* [49]) and J. Mallefet (other species; [36,41,65], Mallefet, unpublished).

3. Ecology of Shark Luminescence

Deciphering the ecological functions of luminescence from elusive animals such as deep-sea sharks is extremely challenging. Indeed, field observations are scarce and unbiased lab experiments proved to be difficult to perform. As a matter of fact, most functions formulated over years regarding the function of shark luminescence remain undemonstrated, mainly due to the difficulties of observing and collecting these rare and elusive organisms to conduct ethological studies on the function of their luminescence. Fortunately, however, detailed analyses of photophore distribution (luminescent "patterns") as well as physical characteristics (intensity, wavelength and angular distribution) and kinetics of luminescence coupled to physical models for pelagic vision and molecular phylogenetic analyses, now allow us to draw a clearer picture of the adaptive benefits of luminescence in sharks.

In this context, counterillumination, i.e., a camouflage technique used by midwater organisms cloaking their silhouette from upward-looking organisms using a ventral glow mimicking downwelling sunlight, e.g., [99–103], is probably the primary function of shark luminescence, for both defensive and predatory purposes [41,65]. Indeed, shark photophores are predominantly situated on the ventral surface area (Figures 2 and 3) and produce a light whose color (wavelength) which is similar to that found in coastal (blue–green) and oceanic (blue) environment (Figure 3b; Table 1). Moreover, the intensity and angular distribution of light emission have been correlated to capture depth residual light parameters in three species: *E. spinax* [103], *E. splendidus* [65], and *S. aliae* [65].

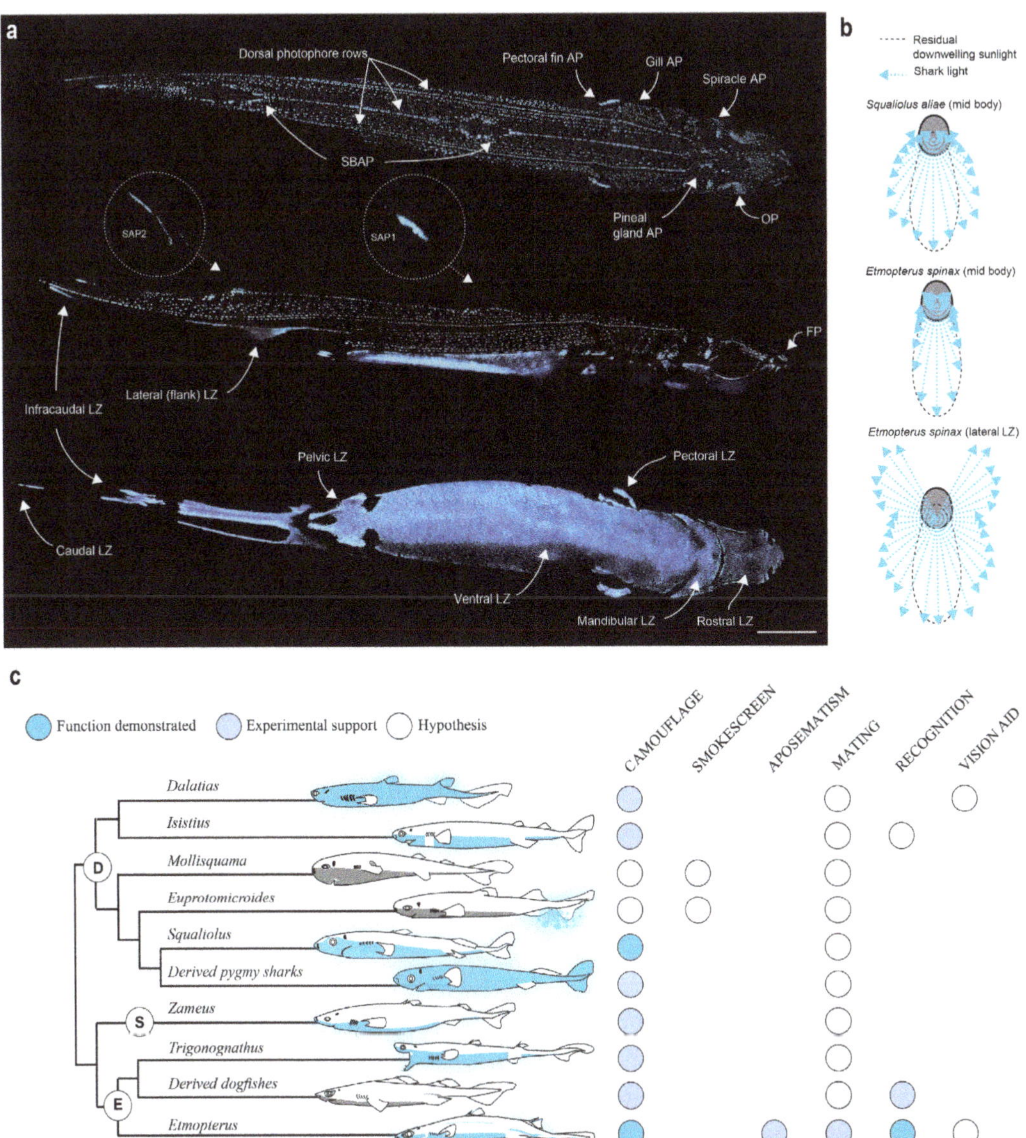

Figure 3. Shark luminescence functions. (**a**) Composite pictures of *E. molleri* and *E. spinax* luminescence patterns. Spontaneous dorsal, lateral and ventral luminescence from *E. molleri*, with key photophores areas highlighted. Circular inserts represent spine-associated photophores (SAPs) of the dorsal fins of *E. spinax* [104]; these photophores are absent in *E. molleri* (which instead has spine base-associated photophores; [105]). (**b**) Angular distribution of dalatiid (*S. aliae*; mid body) and etmopterid (*E. spinax*; mid body and lateral flank marks) luminescence as well as downwelling sunlight; the perfect match observed for *E. spinax* is achieved thanks to a centripetal change in photophore orientation modified from [65,103]. (**c**) Simplified bioluminescent shark phylogeny providing an overview of demonstrated, experimentally supported and putative functions of shark luminescence [27,41,74,75,104–108], Claes and Duchatelet, unpublished observation. AP, associated photophores; FP, frontal photophores; LZ, luminous zone; OP, ocular photophores; SAP1, spine-associated photophores of 1st dorsal fin; SAP2, spine-associated photophores of 2nd dorsal fin; SBAP, spine base-associated photophores. Scale bar, 5 cm. Photographs by J. Mallefet.

Table 1. Shark luminescence color.

Species	Photogenic Structure	Luminescence Color (Wavelength Peak)	References
Isistius brasiliensis	Photophores	Dark blue (455 nm)	[16]
Squaliolus aliae	Photophores	Dark blue (457 nm)	[65]
Euprotomicroides zantedeschia	Pelvic pouch (fluid)	Dark blue	[49]
Dalatias licha	Photophores	Blue	[41]
Euprotomicrus bispinatus	Photophores	Blue	This study
Etmopterus splendidus	Photophores	Blue (476 nm)	[65]
Etmopterus molleri	Photophores	Blue (477 nm)	[65]
Etmopterus bigelowi	Photophores	Blue	[40]
Etmopterus granulosus	Photophores	Blue	[41]
Etmopterus lucifer	Photophores	Blue	[41]
Trigonognathus kabeyai	Photophores	Blue	This study
Etmopterus spinax	Photophores	Blue-green (488 nm)	[65]
Zameus squamulosus	Photophores	Blue-green	[36]
Etmopterus virens	Photophores	Green	[40]
Etmopterus pusillus	Photophores	"Whitish"	[44]

Interestingly, "water box" experiments suggested that *E. spinax*, contrary to other counter-illuminating animals, was not able to adapt rapidly the intensity of its luminescence by more than one order of magnitude in response to changing light regime [103]. Hence an "isolume follower" hypothesis was formulated, stating that sharks move up and down in the water column to remain cryptic [103]. This has been further supported by a large comparative study showing that daytime capture depth of luminescent sharks could be predicted from their ventral photophore cover (e.g., the percentage of the ventral surface area covered with photophores), which varies from 2% in the bareskin dogfish *Centroscyllium kamoharai* to 56% in the viper dogfish *T. kabeyai* [65]. In addition, recent studies show that some etmopterid species have a higher swimming speed and muscular enzymatic activities than their non-luminous counterparts living in the same deep environment [109,110]. These findings could be correlated with a greater physiological demand to perform vertical migration to remain camouflaged. The second most widespread and well-supported function of shark luminescence is intraspecific communication. Although Dalatiidae and Somniosidae present "simple" luminescent patterns where photophores follow a dorsoventral density gradient [33,34,36,41,52] (Figures 2 and 3), etmopterid sharks display complex luminescent photophore aggregations on the ventral area, but also on the flanks, the fins, the tail, around the eyes, the spiracles, the gills, and the epidermal tissue surrounding dorsal spines [34,104–106,111] (Figures 2 and 3). Since these patterns are species-specific and do not show sexual dimorphism (except photophores associated with primary sexual characters, e.g., male claspers, as explained below [112]), they are often used as taxonomic determination tools [34,54,63,68]. Interestingly the shape of lateral luminous areas (flank markings) appears to be clade-specific [38,65] (Figure 2) and their presence correlates with an increased speciation rate [75] and a moderate predation risk [65]. This, and the ability for those sharks to discern the flank marking shape (and other specific luminous zones) at a biologically meaningful distance (1–3 m, or three to ten body lengths, as indicated by visual modeling [111]) strongly support the idea that etmopterid flank markings represent an exaptation of counter-illuminating photophores to facilitate deep-sea communication [65]. Luminescence could also be used as a mating aid, which allows males to identify females from a distance (since photophores cover male claspers of all species [34]; Figure 3a) and visualize their cloaca and pectoral fins (brighter in etmopterid [34]; Figures 2 and 3a) in the darkness of the deep sea (sharks display internal fertilization during which the male stabilizes itself by biting the female's pectoral fin [97]). Sexual dimorphism in the luminescence time course of pharmacologically stimulated pelvic photophores from *E. spinax* further supports this hypothesis [107].

The last luminescence function for which experimental support is available is the aposematism, a mechanism by which an animal advertises potential predators that it is not worth attacking or eating [113,114]. Etmopterid sharks (contrary to dalatiid and somniosid species) have large sharp defensive spines associated with their dorsal fins [97]. In some species, photophores either placed on the edge of the dorsal fins (*E. spinax*; [104]) or around the base of the spines (*E. molleri*; [105]), allow these spines to be seen in the dark from a distance by potential predators, and hence potentially work as an aposematic signal as strongly suggested by experimental data [104,105] (Figure 3a–c).

In addition to these relatively well-established functions, researchers have formulated more speculative functional hypotheses regarding the luminescence of some species (Figure 3c). For instance, the luminescent liquid ejected by the pelvic pouch of *E. zantedeschia* probably works as a defensive "smokescreen" mechanism [49]. Given the morphological similarity of the pectoral glands, an identical function has been suggested for the pocket sharks, *Mollisquama* spp. [55,74]. The ocular and frontal photophores of *E. spinax* [103] and the ventral photophore of *D. licha* [41] could be used as a vision aid (Figure 3c). In addition, the tail of *D. licha*, *S. aliae*, *Z. squamulosus*, *T. kabeyai*, and *Etmopterus* species, which is more mobile and brighter than the rest of the ventral surface area, could work as a distracting lure, and hence be analogous to the caudal photophores of myctophid and tubeshoulder fishes [115–117]. Conversely, the idea that the dark, photophore-free collar of *I. brasiliensis* acts as a lure to attract bigger pelagic fishes or marine mammals on which it feeds [51], now seems at the very least dubious given that numerous common preys of this shark species are either filter feeders or top predators for which such a mechanism is useless [65], and that the closely related species, the largetooth cookiecutter shark, *I. plutodus*, which has a similar diet [97], lacks such a collar [70].

Etmopterid and dalatiid sharks display a set of visual features not found in non-bioluminescent sharks, which strongly supports the idea that their visual system coevolved with their ability to produce light. Aphakic gaps in *S. aliae* [111] and *E. bispinatus* Claes, personal observation] and translucent upper eyelid found in etmopterid sharks (genus *Trigonognathus* and *Etmopterus*; [111]) probably play a role in counterilluminating, e.g., facilitating perception of downwelling light intensity to ensure the perfect match, similarly to what was recently shown for deep-sea bony fishes [118]. On the other hand, the spectral absorbance (485–488 nm) of etmopterid rod photoreceptor as well as the retinal distribution of ganglion cells of those shark (which appear species-specific) appear finely tuned to detect their own bioluminescence, especially from flank markings [111].

4. Photogenic Structures and Specialized Squamation of Bioluminescent Sharks

Sharks display two types of photogenic structures: photophores (Figure 4), with internal luminescence present in all bioluminescent shark species, and secretory glands ejecting a bioluminescent fluid in the environment (external or secretory luminescence; Figure 4) present in *E. zantedeschia* (pelvic pouch) and in *Mollisquama* spp. (pectoral pockets).

Shark photophores are small, reaching a maximum diameter of ~50, ~100, and ~200 μm for Somniosidae, Dalatiidae, and Etmopteridae, respectively [34,36,41,48,52,65,76]. Comparatively, the photophores of bony fishes can reach up to 1 cm [119]. Given their high photophore density (sometimes over ten thousand per square centimeter; [65]), sharks are probably the luminous organisms with the highest number of photophores (a 52 cm TL male *E. spinax* has been estimated to bear about 440,000 photophores; [34]). The general structure of photophores is similar across families (Figure 4a,c,d).

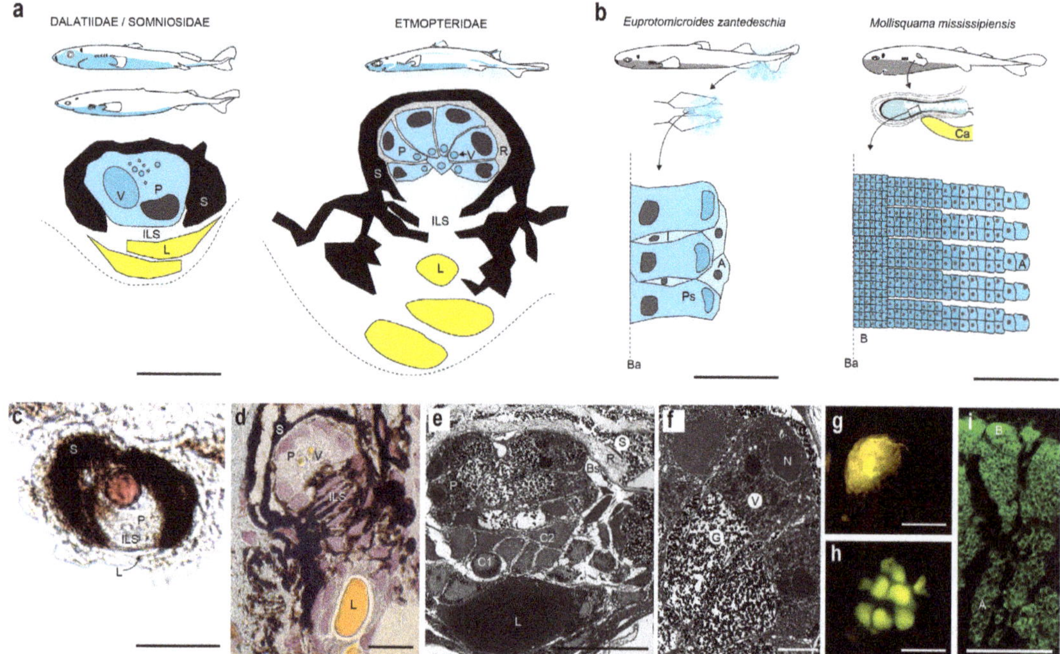

Figure 4. Shark photogenic organs. (**a**) Photophores (modified from [32,34]); (**b**) secretory photogenic structures: light-producing columnar epithelium from pouch of the taillight shark (modified from [34]; left) and putative light-producing cubic tissue from pocket shark's pectoral pockets (modified from [74]; right); (**c**) histological section of *I. brasiliensis* photophore; (**d**) histological section of *E. lucifer* photophore; (**e**) ultrastructure of *E. spinax* photophore; (**f**) ultrastructure of *E. spinax* photocyte; (**g**) autofluorescence from *E. lucifer* photophore; (**h**) autofluorescence from photocyte vesicle from *E. lucifer*; (**i**) autofluorescence from pocket shark's pocket epithelium. A, apical cells; B, basal cells; Ba, basal lamina; Bs, blood sinus; Ca, cartilage rod from pectoral fin. C1, cellular type 1; C2, cellular type 2; E, epidermis; G, granular area; I, inclusion; ILS, iris-like structure; L, lens cell; N, nucleus; P, photocyte; Ps, secretory photocyte; R, reticulated layer (reflector); S, pigmented sheath; V, photocyte vesicles; Va, vesicular area. Indicative scale bars in (**a–e,g,i**), 50 μm; in (**f,h**), 10 μm.

Etmopteridae photophores are composed of 6–14 emitting cells (photocytes) embedded in a cup-shaped pigmented cell sheath, covered by a reflective layer containing guanine crystals and capped by one or several lens cells [16,41,44,76,120]. A multilayer cell zone, called iris-like structure (ILS), is present between the lens cells and the photocytes and is used as a photophore shutter [43,76,77,120–122]. Renwart et al., 2014, described three different cell types constituting the ILS of *E. spinax* photophores: (i) the cellular type I, which contains fibrous material and was assumed to stabilize the lens cells; (ii) the cellular type II, with nucleus only visible via electron microscopy and whose function is unclear, and (iii) the pigmented cells, affiliated to melanophores, presenting pseudopodia-like cellular projections [76] (Figure 4). Photocytes of *E. spinax* were ultrastructurally depicted as containing three distinct areas: the nucleus, the vesicular and the granular areas (Figure 4e,f). The granular area is assumed to be the location of the bioluminescence chemical reaction and associated microsources were named "glowons" [76,120]. Green autofluorescence is observed within photocytes (especially from their granular area) under blue/UV light exposure (Figure 4g,h), which is assumed to be due to the fluorescence properties of the bioluminescent substrate (as it is the case, e.g., for luciferin; [45,47,76,77]). Shark photophores appear to lack intra-organ innervation [46,76], with terminal epidermal nerves reaching only the surrounding of the photophore as demonstrated through acetylated-tubulin labelling [123].

Comparatively, Dalatiidae and Somniosidae harbor less complex and smaller light organs containing a single photocyte, a pigmented sheath and a group of small lens cells [33,34,36,41,52] (Figure 4a). Pigmented cells, attributed to ILS cells, surmounting the photocytes also act as a photophore shutter [41,77]. Photophore development has been studied in *E. spinax* and *S. aliae*, revealing a similar four-phase morphogenesis pattern: (i) apparition of pigmented cells; (ii) formation of the pigmented sheath; (iii) apparition of the protophotocyte (i.e., photocytes not able to produce light); (iv) maturation of the photocyte during which the photocyte acquires its luminescence competence (revealed by the presence of fluorescent vesicles in photocytes; [45,77], Duchatelet, personal observation).

The secretory epithelium from the pelvic pouch of *E. zantedeschia* displays a pseudostratified columnar epithelium with three distinct cell types: columnar cells with a large apical inclusion (which are probably secretory cells), flattened cells present between columnar cells and extending from the basal lamina towards the free surface of the epithelium, and superficial cells forming a thin cover above the apical part of columnar cells [61]. The epithelium of the putatively bioluminescent fluid-producing pectoral pockets of *Mollisquama mississippiensis* shows a surprisingly different structure, since it appears to be stratified, made of over 50 layers of cuboidal cells displaying green autofluorescence and probably involved in the production of a holocrine secretion [74] (Figure 4b,i).

Given that shark photophores are embedded in the shark epidermis, they compete for space with dermal denticles (placoid scales). As a consequence, bioluminescent sharks evolved specific squamation patterns, i.e., pavement-like, cross-, bristle-, hook-, and simple/alveolar leaf-shaped patterns [32,34–36,41,124,125]; Figure 5, which allows the accommodation of photophores between (pavement-like, cross-, bristle-, hook-shaped types) or below (leaf-shaped types) the placoid scales. Histology and light transmission analyses of leaf-shaped type denticles from *Z. squamulosus* recently revealed specific honeycomb structures allowing the transmission of at least 50% of the light produced [36]. On the specific rostral area, the kitefin shark, *D. licha*, also presents highly translucent leaf-shaped placoid scales but without honeycomb structure [41].

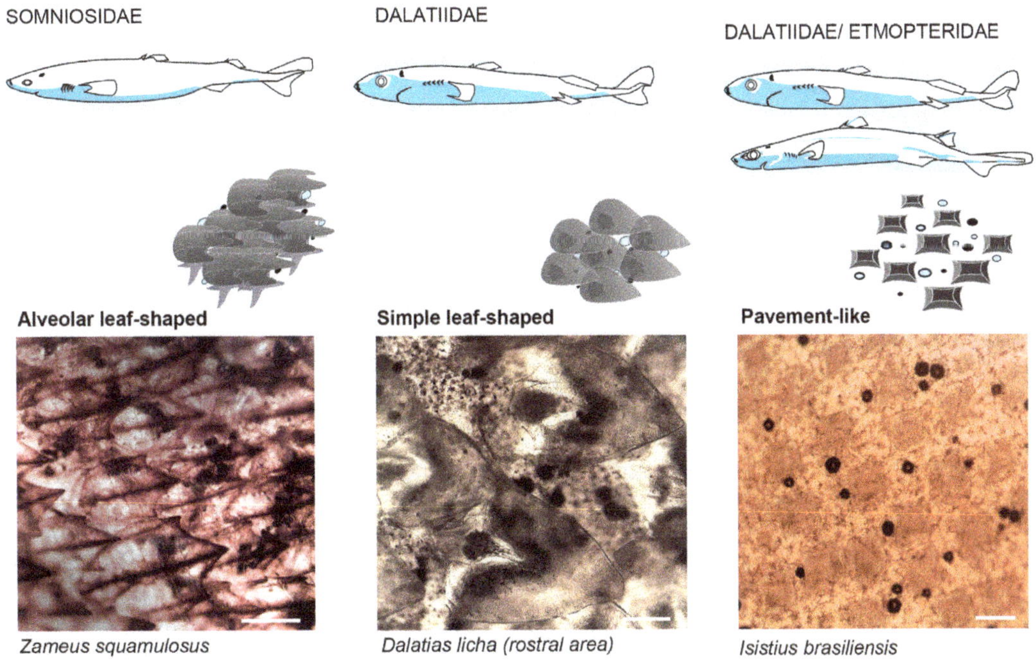

Figure 5. *Cont.*

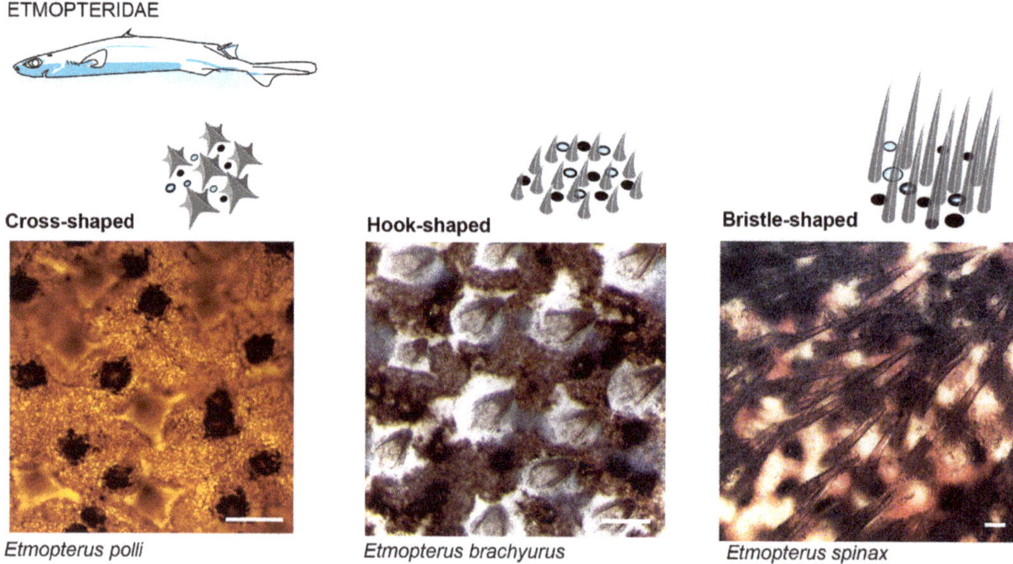

Figure 5. Bioluminescent shark squamation. Placoid scale patterns found in bioluminescent sharks [32,34,36,41,124]. Photophore are represented with blue (open) and dark (closed) dots between placoid scales. Indicative scale bars, 50 μm.

5. Control of Bioluminescence from Shark Photophores

5.1. Hormonal Control

Luminous sharks occupy a peculiar place among bioluminescent organisms regarding the control of their luminescence. Indeed, their photogenic organs are primarily controlled by hormones rather than by nerves, the general condition for intrinsic photogenic organs [42,43,47,48,108]. Although the majority of bony fishes have a nervous control of the light emitted (e.g., via adrenaline and noradrenaline, [126–129]), such classical neurotransmitters have been demonstrated to be inefficient to regulate the shark light emission [16,108]. Intraperitoneal injection of drugs such as adrenalin or acetylcholine in *I. brasiliensis* failed to trigger light emission [16]. Claes and Mallefet (2009) also attempted, without success, to induce light response in the etmopterid *E. spinax* after the application of neural agents (e.g., adrenalin, noradrenalin, serotonin, carbachol, and the classical depolarizing agent KCl) on isolated ventral photogenic skin patches, a technique which became the standard for pharmacological studies of organism luminescence control [108]. The absence of photophore innervations, and the inefficiency of classical neurotransmitters to trigger light emission, led to assumptions of a hormonal control of luminescence in sharks [28,108].

Using *E. spinax* as a model species, and hypothesizing that the shark light emission involves pigment motion within the ILS-melanophores, Claes and Mallefet (2009) have highlighted the implication of melatonin (MT), prolactin (PRL) and alpha melanocyte-stimulating hormone (α-MSH) in the light emission process [108] (Figure 6a). These three hormones are known to be involved in the physiological skin color modulation in elasmobranchs (i.e., sharks and rays): PRL, and α-MSH stimulate melanophore pigment dispersion and, thus, induce skin darkening, and MT regulate melanophore pigment aggregation, leading to a paler skin [130,131]. A few years later, the hormonal control was described in other species such as *E. splendidus* [47], *E. molleri* [42], *E. lucifer* [41] and *E. granulosus* [41], as well as in two Dalatiidae species, *S. aliae* [48] and *D. licha* [41]. Recently, the adrenocorticotropic hormone (ACTH), a melanocortin hormone also involved in vertebrate skin pigment motion within melanophores, was found to inhibit light emission induced by MT in both Etmopteridae and Dalatiidae [41,43] (Figure 6a). Conversely, the melanin-concentrating hormone (MCH), another hormone responsible of the pigment granule aggregation in bony fishes [132–135] had no effect on shark luminescence [136]. Although the pharmacological control of *Z. squamulosus* photophores has not been investigated yet, these organs can be assumed to be under hormonal control as well, given their structural similarity with dalatiid photophores [35,36]. Future work on this subject will allow us to confirm this hypothesis.

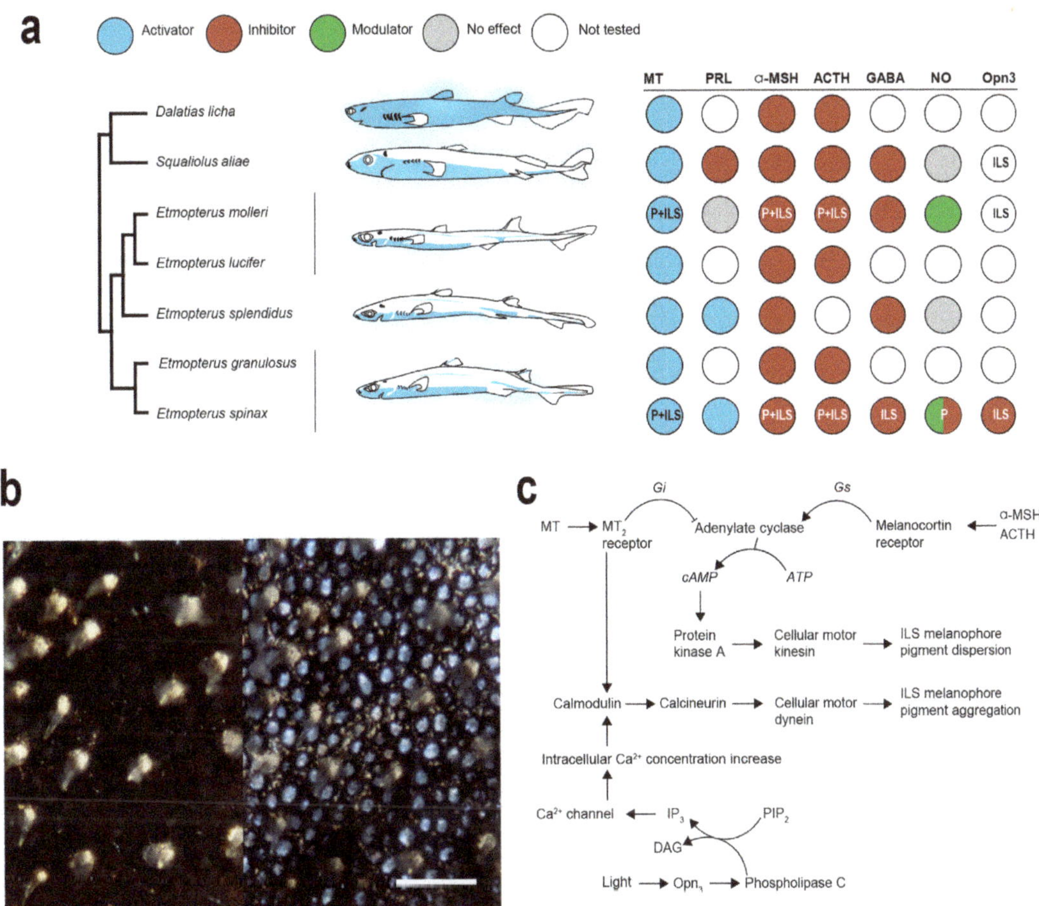

Figure 6. Shark photophore control. (**a**) Overview of shark luminescence control agents mapped against experimentally investigated species over the last 15 years [25,27,41–43,47,48,77,107,108,121,122,137–139]. Letters in circles indicate the identified targets of the agent (ILS, iris-like structure; P, photocytes); (**b**) photophore-containing skin patch before hormonal stimulation (left) and at luminescence peak (right) from *E. spinax* [121]; (**c**) primary hormonal control pathway of shark photophore regulating ILS shutter function (modified from [122]).

Overall, the photophore responses to hormone application appear similar across investigated species, except for PRL (Figure 6a). In Etmopteridae, MT and PRL, both at a concentration of 10^{-6} mol L^{-1}, trigger light emission, while α-MSH at 10^{-6} mol L^{-1} and ACTH at 10^{-5} mol L^{-1} inhibit it [42,43,47,108] (Figure 6a). Both stimulating hormones present different light-emission curves when applied on photogenic ventral skin patches. Although MT triggers a slow increasing and long-lasting luminescence for up to one hour depending on the species, PRL application induces a fast–high response, decreasing rapidly after a few minutes. Nevertheless, it has been shown that both hormones act on the ILS cells pigment movement to open the "iris" and allow the outward light emission [43,121] (Figure 6b). Conversely, α-MSH and ACTH inhibit the light emission with a decrease in the amount of light produced after both MT or PRL applications, and also act on the ILS cells pigment movement by occulting the photocyte, and hence preventing the light from being emitted outside [41,43,108,121,122]. In Dalatiidae, the hormonal control is similar, except for PRL, which was demonstrated to inhibit the light emission process in *S. aliae* [41,48]. The physiological effects of each hormone for each studied shark species are summarized in Figure 6a.

In silico transcriptome analyses and immune localization allowed us to confirm the presence and location of MT and α-MSH/ACTH receptors within the photophore from two Etmopteridae species, *E. spinax* and *E. molleri* [137] (Figure 6a). These G protein-coupled receptors (GPCR), known to be coupled with specific G proteins, are mainly localized within the cell membrane of photocytes and ILS cells [137]. Researchers also highlighted that PRL receptor mRNA sequences are absent from *E. spinax* ventral skin reference transcriptome [137], in agreement with previous works revealing the secondary loss of PRL receptor within the elasmobranch lineages during the growth hormone/prolactin protein family evolution [140]. Therefore, how PRL could cellularly have been perceived by the photophore to trigger luminescence remains enigmatic and assumptions have been made on the involvement of the closely related growth hormone and its specific receptor in the light emission control.

An integrative model of the hormonal control of shark photophore luminescence, highlighting the role of MT, α-MSH and ACTH on both the photocyte and the ILS, has been recently proposed [122] (Figure 6c). MT, through its perception by MT receptor, triggers the release of Gi proteins, inhibiting the adenylate cyclase activity [43,122]. Adenylate cyclase is known to directly act on cellular cAMP concentration [141–143]. Confirming results from Claes and Mallefet, 2009 [108], cAMP concentration assay after MT-induced light emission revealed a drastic decrease in cAMP levels [43,122]. In fish melanophore, a decrease in the cell cAMP concentration triggers the aggregation of melanin pigment toward the nucleus periphery [142,144,145]. Moreover, MT application was demonstrated to activate, through a calmodulin/calcineurin pathway, the cellular motor dynein, which carries pigmented granules towards melanophore-like nucleus periphery [122]. This MT pathway regulates the "opening" of the ILS cells and allows the emitted light to reach the outside of the photophore [122] (Figure 6). The ultrastructure analyses of the ILS structure before and after the MT-induced luminescence confirmed the pigment motion within the ILS-associated melanophores [120]. On the other hand, analysis of the α-MSH/ACTH pathway revealed how these hormones "close" the photophore and, hence, inhibit light emission. Through melanocortin hormone (α-MSH/ACTH) perception, its receptors release a Gs protein, activating the adenylate cyclase, and an increase in the intra-ILS cells cAMP concentrations [43,122]. The final step of this inhibiting pathway involves the cellular motor kinesin, which carries pigment granules on cytoskeleton towards the ILS melanophore pseudopodial-like projection, occulting the light produced [122] (Figure 6c). These two ILS-regulating pathways seem to be conserved across Etmopteridae and Dalatiidae [41,43,47,122] (Figure 6a).

In addition to the hormonal control, pharmacological studies also revealed that γ-aminobutyric acid (GABA) and nitric oxide (NO) can modulate the light emission in certain species [42,48,123,139] (Figure 6a). In the model species, *E. spinax*, GABA inhibits and NO modulates the MT/PRL-induced luminescence [123,139]. Conversely, NO application on *S. aliae* photogenic skin patches, after MT/PRL light emission triggering, had no effects [48] (Figure 6a).

Further research is needed to determine the intracellular interplay between the neuromodulators (NO and GABA) and the hormones as well as to better understand the intracellular events occurring in the photocytes upon hormonal stimulation.

5.2. Extraocular Photoreception and Pigment Motion Regulation

Extraocular photoreception, the ability to detect and respond to light clues outside of the "eyes", has been suggested to be involved in the bioluminescence control in various invertebrate taxa. In the bobtail squid, *Euprymna scolopes*, extraocular photoreceptors and photocytes are colocalized within photophores. Expression of eye-like developing genes in the light "pocket" of *E. scolopes* underlines such a coupling between photoemission and photoreception mechanisms [146,147]. In the comb jelly, *Memniopsis leidyi*, and the burrowing ophiuroid, *Amphiura filiformis*, both photoreceptor and photocyte molecular markers (i.e., opsin and photoprotein/luciferase, respectively) are coexpressed within the same cells. These observations support the existence of a functional link between light perception and bioluminescence control for these species [8,148–150]. By analyzing the opsin and phototransduction genes expression within photophores, light organ photosensitivity was also suggested for a counterilluminating deep-sea shrimp, *Janicella spinicauda*, where it was assumed to play a role in the light emission regulation to ensure a match with the residual downwelling light [151].

As indicated by recent analyses of ventral skin and eye transcriptomes of *E. spinax*, photoreceptors and phototransduction actors were expressed in both tissues [152]. This species displays only two ocular opsins, one rhodopsin and one putatively associated peropsin, highlighting its monochromatic vision [152]. A specific extraocular photopigment, the encephalopsin (Opn3), was also detected in the skin transcriptome of *E. spinax* [152]. The colocation of this extraocular opsin (Es-Opn3) with the ventral skin photophores provides fuel for the putative functional coupling between light emission and light perception in luminous organisms [152]. More precisely, membranes of the ILS cells were shown to be the main expression site of the Es-Opn3, while no expression was found at the photocyte level [152]. Besides, the expression of this opsin was demonstrated to be concomitant with the *in utero* development of the photophores in *E. spinax* and *S. aliae* embryos, appearing with the transformation of protophotocytes to photocytes (i.e., when photocytes can produce luminescence; [77], Duchatelet, unpublished data), which supports the idea that the Es-Opn3 is used to detect embryonic luminescence [77]. The evaluation of the absorbance spectrum of this luminous shark extraocular opsin has added new evidence of the link between the two photobiological processes, since a clear overlap exists between the light emission spectrum of *E. spinax* and the Es-Opn3 photopigment light absorbance [122].

Going further in the phototransduction cascade, Duchatelet et al., 2020, demonstrated an impact of blue light exposure on the intracellular concentration of inositol triphosphate (IP_3) from photogenic ventral skin [122]. This IP_3 concentration modulation confirmed opsin activity and highlighted the first step of the phototransduction cascade occurring at the photophore level. Pharmacological analyses unveiled the next steps of the phototransduction pathway, with the implication of calcium, the Ca^{2+}-dependent calmodulin, and the cytoplasmic motor dynein [122], clearly demonstrating the interconnection between these pathway steps and the pathway leading to the pigment granule aggregation in melanophores. Therefore, Duchatelet et al., 2020, proposed a model of light emission control in *E. spinax* based on the photoperception of its own luminescence that regulates ILS melanophore pigment aggregation and, thus, the aperture of the photophore, which regulates the light output [122] (Figure 6c). Interestingly, while hormones operate at both the photocyte, triggering light emission through an unknown biochemical reaction (i.e., unknown luciferin/luciferase system or photoprotein) and ILS (to regulate the amount of light transmitted as a camera diaphragm) levels, Es-Opn3 appears to be involved only in the pigment motion regulation of the ILS. This dual control is assumed to occur at least in the luminous species belonging to Dalatiidae and Etmopteridae (Duchatelet, unpublished data). The close link between photoemission and photoreception at the light-organ level, putatively regulating the amount of light emitted, seems to have emerged independently in phylogenetically distant luminous organisms: ctenophores, cephalopods, crustaceans, echinoderms, and luminous sharks [8,122,146–152]. It is a fascinating example of convergent evolution.

6. Biochemistry of Shark Luminescence

One of the remaining questions is what type of bioluminescent system is involved in the emission of light in sharks. Two categories of light systems are known to date: (i) luciferase–luciferin systems, and (ii) photoprotein systems. In the former, a substrate, the luciferin, is oxidized by an enzyme, the luciferase, in the presence of oxygen and often other cofactors [5]. In the latter, an enzyme complex as well as a preoxidized luciferin form together a complex protein, the so-called photoprotein, requiring only the contribution of a cofactor (often an ion) to produce light [5,153]. Various luciferin types have been molecularly described in the marine environments to date, such as the coelenterazine (the most phylogenetically widespread luciferin type in the oceanic environment) and its derivatives (dehydrocoelenterazine and enol-sulfate coelenterazine), the dinoflagellate luciferin, or the *Cypridina* luciferin (also called vargulin) [1,13,154]. Associated with these luciferins, different types of luciferases (nine different types) and photoprotein (three different types) have been molecularly depicted [1,5,9,13]. These substrates and enzymes, although initially considered to be species-specific [28], can sometimes be shared by phylogenetically very distant species [5–9]. This has led scientists to hypothesize that certain species can acquire the necessary components for the luminous reaction through their diet [1,5,155–159]. Recently, the golden sweeper fish, *Parapriacanthus ransonnetti*, has been shown to acquire not only its luciferin, but also its luciferase by feeding on luminous ostracods [160]. It is noted that the only species described as being able to synthesize de novo their luciferins are the midwater shrimp *Systellaspis debilis* [161], the ostracod *Metridia pacifica* [162], and the ctenophores *Bolinopsis infundibulum* and *M. leidyi* [163].

Attempts were made to identify the bioluminescent compound responsible for the light emission in *E. spinax* by analyzing the cross-reactivity of known luciferins with extract of shark skin assumed to contain the catalyst (i.e., luciferase). In parallel, the putative diet acquisition of a luciferin through the food chain was hypothesized. Only coelenterazine, the most commonly found luciferin in marine taxa [5,164–167], was found in the digestive tract of *E. spinax*, but none of the tested luciferins (coelenterazine, dinoflagellate/euphausiid luciferin, and *Cypridina* luciferin) has reacted with the shark photogenic skin extract [168]. Similar negative results have been obtained for the cross-reactivity with the respective luciferase: coelenterazine-dependent luciferase (*Renilla* luciferase), cypridinid luciferase, or euphausid luciferase [168]. Analysis of the transcriptomic sequence available for the photogenic skin failed to pinpoint any putative homologs of known luciferase/photoprotein sequences [52,152], leading to assume the involvement of an unknown light-emitting system in luminous sharks, including either an unknown light-emitting system (photoprotein) or a specific active or storage form of a known luciferin using a shark-specific luciferase. Research combining data from transcriptomics, proteomics, and bioinformatics modeling are ongoing, and could allow us to decipher this enigmatic bioluminescent system.

7. Conclusions and Perspectives

Bioluminescent sharks have fascinated humans for almost two centuries. Yet, dedicated research on these enigmatic deep-sea inhabitants involving spectrophotometry, luminometry, pharmacology, light/electron microscopy, biochemistry, molecular analyses, and transcriptomics started only 15 years ago. Results from those studies (over 50 publications in total) have been synthesized in the current review. Overall, findings suggest that luminescence acquisition has been a unique though pivotal evolutionary event for squaliform sharks, which greatly facilitated their radiation in deep-sea habitats and strongly shaped their visual system. From a primary function of camouflage (coopted from the hormonally controlled crypsis mechanism of shallow water elasmobranchs) in Dalatiidae, Somniosidae and basal Etmopteridae, shark bioluminescent patterns progressively became an intra- and interspecific communication tool in derived etmopterid sharks (e.g., *Etmopterus* species), an exaptation that considerably increased their speciation rate and has probably been facilitated by the increase in size and complexity of etmopterid photophores (which allows better orientation of light output, e.g., tangential to the body surface). The secretory luminescence observed in *E. zantedeschia* (and putatively present in pocket sharks, *Mollisquama* sp.), which might have evolved from the invagination and modification of epidermal photophores, represents an example of the high selective pressure occurring in the darkness of the deep sea.

It clearly appears that the future of shark bioluminescence research will also be driven by new molecular data and techniques. The Next-Generation Sequencing methods recently allowed the emergence of transcriptomic studies in non-model organisms such as some selected luminous shark species (i.e., *E. spinax, I. brasiliensis* [52,152]). These studies paved the way for future transcriptomic, proteomic, and genomic studies on luminous sharks. Among an infinite number of fascinating questions, these studies could focus on the identification of the light-emitting molecular toolkit (luciferase, photoprotein, etc.) in luminous sharks.

Up to now, genomic resources for cartilaginous fish are still very limited (i.e., genomes available for the great white shark, *Carcharodon carcharias*, the whale shark, *Rhincodon typus*, the elephant shark, *Callorhinchus milii*, and the little skate, *Leucoraja erinacea*) and absent for luminous sharks. Genome size estimation in different Etmopterid species revealed that these species have among the largest genomes of all investigated Chondrichthyes (e.g., genome size of *E. spinax* might reach 16 Gbp, against 4.63 Gbp in *C. cacharias*, and 3 Gbp in *R. typus*) [169–173]. These impressive genome sizes possibly reveal lineage-specific genome expansion and large-scale alteration events such as gene/genome duplications, transposon insertions or events such as polyploidy [174], and confirm the challenging future of genomic studies on these species.

Further research on these sharks is planned to understand their ecology. In vivo ethological studies could lead to a better understanding of their behavior, notably the "isolumes follower" hypothesis by passive acoustic tagging of these sharks. Another approach could be to follow "in situ" reaction to stimuli mimicking the light emission of conspecifics.

Although recent research allowed to draw a clearer picture on the evolution, ecology and physiology of shark luminescence, gaps in our knowledge of these fascinating animals still exist. In particular, the interplay between hormonal control pathway and neuromodulation as well as the chemistry of shark luminescence remain to be determined. Work is underway to clarify these shadow areas.

Supplementary Materials: The following are available online at https://www.mdpi.com/article/10.3390/oceans2040047/s1, Table S1: Overview table of the luminous species within the luminous Squaliformes (Etmopteridae, Dalatiidae, Somniosidae) according to the literature.

Author Contributions: Conceptualization, L.D., J.M.C. and J.M.; software, L.D.; investigation, L.D., J.M.C., J.D. and J.M.; writing—original draft preparation, L.D., J.M.C. and J.D.; writing—review and editing, L.D., J.M.C., J.D., P.F. and J.M.; supervision, J.M.; project administration, L.D.; funding acquisition, P.F. and J.M. All authors have read and agreed to the published version of the manuscript.

Funding: This research was funded by the Fonds National de la Recherche Scientifique (F.R.S.-FNRS, Belgium), grant number T.0169.20 awarded to the University of Louvain—UCLouvain Marine Biology laboratory and the University of Mons—UMONS Biology of Marine Organisms and Biomimetics laboratory. L.D., J.M.C., J.D., P.F. and J.M. are, respectively a postdoctoral researcher, scientific collaborator of the UCLouvain laboratory, postdoctoral researcher F.R.S.-FNRS, research director F.R.S.-FNRS, and research associate F.R.S.-FNRS.

Acknowledgments: This review is the contribution BRC # 381 of the Biodiversity Research Center (UCLouvain) from the Earth and Life Institute Biodiversity (ELIB) and the "Centre Interuniversitaire de Biologie Marine" (CIBIM).

Conflicts of Interest: The authors declare no conflict of interest.

References

1.   Haddock, S.H.D.; Moline, M.A.; Case, J.F. Bioluminescence in the sea. *Annu. Rev. Mar. Sci.* **2010**, *2*, 443–493. [CrossRef] [PubMed]
2.   Hastings, J.W.; Morin, J.G. Bioluminescence. In *Comparative Animal Physiology. Neural and Integrative Animal Physiology*; Prosser, C.L., Ed.; John Wiley and Sons: Chichester, NY, USA, 1991; Chapter 3; pp. 131–170.
3.   Hastings, J.W. Biological diversity, chemical mechanisms, and the evolutionary origins of bioluminescent systems. *J. Mol. Evol.* **1983**, *19*, 309–321. [CrossRef]
4.   Wilson, T.; Hastings, J.W. Bioluminescence. *Ann. Rev. Cell. Dev. Biol.* **1998**, *14*, 197–230. [CrossRef]
5.   Shimomura, O. *Bioluminescence: Chemical Principles and Methods*; World Scientific Publishing Co. Pte. Ltd.: Singapore, 2012.
6.   Widder, E.A. Bioluminescence in the ocean: Origins of biological, chemical, and ecological diversity. *Science* **2010**, *328*, 704–708. [CrossRef]
7.   Gimenez, G.; Metcalf, P.; Paterson, N.G.; Sharpe, M.L. Mass spectrometry analysis and transcriptome sequencing reveal glowing squid crystal proteins are in the same superfamily as firefly luciferase. *Sci. Rep.* **2016**, *6*, 1–13. [CrossRef] [PubMed]
8.   Delroisse, J.; Ullrich-Lüter, E.; Blaue, S.; Ortega-Martinez, O.; Eeckhaut, I.; Flammang, P.; Mallefet, J. A puzzling homology: A brittle star using a putative cnidarian-type luciferase for bioluminescence. *Open Biol.* **2017**, *7*, 160300. [CrossRef]
9.   Delroisse, J.; Duchatelet, L.; Flammang, P.; Mallefet, J. Leaving the dark side? Insights into evolution of luciferases. *Front. Mar. Sci.* **2021**, *8*, 673620. [CrossRef]
10. Hastings, J.W. Bioluminescence. In *Cell Physiology Source Book*; Sperelakis, N., Ed.; Academic Press: Cambridge, MA, USA, 1995; pp. 665–681. [CrossRef]
11. Oba, Y.; Schultz, D.T. Eco-Evo bioluminescence on land and in the sea. In *Bioluminescence: Fundamentals and Applications in Biotechnology—Volume 1. Advances in Biochemical Engineering/Biotechnology*; Thouand, G., Marks, R., Eds.; Springer: Berlin/Heidelberg, Germany, 2014; Volume 144, pp. 3–36. [CrossRef]
12. Davis, M.P.; Sparks, J.S.; Smith, W.L. Repeated and widespread evolution of bioluminescence in marine fishes. *PLoS ONE* **2016**, *11*, e0155154. [CrossRef]
13. Lau, E.S.; Oakley, T.H. Multi-level convergence of complex traits and the evolution of bioluminescence. *Biol. Rev.* **2020**, *96*, 673–691. [CrossRef]
14. Morin, J.G. Coastal bioluminescence: Patterns and functions. *Bull. Mar. Sci.* **1983**, *33*, 787–817.
15. Martini, S.; Kuhnz, L.; Mallefet, J.; Haddock, S.H.D. Distribution and quantification of bioluminescence as an ecological trait in the deep sea benthos. *Sci. Rep.* **2019**, *9*, 14654. [CrossRef]
16. Herring, P.J.; Morin, J.G. Bioluminescence in fishes. In *Bioluminescence in Action*; Herring, P.J., Ed.; New York Academic Press: New York, NY, USA, 1978; pp. 273–329.
17. Haneda, Y. Further studies on a luminous land snail, *Quantula striata*, in Malaya. *Sci. Rep. Yokosuka City Mus.* **1963**, *8*, 1–9.
18. Rudie, N.G.; Mulkerrin, M.G.; Wampler, J.E. Earthworm bioluminescence: Characterization of high specific activity *Diplocardia longa* luciferase and the reaction it catalyzes. *Biochemistry* **1981**, *20*, 344–350. [CrossRef] [PubMed]
19. Wood, K.V. The chemical mechanism and evolutionary development of beetle bioluminescence. *Photochem. Photobiol.* **1995**, *62*, 662–673. [CrossRef]
20. Broadley, R.A.; Stringer, I.A.N. Prey attraction by larvae of the New Zealand glowworm, *Arachnocampa luminosa* (Diptera: Mycetophilidae). *Invertebr. Biol.* **2001**, *120*, 170–177. [CrossRef]
21. Meyer-Rochow, V.B.; Moore, S. Hitherto unreported aspects of the ecology and anatomy of a unique gastropod: The bioluminescent freshwater pulmonated *Latia neritoides*. In *Bioluminescence in Focus—A Collection of Illuminating Essays*; Meyer-Rochow, V.B., Ed.; Research Signpost: Kerala, India, 2009; pp. 85–104.
22. Oba, Y.; Branham, M.A.; Fukatsu, T. The terrestrial bioluminescent animals of Japan. *Zool. Sci.* **2011**, *28*, 771–789. [CrossRef]
23. Martin, G.J.; Branham, M.A.; Whiting, M.F.; Bybee, S.M. Total evidence phylogeny and the evolution of adult bioluminescence in fireflies (Coleoptera: Lampyridae). *Mol. Phylogenetics Evol.* **2017**, *107*, 564–575. [CrossRef] [PubMed]
24. Bennett, F.D. *Narrative of a Whaling Voyage Round the Globe, from the Year 1833 to 1836*; Рипол Классик: Moscow, Russia, 1840; Volume 2.
25. Claes, J.M. Function and Control of Luminescence from Lantern Shark (*Etmopterus spinax*) Photophores. Ph.D. Thesis, Université catholique de Louvain—UCLouvain, Ottignies-Louvain-la-Neuve, Belgium, 4 May 2010.
26. Renwart, M. Ultrastructure and Biochemistry of the Light-Emitting System of Lantern Shark (*Etmopterus spinax*) Photophores. Ph.D. Thesis, Université catholique de Louvain—UCLouvain, Ottignies-Louvain-la-Neuve, Belgium, 18 December 2014.
27. Duchatelet, L. Extraocular Photoreception and Bioluminescence of Representatives of Two Luminous Shark Families, Etmopteridae and Dalatiidae. Ph.D. Thesis, Université Catholique de Louvain—UCLouvain, Ottignies-Louvain-la-Neuve, Belgium, 3 October 2019.
28. Harvey, E.N. *Bioluminescence. Chapter XVI Pisces*; Academic Press Inc.: New York, NY, USA, 1952; pp. 494–553.
29. Alcock, A. *A Naturalist in Indian Seas: Or, Four Years with the Royal Indian Marine Survey Ship "Investigator"*; Dutton: New York, NY, USA, 1902.
30. de Carvalho, M.R. A synopsis of the deep-sea genus *Benthobatis* Alcock, with a redescription of the type species *Benthobatis moresbyi* Alcock, 1898 (Chondrichthyes, Torpediniformes, Narcinidae). In Proceedings of the 5th Indo-Pacific Fish Conference, Nouméa, New Caledonia, 3–8 November 1997; Séret, B., Sire, J.-Y., Eds.; Societe Francaise d' Ichtyologie: Paris, France, 1999; pp. 231–255.

31. de Carvalho, M.R.; Compagno, L.J.V.; Ebert, D.A. *Benthobatis yangi*, a new species of blind electric ray from Taiwan (Chondrichthyes: Torpediniformes: Narcinidae). *Bull. Mar. Sci.* **2003**, *72*, 923–939.
32. Duchatelet, L.; Moris, V.; Tomita, T.; Mahillon, J.; Sato, K.; Behets, C.; Mallefet, J. The megamouth shark, *Megachasma pelagios*, is not a luminous species. *PLoS ONE* **2020**, *15*, e0242196. [CrossRef]
33. Hubbs, C.L.; Iwai, T.; Matsubara, K. External and internal characters, horizontal and vertical distribution, luminescence, and food of the dwarf pelagic shark, *Euprotomicrus bispinatus*. *Bull. Scripps Inst. Oceanogr.* **1967**, *10*, 1–64.
34. Claes, J.M.; Mallefet, J. Bioluminescence of sharks: First synthesis. In *Bioluminescence in Focus—A Collection of Illuminating Essays*; Meyer-Rochow, V.B., Ed.; Research Signpost: Kerala, India, 2009; pp. 51–65.
35. Straube, N.; Li, C.; Claes, J.M.; Corrigan, S.; Naylor, G.J.P. Molecular phylogeny of Squaliformes and first occurrence of bioluminescence in sharks. *BMC Evol. Biol.* **2015**, *15*, 162. [CrossRef]
36. Duchatelet, L.; Marion, R.; Mallefet, J. A third luminous shark family: Confirmation of luminescence ability for *Zameus squamulosus* (Squaliformes; Somniosidae). *Photochem. Photobiol.* **2021**, *97*, 739–744. [CrossRef] [PubMed]
37. Adnet, S.; Capetta, H. A paleontological and phylogenetical analysis of Squaliform sharks (Chondrichthyes: Squaliformes) based on dental characters. *Lethaia* **2001**, *34*, 234–248. [CrossRef]
38. Straube, N.; Iglésias, S.P.; Sellos, D.Y.; Kriwet, J.; Schliewen, U.K. Molecular phylogeny and node time estimation of bioluminescent lantern sharks (Elasmobranchii: Etmopteridae). *Mol. Phylogenet. Evol.* **2010**, *56*, 905–917. [CrossRef] [PubMed]
39. Compagno, L.J.V.; Dando, M.; Fowler, S. *Sharks of the World*; Harper Collins: London, UK, 2004.
40. Castro, J.I. *The Sharks of North America*; Oxford University Press: New York, NY, USA, 2011.
41. Mallefet, J.; Stevens, D.W.; Duchatelet, L. Bioluminescence of the largest luminous vertebrate, the kitefin shark, *Dalatias licha*: First insights and comparative aspects. *Front. Mar. Sci.* **2021**, *8*, 633582. [CrossRef]
42. Claes, J.M.; Mallefet, J. Comparative control of luminescence in sharks: New insights from the slendertail lanternshark (*Etmopterus molleri*). *J. Exp. Mar. Biol. Ecol.* **2015**, *467*, 87–94. [CrossRef]
43. Duchatelet, L.; Delroisse, J.; Pinte, N.; Sato, K.; Ho, H.-C.; Mallefet, J. Adrenocorticotropic hormone and cyclic adenosine monophosphate are involved in the control of shark bioluminescence. *Photochem. Photobiol.* **2020**, *96*, 37–45. [CrossRef] [PubMed]
44. Ohshima, H. Some observations on the luminous organs of fishes. *J. Coll. Sci. Imp. Univ. Tokyo* **1911**, *27*, 1–25. [CrossRef]
45. Claes, J.M.; Mallefet, J. Early development of bioluminescence suggests camouflage by counter-illumination in the velvet belly lantern shark *Etmopterus spinax* (Squaloidea: Etmopteridae). *J. Fish Biol.* **2008**, *73*, 1337–1350. [CrossRef]
46. Johann, L.; der Organe, I.V. Über eigentümliche epitheliale Gebilde (Leuchtorgane) bei *Spinax niger* Aus dem zoologischen Institut der Universität Rostock. *Von. Z. Wiss. Zool.* **1899**, *66*, 136.
47. Claes, J.M.; Sato, K.; Mallefet, J. Morphology and control of photogenic structures in a rare dwarf pelagic lantern shark (*Etmopterus splendidus*). *J. Exp. Mar. Biol. Ecol.* **2011**, *406*, 1–5. [CrossRef]
48. Claes, J.M.; Ho, H.-C.; Mallefet, J. Control of luminescence from pygmy shark (*Squaliolus aliae*) photophores. *J. Exp. Biol.* **2012**, *215*, 1691–1699. [CrossRef]
49. Stehmann, M.F.W.; Van Oijen, M.; Kamminga, P. Re-description of the rare taillight shark *Euprotomicroides zantedeschia* (Squaliformes, Dalatiidae), based on third and fourth record from off Chile. *Cybium* **2016**, *40*, 187–197.
50. Dickens, D.A.G.; Marshall, N.J. Observations on *Euprotomicrus*. *Mar. Obs.* **1956**, *26*, 73–74.
51. Widder, E.A. A predatory use of counterillumination by the squaloid shark, *Isistius brasiliensis*. *Environ. Biol. Fishes* **1998**, *53*, 267–273. [CrossRef]
52. Delroisse, J.; Duchatelet, L.; Flammang, P.; Mallefet, J. Photophore distribution and enzymatic diversity within the photogenic integument of the cookie cutter shark *Isistius brasiliensis* (Chondrichthyes: Dalatiidae). *Front. Mar. Sci.* **2021**, *8*, 627045. [CrossRef]
53. Ebert, D.A.; Straube, N.; Leslie, R.W.; Weigmann, S. *Etmopterus alphus* n. sp.: A new lanternshark (Squaliformes: Etmopteridae) from the south-western Indian Ocean. *Afr. J. Mar. Sci.* **2016**, *38*, 329–340. [CrossRef]
54. Ebert, D.A.; Leslie, R.W.; Weigmann, S. *Etmopterus brosei* sp. nov.: A new lanternshark (Squaliformes: Etmopteridae) from the southeastern Atlantic and southwestern Indian oceans, with a revised key to the *Etmopterus lucifer* clade. *Mar. Biodiv.* **2021**, *51*, 53. [CrossRef]
55. Grace, M.A.; Doosey, M.H.; Denton, J.S.S.; Naylor, G.J.P.; Bart, H.L.; Maisey, J.G. A new Western North Atlantic Ocean kitefin shark (Squaliformes: Dalatiidae) from the Gulf of Mexico. *Zootaxa* **2019**, *4619*, 109–120. [CrossRef]
56. Burckhardt, R. On the luminous organs of selachian fishes. *Ann. Mag. Nat. Hist.* **1900**, *7*, 558–568. [CrossRef]
57. Bigelow, H.B.; Schoeder, W.C.; Springer, S. New and little known sharks from the Atlantic and from the Gulf of Mexico. *Bull. Mus. Comp. Zool. Harv. Coll.* **1953**, *109*, 213–276.
58. Abe, T. Description of a new squaloid shark, *Centroscyllium kamoharai*, from Japan. *Jpn. J. Ichthyol.* **1966**, *13*, 190–198. [CrossRef]
59. Chan, W.L. New sharks from the South China Sea. *J. Zool.* **1966**, *148*, 218–237. [CrossRef]
60. Dolganov, V.N. A new shark from the family Squalidae caught on the Naska submarine ridge. *Zool. Z.* **1984**, *63*, 1589–1591.
61. Munk, O.; Jørgensen, J.M. Putatively luminous tissue in the abdominal pouch of a male dalatiine shark, *Euprotomicroides zantedeschia* Hulley & Penrith, 1966. *Acta Zool.* **1988**, *69*, 247–251. [CrossRef]
62. Shirai, S.; Nakaya, K. A new squalid species of the genus *Centroscyllium* from the Emperor seamont chain. *Jpn. J. Ichthyol.* **1990**, *36*, 391–398. [CrossRef]
63. Last, P.R.; Burgess, G.H.; Séret, B. Description of six new species of lantern-sharks of the genus *Etmopterus* (Squaloidea: Etmopteridae) from the Australasian region. *Cybium* **2002**, *26*, 203–223.

64. Knuckey, J.D.S.; Ebert, D.A.; Burgess, G.H. *Etmopterus joungi* n. sp., a new species of lanternshark (Squaliformes: Etmopteridae) from Taiwan. *Aqua Int. J. Ichthyol.* **2010**, *17*, 61–72.
65. Claes, J.M.; Nilsson, D.-E.; Straube, N.; Collin, S.P.; Mallefet, J. Iso-luminance counterillumination drove bioluminescent shark radiation. *Sci. Rep.* **2014**, *4*, 4328. [CrossRef]
66. Grace, M.A.; Doosey, M.H.; Bart, H.L.; Naylor, G.J.P. First record of *Mollisquama* sp. (Chondrichthyes: Squaliformes: Dalatiidae) from the Gulf of Mexico, with a morphological comparison to the holotype description of Mollisquama parini Dolganov. *Zootaxa* **2015**, *3948*, 587–600. [CrossRef]
67. Vásquez, V.E.; Ebert, D.A.; Long, D.J. *Etmopterus benchleyi* n. sp., a new lanternshark (Squaliformes: Etmopteridae) from the central eastern Pacific Ocean. *J. Ocean. Sci. Found* **2015**, *17*, 43–55.
68. Ebert, D.A.; Papastamatiou, Y.P.; Kajiura, S.M.; Wetherbee, B.M. *Etmopterus lailae* sp. nov., a new lanternshark (Squaliformes: Etmopteridae) from the Northwestern Hawaiian Islands. *Zootaxa* **2017**, *4237*, 371–382. [CrossRef]
69. White, W.T.; Ebert, D.A.; Mana, R.R.; Corrigan, S. *Etmopterus samadiae* n. sp., a new lanternshark (Squaliformes: Etmopteridae) from Papua New Guinea. *Zootaxa* **2017**, *4244*, 339–354. [CrossRef]
70. de Figueiredo Petean, F.; de Carvalho, M.R. Comparative morphology and systematics of the cookiecutter sharks, genus *Isistius* Gill (1864) (Chondrichthyes: Squaliformes: Dalatiidae). *PLoS ONE* **2018**, *13*, e0201913. [CrossRef] [PubMed]
71. Dolganov, V.N.; Balanov, A.A. *Etmopterus parini* sp. n. (Squaliformes: Etmopteridae), a new shark species from the northwestern Pacific Ocean. *Biol. Morya* **2018**, *44*, 427–430.
72. Ebert, D.A.; Van Hees, K.E. *Etmopterus marshae* sp. Nov, a new lanternshark (Squaliformes: Etmopteridae) from the Philippine Islands, with a revised key to the *Etmopterus lucifer* clade. *Zootaxa* **2018**, *4508*, 197–210. [CrossRef]
73. Dolganov, V.N. On the little-known sharks *Etmopterus villosus* (Etmopteridae) and *Scymnodalatias sherwoodi* (Somniosidae) from the Pacific Ocean. *J. Ichthyol.* **2019**, *59*, 275–279. [CrossRef]
74. Claes, J.M.; Delroisse, J.; Grace, M.A.; Doosey, M.H.; Duchatelet, L.; Mallefet, J. Histological evidence for secretory bioluminescence from pectoral pockets of the American pocket shark (*Mollisquama mississippiensis*). *Sci. Rep.* **2020**, *10*, 18762. [CrossRef]
75. Claes, J.M.; Nilsson, D.-E.; Mallefet, J.; Straube, N. The presence of lateral photophores correlates with increased speciation in deep-sea bioluminescent sharks. *R. Soc. Open Sci.* **2015**, *2*, 150219. [CrossRef] [PubMed]
76. Renwart, M.; Delroisse, J.; Claes, J.M.; Mallefet, J. Ultrastructural organization of lantern shark (*Etmopterus spinax* Linnaeus, 1758) photophores. *Zoomorphology* **2014**, *133*, 405–416. [CrossRef]
77. Duchatelet, L.; Claes, J.M.; Mallefet, J. Embryonic expression of encephalopsin supports bioluminescence perception in lantern-shark photophores. *Mar. Biol.* **2019**, *166*, 21. [CrossRef]
78. Duchatelet, L.; Delroisse, J.; Flammang, P.; Mahillon, J.; Mallefet, J. *Etmopterus spinax*, the velvet belly lanternshark, does not use bacterial luminescence. *Acta Histochem.* **2019**, *121*, 516–521. [CrossRef] [PubMed]
79. Schofield, P.J.; Burgess, G.H. *Etmopterus robinsi* (Elasmobranchii, Etmopteridae), a new species of deepwater lantern shark from the Caribbean Sea and Western North Atlantic, with a redescription of *Etmopterus hillianus*. *Bull. Mar. Sci.* **1997**, *60*, 1060–1073.
80. Straube, N.; Duhamel, G.; Gasco, N.; Kriwet, J.; Schliewen, U.K. Description of a new deep-sea lantern shark *Etmopterus viator* sp. nov. (Squaliformes: Etmopteridae) from the Southern Hemisphere. In *The Kerguelen Plateau, Marine Ecosystem and Fisheries*; Société Française d'Ichtyologie: Paris, France, 2011; pp. 135–148.
81. Fricke, R.; Koch, I. *A New Species of the Lantern Shark Genus Etmopterus from Southern Africa (Elasmobranchii: Squalidae)/With 4 Figures and 1 Table*; Stuttgarter Beiträge zur Naturkunde: Stuttgart, Germany, 1990.
82. Iwai, T. Luminous organs of the deep-sea squaloid shark, *Centroscyllium ritteri* Jordan and Fowler. *Pac. Sci.* **1960**, *14*, 51–54.
83. Schaaf-DaSilva, J.A.; Ebert, D.A. *Etmopterus burgessi* sp. nov., a new species of lanternshark (Squaliformes: Etmopteridae) from Taiwan. *Zootaxa* **2006**, *1373*, 53–64. [CrossRef]
84. Springer, S.; Burgess, G.H. Two new dwarf dogsharks (*Etmopterus*, Squalidae), found off the Caribbean coast of Colombia. *Copeia* **1985**, *1985*, 584–591. [CrossRef]
85. Shirai, S.; Tachikawa, H. Taxonomic resolution of the *Etmopterus pusillus* species group (Elasmobranchii, Etmopteridae), with description of *E. bigelowi*, n. sp. *Copeia* **1993**, *1993*, 483–495. [CrossRef]
86. Dolganov, V.N. Description of new species of sharks of the family Squalidae (Squaliformes) from the north-western part of the Pacific Ocean with remarks of validity of Etmopterus. *Zool. Zhurnal* **1986**, *65*, 149–153.
87. Gubanov, E.P.; Kondyurin, V.V.; Myagkov, N.A. *Sharks of the World Ocean: Identification Handbook. [=Akuly Mirovogo Okeana: Sparvochnik-Opredelitel']*; Agropromizdat: Moscow, Russia, 1986; pp. 1–272.
88. Mochizuki, K.; Ohe, F. *Trigonognathus kabeyai*, a new genus and species of the Squalid sharks from Japan. *Jpn. J. Ichthyol.* **1990**, *36*, 385–390. [CrossRef]
89. Seigel, J.A. Revision of the Dalatiid shark genus *Squaliolus*: Anatomy, systematics, ecology. *Copeia* **1978**, *1978*, 602–614. [CrossRef]
90. Garrick, J.A.F.; Springer, S. *Isistius plutodus*, a new Squaloid shark from the Gulf of Mexico. *Copeia* **1964**, *1964*, 678–682. [CrossRef]
91. Pollerspöck, J.; Straube, N. Bibliography Database of Living/Fossil Sharks, Rays and Chimaeras (Chondrichtyes: Elasmobranchii, Holocephali)—List of Valid Extant Species; List of Described Extant Species; Statistic, World Wide Web Electronic Publication, Version 10/2021; ISSN: 2195-6499. Available online: www.shark-references.com (accessed on 23 November 2021).
92. Denton, J.S.S.; Maisey, J.G.; Grace, M.; Pradel, A.; Doosey, M.H.; Bart, H.L., Jr.; Naylor, G.J.P. Cranial morphology in *Mollisquama* sp. (Squaliformes; Dalatiidae) and patterns of cranial evolution in dalatiid sharks. *J. Anat.* **2018**, *233*, 15–32. [CrossRef] [PubMed]

93. Daley, R.; Appleyard, S.A.; Koopman, M. Genetic catch verification to support recovery plans for deepsea gulper sharks (genus *Centrophorus*, family Centrophoridae)- an Australian example using the 16S gene. *Mar. Freshw. Res.* **2012**, *63*, 708–714. [CrossRef]

94. Stefanni, S.; Catarino, D.; Ribeiro, P.A.; Freitas, M.; Menezes, G.M.; Neat, F.; Stankovic, D. Molecular systematics of the long-snouted deep water dogfish (Centrophoridae, *Deania*) with implications for identification, taxonomy, and conservation. *Front. Mar. Sci.* **2021**, *7*, 588192. [CrossRef]

95. Ziadi-Künzli, F.; Soliman, T.; Imai, H.; Sakurai, M.; Maeda, K.; Tachihara, K. Re-evaluation of deep-sea dogfishes (genus *Squalus*) in Japan using phylogenetic inference. *Deep Sea Res. I* **2020**, *160*, 103261. [CrossRef]

96. White, W.T.; Vaz, D.F.B.; Ho, H.-C.; Ebert, D.A.; de Carvalho, M.R.; Corrigan, S.; Rochel, E.; de Carvalho, M.; Tanaka, S.; Naylor, G.J.P. Redescription of *Scymnodon ichiharai* Yano and Tanaka 1984 (Squaliformes: Somniosidae) from the western North Pacific, with comments on the definition of somniosid genera. *Ichthyol. Res.* **2015**, *62*, 213–229. [CrossRef]

97. Ebert, D.A.; Dando, M.; Fowler, S. *Sharks of the World, a Complete Guide*; Princeton University Press: Princeton, NJ, USA, 2013; pp. 1–624.

98. Ebert, D.A.; Compagno, L.J.V.; De Vries, M.J. A new lanternshark (Squaliformes: Etmopteridae: *Etmopterus*) from Southern Africa. *Copeia* **2011**, *2011*, 379–384. [CrossRef]

99. Clarke, W.D. Function of bioluminescence in mesopelagic organisms. *Nature* **1963**, *198*, 1244–1246. [CrossRef]

100. Denton, E.J.; Gilpin-Brown, J.B.; Wright, P.J. The angular distribution of the light produced by some mesopelagic fish in relation to their camouflage. *Proc. R. Soc. B* **1972**, *182*, 145–158. [CrossRef]

101. Young, R.E.; Kampa, E.M.; Maynard, S.D.; Mencher, F.M.; Roper, C.F.E. Counterillumination and the upper depth limits of midwater animals. *Deep Sea Res. Part I Oceanogr. Res. Pap.* **1980**, *27*, 671–691. [CrossRef]

102. McFall-Ngai, M.; Morin, J.G. Camouflage by disruptive illumination in Leiognathids, a family of shallow-water, bioluminescent fishes. *J. Exp. Biol.* **1991**, *156*, 119–137. [CrossRef]

103. Claes, J.M.; Aksnes, D.L.; Mallefet, J. Phantom hunter of the fjords: Camouflage by counterillumination in a shark (*Etmopterus spinax*). *J. Exp. Mar. Biol. Ecol.* **2010**, *388*, 28–32. [CrossRef]

104. Claes, J.M.; Dean, M.N.; Nilsson, D.-E.; Hart, N.S.; Mallefet, J. A deepwater fish with 'lightsabers'—Dorsal spine-associated luminescence in a counterilluminating lanternshark. *Sci. Rep.* **2013**, *3*, 1308. [CrossRef] [PubMed]

105. Duchatelet, L.; Pinte, N.; Tomita, T.; Sato, K.; Mallefet, J. Etmopteridae bioluminescence: Dorsal pattern specificity and aposematic use. *Zool. Lett.* **2019**, *5*, 9. [CrossRef]

106. Claes, J.M.; Mallefet, J. Ecological Functions of shark luminescence. *Luminescence* **2014**, *29*, 13.

107. Claes, J.M.; Mallefet, J. Functional physiology of lantern shark (*Etmopterus spinax*) luminescent pattern: Differential hormonal regulation of luminous zones. *J. Exp. Biol.* **2010**, *213*, 1852–1858. [CrossRef]

108. Claes, J.M.; Mallefet, J. Hormonal control of luminescence from lantern shark (*Etmopterus spinax*) photophores. *J. Exp. Biol.* **2009**, *212*, 3684–3692. [CrossRef] [PubMed]

109. Pinte, N.; Parisot, P.; Martin, U.; Zintzen, V.; De Vleeschouwer, C.; Roberts, C.D.; Mallefet, J. Ecological features and swimming capabilities of deep-sea sharks from New Zealand. *Deep-Sea Res. I* **2020**, *156*, 103187. [CrossRef]

110. Pinte, N.; Coubris, C.; Jones, E.; Mallefet, J. Red and white muscle proportions and enzyme activities in mesopelagic sharks. *Comp. Biochem. Physiol. B* **2021**, *256*, 110649. [CrossRef]

111. Claes, J.M.; Partridge, J.C.; Hart, N.S.; Garza-Gisholt, E.; Ho, H.-C.; Mallefet, J.; Collin, S.P. Photon hunting in the twilight zone: Visual features of mesopelagic bioluminescent sharks. *PLoS ONE* **2014**, *9*, e104213. [CrossRef] [PubMed]

112. Claes, J.M.; Mallefet, J. Ontogeny of photophore pattern in the velvet belly lanternshark, *Etmopterus spinax*. *Zoology* **2009**, *112*, 433–441. [CrossRef]

113. Santos, J.C.; Coloma, L.A.; Cannatella, D.C. Multiple, recurring origins of aposematism and diet specialization in poison frogs. *Proc. Natl. Acad. Sci. USA* **2003**, *100*, 12792–12797. [CrossRef]

114. Marek, P.; Papaj, D.; Yeager, J.; Molina, S.; Moore, W. Bioluminescent aposematism in millipedes. *Curr. Biol.* **2011**, *21*, R680–R681. [CrossRef]

115. Bolin, R.L. The function of the luminous organs of deep-sea fishes. In Proceedings of the 9th Pacific Science Congress, Bangkok, Thailand, 18 November–9 December 1957; Volume 10, pp. 37–39.

116. Matsui, T.; Rosenblatt, R.H. Review of the deep-sea fish family Platytroctidae (Pisces: Salmoniformes). *Bull. Scripps Inst. Oceanogr. Univ. Calif.* **1987**, *26*, 1–159.

117. Poulsen, J.Y. New observations and ontogenetic transformation of photogenic tissues in the tubeshoulder *Sagamichthys schnakenbecki* (Platytroctidae, Alepocephaliformes). *J. Fish Biol.* **2019**, *94*, 62–76. [CrossRef]

118. Davis, A.L.; Sutton, T.T.; Kier, W.M.; Johnsen, S. Evidence that eye-facing photophores serve as a reference for counterillumination in an order of deep-sea fishes. *Proc. R. Soc. B* **2020**, *287*, 20192918. [CrossRef]

119. Herring, P.J.; Cope, C. Red bioluminescence in fishes: On the suborbital photophores of *Malacosteus*, *Pachystomias* and *Aristostomias*. *Mar. Biol.* **2005**, *148*, 383–394. [CrossRef]

120. Renwart, M.; Delroisse, J.; Flammang, P.; Claes, J.M.; Mallefet, J. Cytological changes during luminescence production in lanternshark (*Etmopterus spinax* Linnaeus, 1758) photophores. *Zoomorphology* **2015**, *134*, 107–116. [CrossRef]

121. Claes, J.M.; Mallefet, J. The lantern shark's light switch: Turning shallow water crypsis into midwater camouflage. *Biol. Lett.* **2010**, *6*, 685–687. [CrossRef]

122. Duchatelet, L.; Sugihara, T.; Delroisse, J.; Koyanagi, M.; Rezsohazy, R.; Terakita, A.; Mallefet, J. From extraocular photoreception to pigment movement regulation: A new control mechanism of the lanternshark luminescence. *Sci. Rep.* **2020**, *10*, 10195. [CrossRef]

123. Claes, J.M.; Krönström, J.; Holmgren, S.; Mallefet, J. Nitric oxide in the control of luminescence from lantern shark (*Etmopterus spinax*) photophores. *J. Exp. Biol.* **2010**, *213*, 3005–3011. [CrossRef] [PubMed]

124. Reif, W.E. Functions of scales and photophores in mesopelagic luminescent sharks. *Acta Zool.* **1985**, *66*, 111–118. [CrossRef]

125. Ferrón, H.G.; Paredes-Aliaga, M.V.; Martínez-Pérez, C.; Botella, H. Bioluminescent-like squamation in the galeomorph shark *Apristurus ampliceps* (Chondrichthyes: Elasmobranchii). *Contrib. Zool.* **2018**, *87*, 187–196. [CrossRef]

126. Christophe, B.; Baguet, F. The adrenergic control of the photocytes luminescence of the *Porichthys* photophores. *Comp. Bioch. Physiol. C Comp. Pharmacol.* **1985**, *81*, 359–365. [CrossRef]

127. Mallefet, J.; Anctil, M. Immunohistochemical detection of biogenic amines in the photophores of the midshipman fish *Porichthys notatus. Can. J. Zool.* **1992**, *70*, 1968–1975. [CrossRef]

128. Zaccone, G.; Abelli, L.; Salpietro, L.; Zaccone, D.; Macrì, B.; Marino, F. Nervous control of photophores in luminescent fishes. *Acta Hictochem.* **2011**, *113*, 387–394. [CrossRef]

129. Mallefet, J.; Duchatelet, L.; Hermans, C.; Baguet, F. Luminescence control of Stomiidae photophores. *Acta Histochem.* **2019**, *121*, 7–15. [CrossRef] [PubMed]

130. Visconti, M.A.; Ramanzini, G.C.; Camargo, C.R.; Castrucci, A.M.L. Elasmobranch color change: A short review and novel data on hormone regulation. *J. Exp. Zool.* **1999**, *284*, 485–491. [CrossRef]

131. Gelsleichter, J.; Evans, A.N. Hormonal regulation of elasmobranch physiology. In *Biology of Sharks and Their Relatives*; Carrier, J.C., Musick, J.A., Heithaus, M.R., Eds.; CRC Press: Boca Raton, FL, USA, 2004; Chapter 11; pp. 314–338.

132. Nagai, M.; Oshima, N.; Fujii, R. A comparative study of melanin concentrating hormone (MCH) action on teleost melanophores. *Biol. Bull.* **1986**, *171*, 360–370. [CrossRef]

133. Oshima, N.; Kasukawa, H.; Fujii, R.; Wilkes, B.C.; Hruby, V.J.; Hadley, M.E. Action of melanin-concentrating hormone (MCH) on teleost chromatophores. *Gen. Comp. Endocrinol.* **1986**, *64*, 381–388. [CrossRef]

134. Mizusawa, K.; Kobayashi, Y.; Sunuma, T.; Asahida, T.; Saito, Y.; Takahashi, A. Inhibiting roles of melanin-concentrating hormone for skin pigment dispersion in barfin flounder, *Verasper moseri. Gen. Comp. Endocrinol.* **2011**, *171*, 75–81. [CrossRef]

135. Mizusawa, K.; Amiya, N.; Yamaguchi, Y.; Takabe, S.; Amano, M.; Breves, J.P.; Fox, B.K.; Grau, E.G.; Hyodo, S.; Takahashi, A. Identification of mRNAs coding for mammalian-type melanin-concentrating hormone and its receptors in the scalloped hammerhead shark *Sphyrna lewini. Gen. Comp. Endocrinol.* **2012**, *179*, 78–87. [CrossRef] [PubMed]

136. Duchatelet, L.; Delroisse, J.; Mallefet, J. Melanin-concentrating hormone is not involved in luminescence emission in the velvet belly lanternshark, *Etmopterus spinax. Mar. Biol.* **2019**, *166*, 140. [CrossRef]

137. Duchatelet, L.; Delroisse, J.; Mallefet, J. Bioluminescence in lanternsharks: Insight from hormone receptor localization. *Gen. Comp. Endocrinol.* **2020**, *294*, 113488. [CrossRef]

138. Claes, J.M.; Mallefet, J. Control of luminescence from lantern shark (*Etmopterus spinax*) photophores. *Commun. Integr. Biol.* **2011**, *4*, 251–253. [CrossRef] [PubMed]

139. Claes, J.M.; Krönström, J.; Holmgren, S.; Mallefet, J. GABA inhibition of luminescence from lantern shark (*Etmopterus spinax*) photophores. *Comp. Biochem. Physiol. C Toxicol. Pharmacol.* **2011**, *153*, 231–236. [CrossRef]

140. Ocampo Daza, D.; Larhammar, D. Evolution of the receptors for growth hormone, prolactin, erythropoietin and thrombopoietin in relation to the vertebrate tetraploidizations. *Gen. Comp. Endocrinol.* **2018**, *257*, 143–160. [CrossRef]

141. Vanacek, J. Cellular mechanisms of melatonin action. *Physiol. Rev.* **1998**, *78*, 687–721. [CrossRef]

142. Busca, R.; Ballotti, R. Cyclic AMP a key messenger in the regulation of skin pigmentation. *Pigment Cell Res.* **2000**, *13*, 60–69. [CrossRef] [PubMed]

143. Sugden, D.; Davidson, K.; Hough, K.A.; The, M.-T. Melatonin, melatonin receptors and melanophores: A moving story. *Pigment Cell Res.* **2004**, *17*, 454–460. [CrossRef]

144. Visconti, M.A.; Castrucci, A.M.L. Melanotropin receptors in the cartilaginous fish, *Potamotrygon reticulatus* and in the lungfish, *Lepidosiren paradoxa. Comp. Biochem. Physiol. C Pharmacol. Toxicol. Endocrinol.* **1993**, *106*, 523–528. [CrossRef]

145. Nery, L.E.M.; Castrucci, A.M.L. Pigment cell signalling for physiological color change. *Comp. Biochem. Physiol. A Mol. Interg. Physiol.* **1997**, *118*, 1135–1144. [CrossRef]

146. Tong, D.; Rozas, N.S.; Oakley, T.H.; Mitchell, J.; Colley, N.J.; McFall-Ngai. Evidence for light perception in a bioluminescent organ. *Proc. Natl. Acad. Sci. USA* **2009**, *106*, 9836–9841. [CrossRef]

147. McFall-Ngai, M.; Heath-Heckman, E.A.C.; Gillette, A.A.; Peyer, S.M.; Harvie, E.A. The secret languages of coevolved symbioses: Insights from the *Euprymna scolopes-Vibrio fischeri* symbiosis. *Semin. Immunol.* **2012**, *24*, 3–8. [CrossRef]

148. Schnitzler, C.E.; Pang, K.; Powers, M.L.; Reitzel, A.M.; Ryan, J.F.; Simmons, D.; Tada, T.; Park, M.; Gupta, J.; Brooks, S.Y.; et al. Genomic organization, evolution, and expression of photoprotein and opsin genes in *Mnemiopsis leidyi*: A new view of ctenophore photocytes. *BMC Biol.* **2012**, *10*, 107. [CrossRef] [PubMed]

149. Delroisse, J.; Ullrich-Luter, E.; Ortega-Martinez, O.; Dupont, S.; Arnone, M.I.; Mallefet, J.; Flammang, P. High opsin diversity in a non-visual infaunal brittle star. *BMC Genom.* **2014**, *15*, 1035. [CrossRef]

150. Delroisse, J.; Ullrich-Lüter, E.; Blaue, S.; Ortega-Martinez, O.; Eeckhaut, I.; Flammang, P.; Mallefet, J. Fine structure of the luminous spines and luciferase detection in the brittle star *Amphiura filiformis. Zool. Anz.* **2017**, *269*, 1–12. [CrossRef]

151. Bracken-Grissom, H.D.; DeLeo, D.M.; Porter, M.L.; Iwanicki, T.; Sickles, J.; Frank, T.M. Light organ photosensitivity in deep-sea shrimp may suggest a novel role in counterillumination. *Sci. Rep.* **2020**, *10*, 1–10. [CrossRef]
152. Delroisse, J.; Duchatelet, L.; Flammang, P.; Mallefet, J. De novo transcriptome analyses provide insights into opsin-based photoreception in the lanternshark *Etmopterus spinax*. *PLoS ONE* **2018**, *13*, e0209767. [CrossRef] [PubMed]
153. McCapra, F. Chemical mechanisms in bioluminescence. *Acc. Chem. Res.* **1976**, *9*, 201–208. [CrossRef]
154. Kaskova, Z.M.; Tsarkova, A.S.; Yampolsky, I.V. 1001 lights: Luciferins, luciferases, their mechanisms of action and applications in chemical analysis, biology and medicine. *Chem. Soc. Rev.* **2016**, *45*, 6048–6077. [CrossRef]
155. Warner, J.A.; Case, J.F. The zoogeography and dietary induction of bioluminescence in the midshipman fish, *Porichthys notatus*. *Biol. Bull.* **1980**, *159*, 231–246. [CrossRef]
156. Frank, T.M.; Widder, E.A.; Latz, M.I.; Case, J.F. Dietary maintenance of bioluminescence in a deep-sea mysid. *J. Exp. Biol.* **1984**, *109*, 385–389. [CrossRef]
157. Thompson, E.M.; Nafpaktitis, B.G.; Tsuji, F.I. Dietary uptake and blood transport of *Vargula* (crustacean) luciferin in the bioluminescent fish, *Porichthys notatus*. *Comp. Biochem. Physiol. A Mol. Integr. Physiol.* **1988**, *89*, 203–209. [CrossRef]
158. Haddock, S.H.D.; Rivers, T.J.; Robison, B.H. Can coelenterates make coelenterazine? Dietary requirement for luciferin in cnidarian bioluminescence. *Proc. Natl. Acad. Sci. USA* **2001**, *98*, 11151. [CrossRef] [PubMed]
159. Mallefet, J.; Duchatelet, L.; Coubris, C. Bioluminescence induction in the ophiuroid *Amphiura filiformis* (Echinodermata). *J. Exp. Biol.* **2020**, *223*, jeb218719. [CrossRef] [PubMed]
160. Bessho-Uehara, M.; Yamamoto, N.; Shigenobu, S.; Mori, H.; Kuwata, K.; Oba, Y. Kleptoprotein bioluminescence: *Parapriacanthus* fish obtain luciferase from ostracod prey. *Sci. Adv.* **2020**, *6*, eaax4942. [CrossRef]
161. Thomson, C.M.; Herring, P.J.; Campbell, A.K. Evidence for *de novo* biosynthesis of coelenterazine in the bioluminescent midwater shrimp, *Systellaspis debilis*. *J. Mar. Biol. Assoc. UK* **1995**, *75*, 165–171. [CrossRef]
162. Oba, Y.; Kato, S.-I.; Ojika, M.; Inouye, S. Biosynthesis of coelenterazine in the deep-sea copepod, *Metridia pacifica*. *Biochem. Biophys. Res. Commun.* **2009**, *390*, 684–688. [CrossRef]
163. Bessho-Uehara, M.; Huang, W.; Patry, W.L.; Browne, W.E.; Weng, J.-K.; Haddock, S.H.D. Evidence for *de novo* biosynthesis of the luminous substrate coelenterazine in ctenophores. *iScience* **2020**, *23*, 101859. [CrossRef]
164. Shimomura, O. Presence of coelenterazine in non-bioluminescent marine organisms. *Comp. Biochem. Physiol.* **1987**, *86B*, 361–363. [CrossRef]
165. Mallefet, J.; Shimomura, O. Presence of coelenterazine in mesopelagic fishes from the Strait of Messina. *Mar. Biol.* **1995**, *124*, 381–385. [CrossRef]
166. Thomson, C.M.; Herring, P.J.; Campbell, A.K. The widespread occurrence and tissue distribution of the imidazolopyrazine luciferins. *J. Biol. Chem.* **1997**, *2*, 87–91. [CrossRef]
167. Duchatelet, L.; Hermans, C.; Duhamel, G.; Cherel, Y.; Guinet, C.; Mallefet, J. Coelenterazine detection in five myctophid species from the Kerguelen Plateau. In *The Kerguelen Plateau: Marine Ecosystem and Fisheries. Proceedings of the Second Symposium*; Welsford, D., Dell, J., Duhamel, G., Eds.; Australian Antartic Division: Kingston, TAS, Australia; pp. 31–41.
168. Renwart, M.; Mallefet, J. First study of the chemistry of the luminous system in a deep-sea shark, *Etmopterus spinax* Linnaeus, 1758 (Chondrichthyes: Etmopteridae). *J. Exp. Mar. Biol. Ecol.* **2013**, *448*, 214–219. [CrossRef]
169. Stingo, V.; Du Buit, M.-H.; Odierna, G. Genome size of some selachian fishes. *Boll. Zool.* **1980**, *47*, 129–137. [CrossRef]
170. Ojima, Y.; Yamamoto, K. Cellular DNA contents of fishes determined by flow cytometry. *La Kromosomo II* **1990**, *57*, 1871–1888.
171. Hardie, D.C.; Hebert, P.D.N. The nucleotypic effects of cellular DNA content in cartilaginous and ray-finned fishes. *Genome* **2003**, *46*, 683–706. [CrossRef]
172. Marra, N.J.; Stanhope, M.J.; Jue, N.K.; Wang, M.; Sun, Q.; Bitar, P.P.; Richards, V.P.; Komissarov, A.; Rayko, M.; Kliver, S.; et al. White shark genome reveals ancient elasmobranch adaptations associated with wound healing and the maintenance of genome stability. *Proc. Natl. Acad. Sci. USA* **2019**, *116*, 4446–4455. [CrossRef]
173. Weber, J.A.; Park, S.G.; Luria, V.; Jeon, S.; Kim, H.-M.; Jeon, Y.; Bhak, Y.; Jun, J.H.; Kim, S.W.; Hong, W.H.; et al. The whale shark genome reveals how genomic and physiological properties scale with body size. *Proc. Natl. Acad. Sci. USA* **2020**, *117*, 20662–20671. [CrossRef]
174. Dufresne, F.; Jeffery, N. A guided tour of large genome size in animals: What we know and where we are heading. *Chromosome Res.* **2011**, *19*, 925–938. [CrossRef]

MDPI

St. Alban-Anlage 66

4052 Basel

Switzerland

Tel. +41 61 683 77 34

Fax +41 61 302 89 18

www.mdpi.com

Oceans Editorial Office

E-mail: oceans@mdpi.com

www.mdpi.com/journal/oceans